Photobiology

Photobiology

The Science of Light and Life

Edited by

Lars Olof Björn

Lund University,
Department of Cell and Organism Biology,
Lund, Sweden

KLUWER ACADEMIC PUBLISHERS

DORDRECHT / BOSTON / LONDON

A C.I.P. Catalogue record for this book is available from the Library of Congress.

ISBN 1-4020-0842-2

Published by Kluwer Academic Publishers,
P.O. Box 17, 3300 AA Dordrecht, The Netherlands.

Sold and distributed in North, Central and South America
by Kluwer Academic Publishers,
101 Philip Drive, Norwell, MA 02061, U.S.A.

In all other countries, sold and distributed
by Kluwer Academic Publishers,
P.O. Box 322, 3300 AH Dordrecht, The Netherlands.

Printed on acid-free paper

Printed in the Netherlands.

(Courtesy of Per Nilsson)

PHOTOBIOLOGY

I am lying on my back beneath the tree,
dozing, looking up into the canopy,
thinking: what a wonder — I can see!

But in the greenery above my face,
an even greater miracle is taking place:
Leaves catch photons from the sun
and molecules from air around.
Quanta and carbon atoms
              become bound.
Life, for them, has just begun.

The sun not only creates life,
            it also takes away
mostly by deranging DNA.
Damage can be, in part, undone
by enzymes using photons from the sun.

Summer nears its end,
           already 'cross the sky
southward aiming birds are flying by.

Other birds for travel choose the night
relying on the stars for guiding light.

Imprinted in their little heads are Gemini,
Orion, Dipper, other features of the sky.
There is room for clocks that measure day
                  and night,
correct for movement of the sky, and tell the
                  time for flight.

Deep into oceans, into caves
the sun cannot directly send its waves.
But through intricacies of foodweb's
               maze,
oxygen from chloroplasts, luciferin,
               luciferase,
at times, in place, where night and
darkness seem to reign,
solar quanta emerge as photons
              once again.

L.O. Björn 2002

CONTENTS

viii

CONTRIBUTORS

LARS OLOF BJÖRN
Lund University
Department of Cell and Organism
Biology, Sölvegatan 35
SE-223 62 Lund, Sweden
Lars Olof Bjorn@fysbot.lu.se

ANDERS JOHNSSON
Department of Physics, NTNU
Norwegian University of
Science and Technology
N-7491 Trondheim, Norway
anders.johnsson@phys.ntnu.no

WOLFGANG ENGELMANN
Universität Tübingen,
Institut für Botanik, Tübingen,
Germany
engelmann@uni-tuebingen.de

RICHARD E. KENDRICK
Laboratory for Plant
Physiology Wageningen
University
Arboretumlaan 4
NL-6703 Wageningen
The Netherlands
dick.kendrick@algem.pf.wau.nl
(lab), kendrick@bos.nl (home)

G. ADRIAN HORRIDGE
Australian National University Research
School of Biological Sciences, Centre
for Visual Sciences, Box 475 PO,
Canberra, A.C.T. 2601 Australia
horridge@rsbs.anu.edu.au

RICHARD L. MCKENZIE
National Institute of Water &
Atmospheric Research (NIWA), Lauder,
PB 50061 Omakau, Central Otago 9182,
New Zealand
r.mckenzie@niwa.cri.nz

PIRJO HUOVINEN
Department of Biological and
Environmental Science, University of
Jyväskylä,
Survontie 9, FIN-40500, Jyväskylä,
Finland
pshuovin@jyu.fi

MARY NORVAL
Department of Medical Microbiology,
The University of Edinburgh Medical
School, Teviot Place,
Edinburgh EH8 9AG, Scotland
M.Norval@ed.ac.uk

JIM L. WELLER
School of Plant Science, University of
Tasmania, GPO Box 252-55,
Hobart, Tasmania 7001, Australia
Jim.Weller@utas.edu.au

PREFACE

My first scientific paper, published 43 years ago, dealt with the effects of red light on the growth of plants. My scientific interest ever since has been focused on photobiology in its many aspects. Because I have been employed as a botanist, my own research has dealt with the photobiology of plants, but all the time I have been interested also in other aspects, such as vision, the photobiology of skin, and bioluminescence. Thirty years ago I wrote a little book called "Light and Life", and it has been my aim to eventually follow this up with a fuller treatise.

Because I always had my hands full, I postponed the project until my retirement. To make it possible for the book to materialize, it has been necessary both to select certain aspects of the discipline, and to recruit coworkers. Many colleagues promised to help, but not all have lived up to the promises. To those who did, and are coauthors to this volume, I direct my thanks; I think that they have done an excellent job.

Living creatures use light for two purposes: for obtaining useful energy, and as information carrier. In the latter case organisms use light mainly to collect information, but in the case of bioluminescence also for sending information, including misleading information to other organisms of their own or other species. Collection of free energy through photosynthesis and collection of information through vision or other photobiological processes may seem to be very different concepts. However, on a deep level, they are of the same kind. They use the difference in temperature between the sun and our planet to evade equilibrium; to maintain and develop order and structure.

As just mentioned, all of photobiology cannot be condensed into a single volume. My idea has been to first provide the basic knowledge that can be of use to all photobiologists, and then give some examples of interesting special topics. A casual reader glancing through the book will easily spot omissions, such as chapters dedicated entirely to, for instance, photosynthesis and vertebrate vision. Sections relevant to photosynthesis appear, however, in chapters 1, 6, 7, and 17, and relevant to vertebrate vision in chapters 1, 7, 8 and 17.

The book is intended as a start, not as the final word. There are several journals specifically dealing with photobiology (Photochemistry & Photobiology, Journal of Photochemistry and Photobiology B: Biology, Photochemical and Photobiological Sciences) or on more specialized photobiology topics (e.g., Photodermatology, Photoimmunology & Photomedicine, Journal of Photosynthesis Research, Photosynthetica). A Comprehensive Series in Photosciences is under publication, and when I am writing this three volumes have appeared in print. There are several photobiology societies arranging meetings and other activities. And least but not least, up-to-date information can be found on the Internet. The most important site, apart from the Web of Science and other scientific databases, is

"Photobiology online", a site maintained jointly by the American and European Societies for Photobiology (ASP and ESP, respectively), at http://169.147.169.1/POL.index.html or http://www.pol-europe.net/ where details about journals etc can be obtained. As for photosynthesis, an excellent site to start Professor Govindjee's at http://www.life.uiuc.edu/PSed_index.htm. But we must never think of photobiology as something isolated from the rest of science, as a science of its own. The title of this book may be somewhat misleading. There is only one science.

I would like to thank the following colleagues for giving advice, providing material, correcting mistakes, and encouraging me: Professor Helen Ghiradella, Professor Adrian Horridge, Dr Michael F. Holick, Dr Almut Kelber, and Dr Michael L. Richardson.

Special thanks go to my wife and beloved photobiologist Gunvor. She has put up with papers and books filling not only tables but also floors for more than a year. To her, and to Professor Hemming Virgin, who started me out on the adventure of photobiology, I dedicate those parts of the book that bear my authorship.

Lund, April 2002-04-19

Lars Olof Björn

LARS OLOF BJÖRN

1. THE NATURE OF LIGHT
AND ITS INTERACTION WITH MATTER

1. INTRODUCTION

The behaviour of light when it travels through space and when it interacts with matter plays a central role in the two main paradigms of twentieth century physics: Relativity and quantum physics. As we shall see throughout this book, it is also important for an understanding of the behaviour and functioning of organisms.

2. PARTICLE AND WAVE PROPERTIES OF LIGHT

The strange particle and wave properties of light are well demonstrated by a modification of Young's double slit experiment. In Young's original experiment (1801), a beam of light impinged on an opaque screen with two parallel, narrow slits. Light passing through the slits was allowed to hit a second screen. Young did not obtain two light strips (corresponding to the two slits) on the second screen, but instead a complicated pattern of several light and dark strips. The pattern obtained can be quantitatively explained by assuming that the light behaves as waves during its passage through the system.

For the experiment to work, it is necessary for the incident light waves to be in step, i.e. the light must be spatially coherent. One way of achieving this is to let the light from a well illuminated small hole (in one more screen) hit the screen with the slits. The pattern produced is a so-called interference pattern, or to be more exact, a pattern produced by a combination of *diffraction* (see next section) in each slit and *interference* between the lights from the two slits. It is difficult to see it if white light is used, since each wavelength-component produces a different pattern. Therefore at least a coloured filter should be used to limit the light to a more narrow wave-band. The easiest way today (which Young could not enjoy) is to use a laser (a simple laser pointer works well), giving at the same time very parallel and very monochromatic light which is also sufficiently strong to be seen well.

It is easy to calculate where the maxima and minima in illumination of the last screen will occur. We can get some idea of this phenomenon of *interference* by just overlaying two sets of semicircular waves spreading from the two slits (Fig. 1), but this does not give a completely correct picture.

1

L.O. Björn (ed.), Photobiology, 1–35.
© 2002 *Kluwer Academic Publishers. Printed in the Netherlands.*

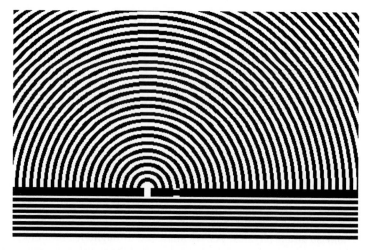

Figure 1a. Light waves impinge from below on a barrier with only one slit open, and spread from this in concentric rings.

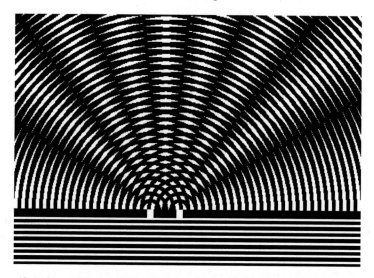

Figure 1b. Light waves impinge from below on a barrier with two slits open. The two wave systems spreading on the other side interfere and in some sectors enhance, in others extinguish one another. The picture is intended only to simplify understanding of the interference phenomenon, and does not give a true description of the distribution of light.

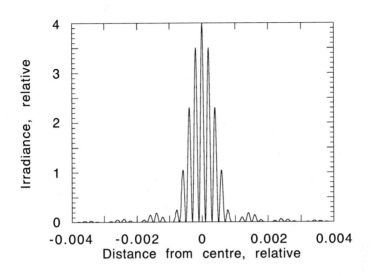

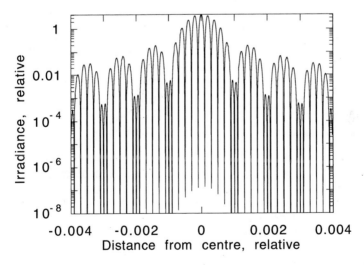

Figure. 2. Interference pattern produced in Young's double slit experiment (computer simulation). The width of each slit is 1 mm, the distance between slit centres 4 mm, and the wavelength 0.001 mm (1 μm). The distance from the centre of the screen is along the horisontal axis and the irradiance ("light intensity") along the vertical axis, both in relative units. Note that the vertical scale is linear in the upper diagram, logarithmic in the lower one.

In a direction forming the angle α with the normal to the slitted screen (i.e. to the original direction of the light), waves from the two slits will enhance each other maximally if the difference in distance to the two slits is an integer multiple of the wavelength, i.e. $d \cdot \sin \alpha = n \cdot \lambda$, where d is the distance between the slits, λ the wavelength, and n a positive integer (0, 1, 2, ...). The waves will cancel each other completely when the difference in distance is half a wavelength, i.e. $d \cdot \sin \alpha = (n+1/2) \cdot \lambda$. To compute the pattern is somewhat more tedious, and we need not go through the details. The outcome depends on the width of each slit, the distance between the slits, and the wavelength of light. An example of a result is shown in Fig. 2

So far so good, light behaves as waves when it travels. But we also know that it behaves as particles when it leaves or arrives (*vide infra*). The most direct demonstration of this is that we can count the photons reaching a sensitive photocell (photomultiplier).

But the exciting and puzzling properties of light stand out most clearly when we combine the original version of Young's experiment with the photon counter. Instead of the visible diffraction pattern of light on the screen, we could dim the light and trace out the pattern as a varying frequency of counts (or, if we so wish, as a varying frequency of clicks as in a classical Geiger counter) as we move the photon counter along the projection screen (Fig. 3a). Since we count single photons, we can dim the light considerably, and still be able to register the light. In fact, we can dim the light

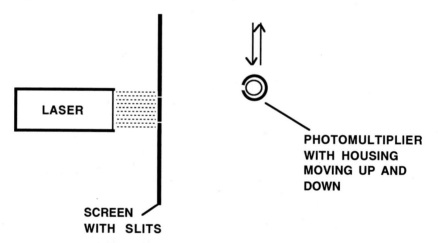

Figure. 3a. Double slit experiment set up to count single photons. The sketch is not to scale. In a real experiment the distance of the photomultiplier from the screen with slits would be greater, and the opening in the photomultiplier housing smaller.

so much that it is very, very unlikely that more than *one photon at a time* will be in flight between our light-source and the photon counter. This type of experiment has actually been performed, and it has been found that a diffraction pattern is still formed under these conditions. We can do the experiment also with an image forming device such as a photographic film or a charge coupled diode (CCD) array as the receiver and get a picture of where the photons hit. A computer simulation of the outcome of such an experiment is shown in Fig. 3b.

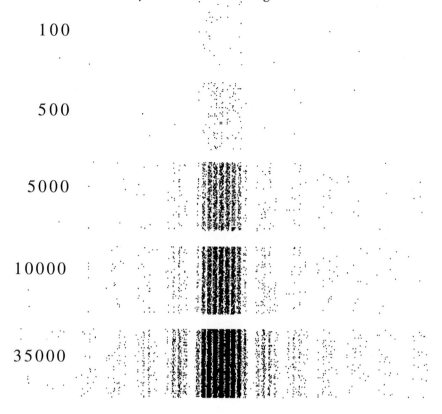

Figure 3b. Simulation of the pattern of photon hits on a screen behind a double slit arranged in the same way as in Figure 2. The number of photons is indicated for each experiment. Although the photon hits take place randomly and cannot be predicted, the interference pattern emerges more and more clearly with increasing number of photons.

If you think a little about what this means, you will be very puzzled indeed. For the diffraction pattern to be formed we need *two* slits. But we can produce the pattern by using only one photon at a time. There can be no interaction between two or more photons, which have travelled different paths, e.g. one photon through one slit and another photon through the other slit. The experiment shows that each photon "must be aware" of both slits, or, in other words, must have travelled through both slits. I know of no other physics experiment that more clearly than this one demonstrates that light *is* not waves *or* particles. The wave and the particle are both *models*, incomplete pictures or imaginations of the nature of light.

The limitations of our senses and our brain prevent us from getting closer to reality than this, simply because it has not made sense during our evolution to get closer to reality. This limitation does not prevent us from using our models very successfully as long as we use them in a correct way.

Let us take one more example to make clear how "weird" (i.e. counterintuitive) the scientific description of the behaviour of light is. When I was younger I used to watch the Andromeda galaxy using my naked eyes (now it is difficult, not only because my vision has worsened, but because there is so much electric light around where I live). I could see the galaxy because atoms in it had emitted light about two million years earlier. The photons after having travelled through empty space interacted with rhodopsin molecules in my eyes. But no photon started on its course following a straight line towards the Earth. It travelled as an expanding wave. Just before interacting with the rhodopsin molecule in my eye, the photon was *everywhere* on a wavefront with a radius of 2 million light years. The energy of the photon was not localized until it interacted with my eye.

3. LIGHT AS PARTICLES AND LIGHT AS WAVES, AND SOME DEFINITIONS

When we are dealing with light as waves, we associate a wavelength to each wave. Visible light has wavelengths in a vacuum in the range 400 to 700 nm (one nm or nanometre equals 10^{-9} m), while ultraviolet radiation has shorter and infrared radiation longer waves.

Photobiologists divide the ultraviolet part of the spectrum into ultraviolet A (UV-A) with 315-400 nm wavelength, UV-B (280-315 nm wavelength) and UV-C (<315 nm wavelength). You may see other limits for these regions in some publications, but the ones supported by the organisation called CIE (Comité Internationale de l'Eclairage), which introduced the concepts. Just as everybody should use the same internationally agreed metre, everybody should honour the definitions of UV-A, UV-B, and UV-C; otherwise there is a risk for chaos in the scientific literature. Plant photobiologists, for which the spectral region 700 to 750 nm is especially important, call this radiation "far-red light". They also call the region 400 to 700 nm "photosynthetically active radiation" or PAR, rather than visible light. Just as radiation outside this band is perfectly visible for some organisms such as some insects, birds and fish (and some light in the range 400 to 700 nm invisible to many animals), so radiation with wavelengths shorter or longer than "photosynthetically active radiation" is photosynthetically active to many organisms.

Natural light never has a single wavelengths, can rather be regarded as a mixture of waves with different wavelengths.

When we characterise light by its wavelength, we usually mean the wavelength in a vacuum. When it travels through a vacuum, the velocity of light is always *exactly* 299792.4562 km s^{-1}, irrespectively of wavelength and the movement of the radiation source in relation to the observer. The reason that this value is exact is that our definitions of the metre and the second are linked by the velocity of light in a vacuum. This velocity is usually designated c, and wavelength λ (the Greek letter lambda). A third property of light which we should keep track of is its frequency, i.e. how many times per time unit the wave (the electric field) goes from one maximum (in one direction) to another maximum (in the same direction). Frequency is traditionally designated ν (Greek letter nu), and in a vacuum we have the following relation between the three quantities just introduced: $c=\lambda \cdot \nu$, or $\lambda=c/\nu$, or $\nu=c/\lambda$. When light passes through matter (such as air or water or our eyes), the velocity and wavelength decrease in proportion, and frequency remains unchanged. Sometimes the wavenumber, i.e. $1/\lambda$, is used for the characterisation of light. It is usually symbolized by n with a line (bar) over it, and a common unit is cm^{-1}.

When we think of light as particles (photons), we assign an amount of energy (E) to each photon. This energy is linked to the wave properties of the light by the relations $E=h \cdot \nu$, where h is Planck's constant, 6.62617636 J s (joule-seconds). It also follows from the preceding that $E=h \cdot c/\lambda$. We can never know the exact wavelength, frequency, or energy of a single photon.

4. DIFFRACTION

We usually think of light travelling in straight lines if there is nothing in its way. We have seen in Young's double slit experiment that it does not always do that. In fact, the great physicist Richard Feynman has shown that its behaviour is best understood if we think of it as always travelling every possible way at the same time, and components travelling those different ways interfering with one another at everyp possible point.

We do not have to have two slits to show how the light "bends" near edges. This "bending" is called diffraction in scientific terminology. It is very important to take diffraction into account to understand some biological phenomena, such a the vision of insects (Chapter 9). Light is diffracted in any small opening and also near any edge. To compute the diffraction pattern we can make use of something called Huygens's principle (sometimes the Huygens-Fresnel principle). It states that we can think of propagating light as a sum of semispherical waves emanating from a wavefront. If the wavefront is flat, the semispherical waves emanating from it add up to a new flat wavefront. But if something stops some of the semispherical waves, the new wavefront is no longer flat. In Fig. 3 we illustrate this in one plane. Flat waves impinge from below on a screen with an opening. Many semicircular waves start out from the opening. Along a line from the middle of the opening the resulting wavefront is flat, but at the edges the semicircular waves produce a bent pattern. We have calculated this pattern more exactly in Fig. 4.

5. POLARIZATION

Light waves are *transverse* , i.e. the oscillation is perpendicular to the direction of wave propagation, the direction of the light (this is in contrast to sound waves, in which particles vibrate in the line of wave propagation). In the case of light, there are no vibrating particles, but a variation in electric and magnetic fields. The electric and magnetic fields are both perpendicular to the direction of propagation, but also perpendicular to one another. When the electric fields of all the components of a light beam are parallel, the beam is said to be *plane-polarized* . The *plane of polarization* is the plane which contains both the electrical field direction and the line of propagation.

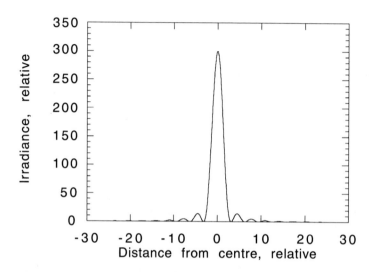

Figure 4. Diffraction pattern in a single slit (the pattern from a round hole looks similar in one dimension, but is slightly different). The horisontal axis shows the sine of the deviation angle in units of the ratio beween wavelength and slit width.

If we add two beams which travel in the same direction and are both plane-polarised and have the same *phase* (i.e. the waves are in step) but different planes of polarisation, the resulting light is also plane-polarised with its plane of polarisation at an intermediate angle.

Light can also be circularly polarized, in which case the electrical field direction spirals along the line of propagation. Since such a spiral can be left- or right-handed, there are two kinds of circular polarization, left-handed and right-handed.

Circularly polarized light can be regarded as the sum of two equally strong plane-polarized components with right angles between the planes of polarizations, and a 90 degree *phase difference* between the components. On the other hand plane-polarized light can be regarded as a sum of equally strong left- and right-handed components of circularly polarized light.

Natural light, such as direct sunlight, is often almost unpolarized, i.e. a random mixture of all possible polarizations. After reflection in a water surface the light becomes partially plane-polarized. Skylight is a mixture of circularly and plane-polarized light which we call elliptically polarized light. We cannot directly perceive the polarization of the light we see. Insects do, and often use the polarization of skylight as an aid in their orientation. Plants in many cases react differently to plane-polarized light depending on its plane of polarization. This holds for chloroplast orientation in seed plants, mosses and green algae, and also for growth of fern gametophytes. A good treatise on the subject (in German) is provided by W. Haupt (1977).

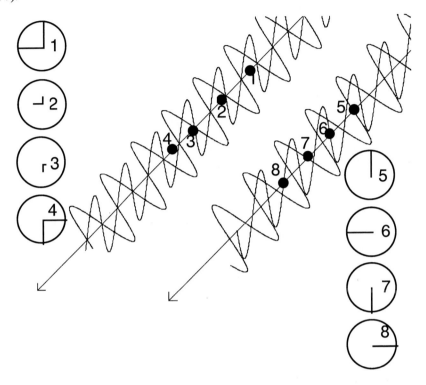

Figure 5. In the upper left part of the figure a plane-polarised light beam, composed of one vertically and one horisontally polarised component, is depicted in perspective, and also "head on" at different points (or at one point at different moments). Numbered points in the perspective drawing correspond to the numbers on the "head on" drawings. Only the electric components of the electromagnetic fields are shown (wavy lines in the perspective drawing, straight lines in the "head on" drawings. In the lower right part of the drawing the same is shwon for a circularly polarised beam.

6. STATISTICS OF PHOTON EMISSION AND ABSORPTION

Usually the members of a population of excited molecules can be expected to emit photons independently of one another, i.e. the time of emission of one photon does not depend on the time of emission of another photon. One exception to this rule occurs when stimulated emission becomes significant, as happens in a laser. Another exception is when there is cooperation between different parts of a cell (e.g., when a dinoflagellate flashes), between different cells in an organism (e.g., when a firefly flashes), or between different individuals in a population (e.g., when fireflies in a tree send out synchronized flashes). The examples in the last sentence are very obvious. However, careful study of the statistics of photon emission offers a very sensitive way of detecting cooperation between different parts of a biological system, and we shall therefore dwell a little on this subject, which also has a bearing on the reliability of measurement of weak radiation in general.

When photons are emitted independently of one another, the distribution of emission events in time is a Poisson distribution, just as in the case of radioactive decay. This means that if the mean number of events in time Δt is x, then the probability of getting exactly n events in the time Δt is $p = e^{-x} \cdot x^n / n!$. In this formula, n! stands for factorial n, i.e. $1 \cdot 2 \cdot 3 \cdot 4 \cdot \ldots \cdot n$. Thus $1! = 1$, $2! = 2$, $3! = 6$, $4! = 24$ and so on. By definition $0! = 1$.

We are familiar with the Poisson distribution of events from listening to a Geiger-Müller counter. That events are Poisson-distributed in time means that they are completely randomly distributed in time. When one event is going to take place does not depend on when a previous event occurred. One might think that there cannot be much useful information to be extracted from such a random process, but such a guess is wrong. The reader is probably already familiar with some of the useful things we can learn from the random decay of atomic nuclei. We can, in fact, use our knowledge of how Poisson statistics works for determining the number of photons that are required to trigger a certain photobiological process. The remarkable thing is that we can do this even without determining the number of photons we shine on the organism that we study.

The principle was first used by Hecht et al. (1942) to determine how many photons that have to be absorbed in the rods of an eye to give a visual impression. Their ingenious experiment was a bit complicated by the fact that our nervous system is wired in such a way that several rods have to be triggered within a short time for a signal to be transmitted to the brain (thereby avoiding false siganalling due to thermal conversion of rhodopsin). We shall demonstrate the principle with a simpler example, an experiment on the unicellular flagellate *Chlamydomonas* (Hegemann & Marwan 1988). This organism swims around with two flagella, and it reacts to light by either stopping ("stop response") or by changing swimming direction ("turning response").

All one has to do is to take a sample of either light-adapted or dark-adapted *Chlamydomonas* cells, subject them to a flash of light, and note which fraction of the cells that either stop or turn. The experiment is then repeated several times with the flash intensity varied between experiments. The absolute fluence in each flash need not be determined, only a relative value. If one possesses a number of calibrated filters no light measurement at all need be performed. Then the fraction of reacting cells for each flash is plotted against the logarithm of the relative flash intensity. It turns out that (for dark adapted cells) the curve so obtained, if plotted on a comparable scale, has the same shape as the curve labelled n=1 in Fig. 6. This holds for both stop response and for turning response, and it means that both responses can be triggered by a single photon. If the experiment is carried out within 20 minutes of removing the cells from strong light, the stop response curve has a shape similar to the curve labelled n=2 in Fig. 6, meaning that in this case two photons are required.

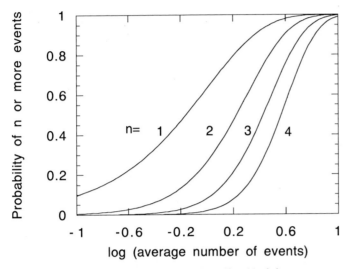

Figure 6. The probability that at least a certain number (n) of absorption events will occur during a sampling time plotted against the logarithm of the average number of events that would occur during a large number of similar samplings. It is seen that the shape of the curves depends on the value of n. If at least n absorption events is necessary for inducing a process, one can determine the number n by plotting the frequency of successful inductions against the logarithm of fluence and compare the shape of the curve obtained with the above diagram.

The curves in Fig. 6 have been computed in the following way (let, as before, x be the average number of events recorded in a large number of trials): The curve for n=1 is the probability (p) of absorption of at least one photon, which is one minus the probability for absorption of no photon, or $p=1-e^{-x}x^1/1!=x/e^x$. The curve for n=2 follows the formula $p=1- e^{-x}x^1/1!-e^{-x}x^2/2!$, the curve for n=3 follows the formula $p=1- e^{-x}x^1/1!-e^{-x}x^2/2!-e^{-x}x^3/3!$ etc.

7. HEAT RADIATION

The term heat radiation is sometimes (erroneously) used synonymously with infrared radiation. We shall use it here as the energy emitted when the energy of the random heat movement of the particles in condensed matter (solids, liquids, or compressed gases) is converted to radiation. It is easiest to think of heat radiation as the glow of a hot body (lamp filament, or the sun), but our own bodies also emit heat radiation, as does, in fact, a lump of ice or even a drop of liquid nitrogen. A body that is cooler than its environment absorbs more radiation than it emits, but still it radiates according to Planck's radiation law to be described below. Heat radiation may be infrared, visible or ultraviolet, and, if we go to exotic objects in the cosmos, even outside this spectral range.

The starting point of the quantum theory was the attempt to explain the spectrum of the radiation emitted by a glowing body. To derive a function which matched the observed spectrum, Planck had to assume that the radiation is emitted in packets (quanta or photons) of energy $h \cdot n$, where n (the Greek letter nu) stands for frequency (which is also the velocity of light divided by the wavelength) and h is a constant, Planck's constant = $6.62620 \cdot 10^{-34}$ J·s (joule-seconds). Planck's radiation law was derived for an ideal black body, best approximated by a hollow body with a small hole in it. With modifications it can be used for other bodies as well. The sun radiates almost as a black body.

Planck's formula can be written in different ways, depending on whether we consider radiation per frequency interval or per wavelength interval, and whether we express the radiation as power (energy per time) or number of photons (per time). Furthermore, we may be interested in the radiation density inside a hollow body (mostly for theoretical purposes), or the radiation flux leaving a body (for most applications).

energy density per frequency interval = $(8\pi h/c^3) \cdot v^3/(e^{hv/kT}-1)$

photon density per frequency interval = $(8\pi/c^3) \cdot v^2/(e^{hv/kT}-1)$

energy density per wavelength interval = $8\pi hc \cdot \lambda^{-5}/(e^{hv/kT}-1)$

photon density per wavelength interval = $8\pi \cdot \lambda^{-4}/(e^{hv/kT}-1)$

These functions are mostly plotted with v or λ as the independent variable and T as a parameter. It should be noted that even for the same T the functions all have maxima at different values of v or λ (see Fig. 7, which shows the plots of energy per wavelength interval and photons per wavelength interval for 5000 K.

These examples are shown merely as an illustration of the fact that the maxima occur at different locations depending on which principle you use for plotting the spectra. This is not only true for heat radiation. It holds for all emission spectra, also for fluorescence emission spectra for instance. The most common sin of people publishing about fluorescence is that they do not understand this. They write "fluorescence, relative" on their vertical axis without further specification, and do not realize, that not even the shape of their spectrum, nor the positions of maxima will be defined in such graphs. The second most common way of sinning is to spell fluorescence incorrectly.

You can see from Fig. 8 that the maxima occur at longer wavelength when the temperature is lower, and also that the total radiation is less in that case. In fact, the wavelength of the maximum is inversely proportional to absolute temperature (Wien's law), while the total photon emission is proportional to the third power of the absolute temperature (i.e. to T^3) and the total energy emission to the fourth power (T^4, Stefan-Boltzmann's law). Wien's and Stefan-Boltzmann's laws can both be derived from Planck's radiation formula, but were found experimentally before Planck did his theoretical derivation.

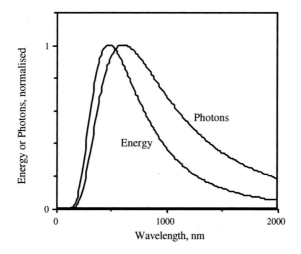

Figure 7. Blackbody radiation (5000 K) plotted as photons per wavelength interval and as energy per wavelength interval.

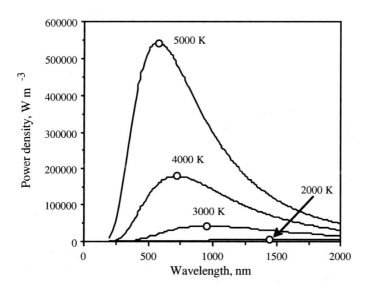

Figure 8. Blackbody radiation plotted as power per wavelength interval for different temperatures. Note that since the graphs show power (i.e, energy per time) per area and per wavelength interval, the dimension is power per volume and the unit W m⁻³. The maximum of each curve is indicated by a circle.

The formulae shown above refer to radiation density in a closed cavity with radiating walls. The *fluence rate*, or amount of radiation per unit of time and unit of *cross-sectional* area falling from all directions on a sphere in this cavity is obtained by multiplying the radiation density by the velocity of light. (Do not worry if you have some difficulty with this here. We shall return later to the concept of fluence rate, which is quite important in photobiology and often misunderstood.)Suppose that the sphere in the cavity is ideally black (absorbing all the radiation falling on it) and of the same temperature as the walls. The second law of thermodynamics states that (assuming that no heat energy is generated or consumed in the sphere) the sphere must stay at the same temperature as the walls, and it must radiate the same amount of radiation (distributed in the same way across the spectrum) as it receives. Therefore its *excitance* (radiation given off per unit of time and unit of *surface* area) is the energy density given by the formulae above multiplied by the velocity of light and divided by 4 (since the *surface* area of the sphere is 4 times the *cross-sectional* area).

To obtain the excitance of a non-black body (such as a glowing tungsten filament in a light bulb, or your own body), the excitance computed for a black body should be multiplied by the *emissivity*. The emissivity varies quite a lot with wavelength, so the multiplication must be carried out separately for each wavelength value in which you are interested. The emissivity also varies somewhat with temperature. The *absorptivity*, or the ability to absorb radiation, is identical to the emissivity; otherwise the second law of thermodynamics would be violated.

It may seem that this is a little too much physics for a biology book, but an understanding of the basic physical principles is very helpful when it comes to the experimental work in photobiology. What has just been described can be used for calibrating measuring equipment in the photobiology laboratory.

8. REFRACTION OF LIGHT

From school you should be familiar with Snell's law. This describes how light is refracted at an interface between two media with different indices of refraction (refractive indices), say n_1 and n_2. Fig. 9, in which we assume $n_1 < n_2$, will serve as a reminder. If you need further explanation you will have to look in other books.

The refractive index can be regarded as the inverse of the relative velocity of light in the medium in question, i.e. it is the velocity in a vacuum divided by that in the medium. It can be shown that Snell's law is equivalent with the statement that the light takes the fastest path possible between any two points on the rays shown. Compared to a straight line (dashed in Fig. 9) between point A on the upper ray and point B on the lower ray, you can see that the light goes a longer distance (solid line) in the medium with refractive index n_1 (lower index, higher velocity) than in the medium with refractive index n_2 (higher index, lower velocity), i.e. AO>AC and OB<CB. The refractive index is a pure number (no unit, as it is the ratio of two velocities). As we have used it here it is a real number (the usual type of number we use in most calculations, represented as a decimal number). In more advanced optical theory the refractive index is a complex ("two-dimensional") number.

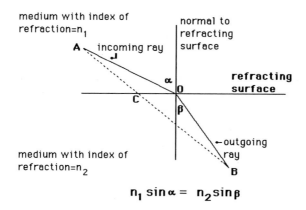

$$n_1 \sin \alpha = n_2 \sin \beta$$

Figure 9. Refraction of light in a plane interface between transparent materials.

As for the values of α and β in relation to one another, the figure looks the same if the light direction is reversed. However, this does not hold any longer when reflection is taken into account, nor when we consider the amount of light in the beams.

Throughout most of the spectrum the refractive index decreases with wavelength, but there are spectral regions (where absorption bands occur) where it increases steeply with wavelength; this phenomenon is, for historical reasons, called *anomalous dispersion* , although it is quite normal. In general, the change in refractive index with wavelength is called dispersion.

In some crystals and many biological materials the refractive index is different depending on direction and plane of polarization of the light. Such a medium is termed *birefringent* . Birefringence occurs in plant cell walls and other structures where elongated molecules are arranged in a certain direction. Measurement of birefringence has been an important method in elucidating the arrangement of molecules in such cases. Media that are originally *isotropic* (with the same properties in different directions, and thus not birefringent) may become birefringent by stretching or squeezing, application of electric fields, or other treatments.

When light passes through a birefringent medium of suitable thickness it becomes circularly or elliptically polarised because of the phase difference that develops between the components of different plane polarisation.

9. REFLECTION OF LIGHT

Reflection may be *specular* (from a shiny, smooth surface or interface) or *diffuse* (from a more or less rough surface or interface). Diffuse reflection is very important in biology, but we shall limit ourselves here to specular reflection at interfaces between dielectric (non-metallic) media.

The angle of incidence is alsways equal to the angle of reflection, but the amount of light reflected (as opposed to refracted) depends on the polarization of the light. The plane in which both the incident and the reflected ray (and the normal to the reflecting surface) lie is called the *plane of incidence* . The component of the light with an electric field parallel to this plane is designated by //, that with an electric field perpendicular to the plane of incidence by +. The fractions, R// and R+, of the irradiance of these components that are reflected are given by Fresnel´s equations, in which α is the angle of incidence (equal to the angle of reflection) and β the angle of transmission (see Fig. 9 in the section on refraction):

$$R// = [\tan(\alpha-\beta)/\tan(\alpha+\beta)]^2$$
$$R+ = [\sin(\alpha-\beta)/\sin(\alpha+\beta)]^2$$

The reflected fraction of unpolarized light is the mean of the two ratios. For normal incidence ($\alpha=\beta=90^0$) another set of equations has to be used, since with the equations above, divisions by zero would occur. In this case there is no distinction between R// and R+:

$$R = [(n_1-n_2)/(n_1+n_2)]^2$$

As an example of use of the last equation, let us consider the reflection in a glass plate ($n_2=1.5$) in air ($n_1=1$). When light strikes the glass plate (perpendicularly), R = $[(1-1.5)/(1+1.5)]^2 = [-0.5/2.5]^2 = 0.04 =4\%$. When the light strikes the second interface (from glass to air) the value of R comes out the same again, because since the expression is squared, it does not matter in this case which of the indices you subtract from the other one. Thus 96% of the 96% of the original beam, or 92.16%, will be transmitted in this "first pass". It can be shown that after an infinite number of passes between the two surfaces the reflected fraction will be R[1+(1-R)/(1+R)] = $2\cdot0.04/(1+0.04) = 7.69\%$ and the transmitted fraction 92.31%. For most practical purposes we may estimate a reflection loss at normal incidence of about 8% in a clean glass plate or glass filter, but if the refractive index is exceptional this value may not hold. If the glass is not clean, it certainly does not.

The multiple internal reflection is not of much effect in a single glass plate, but I wanted to mention it here, because the effect is taken advantage of in so-called interference filters to be described in a later section.

Going back to the case of a<90^0, we find by division, member by member, of the equations above, that R// divided by R+ is $[\cos(\alpha-\beta)/\cos(\alpha+\beta)]^2$. This ratio will always be >1, so R// > R+, or, in other words, the component of light with the electric field perpendicular to the plane of incidence and parallel to the interface will be more easily reflected than the other component. The interface can act as a polarizing device. It can be shown that the reflected beam becomes completely polarised when tan α = n_2/n_1, because none of the light polarised perpendicular to the plane of incidence is reflected (Fig. 10). The angle $\alpha=\arctan(n_2/n_1)$ is called the *Brewster* angle.

If a beam strikes a flat interface obliquely from the side where the refractive index is highest, the outgoing beam will have a greater angle to the normal than the ingoing (according to Snell's law). If the angle of incidence is increased more and more, an angle will eventually be reached when the outgoing beam is parallel to the interface. At greater angles of incidence there will be total reflection, i.e. all light will be reflected, and none transmitted. The smallest angle of incidence at which total reflection occurs is called the *critical angle*.

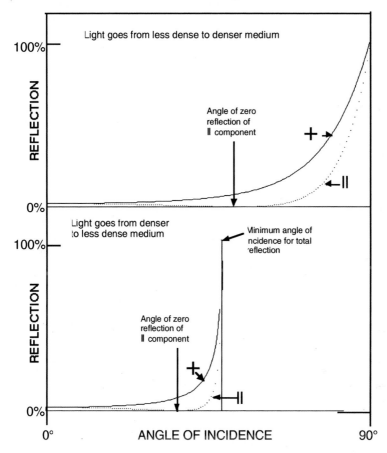

Figure 10. Percent of light reflected for different angles of incidence for light going from air (n=1) to water (n=1.33, top) and from water to air (bottom), and for light polarised with the electric vector in the plane of incidence (II) or perpendicular to the plane of incidence (+). For II-polarised light no light is reflected for a certain angle of incidence (the Brewster angle). For light going from the denser medium (water) to the less dense medium, total reflection occurs for angles of incidence larger than the critical angle.

10. SCATTERING OF LIGHT

Although, strictly speaking, also reflection and refraction are due to scattering (absorption and reemission of electromagnetic energy by material oscillators) we do, in practice, use the term scattering in a more restricted sense for processes which tend to change the propagation of light from an ordered way to a random one. We can distinguish three types of scattering named after three distinguished scientists: Mie scattering, Rayleigh scattering and Raman scattering.

Mie scattering is caused by particles larger than the wavelength of the light and having a refractive index different from that of the continuous phase in which they are suspended. Typical examples are water droplets (clouds, fog) or dust in the atmosphere, or the result of mixing a solution of fat in acetone with water. Also any animal or plant tissue is a strong Mie scatterer, due to the boundaries between cells and between different parts of the cells, and, in the case of plant tissue, between cells and intercellularies. Mie scattering is nothing else than repeated reflection and refraction at numerous interfaces. As we have seen, light of different wavelength is not reflected or refracted in exactly the same way, but most of these differences cancel out in Mie scattering, and there is no strong wavelength dependence of this phenomenon.

Rayleigh scattering is caused by the interaction of light with particles smaller than the wavelength of the light. The particles may even be individual molecules or atoms. In this case there are no interfaces at which reflection or refraction can take place. However, the closer the wavelength of the light is to an absorption band of the scattering substance, i.e. the closer the frequency of the light is to a natural oscillating frequency in the matter, the more strongly the electrons in the matter "feel" the light and the greater is the probability that the electromagnetic field is disturbed when it sweeps by. Most substances have their strongest absorption bands in the far ultraviolet. Therefore, in the infrared, visible and near ultraviolet regions, Rayleigh scattering increases very steeply towards shorter wavelength. Ultraviolet is scattered more strongly than blue, which in turn is scattered more strongly than red. The blue colour of the sky is due to more blue than red light being scattered out of the direction of the direct sunlight.

To be more precise, Rayleigh scattering is inversely proportional to the fourth power of the wavelength, i.e. proportional to $1/\lambda^4$.

In Rayleigh scattering the direction of the electrical field is not changed. If, for instance, a horisontal beam, vertically polarized (i.e. with the electric field vertical), is scattered, the electric field remains vertical. But light can never propagate in the direction of its electric field (remember, it is a transverse wave). This means that the light is not scattered up or down, only in horizontal directions. If, on the other hand, the incident light is not polarized, it is scattered in all directions, but with different polarizations.

In both Mie and Rayleigh scattering the wavelength of the light remains unchanged. In Raman scattering, on the contrary, either part of the photon energy is given off to the scattering particles (which in this case are molecules), or some extra energy is taken up from the particles. The amount of energy taken up or given off corresponds to energy differences between vibrational levels in the molecule. Raman scattering can be used as an analysis method, and is also a source of error in fluorescence analysis, but we do not need to consider it in photobiological phenomena, since it is always very weak.

11. PROPAGATION OF LIGHT IN ABSORBING AND SCATTERING MEDIA

We shall consider here first the simplest case: A light beam (irradiance I_0) that perpendicularly strikes the flat front surface of an homogeneous non-scattering but absorbing object. The most common objects of this kind that we deal with in the laboratory are spectrophotometer cuvettes and glass filters. A small fraction of the incident light is specularly reflected at the surface according to Fresnel's equation (see section 9). For simplicity we disregard this in the present section. In spectrophoto- metry reflection is taken care of by comparing a sample with a reference cuvette having approximately the same reflectivity as the sample cuvette.

At depth x within the object the irradiance (see Chapter 2 for definitions of irradiance and other terms) will be $I_x = I_0 \cdot e^{-Kx}$, where K is the linear absorption coefficient. The relationship is known as Lambert's law, and follows mathematically from the conditions that (1) the light is propagated in a straight line and (2) that the probability of a photon being absorbed is the same everywhere in the sample.

In spectrophotometry we also make use of Beer's law, which states that, under certain conditions, K is a product of the molar concentration of the absorbing substance and its molar absorption coefficient (or, in the case of several absorbing substances, the sum of several such products).

However, we are concerned now not with spectrophotometry, but with the propagation of light in living matter. Almost invariably we will be facing complications caused by intense scattering. A general quantitative treatment of scattering is so complicated as to be useless for the photobiologist. All it would lead to would be a system of equations with mostly unknown quantities.

A simplified theory, which has been found very useful as a first approximation is the Kubelka-Munk theory (Kubelka & Munk 1931). It should be observed that this theory is valid only for "macrohomogeneous" objects, i.e. those that on a macroscopic scale are uniform and isotropic, with absorption and scattering coefficients that can be determined. Seyfried and Fukshansky (1983) have shown how the theory can be modified for an object consisting of several macro-homogeneous layers. Specular reflection at the surfaces has to be dealt with separately. Uncertainty in the specular reflection leads to uncertainties in the absorption and scattering coefficients if they, as proposed by Seyfried and Fukshansky, are determined from overall reflection and transmission by the object. In any case the method is good enought to demonstrate here the general features of light propagation in media which both absorb and scatter light.

Suppose we can determine, with sufficient confidence, the reflectance R (except for specular reflectance) and transmittance T of our sample. The linear absorption coefficient K and the linear scattering coefficient S, as well as the fluence rate at any point inside the sample can then, with some effort, be computed from the system of equations

$R = 1/[a+b\cdot coth(bSd)]$

$T = b/[a\cdot sinh(bSd)+b\cdot cosh(bSd)]$

$a = (S\underline{+K})/S$

$b = \sqrt{(a^2-1)}$

$I_x = I_o\cdot T\cdot[\{(a+1)/b\}\cdot sinh\{bS\cdot(d-x)\}+cosh\{bS\cdot(d-x)\}]$

Here I_o is the fluence rate incident from one side and I_x the fluence rate at depth x of a sample of overall thickness d. The so-called hyperbolic operators sinh, cosh and coth are defined by the following relationships: $sinh(y) = (e^y-e^{-y})/2$; $cosh(y) = (e^y+e^{-y})/2$; $coth(y) = cosh(y)/sinh(y)$. If light is incident from both sides, the last equation has to be modified.

To demonstrate, without too much computation, the effect of scattering, we shall assume that we have determined K and S. For any sample thickness, d, we can then compute I_x as a function of x.

Note the following features in the examples of computer outputs below Fig. 11-12:

1. When S is given a low value (0.01) the Kubelka-Munk curve coincides with the Lambert curve (and is therefore invisible).

2. When S has a value similar to, or higher than K, i.e. when scattering is appreciable compared to absorption, the fluence rate in the sample near the illuminated side is higher than the incident fluence rate. This is no violation of the law of energy conservation; the sample does not create any new light. However, the concentration of photons is increased by their bouncing back and forth.

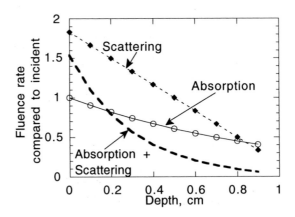

Figure. 11a. Decrease of fluence rate with depth. See Fig. 11b for details.

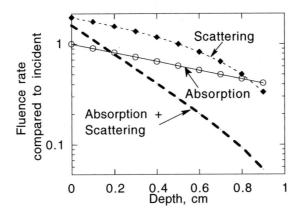

Figure 11b. The decrease of fluence rate with depth of penetration in a 1 cm thick slab of a medium with absorption only (linear absorption coefficient 1 cm⁻¹), scattering only (linear scattering coefficient 5 cm⁻¹), and one with both absorption (1 cm⁻¹) and scattering (5 cm⁻¹). The values were computed using Kubelka-Munk theory and assuming isotropic incident light. In the upper frame the fluence rate scale is linear, in the lower one logarithmic. Note that in a scattering medium fluence rate can exceed the fluence rate of the incident light. Fig. 11a is a linear plot, Fig. 11b has a logarithmic vertical scale.

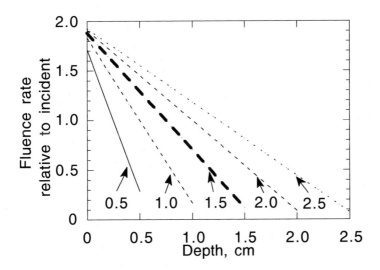

Figure 12. Decrease of fluence rate in layers of the indicated thickness (in cm) of a purely scattering medium (no absorption); the linear scattering coefficient is 5 cm⁻¹ in all cases. Note that the fluence rate decreases more quickly in a thin scatterer, because there is less backscatter of light.

The expediency with which the Kubelka-Munk equations can be evaluated using a computer must not cause us to forget the limitations of the Kubelka-Munk theory. One severe restriction is that only diffuse incident light is considered. We need only enter 3 constants, K, S and thickness of the scattering medium, to describe the scattering object. A more complete description would give more realistic results, but apart from the difficulty in choosing the correct constants, the equations and algorithms would rise in complexity very fast. More complete theories are described by Star et al. (1988) and Keijzer et al. (1988).

12. SPECTRA OF ISOLATED ATOMS

We shall deal in this section with isolated atoms (which are not part of di- or polyatomic molecules, and also not close to one another for other reasons, sucn as high pressure). They can increase their internal energy by absorbing photons, and also give off energy by emitting photons. They can absorb or emit only very particular photons, whose energy corresponds very exactly to differences between energy levels in the atom. The simplest case is the hydrogen atom, and it has been found that its energy levels are inversely proportional to $1/n^2$, where n represents positive integers (1, 2, 3...). The possible energy jumps are then proportional to the energy differences $1/n^2 - 1/m^2$, where n=1,2,3.... and m= n+1, n+2.... Since, according to the relationships $E = h \cdot v$ and $\lambda = c/v$, the energy (E) of a photon is inversely proportional to wavelength (λ), the wavelengths of light which can be absorbed or emitted by a hydrogen atom are given by $1/\lambda = R*(1/n^2 - 1/m^2)$, see Fig. 13. The proportionality constant is called the Rydberg constant.

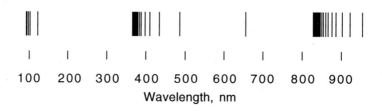

Figure 13. Spectrum of atomic hydrogen (computer simulation).

In ordinary hydrogen gas the atoms are combined in pairs. However, when an electric current runs through the gas, the pairs are split, and photon emission from energized (excited) free hydrogen atoms takes place. In the laboratory we use lamps containing heavy hydrogen (deuterium), for instance in the spectrophotometer. We use the continuous part of the spectrum in the ultraviolet (as well as continuous emission in the ultraviolet arising from molecular deuterium) for measuring ultraviolet absorption of samples. We use the two first lines of the Balmer series (H_α at 656 nm and H_β at 486 nm) for wavelength calibration. In the program above an approximate Rydberg constant in between that for light and heavy hydrogen was used. You can enter the more exact value for heavy hydrogen if you want to get accurate values for a deuterium lamp.

In Nature the H_α, H_β and some other hydrogen lines appear in the spectrum from the sun as absorption lines (Fraunhofer lines), because of the presence of non-excited hydrogen atoms in the atmosphere of the sun outside the glowing photosphere. Light of these particular wavelengths is therefore absent in the daylight spectrum. The absence of H_α light from daylight should make it possible to measure other light (e.g., fluorescence) at this wavelength in full daylight. However, the chlorophyll fluorescence from plants is weak at such a short wavelength.

One other case where the photobiologist is concerned with the spectrum of isolated atoms is when he uses low-pressure mercury lamps. We shall return to this in Chapter 3.

13. ENERGY LEVELS IN DIATOMIC AND POLYATOMIC MOLECULES

The energy relations immediately become much more complex when we proceed from single atoms to molecules consisting of two atoms each, i.e. diatomic molecules. The simplest example of such a molecule (if we disregard the hydrogen molecular ion H_2^+) is the hydrogen molecule, H_2.

In the molecule we have, in addition to the electronic energy described for the atom, also vibrational and rotational energy. In diatomic molecules the bond between the atoms, mediated by the electrons, can be regarded as an elastic string or spring, which stretches and contracts. At one instant the nuclei of the two atoms move towards one another. When the positively charged nuclei come close enough, their mutual electric repulsion becomes strong enough to reverse the motion, and the distance between the nuclei starts to increase. The nuclei move apart until the attractive force from the negatively charged electrons becomes strong enough to reverse the motion again.

This oscillating movement of the nuclei has some resemblance to that of a pendulum, but one difference is that it is asymmetric. The force on the nuclei is not proportional to the distance from a symmetry point, and therefore the molecule is an inharmonic oscillator rather than an harmonic one.

The changes in energy due to changes in oscillating movement are smaller than (the largest) energy jumps due to electronic transitions (changes in electronic energy).

In molecules consisting of more than two atoms each there are also oscillations due to the bending of bonds, but we shall disregard this in the following.

The molecule can also absorb or emit energy by changing its state of rotation. In diatomic molecules only rotation around an axis perpendicular to the bond contributes to the rotational energy, but in more complicated molecules we must consider three axes of rotation, all perpendicular to one another.

All these energy changes are quantized, i.e. only certain energy changes are possible. However, because the vibrational and rotational energy amounts are much smaller than the electronic energy amounts, and are combined with them, the molecules have apparently continuous absorption and emission bands rather than lines. At equilibrium, the number of molecules (N_x, N_y) "occupying various energy states" as the jargon goes, i.e. having various amounts of energy (E_x, E_y), is related to the energy differences between the states by the formula $N_x/N_y = e^{(E_y-E_x)/(kT)}$.

We shall now restrict the discussion to the stretching vibrations and their interaction with the electronic energy transitions. At one point in the stretching oscillation the force acting on the nuclei is zero (the repulsive and attractive force compensate one another exactly). All the vibrational energy is then kinetic (translational) energy. In contrast, when the distance between the nuclei is either minimal or maximal, i.e. at the inner and outer turning points, the velocity is zero, and therefore the kinetic energy is zero. All the vibrational energy is then potential (positional) energy. In between, the kinetic and potential parts of the energy change in such a way that their sum is constant.

In the diagram below (Fig. 14) the distance between the atomic nuclei is plotted in the horizontal direction (lowest values to the left) and the energy of the molecule in the vertical direction (lowest values at the bottom). The curved lines show the potential energy for various distances and for two different electronic states of the molecule. The various horizontal lines within the curved lines show the total energy for various vibrational states and for the two electronic states. The turning points in the oscillating movement of the nuclei is where these horizontal lines reach the curves. For these interatomic distances the kinetic energy is zero.

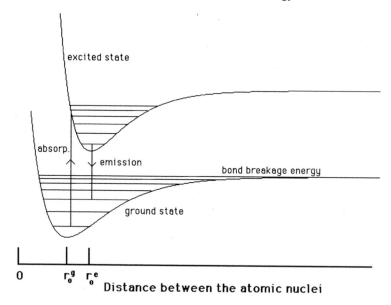

Figure 14. The potential energies (vertical coordinate) of the electronic ground state and the first excited state are shown by the curves as functions of the distance between the atomic nuclei in a diatomic molecule (horizontal coordinate). The equilibrium distances (lowest potential energy) for the ground and excited states are denoted by rog and roe, respectively. At this distance the potential energy is minimal, and the kinetic energy (distance between curves and horizontal lines) is maximal. However, the molecule never comes to rest at this position. Even at zero absolute temperature the vibrations continue (lowest horizontal lines on the curves).

Looking at Fig. 14 and the lengths of the vertical lines in it we get some understanding of why absorption maxima occur at shorter wavelength (higher photon energy) than emission maxima (this difference is referred to as the "Stoke's shift"). The maxima, which we can determine experimentally, of course correspond to the wavelengths and photon energies of the most likely transitions. Later we will see how one can look at the same phenomenon from quite a different point of view.

A macroscopic pendulum moves most slowly near the turning points. If it were possible to get snapshots of the molecule at random times, one would therefore expect most of the snapshots to show the atoms near the turning points. However, the quantum physics is more complicated than that. For the lowest vibrational state, the zero state, it is quite the opposite, and the probability is greatest that the molecule will be near the state of zero potential energy and maximum kinetic energy. Thus, when for some reason a molecule changes electronic state, in most cases the transition will occur from near the mid-point of the line for the lowest vibrational state. The vertical line to the left in Fig. 14 shows a likely transition from the lower electronic state to the higher electronic state, and the line to the right shows a likely transition from the higher electronic state to the lower one. The upward transition could be associated with absorption of photons, and the downward one by emission of photons.

Apart from the changes in vibrational and rotational energy there are other causes of the "broadening" of spectra mentioned above (from line spectra to band spectra). More complicated molecules are usually (and the biomolecules always) in a condensed phase (liquid or solid) rather than in a low pressure gas. The different molecules in the phase affect one another in complicated ways so that the energy levels of are not the same as those of its neighbours. Finally, the different molecules are not identical (as a collection of isolated atoms of the same kind are) since they, even if they correspond to a single chemical formula, may have different *conformation* , e.g. an extended or folded chain of atoms. This results in continuous absorption and emission spectra.

Because, at ordinary temperatures, transitions between different conformational states take place readily, we do not experience molecules with different conformations as different kinds of molecules. By greatly lowering the temperature we may prevent the transitions between different conformational states as well as between different vibrational and rotational states, and it becomes possible to selectively deplete one state by monochromatic light from a laser. In this way one may "burn holes" in an absorption spectrum and see which portion of a spectrum is associated with a particular state (Fig. 15).

Even at a temperature of absolute zero the oscillation continues, with all molecules in the lowest vibrational state, the zero state. In fact, even at room temperature the majority of the molecules are in this state, but a substantial fraction are in higher states.

Even at moderately lowered temperatures absorption spectra (as well as fluorescence spectra) are sharpened (Fig. 16). This effect is often taken advantage of in spectroscopic investigations of biological samples containing several substances with similar spectra, such as cytochromes or chlorophyll proteins.

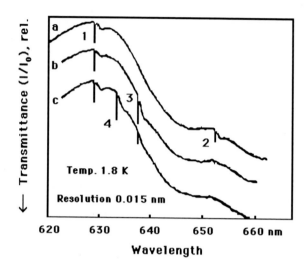

Figure 15. A solution of C-phycocyanin was irradiated with a strong laser beam. Various wavelengths were used in the order indicated by the numbers and the vertical lines on the curves. These lines are, in fact, narrow dips or "holes" in the absorption spectra caused by depopulation of specific molecular states by the laser light. Note that irradiation with light of shorter wavelength following one with light of longer wavelength causes a repopulation of the less energetic state (corresponding to the longer wavelength). For instance, irradiation 3 practically cancels the effect of irradiation 2 (but not that of irradiation 1 at an even shorter wavelength than 3). Similarly irradiation 4 partly cancels the effect of irradiation 3. From Friedrich et al. (1981), modified.

It should be understood that a molecule can appear in an electronically excited state for reasons other than having absorbed light or ultraviolet radiation. In rare cases the collisions with other molecules can give a molecule sufficient energy for transition to an excited state. Chemical reactions may also produce reactants in electronically excited states, which can lose their energy by emission of light. This is how chemiluminescence works, and bioluminescence, which will be described later, is biochemical chemiluminescence.

14. QUANTUM YIELD OF FLUORESCENCE

We have mentioned before that a molecule in the excited state may lose its energy in different way. It can send it out as light: radiative de-excitation. This process is called fluorescence if the transition is from a singlet excited state to a singlet ground state, and phosphorescence if the transition is from a triplet excited state to a singlet ground state or from a singlet excited state to a triplet ground state (the most important example of the latter is phosphorescence of singlet oxygen, see the section on experiments). It can also lose the energy as vibrations to neighbouring molecules (thermal de-excitation). Singlet excited states can disappear by "intersystem crossing" to produce triplet states; this happens, e.g., sometimes with chlorophyll molecules. And finally, energy may be lost through chemical reactions. Thus the total rate of disappearance of singlet excitation can be described

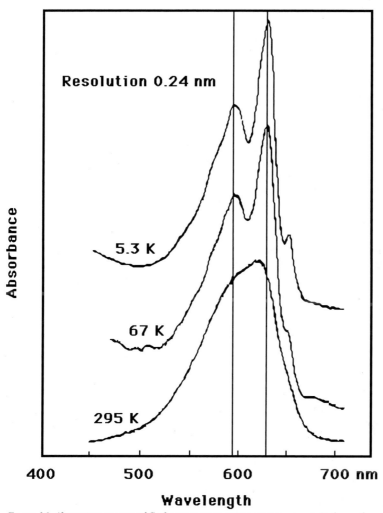

Resolution 0.24 nm

5.3 K

67 K

295 K

Absorbance

400   500   600   700 nm

Wavelength

Figure 16. Absorption spectra of C-phycocyanin at various temperatures. Redrawn from Friedrich et al. (1981).

as the sum of the rates for the different de-excitation "pathways". In most cases each molecule "acts on its own", so the kinetics of disappearance of singlet excitate states is of first order (in contrast, deexcitation of singlet oxygen is mixed first and second order at higher concentrations). It can thus be described by a first order rate constant k, which is a sum of the rate constants for the different pathways:

$$k = k_f + k_{th} + t_{ic} + k_{ch},$$

where f stands for radiative de-excitation (usually fluorescence), th for thermal de-excitation, ic for intersystem crossing, and ch for chemical de-excitation.

Under steady illumination a steady state develops, so the rate of excitation by absorption of photons equals the total rate of de-excitation. Therefore, the ratio of the number of photons emitted as fluorescence to the number of photons absorbed will be

$$f_f = k_f/k = k_f/(k_f + k_{th} + t_{ic} + k_{ch}).$$

The quantity f_f is called the quantum yield of fluorescence. In the same way we have a quantum yield for each de-excitation path, for example also for the chemical deactivation:

$$f_{ch} = k_{ch}/k = k_{ch}/(k_f + k_{th} + t_{ic} + k_{ch}).$$

The different pathways compete with one another. Therefore the chlorophyll fluorescence from a plant, which is usually weak and invisible, increases if we add a poison which stops photosynthesis, the main pathway for chemical de-excitation. The fluorescence from chlorophyll becomes even stronger and clearly visible if we extract the chlorophyll and illuminate it dissolved in an organic solvent as acetone. Then we have not only completely stopped chemical de-excitation, but also decreased thermal de-excitation.

Studying the changes of fluorescence from chlorophyll is an important way for investigating the functioning of the photosynthetic apparatus. In this context one often uses the terms photochemical quenching and non-photochemical quenching, respectively, for the chemical and thermal de-excitations competing with fluorescence.

15. THE RELATION BETWEEN ABSORPTION AND EMISSION SPECTRA

A simple relationship between absorption and emission spectra of even very complicated molecules was first hinted at by E.H. Kennard and later elaborated mostly by B.I. Stepanov. The relation is most commonly referred to as the Stepanov relationship. The basic idea is that it is of no consequence to the future behaviour of a molecule in what way it reached a certain state. From this it follows that any emission from an excited (energy-rich) electronic state of any kind of molecule *in thermal and conformational equilibrium with its surroundings* must have the same spectrum, whether the molecule reached the energy-rich state by collisions with its neighbours, or by absorbing a photon, or as a result of a chemical reaction. More specifically, the shape of the fluorescence spectrum (excited state reached by absorption of photons) is identical to the shape of the heat radiation spectrum from that kind of molecule. The heat radiation spectrum follows Planck's law for a blackbody, modified by the emissivity of the substance. But the emissivity, as was already mentioned, is the same as the absorptivity, which has the same spectral dependence as the experimentally measured absorption coefficient. Thus (fluorescence spectrum) = (absorption spectrum) x (blackbody spectrum). The multiplication sign here stands for *convolution* , i.e. multiplication of pairs of values throughout the spectrum. The fluorescence spectrum will then be expressed as photons per wavelength interval, energy per frequency interval etc. depending on how the blackbody spectrum is entered into the equation.

The Stepanov relationship breaks down when heat energy cannot easily diffuse away from the emitting molecule. This happens in solid media or liquids of high viscosity, and is always the case at low temperatures. Conversely, by comparing a fluorescence and an absorption spectrum it can be found out whether or not the molecules in the excited state are in thermal and conformational equilibrium with the surroundings at the time of photon emission. An example of the application of these ideas to a biological system is provided by Björn & Björn (1986).

16. THE MOLECULAR GEOMETRY OF THE ABSORPTION PROCESS

In a molecule the center of positive charge (associated with the nuclei) may, or may not, coincide with the center of negative charge (associated with the electrons). If the center of positive charge does not coincide with the center of negative charge, the molecule is a dipole. Unless molecules are symmetric, as CCl_4 or CH_4, they are more or less strong dipoles.

A dipole is characterized by a dipole moment. This is a vector, having direction and magnitude. The magnitude is the distance between positive and negative charge centres times the amount of charge. The direction of the dipole moment is from the negative to the positive charge centre. As other vectors, the dipole moment is often symbolized by an arrow (Fig. 17).

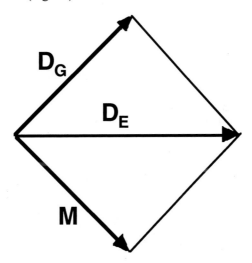

Figure 17. The transition moment M for a transition from the ground state to an excited state is the vectorial difference between the dipole moments of the molecule in the ground state (D_G) and in the excited state (D_E).

When a molecule is electronically excited, the negative charge is generally displaced in relation to the positive charge, i.e. there is a change in dipole moment. This change in dipole moment is called the transition moment, and often symbolized by **M**. Like the dipole moment it is a vector; in fact it is the dipole moment of the excited state (D_E) minus the dipole moment of the ground state (D_G). Symbolizing the vectors by arrows we may describe the subtraction as shown in Fig. 17.

The magnitude of the transition moment can be estimated from the absorption spectrum of the compound in question. Denoting the molar absorption coefficient by ε and light velocity divided by wavelength (frequency, c/λ) by ν, the oscillator strength (the square of the magnitude of the transition moment), is approximately $4.32 \cdot 10^{-9} \int \varepsilon d\nu$, where integration is carried out over the absorption band.

In most cases of excitation by light absorption, the probability of absorption is proportional to the square of the component of the transition moment in the direction of the electric field of the light. (There are cases of interaction between electrons and the magnetic field of the light rather than the electric field, but these cases are of little interest in photobiology.) Expressed in another way, the probability is proportional to $|M|^2 \cos^2 \alpha$, where α is the angle between the transition moment and the direction of the electric vector of the light wave, i.e. the direction in the plane of polarization which is perpendicular to the direction of light propagation.

To those of you who think this is hard to follow: Remember that the probability of absorption depends on how the molecule is oriented in relation to the direction of the light and (for plane polarized light) the plane of polarization. For absorption of light by molecules in ordinary solutions this is of no consequence, since the molecules (except in special cases) have random directions. For absorption of light by molecules in living cells it is sometimes very important, since these molecules may be very accurately aligned. In such cases light polarised in a direction parallel to the transition moment of the absorbing molecule is more strongly absorbed than light polarised in other directions. This phenomenon is called (absorption) dichroism.

In the same molecule there may be transition moments with different directions. For example, in the chlorophyll molecule, two transition moments are nearly at right angles to one another. The transition moment for emission of fluorescence may have a direction different from that for excitation of fluorescence. By measuring the polarisation of fluorescence from molecules irradiated by polarised light, one can gain information about the angle between the transition moments.

17. TRANSFER OF ELECTRONIC EXCITATION ENERGY BETWEEN MOLECULES

Transfer of electronic excitation energy from compound A to compound B may be symbolized as $A^* + B \Longrightarrow A + B^*$. The energy quantum to be transferred must have a size such that it can be given off by A, i.e. correspond to the enrgy of a photon within the fluorescence band of A. Furthermore, it must be of a size that can be taken up by B, i.e. correspond to the energy of a photon within the absorption band of B.

There are a few photobiological phenomena in which this energy transfer is actually mediated by a photon. As an example we may mention the transfer of energy from luciferin in the lantern of a firefly female to the rhodopsin in the eye of a firefly male. However, in the majority of cases, the radiation transfer is radiationless, a process which is much more efficient at short range. Very few of the photons emitted by firefly females happen to be absorbed in rhodopsin molecules of firefly males. The advantage of energy transfer by photons is that it can take place over distance. We also all depend on the energy transfer taking place directly between atoms in the sun and chlorophyll molecules in plants, but also this is a very wasteful process in the sense that a very small fraction of the photons emitted by the sun end up in chlorophyll molecules. On the other hand, once the quantum has been caught by a chlorophyll molecule (or a molecule of phycoerythrin or phycocyanin), it is channeled from molecule to molecule with an efficiency of practically 100% by radiationless energy transfer.

There are two main mechanisms for radiationless energy transfer, *exciton coupling* and the *Förster mechanism*. Exciton coupling occurs in a pure form in the photosynthetic antennae of green photosynthetic bacteria like *Chloroflexus*. The so called chlorosomes of these bacteria contain rods made up of bacteriochlorophyll and carotenoid molecules. The pigment molecules are so tightly packed together that the whole rod behaves almost as a single pigment molecule; the energy is delocalised. This phenomenon, called *exciton coupling*, provides very fast transfer of the energy to the reaction centre.

In other cases chromophores may just pairwise be close enough to share energy, and form what is called *exciplexes*. When exciplexes are formed, the energy levels split up.

The other mechanism, the Förster mechanism or resonance transfer, or dipole-dipole interaction, is exemplified in a rather pure form in the phycobilisomes, pigment antennae of cyanobacteria and red algae. Only a few of the chromophores in the phycobilisomes are close enough to form exciplexes. Both the Förster mechanism and the phycobilisomes are so important that we have chosen to devote special sections to them.

18. THE FÖRSTER MECHANISM FOR ENERGY TRANSFER

Thus, the transition of the molecule from one electronic energy state to another one causes a change in the electrical field around it. Conversely, a change in the electric field can cause the transition from one energy state to another one. The field change caused by the transition in one molecule can cause the opposite transition in a neighbouring molecule. This is the essence of energy transfer by dipole-dipole interaction, the Förster mechanism for energy transfer.

Just as the field change from the transition taking place in one molecule (the donor) drops off with the third power of the distance, so the sensitivity of the other molecule (the acceptor) to a field change drops off with the third power of the distance. The combined effect is a sixth power relationship: The rate of dipole-dipole energy transfer between two molecules is inversely proportional to the sixth power of the distance.

We are now ready to have a look at a simplified Förster's formula:

Energy transfer rate = factor $\cdot$ F $\cdot$ (overlap integral) $\cdot$ $\cos^2\alpha/(r^6)$.

Here F is the fluorescence quantum yield in the absence of the acceptor (cf. section 15), α the angle between the transition moments of the molecules, r is the distance, and the overlap integral is the convolution (point- wise product) of the donor fluorescence spectrum by the acceptor absorp- tion spectrum, integrated over the whole spectral region in common.

19. TRIPLET STATES

Our description of molecular energy states so far has been aimed primarily of explaining the properties and processes associated with so-called singlet states. A molecule is said to be in a singlet state when all its electrons are grouped in pairs, so that the two electrons in each pair have opposite *spins*. Spin is a property of en electron or other charged particle that makes it act like a small magnet, to produce a magnetic field. Also positive charged particles like atomic nuclei have spin.

Because all electrons in a molecule in a singlet state occur in pairs, and the electrons in each pair have opposite spins, the electrons produce no net magnetic field. Most molecules like to be in a singlet state, so usually the ground state, the most stable state, having the lowest electronic energy, is a singlet state. A notable exception is the dioxygen molecule (making up ordinary oxygen in the air), which we shall come back to later.

However, it can occasionally happen, that when a molecule has been excited from its ground (singlet) state to an excited singlet state, an electron "flips over", i.e. changes spin. Let us take a concrete and important example: the chlorophyll a molecule (Fig. 18). Like other chlorophyll forms, chlorophyll a has two prominent absorption bands, corresponding to two electronic transitions with high probability. For chlorophyll a these absorption bands are in the blue and red parts of the spectrum. In a collection of chlorophyll a molecules, be it in the plant or in solution, most of the molecules are in the ground state. Absorption of a photon of red light transforms a ground-state molecule to the first excited singlet state. Absorption of a photon of blue light transforms a molecule from the ground state to the second excited singlet state. A molecule in the second excited singlet state very rapidly transfers some of its electronic energy to vibrational energy (heat), and lands in the first excited singlet state. Then various things can happen. The most "exciting" (pardon the expression) of the possibilities is that an electron completely leaves the molecule. This is the key step in photosynthesis and the key step in the whole living world. Another possibility is that the molecule "shakes off" more energy, and heats its environment even more, and returns to the ground state. A third possibility is that it emits a photon, which carries away the excess energy and also returns the molecule to the ground state. A fourth possibility, which is realised in only a small fraction of the cases, is that the molecule is transferred from the first excited singlet state to the first excited triplet state. Although this happens after only a small fraction of the excitations it is important, and if plants would not be specially equipped to handle such events, they would not survive.

A change from a singlet to a triplet states, which involves a spin change, a "flip" of an electron, is sometimes referred to as "intersystem crossing", because singlet and triplet states can be considered to be two systems of states. Intersystem crossings are still sometimes also called "forbidden transitions", because early theories did not include the rare occasions when they occur.

Also the change from the excited triplet state to the (singlet) ground state is "forbidden". In fact it does occur (as many forbidden things do also in our society). In any case it does not take place quickly or, in other words, the excited triplet state has a long lifetime. A triplet molecule does not easily react with a singlet molecule, but if it meets another triplet molecule things are different. The magnetic fields created by the unpaired electrons interact. Even if the triplet molecules should not react chemically, they can exchange energy and become two singlet molecules. But because creation of a triplet state is in most cases a rare event, most triplet molecules are in low concentration, and the chance that two meet is not great. We shall now come to a very important exception to this rule, already mentioned above.

20. THE DIOXYGEN MOLECULE

The molecules of ordinary oxygen which we breathe have very remarkable properties. The most remarkable, important and unusual of them all is that dioxygen molecules have a triplet ground state. From what will follow, the reader might get the impression that this is very unfortunate, because it makes oxygen a bit difficult to handle for organisms, and imposes many threats. We shall deal with some of these in the chapters on photoxicity and photsynthesis. But as is the case with many properties of the rest of the surprising and exciting world which we inhabit, if things would not be exactly as they are, we would not be around. Just think for a moment that the dioxygen in the air would be in the singlet state instead, what would happen? We all know that oxygen under certain circumstances can react with organic matter such as wood or our own bodies. Not only single houses or trees, but whole towns and forests have sometimes burnt down. When oxygen oxidises organic matter, large amounts of energy are released as heat. Processes which release energy usually take place quite easily. But for a house to catch fire, something has to get hot to start with. When fire once has started, other things get hot, and the fire is not easy to extinguish. Why is it that the fire does not start spontaneously?

The answer is that dioxygen consists of triplet molecules, and triplet molecules do note easily react with singlet molecules. Only after things get hot and some of the organic molecules get into states with lone electrons, does a reaction with oxygen take place. When this happens, heat is released, more organic molecules acquire unpaired electrons and can react with oxygen, and so on.

Since we know that electrons like to join to pairs, one would expect that two oxygen atoms combine to a dioxygen molecule by joining two unpaired electrons to a pair. Instead, they combine to form a molecule with two lone electrons, a diradical.

34

21. SINGLET OXYGEN

Also singlet dioxygen exists, but its lowest electronic state has more energy than the triplet ground state. Singlet oxygen can be produced by reaction of ordinary triplet oxygen with another compound in the triplet state, provided the energy of that other molecule is high enough. As we can see from Fig. 18, the energy of triplet chlorophyll (in relation to the singlet ground state of chlorophyll) is high enough to transfer oxygen from its triplet ground state to an excited singlet state called $^1\Delta$g, according to the following scheme:

chlorophyll(excited triplet) $^+$ O$_2$(ground state) ==>
chlorophyll(ground state) $^+$ O$_2$(singlet)

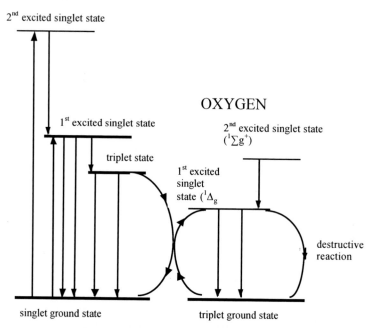

Figure 18. The various energy states (horisontal lines) of chlorophyll (left) and molecular oxygen (right) and their energy transitions (arrows). Energy is plotted upwards, i.e. a high horisontal line depicts a high energy state. Only the most important electronic levels are indicated, and the vibrational levels have been omitted. Thicker lines depict energy transitions associated with absorption (upward arrows) or emission (downward arrows) of light. The long upward arrow from the ground state of chlorophyll to the second excited state represents absorption of blue light; the shorter upward arrow from the ground state to the first excited state absorption of red light. Thin arrows represent radiationless transitions, in which energy is either transformed to heat (straight arrows), or reaction with another molecules (curved arrows) takes place. Emission of light can take place either as fluorescence (rapid light emission from singlet to singlet state) or as phosphorescence (slow light emission associated with change from singlet to triplet (as in chlorophyll at low temperature) or from triplet to singlet (in oxygen gas even at room temperature).

The singlet so created is very reactive, and can attack various other singlet molecules in the cell. If the plant did not have special systems both for preventing as much as possible the formation of singlet oxygen (this is the role of carotene in the plant cell) and for ameliorating the effects of it if it is formed, the plant could not survive for long, as shown with mutants lacking these protective systems.

In addition to chlorophyll many other pigments, when illuminated, can form triplet states and generate singlet oxygen. We shall deal with this further in the chapter on phototoxicity.

Figure 19 shows the configuration of the outermost electrons in various forms of "ordinary" and "reactive" oxygen.

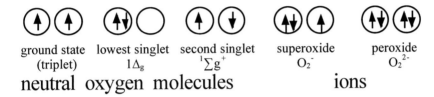

| ground state | lowest singlet | second singlet | superoxide | peroxide |
| (triplet) | $1\Delta_g$ | $^1\Sigma g^+$ | O_2^- | O_2^{2-} |

neutral oxygen molecules   ions

Figure 19. The electronic configurations in various forms of neutral dioxygen (left) and dioxygen ions of biological importance (right). Only the "antibonding" (π^*) electrons are shown, since all lower orbitals are similar (completely filled) for all the species. Arrows of opposite directions represent electrons of opposite spin.

REFERENCES

Alonso, M. & Finn, E.J. 1967. Fundamental university physics, Vol. II: Fields and waves. Reading, Mass.: Addison-Wesley. Library of Congress Catalog Card Number 66-10828.

Björn, L.O. & Björn, G.S. (1986). Studies on energy dissipation in phycobilisomes using the Kennard-Stepanov relation between absorption and fluorescence emission spectra. *Photochem. Photobiol., 44*, 535-542.

Friedrich, J., Scheer, H., Zickendraht-Wendelstadt, B. & Haarer, D. (1981). High-resolution optical studies on C-phycocyanin via photochemical hole burning. *J. Am. Chem. Soc., 103*, 1030-1035.

Haupt, W. (1977). *Bewegungsphysiologie der Pflanzen*, Georg Thieme Verlag, ISBN 3-13-509201-1.

Hecht, E. (1987) . *Optics.* 2nd ed., pp. x+676. Reading (Mass.): Addison Wesley. ISBN 0-201-11611-1.

Hecht, S., Shlaer, S. & Pirenne, M.H. (1942). Energy, quanta and vision. *J. Gen. Physiol., 25*, 819-840.

Hegemann, P. & Marwan, W. (1988). Single photons are sufficient to trigger movement responses in *Chlamydomonas reinhardtii*. *Photochem. Photobiol., 48*, 90-106.

Keijzer, M., Star, W.M. & Storchi, P.R.M. (1988). Optical diffusion in layered media. *Appl. Optics, 27*, 1820-1824.

Kennard, E.H. (1918). On the thermodynamics of fluorescence. *Phys. Rev., 11*, 29-38.

Kublelka, P. & Munk, F. (1931). Ein Beitrag zur Optik der Farbanstriche. *Phys. Rev., 11*, 672-683.

Star, W.M., Marijnissen, J.P.A. & van Gemert, M.J.C. (1988). Light dosimetry in optical phantoms and in tissues: I. Multiple flux and transport theory. *Phys. Med. Biol., 33*, 437-454.

Stepanov, B.I. 1957. A universal relation between the absorption and luminescence spectra of complex molecules. Dokl. Akad. Nauk SSSR 112:830-842 (Engl. transl. *Soviet Phys. Doklady, 2*, 81-84).

LARS OLOF BJÖRN

2. PRINCIPLES AND NOMENCLATURE FOR THE QUANTIFICATION OF LIGHT

1. INTRODUCTION: WHY THIS CHAPTER IS NECESSARY

This chapter will not deal with measuring equipment or measuring techniques, but with basic concepts of light quantification. This topic seems confusing, not only to the layman and the student, but also to the expert. Some reasons for this confusion are as follows:

1. The layman and beginnig student erronously regard the "amount" or "intensity" of light as something that can be completely described by a number. Such a view disregards the facts that

a) Light consists of components with different wavelength. A full description of the light would thus give information about the "amount" of light of each wavelength.

b) Light has direction. The simplest case is that all the light we are considering has the same direction, i.e. the light is collimated; the rays are all parallel. Another case is that light is isotropic, i.e. all directions are equally represented. Between these extremes there is an infinite number of possible distributions of directional components.

c) Light may be polarized, either circularly polarized or plane polarized. In the rest of this chapter we shall disregard this complication, but one should always be aware of the fact that a device such as a photocell may be differentially sensitive to components of different polarization, and polarization may be introduced by part of the experimental setup, such as a monochromator or a reflecting surface.

d) Light may be more or less coherent, with light-waves "going in step". Also this complication we shall neglect in this chapter.

e) Finally, people often disregard or neglect or confuse the concept of time. We must make up our mind whether we want to express an *instantaneous* or a *time-integrated* quantity, e.g. *fluence rate* or *fluence, power* or *energy*. Power means energy per time unit.

2. Light is of interest for people investigating or working with widely different parts of reality. Experts in different fields have used different concepts and different nomenclature, partially depending on what properties of light that have been interesting for them, and partly due to the whims of historical development. Only rather recently have there been serious attempts to achieve a uniform nomenclature, and the process is not yet complete.

L.O. Björn (ed.), Photobiology, 37–44.

2. THE WAVELENGTH PROBLEM

As we cannot always quantify light by giving the complete spectral distribution, we have to quantify it in some simpler way. From the purely physical viewpoint, there are two basic ways. Either we express a quantity related to the number of photons, or a quantity related to the energy of light. For a light of single wavelength the energy of a photon is inversely proportional to the wavelength, and the proportionality constant is Planck's constant multiplied by the velocity of light.

There is an ultimate way of calibration only for the energy of light. For this we can use a hollow heat radiator of known temperature, which will radiate in a way predictable by basic physics. Using such a radiator a photothermal device, such as a thermopile or a bolometer (see the chapter on light measurement) can be calibrated, and then any kind of light can be measured with it and expressed in energy or power units. We can use it for measuring a series of "monochromatic" (i.e. narrow-band) light beams, and then they, in turn, can be used for calibrating other measuring devices in either energy or photon units.

Actinometers, i.e. photochemical devices, seem to count photons, but also in this case the ability of photons to cause a response (the quantum yield) varies with wavelength.

We can also use photomultipliers as photon counters, but we should be aware that they do not, strictly speaking, count photons, but impulses caused by photons. Some impulses are not caused by incident photons, but by electrons knocked out from the photocathode by the heat vibration of the atoms in it. We try to minimize this by cooling the photocathode. Furthermore, all photons do not have the same ability to knock out electrons from the photocathode and cause pulses to be counted. This ability is wavelength-dependent. Therefore we cannot use photon counting as an independent calibration method.

The units for expressing light as photons are

1) photons (number of photons)

2) mole of photons (the symbol is mol), which is $6.02217 \cdot 10^{23}$ photons, or a unit derived from this, such as micromole of photons ($6.02217 \cdot 10^{17}$ photons). The symbol for the latter is μmol.

Either of these (1 or 2) can be expressed per time and/or per area, or (rare in biological contexts) per volume.

The unit for energy is joule (J). Energy per time is power, and joule per second is watt (W). Both can be expressed per area (or, rarely in biological contexts, per volume, i.e. energy density or power density).

You should note that simply giving a value followed by "W/m^2" without further qualification is not defined, since one cannot be sure what kind of area you are expressing with m^2. Is it a flat area or a curved one? If it is flat, what is its direction? This brings us to the topic of the next section.

3. THE PROBLEM OF DIRECTION AND SHAPE

Most light measuring systems are calibrated using light of (approximately) a single direction, i.e. collimated light. However, light in nature, where most plants live, is not collimated. If the sky is cloudless and unobstructed the rays coming directly from the sun are rather well collimated, but in addition there is skylight and light reflected from the ground and various objects. A plant physiologist who wants to understand how plants use and react to light has to take this into account.

Traditionally, most measuring devices can be regarded as having a flat sensitive surface, and when we calibrate the instrument we generally position this surface perpendicularly to a collimated calibration beam. A plant leaf is also flat, so in the first approximation we can measure light in single-leaf experiments with a flat device with the same direction as the leaf. But a whole plant is far from flat (except in very special cases). Different surfaces on the plant have different directions. Ideally we should know the detailed directional (and spectral) distribution of the light impinging on the plant, but this is not possible in practice. Since a plant is a three-dimensional object, it would in most cases be better to determine the light using a device having a spherical shape and equally sensitive to light from any direction. This brings us to the distinction between

1) Irradiance, i.e. radiation power incident on a flat surface of unit area, and

2) Energy fluence rate (or fluence rate for short), i.e. radiation power incident on a sphere of unit cross section. The term fluence rate was introduced by Rupert (1974),.

Both these concepts have their correspondences in photon terms. For case 1 the nomenclature is not settled, but it would be logical to use the term photon irradiance. Many people, especially in the photosynthesis field, use the term photon flux density, and the abbreviation PFD (PPFD for photosynthetic flux density, see below). For case 2 the term photon fluence rate is well accepted among plant physiologists, but hardly among scientists in general.

Energy fluence is the energy fluence rate integrated over time. By fluence the same thing as meant as by energy fluence.

We shall now compare irradiance and energy fluence for different directional distributions of light (cf. Figure 1, next page):

1) Collimated light falling perpendicularly to the irradiance reference surface. In this case the flat surface of unit area and the sphere of unit cross sectional area will intercept the light equally, and irradiance will be the same as fluence rate.

2) Collimated light falling at an angle x to the normal of the irradiance reference plane. In this case the light intercepted by the flat surface of unit area will be less than that intercepted by the sphere of unit cross sectional area. The irradiance will be $\cos(x)$ times the fluence rate. Since $\cos(x)$ is less than unity, the irradiance in this case will be lower than the fluence rate.

3) Completely diffuse light falling from one side only. The ratio of irradiance to fluence in this case will be an average of $\cos(x)$ for all angles x from 0 to $+\pi/2$ weighted by $\sin(x)$, i.e. $\int\sin(x)\cdot\cos(x)\cdot dx/\int\sin(x)\cdot dx$ with the integral running from 0 to $+\pi/2$, and this is equal to 1/2. Thus the irradiance in this case is half the fluence rate. The reason we have to weight $\cos(x)$ by $\sin(x)$ is that all values the of x are not equally "common", do not have the same probability. The various directions may be thought of as corresponding to points on a big sphere, the center of which is the point of measurement. The sphere can be thought of as divided into a pile of rings,

and each ring (corresponding to a value of x) has a radius, and hence a circumference proportional to sin(x).

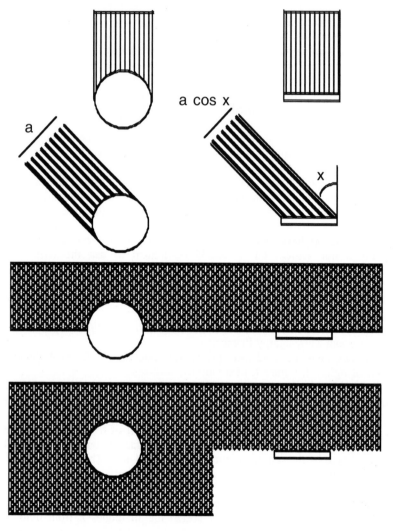

Figure 1. The concepts of irradiance and fluence rate. In the top diagramme the incident light is perpendicular to the surface of the flat irradiance sensor. In this case fluence rate and irradiance are equal. In the next case the incidence angle is x. The irradiance sensor then intercepts only the fraction a cos x of what the fluence rate sensor does. In the third case the light is diffuse, but incident only from above. Then the fluence rate is twice the irradiance. In the fourth case the sensors are immersed in diffuse radiation from all directions, but the irradiance sensor senses radiation only from above. In this case the fluence rate is four times the irradiance.

4) Completely diffuse light from both sides, i.e. isotropic light. The sphere is then hit by light over its whole surface, but for the flat receiver we still count only one surface (irradiance is defined in this way), so irradiance is one quarter of the fluence rate in this case. We can easily remember this if we think that the area of a circle is one quarter of the area of a sphere with the same radius.

We may now make the following table of various quantities associated with light measurements:

	Flat receiver	Spherical receiver
Instantaneous values:		
Energy system	(energy) irradiance	(energy) fluence rate
	unit: W/m^2	unit: W/m^2
Photon system	photon irradiance (=photon flux density)	photon fluence rate
	unit: $mol\ m^{-2}\ s^{-1}$	unit: $mol\ m^{-2}\ s^{-1}$
Time integrated values:		
Energy system	(energy density)	(energy) fluence
	unit: $J\ m^{-2}$	unit: $J\ m^{-2}$
Photon system	-	photon fluence
		unit: $mol\ m^{-2}$

In all the above cases we add the word "spectral" before the various terms if we wish to describe the spectral variation of the quantity. We may thus write, e.g., spectral fluence rate on the vertical axis of a spectrum of light received by a spherical sensor.

What we here have called fluence rate is termed *actinic flux* by atmospheric scientists. The term space irradiance has been introduced by Grum & Becherer (1979). Two more terms with the same meaning are *space irradiance* and *scalar irradiance*. The term *spherical irradiance* has been used in similar contexts, but means one quarter of the fluence rate. *Vectorial irradance* is just the same as irradiance. The reader should use just one system of terms, preferabley irradiance and fluence rate, but may encounter all these other terms in the literature.

Few instruments on the market, and very few spectroradiometers, are designed for direct measurement of fluence rate. Most of them are constructed for irradiance measurements and a few for measurement of radiance (see below9. But Björn (1995) and Björn & Vogelmann (1996) have shown how irradiance meters can be used for estimation of fluence rate.

So far we have been dealing with light falling on a surface, either a flat or a spherical one. But we may need to express also other quantities, for instance the total power (energy per time unit) output of a light source. The unit for this is, of course, W (watt). The power emission per unit area of the source is called the *radiant excitance* and is measured in $W\ m^{-2}$ (just like irradiance and fluence rate, so beware of confusing them). The power emission takes place in different directions; in total there is a solid angle of 4π steradians (sr) surrounding a source. Usually the emission is not equally distributed in all directions, so for a certain direction we might like to specify the power emission per steradian. This quantity is called the *radiant intensity* in that direction, and the unit is W sr^{-1}. Note that the term intensity is often (erroneously) used in another and usually not well defined sense). The radiant intensity per area unit on a plane perpendicular to the light is called *radiance*, and the unit is W sr^{-1} m^{-2}. If we integrate the radiance over all 4π radians, we get back to the fluence rate we are already familiar with.

Fortunately the average photobiologist need not keep all these concepts in his or her head at all times. You can look them up when needed. However, you must be clear over the meaning of irradiance and fluence rate, and not confuse these two concepts. As a practical illustration of how the ratio of fluence rate to irradiance can vary under natural conditions see Fig. 2.

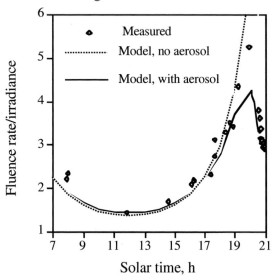

Figure 2. The variation of the ratio of fluence rate to irradiance over a clear summer day in southern Sweden. Presence of small particles in the air (aerosol) dampens the variation. The graph is for cloudless conditions and 400-700 nm (the "PAR" spectral band, c.f. Fig. 3). For ultraviolet radiation, especially for UV-B radiation, the variation is smaller, because even clean air scatters this radiation to make it to a large extent diffuse. For overcast conditions the variation is also smaller than for clear skies. Measurements and calculations by the author.

4. BIOLOGICAL WEIGHTING FUNCTIONS AND UNITS

Section 3 concludes the physical quantification of light. However, there has been a need for additional concepts in connection with organisms and biological problems. Traditionally there has been a special system related to human perception of light. We can here limit ourselves to *illuminance*, which is expressed in *lux* (lx). Neglecting the historical development, we can say that lux is the integrated spectral irradiance weighted by a special weighting function. This weighting function is precisely described mathematically, but can be thought of as the average (photopic, i.e. related to strong light vision mediated by cones) eye spectral sensitivity for a large number of people. (Rarely we also see the expression "scotopic lux", which is the corresponding using the scotopic visibility weighting function). The photopic visibility function has its maximum at 555 nm, and for this wavelength one W m^{-2} equals 683 lux. For all other wavelengths one W m^{-2} is less than 683 lux. Illuminance integrated over receiving flat area is called *luminous flux*, and the unit is *lumen* (lm). Thus lux is lumen per square meter. In older American literature the unit foot candle (f.c.) is used instead of lux. Foot-candle equals lumen per square foot, and since there are 3.2808399 feet in a meter, there are 3.2808399^2 = 10.763910 square feet in a square meter, and also 10.763910 lux in a foot candle. There are also a number of other photometric concepts and units, which we seldom need in photobiology. Many of them are defined in the Handbook of Chemistry and Physics.

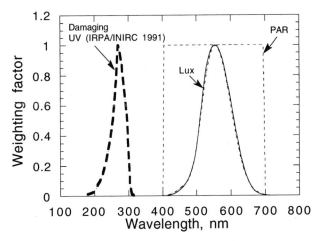

Figure 3. Examples of weighting functions. The "Damaging UV" function is one devised by the International Radiation Protection Association (IRPA) and the International Commission on Non-ionizing Radiation Protection (ICNIRP) and thus having a certain official status. "PAR" stands for photosynthetically active radiation, and this weighting factor is unity from 400 to 700 nm and zero outside this interval. This weighting function is applied more often to photon irradiance or photon fluence rate than to energy irradiance and energy fluence rate. The "Lux" function is that used for conversion of W m^{-2} to lux (or W to lumen). The maximum is here made unity to allow plotting together with the other functions, but in absolute units it corresponds to 683 lux per W at 555 nm. The "Lux" graph in fact consists of two plots, so close that they hardly can be distiguished in the diagram, both the official values from a table and an analytical approxomation consisting of the difference between two Gauss functions: Weighting factor = exp[-(555-λ)/63.25]^2-exp[-(495-λ)/30]^2/6.8, where λ stands for wavelength in nm.

Similarly we may, for purposes other than vision (reading light, working light) weight the spectral irradiance by other functions. These functions approximate various photobiological action spectra (see the chapter on action spectra). One special function is zero below 400 nm and above 700 nm, and unity from 400 to 700 nm. This describes, by definition the "photosynthetically active radiation", PAR. Usually one uses the spectral photon irradiance to weight by this function, and this is the meaning of the often used term PPFD, "photosynthetic photon flux density". To assume photosynthetic zero action outside the range 400-700 nm, and the same action for all components within the range is, of course, physiologically speaking an approximation, but an approximation that people have agreed upon, just as the definition of lux involves an approximation which holds well only for scotopic (rod) vision.

Other weighting functions are used for "sunburn" meters to yield "sunburn units", but in this field we have to watch out for various "units" used by various people. One kind of sunburn meter used much in the past is the Robinson-Berger meter, but recently a new agreement has been used for using an weighting spectrum more closely resembling the true sunburn action spectrum (of caucasian skin). One weighting function to determine safe working conditions is shown in Fig. 3 as "Damaging UV". This particular function was agreed upon in 1991 by the International Radiation Protection Association (IRPA) and the International Commission on Non-ionizing Radiation Protection (ICNIRP). Another, slightly different function for similar purposes was devised by ACGIH (American Conference of Governmental Industrial Hygienists). In reality, of course, different kinds of damage, such as damage to the cornea and to the lens in the eye, and to skin of persons with different pigmentation, have different action spectra. Also the standard "PAR" (photosynthetically active radiation) is an approximation to reality, since different plants, and even the same plant in different states, have different action spectra for photosynthesis.

There are also numerous other weighting spectra in use for estimating radiation with other biological actions. We shall come in contact with some of them in the chapter on ultraviolet radiation effects.

Some meters, such as lux meters, sunburn meters, and meters for PAR, are constructed with spectral responses approximating the weighting functions and can therefore directly yield the values we want without spectral decomposition of the light. For more precise work, and in the case of, e.g., ulktraviolet inhibition of plant growth, it is necessary to measure (using a spectroradiometer) each wavelength component separately, and weight by the weighting function using arithmetics (usually computers are used).

REFERENCES AND FURTHER READING

Björn, L.O. (1995). Estimation of fluence rate from irradiance measurements with a cosine corrected sensor. *J. Photochem. Photobiol. B. Biol., 29,* 179-183.

Björn, L.O. & Vogelmann, T.C. (1996). Quantifying light and ultraviolet radiation in plant biology. *Photochem. Photobiol., 64,* 403-406.

Björn, L.O. (1997). Quantification of light in biology. - Physiol. Plant. (in press).

Grum, F. & Becherer, R.J. (1979). *Optical radiation measurements,* 1, pp. 14-15. New York: Academic Press.

Handbook of Chemistry and Physics (many editions). Cleveland (OH): CRC Press.

Rupert, C.S. (1974). Dosimetric concepts in photobiology. *Photochem. Photobiol. , 20,* 203-212.

LARS OLOF BJÖRN

3. GENERATION AND CONTROL OF LIGHT

1. INTRODUCTION

Any photobiological experimental setup consists of three main parts: A light source, a light path, and a target. The biological object under investigation may form the light source, part of the light path, or, as the most common case, the target.

In the following we shall treat the non-biological components of the experimental setup.

2. LIGHT SOURCES

2.1. The sun

Almost all the natural light at the surface of the earth comes from the sun (this holds, of course, also for moonlight). The sun, on the whole, radiates as a glowing blackbody at a temperature of about 6000 K. We have already mentioned the absence of some wavelength components from sunlight due to absorption in the outer cooler layers of the sun. Sunlight is further modified by the earth's atmosphere before it reaches ground level. More about this and other natural light conditions will follow in Chapter 5.

2.2 . Incandescent lamps

The light from an incandescent lamp originates at the surface of a glowing filament, which nowadays is, almost invariably, made from tungsten. It is heated by an electric current flowing through it. In order not to be destroyed (oxidized) by the oxygen in the air, the filament is enclosed in an envelope made of glass or quartz. The envelope is either evacuated (most commonly for small lamps) or filled with an inert gas, or iodine vapour.

The spectral composition of the emitted light is strongly dependent on the temperature of the filament. As a first approximation the spectrum varies with temperature as does that of blackbody radiation (Planck's law). However, due to the wavelength-dependence of the emissivity of tungsten, more short-wave radiation is emitted than from a blackbody of the same temperature. The filament is, in most cases, coiled, which makes it somewhat "blacker" (more like a glowing cavity) than a smooth tungsten surface would be.

L.O. Björn (ed.), Photobiology, 45–63.

In most cases it is desirable to obtain as much as possible of the radiation towards the short wavelength end of the spectrum, i.e. to operate the lamp with as high a filament temperature as possible (the temperature can be increased by increasing the voltage). However, this results in shorter life of the lamp, because the tungsten evaporates more quickly (and condenses again on the envelope, which blackens it). As a rule of thumb an increase of the voltage by 10% above the recommended voltage will decrease lamp life to 1/3 (i.e., by two thirds).

The lamp life at a certain filament temperature can be increased by making the filament thicker. Thereby a certain power (wattage) is reached at a lower voltage (using a higher current). A further advantage of low-voltage lamps is, in many cases, that the filament is shorter. Quite often a small ("point-like") light source is desirable in optical systems.

A gas-filled envelope permits a higher filament temperature than an envelope with vacuum. A mixture of argon and nitrogen is the most common choice (ordinary household bulbs). Addition of iodine ("halogen lamps" or "quartz-iodine lamps") permits an even higher temperature. This is because the iodine vapour combines with the tungsten vapour. The compound is decomposed again when the molecules hit the hot filament, which is thereby continuously regenerated. For the regeneration to work properly, the temperature of the envelope must be so high that the tungsten iodide cannot condense on it. Therefore such lamps are manufactured with a very small envelope made of quartz, which can stand high temperature.

The advantages of incandescent lamps are low price, no requirement for complex electrical circuitry, and stability. The main disadvantage is that the emission in the short-wavelength end of the spectrum is low, and that the spectrum is strongly dependent on the current through the lamp, and therefore difficult to keep absolutely constant. A method to estimate the spectrum of an incandescent lamp fed with direct current (within the spectral range where the glass or quartz envelope does not absorb appreciably) has been devised by Björn (1971). The principle is as follows: Using a small current and measuring both the current through and the voltage across the lamp, the room temperature resistance of the lamp is measured (this is essentially the resistance of the tungsten filament). After the lamp has been powered up (with direct current) the current and voltage are measured again. From this the "hot" resistance is computed. The temperature of the glowing filament can be computed from the ratio between room temperature resistance and "hot" resistance, and from this (using the Planck formula for a blackbody radiator, and the known temperature- and wavelength-dependent emissivity of tungsten) the spectrum can be computed.

Using these principles emission spectra for a lamp with a 24 V rating operated at various voltages were computed (Tab. 1, Fig. 1a, 1b). Seeing the values on a linear scale only may be misleading, and the values are therefore also plotted on a logarithmic scale for spectral irradiance.

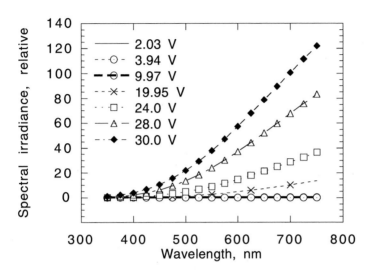

Figure 1a. Emission spectra of an incandescent lamp with a nominal rating of 24 V run at different voltages. The corresponding currents and filament temperatures are shown in Table 1.

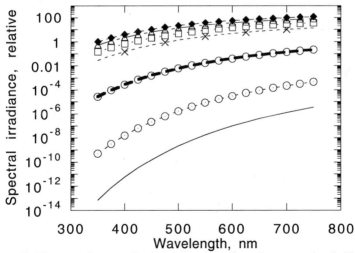

Figure 1b. The same values as in Fig. 1a but plotted on a logarithmic vertical scale. The symbols stand for the same voltages as in Fig. 1 a.

Table 1

Voltage, V	Current, mA	Abs. temp., K
2.03	22.0	1013
3.94	33.4	1243
9.97	58.5	1690
19.95	87.0	2164
24.0	97.0	2306
28.0	106	2435
30.0	110	2500

2.3. Electric discharges in gases of low pressure

In a gas discharge lamp an electric current is flowing through a gas. The gas emits light, the spectral composition of which depends on the gas. When the gas has a low pressure it emits a line spectrum, i.e. only light of certain wavelengths is represented (in contrast to the continuous spectrum emitted by an incandescent lamp).

The basic parts of a gas discharge lamp are a gas enclosed in a transparent envelope and the two electrodes necessary for the conduction of current to and from the gas. In addition to this it is often necessary with other parts, such as heating filaments, to release enough electrons to start the current through the gas or to vapourise, e.g., mercury or sodium when vapours of these metals are to be used as emitting gas.

The electric resistance of an incandescent lamp increases when the current through it increases, since tungsten has a higher resistivity the higher the temperature. Thus an incandescent lamp is self-regulating and burns in a stable way as long as the voltage is constant. In a gas discharge lamp the reverse holds: Its electric resistance decreases with increasing current. Therefore such a lamp has to be connected to some kind of circuitry limiting the current. In the case of direct current a series resistor is often used, in the case of alternating current, a choke.

Gas discharge lamps containing mercury vapour of very low pressure emit most of the energy as ultraviolet radiation of wavelength 253.7 nm. This wavelength is close to the absorption maximum of nucleic acids, and the radiation is also absorbed by the aromatic amino acids in proteins, and many other biological molecules. The photons are also energetic enough to initiate many chemical reactions, and therefore this kind of radiation is very destructive for living matter. Low pressure mercury lamps with quartz envelopes (which transmit this kind of radiation, in contrast to glass envelopes) are therefore used as sterilization lamps.

Fluorescent lamps are similar lamps, but with glass envelopes which on the inner surfaces have a fluorescent layer converting the UV radiation to visible light (or, in certain cases, to UV radiation of longer wavelength than the original emission).

Glow lamps are a kind of low pressure gas discharge lamp, usually containing neon. Lamps containing the element to be measured are used for atomic absorption spectroscopy. Low pressure sodium lamps, emitting almost monochromatic light at 589 nm, have been used extensively as sources of outdoor working light, because they give more visible light per unit of energy input than any other type of lamp. To a large extent they have been abandoned, mostly because our colour vision cannot be used with monochromatic light.

Microwave radiation is used for the energy input in some other gas filled lamps, such as "electrodeless" high pressure xenon lamps.

2.4. Medium and high pressure gas discharge lamps

If the vapour pressure in a mercury lamp is increased, more and more of the emission at 253.7 nm is reabsorbed, and finally very little of this radiation escapes from the vapour. Instead spectral lines of longer wavelength emerge (medium pressure mercury lamps). At even higher pressures the spectral lines are broadened to bands (high pressure mercury lamps), and finally a continuous spectrum results (super high pressure mercury lamps).

Deuterium (heavy hydrogen) lamps of medium pressure are used as light sources for spectrophotometry in the ultraviolet region.

Lamps containing xenon of high pressure are used to obtain a strong continuous emission from 300 nm and into the infrared. Depending on the composition of the envelope, more or less of shorter wavelength ultraviolet also escapes. Xenon lamps come in a great variety of types. We use lamps running on about 24 volt direct current (but ignited with about 2000 volts= danger!!) and wattages (rated powers) from 150 to 900 W. Xenon lamps of higher wattage are often water-cooled. Electrodeless xenon lamps are also manufactured. They are powered by microwave radiation.

Because xenon lamps with UV-transparent envelopes cause conversion of oxygen to ozone, such lamps must be provided with exhausts to transport the ozone out of the building. The same holds for high-pressure mercury lamps with UV-transparent envelopes.

All high pressure lamps are dangerous because they can explode. They must therefore never be operated without protective housing. Also when cold they should be handled with care using eye protection and other appropriate safety measures.

2.5. Flashlamps

Electronic flashes are xenon lamps through which a capacitor is discharged when a special triggering pulse has ionized the gas. The energy available is proportional to the capacitance of the capacitor and to the square of the voltage to which it has been charged.

In many cases it is desirable to have a short flash duration. This requires that the impedance of the circuit is low (short leads) and the capacitance low (high voltage has to be applied to the capacitor to get enough energy with a low capacitance). It is also necessary to prevent the circuit from oscillating and causing multiple flashes.

Ordinary photographic flashes have a flash duration of about one millisecond. If they are "automatic", i.e. combined with a light-sensing photodiode and appropriote circuitry, the flash will be cut off when a certain amount of light energy has been emitted.

2.6. Light emitting diodes

Light emitting diodes (LEDs) are used in applications where not very strong light is needed, for instance as indicator lights and displays. However, the maximum output power available from LEDs is increasing, and LEDs are the cheapest devices which can be modulated very rapidly: using an appropriate circuit they can be switched on and off in a few nanoseconds.

For this reason they have been used as light sources for measuring variable fluorescence in plants. The trouble with this is that only red emitting diodes are intense enough, and their red light is not easy to efficiently separate from the chlorophyll fluorescence.

LEDs of several spectral emission types are presently manufactured: ultraviolet A, blue, green, yellow, red and infrared (the latter emitting at about 900 nm). It should be noted that they are not monochromatic light sources, and especially the UV-A and blue emitting LEDs have a broadband emission of longer wavelength than the nominal emission. For some types the emission spectrum changes with operating current. LEDs are powered by a low voltage source (e.g. a 1.5 volt battery; some types need up to 5 V) in series with a resistor limiting the current to the rated value. Proper polarity should be observed.

Traditional LEDs contain inorganic semiconductors such as GaN, InGaN, SiC and GaAsP (see table). Very recently several laboratories and companies have started to develop organic light emitting diodes (OLEDs), which will probably widen the range of spectral types available.

Table 2.
Examples of LEDs and from where to obtain them

Peak wavel., nm	Semiconductor(s)	Company and web address
370-390	GaN	Nichia America
460	GaN	www.nichia.com
470, 505, 525	SiC/GaP	Ledtronics
574, 595, 611	InGaAlP	www.ledtronics.com
630	GaAsP/GaP	
660	GaAlAs/GaAlAs	
660, 700, 720	GaAlAsP	Roithner Lasertechnik
780, 810, 905	GaAlAsP	

Roithner also markets a range of infrared emitting diodes with emission peak wavelengths to over 4.5 micrometres.

An interesting new development is the construction of a LED which can generate a single photon at a time (Yuan et al. 2002).

2.7. Lasers

Laser is an acronym for Light Amplification by Stimulated Emission of Radiation. Stimulated emission occurs when a photon causes a molecule in an excited state to emit a second photon. Stimulated emission as such requires no special equipment. It occurs regularly when photons of the proper energy encounter excited molecules. However, as a rule, excited molecules are very rare compared to molecules in the ground state (remember the formula for the equilibrium concentration, $N_y/N_x = e^{-(E_x-E_y)/kT}$). To get light amplification by stimulated emission we must have more stimulated emission than light absorption, which means that we must have more molecules in the proper excited state than in the ground state. This can be achieved in different ways; however, never by "direct lift" from the ground state by absorption of light. Various lasers employ indirect "optical pumping" (sometimes by another laser), electrical energy, or chemical reactions. For a laser to work, photon losses must also be minimized by a suitable optical configuration, often involving mirrors.

Laser light has some unusual properties:

1) Laser light is coherent in the sense that the light constitutes very long wave-trains, contrary to ordinary light, where each photon can be regarded as a limited wave-packet independent of other photons.

2) Laser light can be made very collimated (all rays very parallel).

3) Laser light is usually very monochromatic (very narrow spectral bandwidth) or consists of a small number of such very narrow bands.

4) Laser light may be (but is not necessarily) plane polarized.

5) The light from some types of lasers is given off in extremely short pulses of extremely high power (energy per time unit). However, this does not hold for all lasers. Some lasers emit light continuously and have a very feeble power.

Even lasers of low power, such as the helium-neon laser, should be handled with some caution. This is because the beam is so narrow, parallel and monochromatic that if it hits your eye all its power will be focused onto a very small area of your retina and blind that particular spot.

One kind of laser that is in everyday use is the continuous *helium-neon laser* , emitting at 632.8 nm and a few infrared wavelengths. *Dye lasers* are advantageous in many cases because the wavelength can be selected over a wide range (within the fluorescence band of the dye used, and the dye can be changed if necessary). You may sometimes encounter a YAG laser. YAG is the acronym for Yttrium Aluminium Garnet, containing trivalent neodymium ions in $Y_3Al_5O_{12}$. They are very powerful emitters of infrared radiation of 1,060 nm wavelength. For photobiological purposes they are sometimes combined with *frequency doublers* made of potassium phosphate crystals, so that green light of 530 nm wavelength is obtained. The wavelength can be further changed either by letting the light undergo Raman scattering or using it as a power source for a dye laser. *Diode lasers* are photodiodes emitting coherent light. They are now the most common lasers, used in CD players and other optical readout devices and laser pointers. They are available from 370 nm in the UV-A band to the long-wavelength red.

3. SELECTION OF LIGHT

In many cases you do not want to use the light as it comes from your light source. You may wish to remove some parts of the spectrum or select just a narrow spectral band, or select light with a certain polarization, or you might wish to modulate the light in time, for instance quickly change from darkness to light or obtain a series of light pulses. The first three sections below will deal with wavelength selection.

3.1 Filters with light absorbing substances

The simplest devices for modifying the spectral composition of light are filters with light absorbing substances which remove certain components from the spectrum. These colour filters may be solid or liquid.

Cheap filters, which are quite useful for some purposes, consist of coloured plastic (for instance Plexiglas or cellulose acetate). Coloured cellophane is not recommended as it is rather unstable, and also cellulose acetate has to be used with great caution since it bleaches with time. It must also be realized that all these substances transmit far-red light and infrared radiation freely. Thus a piece of green Plexiglas does not transmit just green light. It is very instructive to put one, two, three etc. sheets of green Plexiglas on an overhead projector and watch the effect (or look through the sheets towards an incandescent lamp). One or two sheets look clearly green, then the colour becomes indescribably dirty, and finally shifts to deep red. This is because far-red light is transmitted even more freely than green light, and when practically all the green light (to which our eyes are most sensitive) is gone, we see the remaining far-red, which is otherwise hidden by the green. Remember that plants, contrary to our eyes, are more sensitive to far-red light than to green!

Undyed cellulose acetate can be used as a cut-off filter to remove ultraviolet-C radiation but retain UV-A and UV-B. The exact absorption depends on the thickness and must be checked regularly, as the filter changes, especially in front of a UV radiation lamp. A more stable alternative is a special type of uncoloured Plexiglas, number FBL.2458 (from Röhm GmbH, ordinarily used in front of the UV radiators in solaria).

A great variety of gelatin filters for photographic use are manufactured by Ilford, Kodak and other companies. They may also be very unstable in strong light, and also freely transmit far-red light.

From an optical viewpoint, some solutions are much better filters than anything that can be made in solid form, but solutions are, in many cases inconvenient to use. Thick layers may be required for the optical properties you want, and the liquid filters may therefore become bulky. Furthermore, some of the most useful coloured substances are, unfortunately, cancerogenic or toxic in other ways.

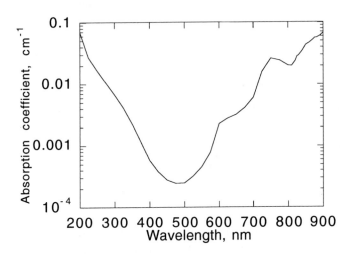

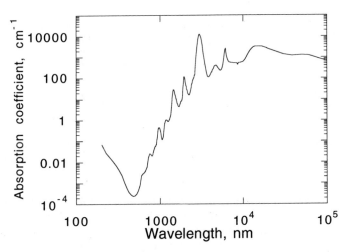

Figure. 2. The absorption spectrum of pure water plotted in two different ways. Data from Hale & Querry (1973), replotted.

One useful substance, which is not particularly dangerous, is water. It can be used for removing infrared radiation and thus avoid heating by light from incandescent and xenon lamps. Addition of copper sulfate increases the absorption of far-red. Copper sulfate should be used only with distilled water, and the solution should be acidified with sulfuric acid to avoid precipitation of cupric carbonate.

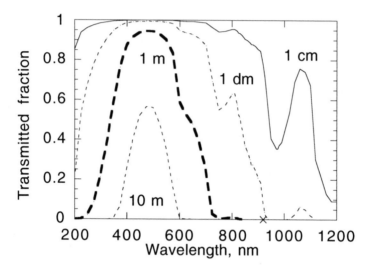

Figure 3. The fraction of light of various wavelength that is absorbed by (pure) water layers of the thicknesses indicated. Ten meter filters are not very practical in the laboratory, but 10 and even 100 m of water are natural light filters for many organisms. The transmission in natural waters differs from that in chemically pure water and will be dealt with in a later chapter. Calculated from data of Hale & Querry (1973).

Solutions of potassium dichromate (cancerogenic!) are very good for removing light of wavelength shorter than about 500 nm, as one might want to do in, e.g., studies of fluorescence.

For this purpose it is far superior to glass filters or any filters containing organic compounds, because of the total lack of fluorescence. We use it for filtering away the blue excitation light when we study the red chlorophyll fluorescence from plants.

Containers (cuvettes) for solutions of copper sulfate or potassium dichromate can be made by gluing together pieces of ordinary clear Plexiglas. Be sure that the glue has dried thoroughly and throughout before you pour your solution into the cuvettes, or you will lose more time than you are trying to save. It is best to test for leaks using distilled water before you put your solutions in. Distilled water can be removed again from the leaks, but crystals cannot!

Cobalt chloride and nickel sulfate (nasty, cancerogenic, toxic, in the case of nickel allergenic) dissolved in water or aqueous ethanol make very good broadband filters for the UV-B region, but because the substances are so dangerous I do not recommend them unless you really know what you are doing and are sure that your cuvettes will not leak. Detailed descriptions of solution filters, especially for isolationg lines of the mercury spectrum, are found in Calvert & Pitts (1966) and Rabek (1982).

Coloured glass filters are made by several companies. My personal experience is mainly with filters manufactured by Schott & Gen. (Mainz, West Germany) and Corning (USA). There is a large assortment of such filters available (Fig. 4), so you will have to consult a catalogue before ordering. Of the filters from Schott I have most often found BG12 and BG18 useful for isolating broad-band blue light, and a series of cut-off filters, which cut off the short end of the spectrum (cut-off wavelengths from 250 to 780 nm) but transmit light of longer wavelength. For a particular kind of glass the cut-off point can be varied by having filters of different thickness, or using several filters in tandem.

All the filters absorbing the unwanted light convert the absorbed energy to heat. If the light to be filtered is strong, the filters may become overheated and be destroyed. Organic substances may be decomposed, plastic may melt or burn, solutions may boil, and glass may crack. The risk for these unwanted effects is considerably less in the case of interference filters.

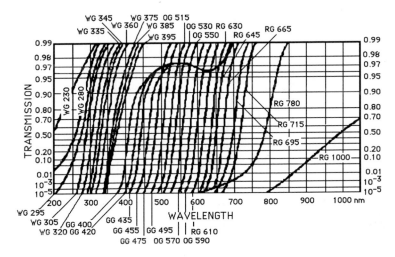

Figure 4. Cut-off filters made of glass, manufactured by Schott & Gen., Mainz. Note that the vertical transmission scale is not linear. The filters absorb short wavelength and transmit long wavelength light. The diagram, drawn for 1 mm WG 230 to GG 395) or 3 mm (GG 400 to RG 1000) thick filters, does not take reflection (roughly 0.08 or 8%) into account.

3.2 Interference filters

An interference filter removes the unwanted radiation from a beam not by absorption, but by reflection. It does not contain any coloured substances, but instead a number of partially reflecting and partially transmitting interfaces. Some interference filters contain very thin metal films, others are made from alternating layers of transparent compounds of high and low refractive indices. The complete theory for interference filters is complicated. However, its essence is that when the spacing between the layers is a quarter of a wavelength, destructive interference will occur in the reflected beam, so no light is reflected for this particular wavelength. Instead light of this wavelength is transmitted. Light with twice or three times (or any integer times) the basal wavelength will also be transmitted, since there will be layer distances corresponding to a quarter of these wavelengths also. The reader is referred to Chapter 7, section 10, which deals with a very similar topic.

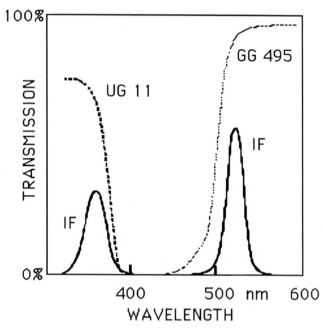

Figure 5. The transmission spectrum for an interference filter (IF) with two transmission peaks. Either one of the transmission bands can be selected by combining the interference filter with suitable glass filters: UG 11 for the short wavelength band or GG 495 for the long wavelength band.

Interference filters of the type just described will thus allow several narrow spectral bands to pass through, with wavelengths 4, 8, 12 ... times the distance between interfaces. By combining interference and glass filters, one of the bands can be selected (Fig. 5). When using such combination filters, it is essential (at least if strong light is to be filtered) that the interference part of the filter faces the incident light. If the absorbing glass part is hit by the unfiltered light, the filter might become overheated.

For the filter to function properly, the light to be filtered must be nearly perpendicular to the filter, or the transmitted band will be broadened and shifted to longer wavelength. Thus only collimated (parallel) light, not diffuse light or light from an extended light source (e.g. a fluorescent tube) can be efficiently filtered by an interference filter.

Even if interference filters do not heat up as easily as absorbing filters, care should be taken so that their temperature does not rise by conduction from other parts of your apparatus. They should also be protected from moisture. When not in use they should be kept in a desiccator with dry silica gel.

The half-band width of a spectral band is defined as the difference in wavelength (or frequency) between the two points in the spectrum where the band is half the maximum height. Photobiologists often use interference filters with half-band widths of about 15 nm. This gives a reasonable compromise between spectral purity and amount of light transmitted. For some purposes filters with half-band widths up to 50 nm are useful. There are also interference filters with half-band widths as small as a fraction of a nanometer. They are used, e.g., by astronomers for photographing the sun using light emitted by a single kind of atom.

Continuous interference filters (also called spectral wedges) transmit light of different wavelength at different points on the filter. They are usually oblong, with different wavelengths along their length. Also circular spectral wedges have been manufactured.

There are also interference filters other than the narrow band type. One useful type is Calflex, manufactured by Schott & Gen. One version of it transmits almost the full visible range and reflects ultraviolet and infrared (including far-red).

3.3 Monochromators

For high spectral purity of light, yet flexibility in the choice of wavelength, a monochromator is the device of choice. A very simple monochromator can be made from a continuous interference filter between two slits (Fig. 6). Light of different wavelength is obtained simply by sliding the interference filter. However, this arrangement is not suitable when a small half-band width (high purity light) is required.

In earlier times, most monochromators contained a prism as the dispersing element (i.e. the component deflecting the light differently depending on wavelength). Gratings were too difficult to make, and hence expensive. New methods, however, allow the mass production of high quality gratings, and nowadays practically all monochromators for the near infrared, visible and ultraviolet regions use a reflection grating as the dispersing element.

The basic theory for a grating is best understood as an extension of Young's double slit experiment. Using a computer we can investigate the effect of increasing the number of slits more and more. The essential part of a programme for this are the three equations relating wavelength (l), deflection angle (θ) and relative fluence rate (I) to the width (b), number (n), and distance (a) of the slits:

$\alpha = (a*\pi/\lambda)*\sin \theta$
$b = (b*\pi/\lambda)* \sin \theta$
$I = 4*[\sin \beta * \sin(n*\alpha)/(\alpha*\beta)]^2$

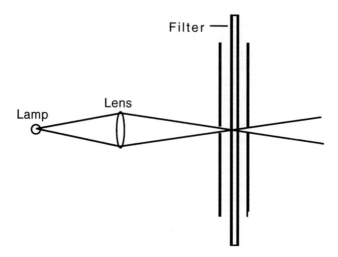

Figure 6. Simple monochromator consisting of an interference filter between two slits. It is here combined with a simple illuminator consisting of a lamp and a lens, providing light which is almost perpendicular to the filter, which is essential for proper function.

In the outputs of the programme below distance between and width of slits is in mm, wavelength in micrometre.

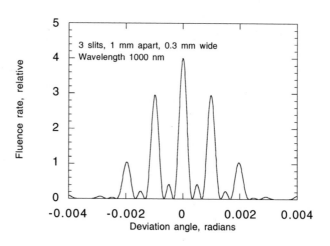

Figure 7a. Output from the multislit diffraction program.

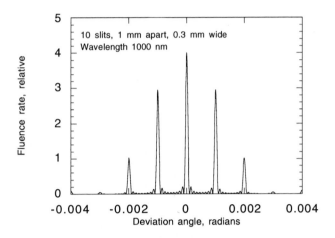

Figure 7b. Printout from the multislit diffraction program. 10 slits instead of 3 as in Fig. 7a.

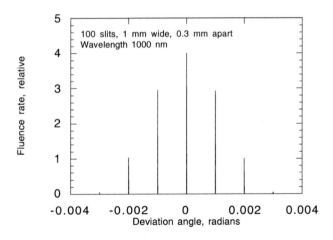

Figure 7c. Output from the multislit diffraction program. 100 slits.

Comparing the first three printouts (Figures 7a-c) and looking back at the double slit experiment, you can see that, when increasing the number of slits, the light becomes more and more concentrated to certain peaks (remembering that the horizontal axis represents deflection angle, we may say that the light becomes more and more concentrated to certain angles as the number of slits increases).

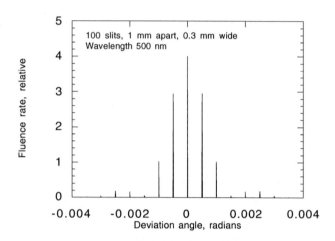

Figure 7d. Output from the multislit diffraction program. The wavelength is only half that of the previous three examples, while the number of slits is kept at 100.

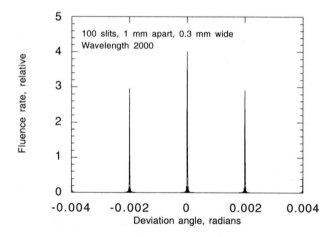

Figure 7e. Output from the multislit diffraction program. The wavelength has been doubled compared to the first three examples, while the number of slits is kept at 100.

The highest, central peak (non-deflected light) is called the first order light. With mixed input light this includes all wavelengths. The two peaks on each side of it is the first order light, and so on.

Comparing the last three printouts (Figures 7c-e), we see that the deflection angles increase with wavelength. We may think of the length unit as 1 mm, which means that we have 0.3 mm wide slits with their centres 1 mm apart, and the wavelength is 1000 nm in the first 3 printouts. We can see that the second order light with wavelength 500 nm is deflected to the same angle as the first order light with wavelength 1000 nm. Likewise, fourth order light of 500 nm wavelength, second order light of 1000 nm wavelength, and first order light of 2000 nm wavelength are all deflected to the same angle. As will be explained later, this has important consequences for photobiological experimentation.

Transmission gratings in the form of multiple slits are seldom used, except as here for teaching the theory of gratings. Instead mirrors with grooves, called reflection gratings, are made in many variants, usually as replicas, i.e. copies molded in plastic from very expensive originals. The principle of operation is essentially the same, except that the light exits on the same side of the grating as it enters.

Modern reflection gratings have the added refinement of blaze. This means that the grooves are not symmetrical but shaped in such a way that the direction of specular reflection coincides with the diffraction direction for a certain wavelength, the blaze wavelength. That makes a grating particularly efficient for this wavelength. If you have a choice, you should select a grating with a blaze wavelength near the wavelength you are particularly interested in, or for which it is difficult to get sufficiently strong light. For a lamp-monochromator combination like the one shown below it is usually preferred to have a low blaze wavelength to compensate for the lower lamp output in the UV and blue regions. On the other hand, in equipment for the analysis of light (monochromator- photomultiplier combinations in spectroradiometers or the emission units of spectrofluorometers) it may be advantageous to use a high blaze wavelength to compensate for the lower sensitivity of the photocell for long-wavelength light.

A grating monochromator, in addition to the grating, consists of an entrance slit, an exit slit, and optics which forms an image of the entrance slit at the exit slit via reflection in the grating. In Figs 8 and 9 we see a schematic diagram and the external appearance of one type of monochromator suitable for photobiological use. In this case a plane grating is combined with a concave mirror. Another

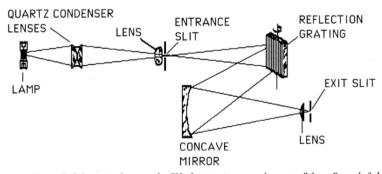

Figure 8. Schematic diagram for "High intensity monochromator" from Bausch & Lomb.

solution is to use a concave grating, which focuses the light without any concave mirror. The wavelength of light leaving the exit slit is changed by rotating the grating. The monochromator is shown in combination with an illumination unit consisting of a lamp and a lens.

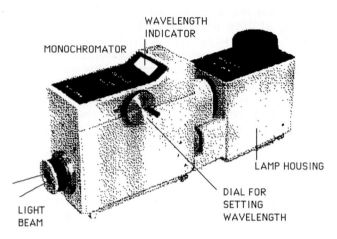

Figure 9. External appearance of "High intensity monochromator" from Bausch & Lomb.

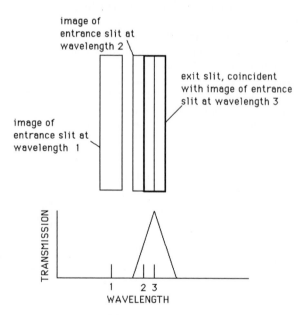

Figure 10. Images of the entrance slit for different wavelengths in relation to the exit slit (above), and the ideal transmission function, neglecting diffraction in entrance and exit slits.

When using a grating monochromator, one should be aware of the fact that despite the name, with a certain setting, it transmits light of more than one wavelength:

First, we have the problem of light of various orders (first, second etc.) mentioned above. The wavelength scale on the monochromator is valid for only a certain order, usually the first. When the dial is set at 300 nm, also second order light of 600 nm wavelength may be transmitted.

Second, the spectral composition of the light within one spectral transmission band depends on the size of the slits. A compromise has to be made in choosing the slits: The wider they are, the more light is transmitted (Fig. 10), but the more narrow they are, the higher the spectral purity (which is, in many cases, important). The best compromises are found when the image of the entrance slit (in monochromatic light) just covers the exit slit. In this case, the spectral transmission band shape would be triangular, were it not for diffraction at the slits and imperfections in the construction. In reality, the spectral band transmitted will be of a somewhat rounded triangular shape (cf. Chapter 4).

An interesting new development is the construction of acousto-optic tunable interference filters (see Tran 1997 for a review). Such a filter, is in a way intermediate between an interference filter and a grating. The basic principle is that a sound-wave in a crystal creates regions of alternating low and high refractive index, which causes diffraction of a light-beam. In this way the unit functions as a grating. But unlike a grating it need not be rotated for changing the wavelength of light exiting in a certain direction. This can be done by changing the frequency of the sound wave, which is generated piezoelectrically. This makes possible very rapid scanning over a spectral range by changing the frequency of the driving voltage.

REFERENCES

Calvert, J.G. & Pitts, J.N., Jr. (1966). Photochemistry. New York: Wiley.

Hale, G.M. & M. R. Querry,M.R. (1973). Optical constants of water in the 200 nm to 200 μm wavelength region. *Appl. Opt., 12,* 555-563.

Rabek, J.F. (1982). Experimental methods in photochemistry and photophysics, part 2. Chichester. ISBN 0-471-90030-3.

Tran, C.D. (1997). Principles and analytical applications of acousto-optic tunable filters, an overview. *Talanta, 45,* 237-248.

Yuan, Z., Kardynal, B., Stevenson, R.M., Shields, A.J., Lobo, C.J., Cooper, K., Beattie, N.S., Ritchie, D.A. & Pepper, M. (2002). Electrically driven single-photon source. *Science, 295,* 102-105.

LARS OLOF BJÖRN

4. THE MEASUREMENT OF LIGHT

1. INTRODUCTION

There are three principal types of light sensitive devices in common use, based upon three different effects of light on matter: photothermal, photoelectric and photochemical devices. We shall describe these and their uses, and then go on to describe a more complex device, the spectroradiometer.

2. PHOTOTHERMAL DEVICES

Photothermal devices have slow response and low sensitivity. Their great advantage is that, contrary to photoelectric and photochemical devices, they have the same response per energy unit throughout a very wide spectral range. Their principle of operation is that the light to be measured is allowed to be absorbed by a target. The temperature of the target is raised by the absorbed energy and the temperature rise is taken as a measure of the amount of energy absorbed.

Although other photothermal devices exist, the two most important ones are the bolometer and the thermopile.

2.1. The bolometer

In a bolometer the target is a temperature-dependent resistor. The resistivity of all materials are temperature dependent. In the first bolometers thin platinum foils, blackened with colloidal platinum for efficient absorption of light, were used. The resistivity of platinum rises with temperature. The platinum foils were freely suspended in the air, and these bolometers were very sensitive to air currents.

In the bolometers commonly used in photobiology laboratories today, the targets are thermistors, i.e. semiconductor resistors with a large negative temperature coefficient. They are protected by a window made from sapphire, lithium fluoride, or other material with a wide spectral transmittance range. The light target is part of a Wheatstone bridge, so that small changes in resistance can be recorded.

The setup is schematically depicted in Figure 1. Of the four resistor arms of the bridge, one is variable, so that the bridge can be balanced (same potential at the top as at the bottom, and no current flowing through the meter) with the target resistor shielded from light. In general the potential difference between top R_1/R_2 junction) and bottom (R_3/R_4 junction) will be $U[R_2/(R_1+R_2)-R_3/(R_3+R_4)]$, so that when the

65

L.O. Björn (ed.), Photobiology, 65–85.
© 2002 *Kluwer Academic Publishers. Printed in the Netherlands.*

bridge is balanced $R_2/(R_1+R_2)=R_3/(R_3+R_4)$. We now remove the light shield and allow the light to be measured to fall on R_2. This resistor now heats up and changes its resistance by the amount ΔR_2. The concomitant change in potential between "up" and "down" will be $U \cdot \Delta R_2/(R_1+R_2)$, and the change in current flowing through the meter will be proportional to the resistance change of the target.

Disregarding heating by the current flowing through it, the energy taken up by the target consists of the light to be measured plus heat radiation from the surroundings, assumed to be at absolute temperature T_a. The heat radiation received is proportional to T_a^4 (Stefan-Boltzmann's law). The radiation energy given off by the target (at absolute temperature T_t) is proportional to T_t^4. When equilibrium has been reached we thus have the relationship for the irradiance of the light to be measured (k_1 a constant):

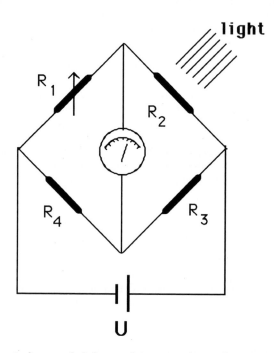

Figure 1. Schematic diagram of a bolometer. It is connected to a voltage source with voltage U.

irradiance $+ k_1 \cdot T_a^4 = k_1 \cdot T_t^4$, or

irradiance $= k_1 \cdot (T_t^4 - T_a^4) = k_1 \cdot (T_t^2 + T_a^2) \cdot (T_t + T_a) \cdot (T_t - T_a) \approx k_2 \cdot (T_t - T_a) = k_2 \cdot DT$. Thus the irradiance is proportional to the temperature change of the target resistor and thus, as shown previously, proportional to the current flowing through the meter.

Irradiances down to about 1 W m^{-2} can be measured with a standard bolometer. At lower irradiances the drift problems become serious. When a low irradiance is to be measured, it is best to connect the bolometer to a strip-chart recorder, to keep track of the drift of the base-line. A suitable procedure is to expose the bolometer to the light for 30 seconds, then shield it for the same period for recording of the base line, then expose it again, etc. Prolonged exposure decreases the reading, because the balancing resistors (not directly exposed to the light) heat up by heat conduction.

The calibration of a bolometer can be easily checked as described by Björn (1971). For highest accuracy a special standard lamp should be used in the way specified in the directions supplied with it. Regarding standard lamps see the section on spectroradiometers.

2.2. The thermopile

A thermocouple is a couple of junctions between two metals. Wherever two metals are in contact, a temperature dependent potential difference exists. A thermopile consists of several thermocouples (each one with two junctions) connected in series as shown in Figure 2. Of each couple of junctions, one is shielded from and the other one exposed to the light to be measured. The sensitivity and speed of response are increased by attaching little light absorbing (and heat radiating) shields to the junctions, and these shields should be blackened for efficient absorption (and reradiation).

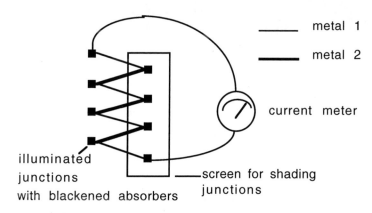

Figure 2. Schematic diagram of thermopile with current measuring meter.

For optimal results the input resistance of the current measuring meter should be matched to the resistance of the thermopile, which is of the order of 10 to 100 ohm. A thermopile is usable down to about the same irradiance as a bolometer. As for the bolometer, the output current is proportional to the irradiance.

3. PHOTOELECTRIC DEVICES

A great number of photoelectric devices exist. They can be divided into two main categories, depending on whether they exploit an outer photoeffect (at a metal surface in vacuum or gas), or an inner photoeffect inside a solid semiconductor.

3.1. Device based on outer photoelectric effect: the photomultiplier

Although there are many kinds of photocells (as well as television camera tubes, image intensifiers etc.) utilizing the outer photoeffect, we shall limit ourselves here to a description of the photomultiplier, a device extensively used by photobiologists.

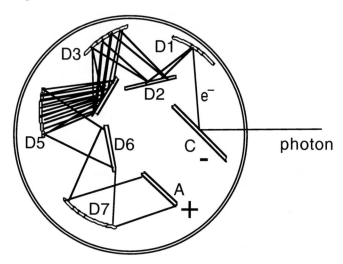

Figure 3. Diagram of (side-on) photomultiplier.

Fig. 3 shows the basic principle. Inside an evacuated envelope of glass or quartz there are a number of metal plates. The one marked C is the photocathode. It is held at a large negative potential relative to ground (wires connected to electrical circuitry not shown). The surface of the photocathode exposed to the light to be measured is covered with a layer of special metals. Usually a mixture of several metals, some of which are alkali metals, is used. Depending on the particular metal alloy, photomultipliers have different spectral responses.

When a photon hits the photocathode, an electron is released from the metal (as is known from chemistry, alkali metals are especially prone to losing electrons; they have a low *work function*). Because the photocathode is at a low electric potential, the electron does not return to the surface. Instead it is accelerated towards another metal plate nearby, which is held at a higher potential, dynode 1 (D1). In flight the electron acquires such a velocity that when it hits the dynode it releases 2 or 3 electrons from it. They travel on to dynode 2 (at an even higher potential) where further electrons are released. Photomultipliers are constructed with up to 12 dynodes in series, and at each dynode more electrons are added. Finally the electron swarm is collected at the final plate, the anode (A in Fig. 3). This is usually maintained close to ground potential. The electrons flowing to the anode represent an electrical current which also flows through the wire connected to the anode (not shown), and this can be recorded and used as a measure of the light incident on the photocathode.

Contrary to the thermoelectric devices described in the previous section, photomultipliers have different sensitivities to different kinds of light. Furthermore, they are rather unstable. Their great advantage lies in the high light sensitivity: even individual photons can be recorded by some photomultipliers under suitable conditions. Provided the electronic circuitry to which they are connected has a low time constant, photomultipliers also have a short response time (although different photomultipliers differ in this respect).

The diagram shows a so-called side-on photomultipler. There are also other designs. Another common one is the end-on photomultipler, where the photocathode consists of a thin, semitransparent metal film on the inside of the flat end of the cylindrical envelope. In this type the spectral response is also dependent on the thickness of the film, which usually varies somewhat over the surface.

Photomultipliers require an operating voltage of 500 to 5000 volt; in many cases about 1000 V is used. The different electrodes are given their proper voltages by a chain of resistors between the negative high voltage lead and ground. The output current is very strongly dependent on the operating voltage, which must therefore be held very constant (with N dynodes and operating voltage U, the output current is roughly proportional to U^N).

The output is not always measured as a current. As the incident irradiance is lowered, the discontinuous nature (quantization) of light becomes more and more apparent. At very low light levels it becomes advantageous to record individual photons by counting the pulses of current flowing to the anode. The measurement of very weak light is further dealt with in section 7 below.

Photomultipliers are made with different cathode layers for different spectral ranges from the ultraviolet to the near infrared (to about 900 nm). Photomultipliers sensitive to light of long wavelength generally have a higher dark current (dark noise) than others. The noise level can be decreased (to achieve a better signal to noise ratio, S/N ratio) by lowering the temperature. Liquid nitrogen, dry ice or Peltier coolers are generally used for this. Cooling must be combined with precautions to avoid dew on the optical components.

3.2. Devices based on semiconductors (inner photoelectric effect)

In the inner photoelectric effect absorption of a photon inside a solid semiconductor results in the separation of a positive charge from a negative one.

Photoconductive cells with a semiconductor as the light sensitive element have been used extensively in the past. Among these can be mentioned lead sulphide cells for measurements in the near infrared, for which no photomultipliers or other suitable photocells were available, and cadmium sulphide cells for photographic exposure meters. Photoconductive cells are variable resistors, and require an external source of voltage for creation of a current that can be measured. They have a slow response and are not used much for scientific measurements.

Photodiodes and phototransistors, on the other hand, can be made to have a very rapid response. For very long wavelengths (above 1000 nm) they also compete with photomultipliers in terms of sensitivity. The so-called avalanche photodiode can be regarded as a semiconductor equivalent to the photomultiplier. Photodiodes are often preferred in applications where photomultipliers ould also be used, because photodiodes are small, cheap, rugged and do not require high voltages. In our laboratory we use them for measuring variable chlorophyll fluorescence in plants, and, in combination with light emitting diodes, for indicating position of rotating shafts.

Photodiode arrays have become popular for recording a whole spectrum at one time (each diode in the array measures one spectral band).

Two main types of barrier layer cells are in current use: The selenium cell and the silicon cell. They are similar in principle, but differ in their spectral response: The selenium cell is most sensitive to green and blue light, the silicon cell to red and far-red light. The general principle is shown in Figure 4.

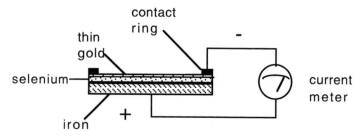

Figure 4. Selenium barrier layer cell with required circuitry. Note that the device generates its own operating voltage and does not require any battery or other external voltage source.

Because selenium barrier layer cells have a spectral sensitivity somewhat resembling that of the human eye, they are used in photometers (lux meters) for measurement of visible light at working places etc. By combination with suitable filters the spectral sensitivity curve can be made to almost completely match the curve for scotopic vision (which defines illuminance, mor on this later). Earlier such cells were used also for photographic exposure meters, where they have now been replaced by cadmium sulphide photoconductive cells, and more recently by photodiodes.

4. PHOTOCHEMICAL DEVICES: ACTINOMETERS AND DOSIMETERS

Chemical systems for measurement of light and ultraviolet radiation are called actinometers.

The best-known photochemical device for recording light is the photographic film. This has been used also for quantitative measurements, but it will not be further discussed here. Other chemical systems are usually better suited for quantitative measurements of radiation.

Actinometers have the advantage of not having a need for calibration by the user, and thus do not require the purchase of an expensive standard lamp with an expensive power supply. Standardisation has usually been taken care of by those who have designed the actinometer. Another advantage is that the geometry can more easily be adjusted to the measurement problem. The shape of a liquid actinometer can easily be made to correspond to the overall shape of the irradiated object under study. In many cases it is of interest to study a suspension or solution that can be put in an ordinary cuvette for spectrophotometry or fluorimetry, and the actinometer solution can be put into a similar cuvette.

A large number of actinometers have been devised. Kuhn et al. (1989) list and briefly describe and give references to 67 different systems, involving gaseous, liquid and solid phases. Of these they recommend five. In general actinomters are sensitive to short-wave radiation and insensitive to long-wave radiation. Insensitivity to long-wave radiation can be both a drawback and an advantage, but by choosing the best actinometer for the purpose one can avoid the disadvantages. One advantage of using an actinometer insensitive to long-wave radiation is that one can work under illumination visible to the human eye, without disturbing the measurement. We shall give an introduction below to a few different actinometers, all of which are not mentioned by Kuhn et al. (1989), and then go on to describe more in detail the most popular one for ultraviolet radiation, the potassium ferrioxalate or potassium iron(III) oxalate actinometer.

1. The potassium iodide actinometer (Rahn 1997) is sensitive primarily to UV-C radiation (wavelength below 280 nm), and with slight sensitivity also for short-wave UV-A. It is suitable for determining the 253.7 nm radiation from low pressure mercury lamps (bactericidal lamps), since the contribution from other spectral lines of the lamp will be negligible (but the ferrioxalate actinometer works almost as well for this purpose). It can be handled in ordinary incandescent light (not light from unshielded quartz-iodine lamps). The reaction on which this actinometer is based is the oxidation of iodide ion by iodate ion to form iodine, or rather triiodine ion (I_3^-).

2. An actinometer sensitive to visible light (photosynthetically active radiation) has been described by Wegner & Adamson (1966). It works up to above 700 nm, and is based on potassium tetrathiocyanatodiamm inechromate (III), $K[Cr(NH_3)_2(SCN)_4]$. The latter can rather easily be prepared from the commercially available Reinecke's salt, i.e $NH_4[Cr(NH_3)_2(SCN)_4]$. Irradiation causes the uptake of water and release of thiocyanate, which can be measured spectrophotometrically after addition of an iron(III) salt.

3. Two more actinometers which have recently been used in biological contexts is the 2-nitrobenzaldehyde actinometer (Allen et al. 2000) and the oxalic acid/uranyl sulfate actinometer (Mirón et al. 2000). A different version of the latter one preferable for small radiation doses is among those recommended by Kuhn et al. (1989).

4. As already mentioned, the most popular actinometer for ultraviolet radiation (and for violet and blue radiation as well) is the ferrioxalate actinometer. The description below will be sufficient for the experimenter starting in the field. For

more detailed information you should consult Parker (1953), Hatchard & Parker (1956) and Lee & Seliger (1964). Complete recipies have also been published by, e.g., Seliger & McElroy (1965) and Jagger (1967).

In the ferrioxalate actinometer the following photochemical reaction is exploited:

$1/2 \, (COO^-)_2 + Fe^{3+} + photon \Longrightarrow CO_2 + Fe^{2+}$, or

oxalate ion + Fe(III) ion + light $\Longrightarrow$ carbon dioxide + Fe(II) ion. The quantum yield for this reaction (i.e. the number of iron ions reduced per photon absorbed) is slightly wavelength dependent but close to 1 in the spectral region, 250-500 nm, where the ferrioxalate actinometer is used. Usually a 1 cm layer of 0.006 M ferrioxalate solution is used. Quantum yield, and the fraction of the radiation (perpendicular to the 1 cm layer) absorbed are shown in the following table and graph:

Wavelength	Quantum yield	Fraction absorbed	Q. yield x Fract. abs.
253.7	1.26	1	1.260
300.0	1.26	1	1.260
313.3	1.26	1	1.260
334.1	1.26	1	1.260
365.6	1.26	1	1.260
404.7	1.16	0.92	1.067
435.0	1.11	0.49	0.544
509.0	0.85	0.02	0.017

The quantum yields for 0.15 M actinometer solution are 0.952 of the above values. Irradiation of the side walls of the cuvette should be avoided, i.e. the beam should be smaller than the cross section of the cuvette.

The amount of iron(II) formed can be measured spectrophotometrically after addition of phenanthroline, which gives a strongly absorbing yellow complex with iron(II) ions.

The ferrioxalate for the actinometer is prepared by mixing 3 volumes of 1.5 M $(COOK)_2$ with 1 volume of 1.5 M $FeCl_3$ and stirring vigorously. This step and all the following involving ferrioxalate should be carried out under red light (red fluorescent tubes). The precipitated $K_3Fe(C_2O_4)_3 \cdot 3H_2O$ should be dissolved in a minimal amount of hot water, and the solution allowed to cool for crystallization (this crystallization should be repeated twice more).

Below is a recipe for the three solutions required for carrying out actinometry.

Solution A. Dissolve 2.947 g of the purified and dried iron(III) oxalate in 800 ml distilled water, add 100 ml 0.5 M sulphuric acid, and dilute the solution to 1 litre. This gives 0.006 M actinometer solution, which is suitable for measurement of ultraviolet radiation. For visible light, which is only partially absorbed, it may be advantageous to use 0.15 M iron(III) oxalate instead, i.e. 73.68 g per litre solution.

Solution B. The phenanthroline solution to be used for developing the colour with iron (II) ions should be 0.1% w/v 1:10 phenanthroline monohydrate in distilled water.

Solution C. Prepare an acetate buffer by mixing 600 ml of 0.5 M sodium acetate with 360 ml of solution B.

Solution A is irradiated with the light to be measured. The geometries of the container and of the light are important, and must be taken into account when evaluating the result. The simplest case is when the light is collimated, the container a flat spectrophotometer cell, the light strikes one face of the cell perpendicularly, and no light is transmitted. Even in this case one has to distinguish whether the cell or the beam has the greater cross-section, and correct for reflection in the cell surfaces. The irradiation time should be adjusted so that no more than 20% of the iron is reduced.

After the irradiation two volumes of the irradiated solution are mixed with two volumes of solution B and one volume of solution C, and then diluted to 10 volumes with distilled water. After 30 minutes the absorbance at 510 nm is measured against a blank made up in the same way with unirradiated solution A.

Example of calculation: Four milliliters of 0.006 M actinometer solution are irradiated in a flat quartz container by parallel rays of UV-B impinging at right angles to one surface (and not able to enter any other surface). The radiation cross section intercepted by the solution is 2 cm^2. Five minutes of irradiation produces 0.6 µmol iron(II).

Throughout the UV-B region the quantum yield is 1.26. Reflection from the surface is estimated to be 7% (by application of Snell's law). None of the radiation penetrates the solution to the rear surface, since the solution thickness is well over 1 cm. Therefore 0.6 mmol corresponds to $0.6/(1.26 \cdot 0.93)$ µmol $= 0.512$ µmol radiation incident on 2 cm^2 in 5 minutes, and the photon irradiance (quantum flux density, in this case equal to the photon fluence rate, since the rays are parallel and at right angles to the surface) is $0.510/(2 \cdot 5)$ µmol cm^{-2} min^{-1} = 5.10 nmol cm^{-2} min^{-1} or $5.10 \cdot 10^4/60$ nmol m^{-2} s^{-1} = 850 nm m^{-2} s^{-1}.

The great limitation of the ferrioxalate actinometer is that it is not sensitive to long wavelength light (in many cases this is also an advantage; one reason being that red working light can be used without interference with the measurements). Several actinometers sensitive to longer wavelengths have been designed. Warburg, for instance, used one based on chlorophyll. A modern, red sensitive actinometer has been described by Adick et al. (1989).

Chemical or biological systems, mostly in the solid state, for recording light, and ultraviolet radiation in particular, are widely employed for estimating exposure of people, leaves in a plant canopy and other objects which for various reasons are not easily amenable to measurements with electronic devices. These chemical devices are generally referred to as *dosimeters* rather than actinometers, even if there is no defined delimitation between these categories. As the construction, calibration and use of chemical and other dosimeters have been the subject of frequent reviewing (Bérces et al. 1999, Horneck et al. 1996, Marijnissen & Star 1987), I shall not dwell on them here, only stress that their radiation-sensitive components can be either chemical substance (natural like DNA or provitamin D, or artificial ones), or living cells (as various spores and bacteria).

5. FLUORESCENT WAVELENGTH CONVERTERS ("QUANTUM COUNTERS")

As stated earlier, photomultipliers have the advantage of being very sensitive, but also the disadvantage of having wavelength-dependent sensitivity. Fluorescent wavelength converters or "quantum counters" are solids or solutions, usually used in conjunction with photomultipliers, to obtain devices which are sensitive, yet have a sensitivity per photon which is independent of wavelength over a certain interval. The idea is to use a solution which has an absorbance high enough that all photons (except those reflected) will be absorbed, and which has a high fluorescence yield. Incident light of any wavelength distribution within certain limits is then converted, photon for photon, to light of a fixed wavelength distribution (the fluorescence spectrum of the "quantum counter"), to which the photomultiplier has a fixed sensitivity. One of the major uses of "quantum counters" is to calibrate the excitation units of spectrofluorimeters. The "quantum counter" most widely used consists of a concentrated solution of rhodamine B in ethylene glycol. It is useful for wavelengths up to 600 nm.

6. SPECTRORADIOMETRY

6.1. General

A spectroradiometer is an apparatus with which you can measure the spectrum of light, i.e. either the spectral irradiance, the spectral fluence rate, or the spectral radiance as a function of wavelength (or frequency, which is equivalent, but less commonly used by biologists). It consists of three main parts: (a) input optics, different for spectral irradiance, spectral fluence rate, or spectral radiance, (b) a monochromator or, preferably, a double monochromator, and (c) a transducer for converting the light signal to an electrical signal. The latter may be, in some cases, a photodiode, but is usually a photomultiplier. In some spectroradiometers instead of a monochromator there is a spectrograph which projects a whole spectrum, and the transducer is a diode array, charge coupled device (CCD) or a multichannel plate which samples the whole spectrum at once. The latter arrangement has the advantage of speed and synchronous sampling of all spectral channels, but is not always suitable. In particular it is very unsuitable for measuring ultraviolet radiation in daylight, in which case straylight problems must be minimized by use of a double monochromator.

A complete spectroradiometer system also requires some facility for frequent recalibration, as especially photomultipliers have very bad long-term stability.

6.2. Input optics

Before deciding on input optics, we need to decide what quantity we wish to measure. For spectral radiance we need input optics with which we can sample a very narrow solid angle. This can be some kind of telescope, but for most purposes it is sufficient to have a tube with a terminal stop which determines the sampling angle, and a few internal baffles and an inner matte black surface to avoid reflections inside the tube to reach the monochromator entrance slit (Fig. 5).

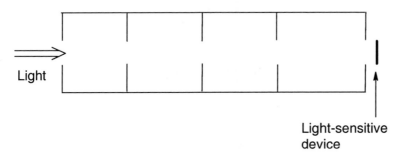

Figure 5. Input optics for radiance (narrow angle) measurements, consisting of a tube with stops and internal baffles, and inside painted dull black to prevent internal reflections.

For spectral fluence rate we need a device which samples all directions with equal sensitivity. This is an ideal which cannot really be fulfilled, but it can be approached. Using input optics for irradiance, it is possible to combine several measurements to obtain the fluence rate (Björn 1995, Björn & Vogelmann 1996).

For spectral irradiance measurements we need input optics which has a "cosine response", meaning that the sensitivity, or the efficiency of sampling, for a certain direction should be proportional to the cosine of the angle between that direction and the optical axis. The concept of "cosine response" is graphically explained in Fig. 6.

As a first crude device to reach this situation one could let the light to be measured strike a strongly scattering but translucent plate above the entrance slit of the monochromator. Suitable materials for this is ground quartz or fused silica or teflon, depending on the spectral range. A flat plate is, however, not very satisfactory, especially for large deviations from the optical axis. For measuring irradiance of light from an extended light source, for instance an overcast sky, light at these large angles is very important, since the "amount of sky" corresponding to a certain angular deviation from the optical axis is proportional to the sine of the angle. Somewhat better results are obtained using a hemispherical scattering dome over the slit. The only device which works real well is an integrating sphere (Fig. 7), and to work well it must be well designed. Details on this subject are provided by Optronics Laboratories (1995, 2001).

In some cases we are more interested in the shape of a spectrum than in the absolute light level, and then the angular sensitivity function is less important. We may also be interested in measuring light in places where it is not easy to put the spectroradiometer itself (such as underwater or inside your mouth). In that case the best choice may be to use fibreoptics at the input end of the spectroradiometer. Even then one may add, for instance, a small scattering device at the tip of the fibre optic conductor to collect light from different directions. Single light-conducting fibres may even be used to measure light inside plant or animal tissues (Vogelmann & Björn 1984).

Figure 6.. Cosine sensitivity of a receiver. The sensitivity is greatest straight up in this case, and decreases proportionally to the cosine of the angle to the vertical. In the horisontal direction the sensitivity is zero.

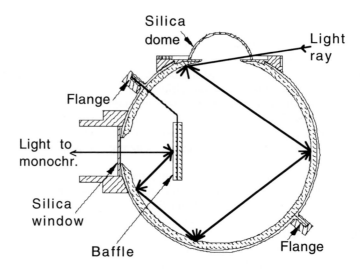

Figure 7. Integrating sphere as used as input optics for spectroradiometric irradiance measurements. On top is a silica dome, which can be omitted if the instrument is to be used only indoors or in good weather conditions. Below that is the opening in the sphere which defines the area over which the irradiance measurement is taken, and the direction of the reference surface. A light ray has been drawn which strikes this at a low angle. It is important that the walls of the sphere taper off to an edge to allow rays at such low angles to enter the sphere. The ray strikes the inner diffusely reflecting wall. From the point where it strikes light is scattered in all directions. We have followed one possible path through the sphere, but the little "brushes" at each scattering point are meant to indicate that there are many possibilities. Eventually the light strikes the back side of the baffle, which serves as the direct light source for the monochromator (cf. Fig. 8).

6.3. Example of a spectroradiometer

We show here (Fig. 8), as an example of a spectroradiometer often used by biologists, the construction of Model 754 from Optronic Laboratories.

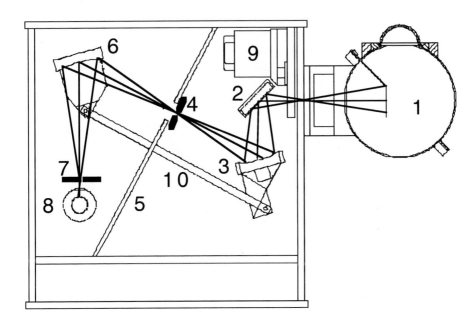

Figure. 8. Diagram of spectroradiometer (Optronic Laboratories model 754), simplified to enhance the important optical parts. Light from the integrating sphere (1, cf. Fig. 6) enters the light-tight box through a slit (entrance slit for the first monochromator, not specifically shown) and is deflected by the mirror (2) to the first grating (3). The chosen wavelength band leaves this first monochromator unit through the exit slit (4) in the wall (5) separating the two monochromators. This slit (4) is also the entrance slit for the second monochromator. The second grating (6) again disperses the light and deflects the chosen wavelength band onto a slit (7) in front of the photomultiplier (8). A stepper motor (9) turns the grating to change the wavelength. The rotating grating supports are connected by a bar (10), thus assuring that they follow one another with high fidelity.

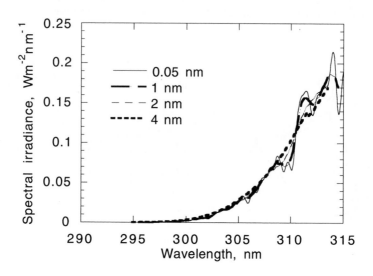

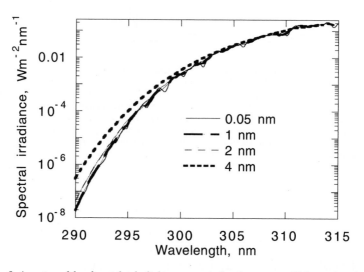

Figure 9. A portion of the ultraviolet daylight spectrum in Lund at noon on 15 June, plotted with three different half-bandwidths (modelled values). Upper frame linear, lower frame logarithmic spectral irradiance scale. In the linear plot we can see how the fine-structure is gradually smoothed out, while the semilogarithmic plot shows more clearly how systematic positive errors develop in the short-wavelength part. With 4 nm bandwidth the error is more than an order of magnitude for part of the spectrum.

Spectroradiometers for measurement of UV-B radiation in daylight do not work well with a single monochromator, mainly because spectral irradiance in the UV-B region changes very rapidly with wavelength. Very small amounts of radiation outside the intended band can therefore ruin the measurement. When two monochromators are used in tandem, it is of course very important that their wavelength settings agree throughout the scan. The only way to achieve this is to have both built on a single base plate and driven by a single wavelength drive mechanism.

Because biological chromophores have rather broad spectra, the fine-structure of the daylight spectrum or the spectrum of any other light is not of great importance. Still. it is important that the bandwidth of the measuring system is not too great, i.e. that the slits in the monochromator are not wide when daylight ultraviolet-B radiation is to be measured. This is because the spectrum is so steep in this region. Fig. 9 illustrates this: when the bandwidth is increased above 1 nm, values in the short-wavelength part of the UV-B band start to Errors are also introduced in the calibration (see below) if the bandwidth is too great.

6.4. Calibration of spectroradiometers

6.4.1. Wavelength calibration with lamps.. This is the simple part, but important. If the wavelength is not correct, then everything else will be wrong, too. The wavelength error should be less than 1 nm; for measurements of daylight ultraviolet-B radiation preferably much less.

For such calibration any medium pressure mercury lamp works well, even an ordinary fluorescent lamp. Easily recognizable spectral lines occur at 253.7 nm, 265.2 nm, 312.6+313.2 nm, 334.2 nm,365.0 nm, 365.4 nm, 366.3 nm, 404.7 nm, 435.8 nm, 491.6 nm, 496.0 nm, 546.1 nm, 577.0 nm, 579.1 nm, 623.4 nm and 690.7 nm. The short-wavelength bands do not penetrate the envelope of ordinary fluorescent lamps.

6.4.2. Irradiance calibration with standard lamps. Usually a spectroradiometer is calibrated using a lamp with a known output at different wavelengths. The most commonly used lamp is a 1 kW tubular quartz-iodine incandescent lamp. The reason to use such a powerful lamp is to obtain sufficient output in the ultraviolet region, and a lamp of this power can be used down to 250 nm. (For shorter wavelengths usually deuterium standard lamps are used).

Using transfer standards, these standard lamps are ultimately calibrated against cavity radiators held at a well-determined temperature, and designed so they follow as closely as possible the theoretical Planck blackbody radiation formula (see Chapter 1). When you purchase such a lamp you will obtain a table of the spectral irradiance obtainable at certain wavelengths at certain distance using a specific geometry, and also information about the accuracy. You will be surprised at (a) wide uncertainty limits compared to most other kinds of physical measurements (typically 3.5% at 250 nm) and (b) the high price of the lamp. After all it looks almost like the lamp you have in the overhead projector in your lecture hall. What you should consider at this moment of surprise is (a) that the lamp has been preburnt and specially selected among several to be particularly stable, (b) the effort and cost put down in calibrating it as accurately as possible. You should buy a second similar but uncalibrated lamp at a much lower cost, calibrate this as a working standard against your expensive lamp, and only occasionally use your expensive lamp to check your working standard.

The disadvantage in using a 1 kW lamp (apart from the heat it produces in your usually small calibration room) is that it requires a direct current of 8 A to run at 125 V, and that quite a big power supply is needed to produce this with good accuracy. It is very important that you really run it at the specified current at which it was calibrated. An error of 0.1% in the current produces a 1.2% error in the spectral irradiance at 250 nm, and a 0.6% error at 400 nm. It is important that the current is as ripple-free (has as little ac component) as possible. The best way of measuring the current is to measure the voltage across a precision resistor of, say, 0.1 ohm in the lamp circuit with a good digital voltmeter.

Calibration of a spectroradiometer is a tricky thing. When you calibrate it with your standard lamp you usually put the standard lamp on the optical axis. Suppose that you calibrate two spectroradiometers in this way with the same standard lamp in the same setup, so that they show the same result when you measure light from a lamp in the laboratory. Then you take the spectroradiometers outdoors and measure the daylight. You will then likely find that the two spectroradiometers show different results. This can have different causes, but the two dominating ones are probably that (a) the temperature is different and the two spectroradiometers have different temperature temperature dependencies , and (2) they have different off-axis sensitivities, and you are now measuring a very extended light source rather than an almost point-like standard source.

It is recommended that you recalibrate a spectroradiometer about once a month. This interval can be modified according to the experience you obtain over time with your particular instrument under your particular working conditions. If you move your instrument around, you should perform a rather easy wavelength check at each new location.

6.4.3. Irradiance calibration with an improvised standard lamp. Björn (1971) devised a method to use an ordinary tubular incandescent lamp as a standard lamp without prior optical calibration of the lamp, just relying on electrical measurements. The basic idea is that the temperature of the lamp filament is calculated from the increase in electrical resistance when it heats up from room temperature, and calculate the spectrum of the glowing filament from its temperature using Planck's radiation formula with appropriate corrections for the (temperature and wavelength dependent) emissivity of tungsten. This method is not recommended if calibrated standard lamps are obtainable for the experimentor.

6.4.4. Calibration without a standard lamp. Considering the cost of standard lamps, their instability, and the difficulty of ensuring the same standard everywhere in the world, it would be good if there would be a radiation source available free of charge such that everyone could use the same radiation source. There is such a radiation source: the sun.

6.4.5. Wavelength calibration using daylight. The sun is essentially a heat radiator, and thus radiating an essentially continuous spectrum, while a line spectrum is more suitable for wavelength calibration. However, some of the outer layers of the sun have a temperature that is low enough to reabsorb light from the inner layers, and still hot enough to contain single atoms, not united to molecules. They produce the so-called Fraunhofer lines in daylight, which are absorption spectral lines of these free atoms. The wavelengths of these Fraunhofer lines can be looked up in various sources. Here we just mention the two, due to hydrogen, which are most suitable for wavelength calibration in the visible region: 486.1344 nm (Fraunhofer F line) and 686.9955 nm (Fraunhofer C line).

6.4.6. Irradiance calibration using the sun. Surprisingly, the sun can be used also for irradiance calibration. This is surprising because we have the variable terrestrial atmosphere between ourselves and the sun, and this atmosphere is not the same everywhere. If we are on a mountain there is less atmosphere between us and the sun than at sea-level, and other factors also contribute to different attenuation of the sunlight at different places and times. However, these difficulties can be circumvented, provided the atmosphere is reasonably clear and stable over a day.

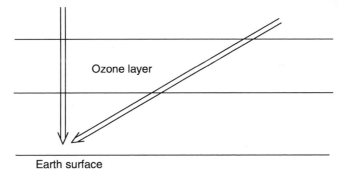

Figure 10. Effect of atmosphere on sunlight from different directions. If scattered light is excluded, so only the direct beam is considered, the length of the path of light through the atmosphere is longer for obliquely incident light, in proportion to the cosine of the incidence angle.

Consider first two (unrealistically) simple cases (Fig. 10). We have no clouds in the sky. In the first case the sun is direcly overhead, and the light is attenuated by, as the jargon in the field goes, one air mass. At a certain wavelength the spectral irradiance from the direct sunlight can be written as $I_1 = I_e e^{-a}$, where I_e is the corresponding extraterrestrial spectral irradiance, i.e. the spectral irradiance just outside the terrestrial atmosphere. The spectral irradiances we consider here are in the direction of the sun (light falling on a plane perpendicular to the direction to the sun). We consider only the direct sunlight, i.e. not that scattered in the atmosphere or reflected from the ground. In the second case the sun is at an angle of 60° to the vertical, i.e. the zenith angle of the sun is 60°. In this case the light path through the atmosphere is twice as long as in the first case. The irradiance then must be (provided the Lambert-Beer law is valid) $I_2 = I_e.e^{-2a}$, since the light is now attenuated by two air masses. Only when the scattered light is excluded by a narrow-angle input does the Lambert-Beer law hold (c.f. Chapter 1, section 11). The ratio between the two irradiances below the atmosphere is then $I_2/I_1 = e^{-a}$, and this ratio we should be able to determine without absolute calibration of the spectroradiometer. But knowing e^{-a} we can now calibrate the spectroradiometer against the extraterrestrial irradiance using the relationship $I_1 = I_e.e^{-a}$.

This is the principle of the Langley calibration method. In practice it is a bit more complicated than described here. One has to take spectra of the sun over at least one day, and preferable over several days when the weather is stable and the sky without clouds. For the Lambert-Beer law to be valid one has to measure the *direct* sunlight and exclude skylight as well as possible. One way of doing this is to measure the sunlight through a narrow baffled tube (Fig. 5) following the sun. Another way is to take a difference reading, the difference between the total daylight and the daylight when shadowing the sun. One then plots the logarithm of the reading against the air mass (the air mass is proportional to one over the cosine of the zenith angle). This gives a nearly straight-line relationship which can be extrapolated to zero air mass, corresponding to the reading which would have been obtained outside the atmosphere. For high accuracy one must apply various corrections (especially for large zenith angles), for instance for the refraction (bending) of the light rays in the atmosphere and for the curvature of the earth, and for the fact that different attenuators in the atmosphere do not have the same height distribution (especially for the fact that ozone, absorbing at short wavelengths, is higher than most attenuators). One must also take into account the variation of the sun-earth distance over the year, but this is easy. With decreasing wavelength the difficulties increase (for instance due to the rapidly changing ozone attenuation, the strong wavelength dependence of ozone absorption, less constant output of the sun at short wavelengths, and the smaller signal in relation to instrument noise). Below 300 nm this method cannot be used at all.

7. SPECIAL METHODS FOR MEASUREMENT OF VERY WEAK LIGHT

7.1 Introduction

Only methods based on photomultipliers will be reviewed here, but photomultipliers can be used in different ways. We shall not touch upon imaging of very weak light, which is important in many contexts from astronomy to biology. Photomultipliers can be used in the following main ways.

7.2 Direct current mode

This is the "classical way" described in section 3.1 above. The d.c. component of the anode current is measured. If the light is steady or varies only slowly, an amplifier with a long time constant can be used to obtain a "smoothed" or averaged value of the current; alternatively this averaging can, of course, be achieved using a computer. Then the light is shut off and the dark current is measured in the same way. With the photomultiplier connected to suitable electronic circuitry the difference between "light" and "dark" currents is proportional to the light to be measured.

If the light is very weak this does not work well. This is because the difference between "light" and "dark" curren ts will be a small difference between two larger terms, and a small relative error in any of them will result in a larger relative error in the difference. Furthermore, for weak light a long time will be required to get a reliable value of the light current, and in the meantime the dark current might drift.

7.3. Chopping of light and use of lock-in amplifier

In this mode for instance a rotating shutter or mirror is used to cut the light into short pulses separated by darkness of similar duration. A special amplifier amplifies the current during the light and dark periods separately, and the difference is continuously computed, or electric charges from the two sets of periods are stored for integration of the difference over time. In this way the effect of drift with time of the dark current is minimized. In this way much weaker light can be measured than in the direct current mode. It is, e.g., easy to measure the light emitted from a plant leaf (as a reversal of the photosynthetic process, see Chapter 16) for tens of minutes after the leaf has been placed in "darkness". However, "ultraweak luminescence" can hardly be measured with this method.

7.4. Measurement of shot noise

Shot noise is the "noise" of the d.c. signal from the photomultiplier arising from the quantized nature of light, i.e. the current pulses generated by the single photons. Yoh-Han Pao et al. (1966) suggested that the shot noise, treated as an a.c. signal, could be used as a measure of the light. Later experiments indicate that the signal-to-noise ratio for this method is somewhat better than for the lock-in method. However, because of the advantages of the method to be described below this method has not been used much.

7.5. Pulse counting

The dominating technique today for measuring very weak light is to count the current pulses generated in a photomultiplier by single photons. All pulses of the anode current, however, are not due to photons. Electrons are released both from the photocathode and from the dynode surfaces also by thermal energy. This is what gives the dark current in the d.c. mode. One great advantage with the pulse counting method is that pulses due to thermal emission from the dynodes can be filtered off, because they are smaller than those coming from the photocathode, since they have gone through fewer amplification steps. To achieve this the pulses from the anode go to a pulse discriminator , which allows only pulses above a certain amplitude to pass. The pulses passing through are then shaped to a uniform amplitude and pulse shape, so they can be more accurately counted. However, this cannot be done in a perfect way. Some photon induced pulses are discarded, and some spurious pulses are passed. Some photons do not give rise to any pulses at all, because the quantum efficiency of the photocathode is lower than one (even in the spectral region where it is highest). And there is no way by which thermal pulses arising in the photocathode can be distinguished from light-induced pulses. Therefore a photomultiplier can never be used as an absolute photon counting device. To estimate the true number of photons arriving at the photocathode the photomultiplier has to be calibrated, and a correction has to be made for "dark pulses".

Just as in a geiger counter there is a certain minimum time necessary for two pulses to be counted separately. Since the pulses are poisson distributed in time, the counting efficiency starts to decline gradually when the photon absorption rate increases over a certain limit. The absolute standard error of the number of pulses recorded, according to poisson statistics, is proportional to the square root of the number of pulses counted, and the relative error proportional to one over this square root. With absolute standard errors of e_l for the light count and e_d for the dark count, the standard error of the difference between light and dark counts is the square root of $(e_l^2 + e_d^2)$.

REFERENCES AND LITERATURE FOR FURTHER READING

Adick, H.-J., Schmidt, R. & Brauer, H.-D. (1989). A chemical actinometer for the wavelength range 610-670 nm. *J. Photochem. Photobiol., 49A,* 311-316.

Alfano, R.R. & Ockman, N. (1968). Methods for detecting weak light signals. *J. Optical Soc. Am., 58,* 90-95.

Allen, J.M., Allen, S.K. & Baertschi, W.W. (2000). 2-Nitrobensaldehyde: a convenient UV-A and UV-B chemical actinometer for drug photostability testing. *J. Pharmaceutical Biomedical Anal., 24,* 167-178.

Bérces, A., Fekete, A., Gáspár, P., Grof, P., Rettberg, P., Hornedk, G. & Rontó, G. (1999). Biological UV dosimeters in the assessment of biological hazards from environmental radiation. *J. Photochem. Photobiol. B: Biology, 53,* 36-43.

Björn, L.O. (1971). Simple methods for the calibration of light measuring equipment. Physiol. Plant., 25, 300-307.

Björn, L.O. (1995). Estimation of fluence rate from irradiance measurements with a cosine corrected sensor. *J. Photochem. Photobiol. B. Biol., 29,* 179-183.

Björn, L.O. & Vogelmann, T.C. (1996). Quantifying light and ultraviolet radiation in plant biology. - *Photochem. Photobiol., 64,* 403-406.

Hale, G.M. & M. R. Querry,M.R. (1973). Optical constants of water in the 200 nm to 200 µm wavelength region. *Appl. Opt., 12,* 555-563.

Hatchard, C.G. & Parker, C.A. (1956). A new sensitive actinometer. II. Potassium ferrioxalate as a standard chemical actinometer. *Roy. Soc. (London), Proc., A235,* 518-536.

Jagger, J. (1967). *Introduction to research in ultraviolet photobiology.* Englewood Cliffs (N.J.): Prentice Hall.

Kuhn, H.J., Braslavsky, S.E. & Schmidt, R. (1989). Chemical actinometry. *Pure. Appl. Chem., 61,* 187-210

Lee, J. & Seliger, H.H. (1964). Quantum yield of the ferrioxalate actinometer. *J. Chem. Phys., 40,* 519-523.

Marijnissen, J.P.A. & Star, W.M. (1987). Quantitative light dosimetry in vitro and in vivo. *Lasers in Medical Science, 2,* 235-242.

Mirón, A.S., Grima, E.M., Sevilla, J.M.F., Chisti, Y. & Camacho, F.G. (2000). Assessment of the photosynthetically active incident radiation on outdoor photobioreactors using oxalic acid/uranyl sulfate chemical actinometer. *J. Appl. Phycol., 12,* 385-394.

Optronics Laboratories (1995). *Improving integrating sphere design for near-perfect cosine response.* Application Note (A9). Downloadable from http://www.olinet.com

Optronic Laboratories (2001). *Standard spheres and sphere standards.* Application Note (A15). Downloadable from http://www.olinet.com

Rahn, R.O. (1997). Potassium iodide as a chemical actinometer for 254 nm radiation: Use of iodate as an electron scavenger. *Photochem. Photobiol., 66,* 450-.

Rolfe, J. & Moore, S.E. (1970). The efficient use of photomultiplier tubes for recording spectra. *Appl. Optics, 9,* 63-71.

Ryer, A.D. (1997) *Light measurement handbook.* Newburyport: International Light, Inc., Technical Publications Dept. ISBN 0-9658356. Downloadable from http://www.intl-light.com/handbook/

Schmid, B. & Wehrli, C. (1995). Comparison of sun photometer calibration by use of the Langley technique and the standard lamp. *Appl. Optics, 34,* 4500-4512.

Schmid, B., Spyak, P.R,, Biggar, S.F., Wehrli, C., Sekler, J., Ingold, T., Mätzler, *C. & Kämpfer, N. (1998). Evaluation of the applicability of solar and lamp radiometric calibrations of a precision sun photometer operating between 300 and 1025 nm. *Appl. Optics, 37,* 3923-3941.

Schneider, W.E. & Young, R. *Spectroradiometry methods.* Application Note (A14), pp. 49. Orlando (Fla): Optronics Laboratories. Downloadable from http://www.olinet.com

Seliger, H.H. & McElroy, W. (1965). *Light: Physical and biological action.* New York: Academic Press.

Vogelmann, T.C. & Björn, L.O. (1984). Measurement of light gradients and spectral regime in plant tissue with a fiber optic probe. *Physiol. Plant., 60,* 361-368.

Wilson, S.R. & Forgan, B.W. (1995). *In situ* calibration technique for UV spectral measurements. *Appl. Optics,* 34, 5475-5484.

Wegner, E.E. & Adamson, W.W. (1966). Photochemistry of complexions. III. Absolute quantum yields for the photolysis of some aqueous chromium (III) complexes. Chemical actinometry in the long wavelength visible region. *J. Am. Chem. Soc., 88,* 394-404.

LARS OLOF BJÖRN

5. NATURAL LIGHT

1. INTRODUCTION

Natural light at the surface of the earth is almost synonymous with light from the sun. Light from other stars has, as far as is known, photobiological importance only for the navigation by night-migrating birds.

Moonlight, which originates in the sun, is important for the setting of some biological rhythms. A full moon may perturb the photoperiodism of some short-day plants, and also synchronize rhythms in some marine animals.

However, the majority of photobiological phenomena are ruled by daylight, and we shall devote the remainder of this chapter to this topic. We shall treat the shortest wavelength components of daylight, the ultraviolet-B radiation at the end of the chapter, as special problems are involved with this waveband.

2. PRINCIPLES FOR THE MODIFICATION OF SUNLIGHT BY THE EARTH'S ATMOSPHERE

As mentioned earlier, the radiation from the sun is spectrally very similar to blackbody radiation of 6000 K. There are, however deviations both in the basic shape, and due to reabsorption (Fraunhofer lines) of some light by gases in the higher, cooler, layers of the sun.

The earth's atmosphere reflects, refracts, scatters, and partially absorbs the radiation from the sun, and thereby changes its spectral composition considerably. Part of the absorption and Rayleigh scattering is due to the main gases in the atmosphere, the concentration of which can be regarded as constant. Another part is due to ozone and water vapour, which occur in highly variable amounts. A third part is due to aerosol, which is also highly variable. The absorption causes loss of light, while scattering causes some light to be lost to space, while another part appears as diffuse light (skylight). Light is also reflected by clouds, and thereby mostly lost to space, but this will not be considered in detail here. Light reflected from the ground is partly scattered downwards or reflected from clouds, and again appears at the surface as diffuse light, and for this reason the ground reflectivity has some effect on skylight.

Daylight is also strongly dependent on the elevation of the sun above the horizon (90° minus the solar elevation is called the zenith angle of the sun, and often symbolized by the Greek letter theta, θ), because the lower the sun, the more air the rays must pass before they reach the ground. Daylight can be considered as

87

L.O. Björn (ed.), Photobiology, 87–95.

composed of two components, direct sunlight and scattered light. The scattered light is in most cases dominated by skylight, while some may reach the observer as scattered from the ground, trees etc.

3. THE UV-A, VISIBLE AND INFRARED COMPONENTS OF DAYLIGHT IN THE OPEN TERRESTRIAL ENVIRONMENT UNDER CLEAR SKIES

Accurate methods for computational modelling of daylight depart from the radiative transfer theory described by Chandrasekhar (1950, 1960). However, the fundamental formulae can usually not be used directly; different approximations have to be used for various cases, and this is nothing for the average biologist to work with. However, as long as we are dealing with clear skies (no clouds), relatively long wavelengths (near infrared, visible and ultraviolet-A radiations), and as long as we stay above water and vegetation, daylight can be well described by methods that are more easily handled.

A simple and for most purposes adequate procedure for this has been published by Bird & Riordan (1986). Their model has become very popular and their paper had been cited well over a hundred times when this book is being written. An alternative approach for this part of the spectrum is that of Green & Chai (1988). We shall use the approach of Bird & Riordan (1986) here to show how the direct component (sunlight) and the component scattered by the atmosphere (skylight) vary with the solar elevation (i.e. with the zenith angle). The same algorithm can be used also for visualizing how other factors, such as air pressure, air humidity, aerosol, ozone column and ground albedo affect daylight. We show the result only from 300 to 800 nm, but the paper by Bird & Riordan (1986) can be used to model radiation up to 4 micrometres, i.e. 4000 nm.

Fig. 1 shows three spectra, representing the direct sunlight, the skylight (diffuse radiation), and their sum, the so called global radiation (the total daylight). On top of the figure the vertical scale is indicated by a short horizontal line on the vertical axis and a value of spectral irradiance in $W\ m^{-2}\ nm^{-1}$.

Note that the skylight has its maximum moved towards shorter wavelengths compared to the direct sunlight. This corresponds to the fact that the sky appears blue in colour, and also to the fact that Rayleigh scattering is inversely proportional to the fourth power of the wavelength.

Figure 1 is for the irradiance on a horisontal plane. We can now do the corresponding computation for the irradiance on a vertical plane in the compass direction (azimuth) toward the sun (Fig. 2).

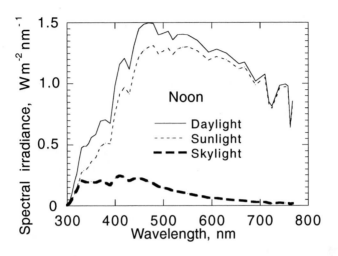

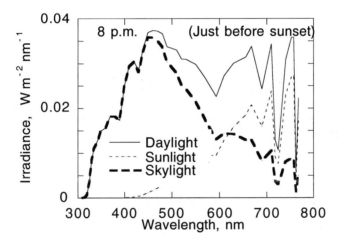

Figure 1. Irradiance at noon (top frame) and just before sunset (bottom frame) from above on a horisontal plane in Lund (south Sweden, 55.7°N, 13.4°E) on 15 July 2002, as computed with the algorithm of Bird & Riordan (1986). The ozone column was assumed to be 300 dobson units and the ground albedo 0.2, aerosol 0, and air pressure 1000 millibar.

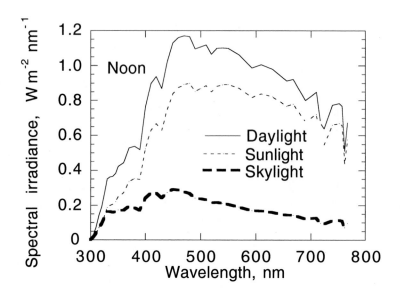

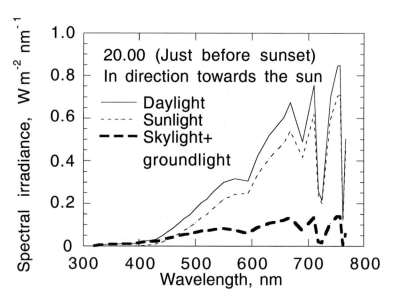

Figure 2. The same as Fig. 1, except that the plane is vertical and pointed in the compass direction of the sun.

These spectra are rather different. The sunset sunlight is of course much stronger in the horisontal direction (on a vertical plane). The scattered light is now not only skylight, but also light scattered from the ground, and therefore it contains much more long-wave components. Note also how deep the absorption bands for water vapour and oxygen have become, because the light must pass so much air when the sun is so low in the horizon. We can see from this that the concept "daylight spectrum" has no meaning if the geometry of measurement is not specified. We would get a third set of spectra for the fluence rate. The fluence rate can also be readily calculated using the algorithm of Bird & Riordan (1986) slightly modified: The diffuse component should be doubled, and the factor cosinus of incidence angle should be dropped in the expression for the direct component (sunlight).

Björn (1989) describes how, instead of tilt angle and incidence angle, one can enter tilt angle and compass direction of tilt in the computation of daylight irradiance.

The algorithm of Bird & Riordan (1986) assumes the skylight to come equally from all over the sky, or in other words, the sky *radiance* is uniform. This is an approximation, and other more accurate descriptions exist. A model based on radiative transfer theory has been published by Liang & Lewis (1996), while the group of R.H. Grant (Grant & Heisler 1997, Grant et al. 1996a,b, 1997), based on measurements, has developed a set of very simple models for various cloud conditions.

The paper by Bird & Riordan (1986), as stated in its title, deals only with clear skies. In modelling the diffuse skylight it assumes the sky to have uniform (isotropic) radiance. This latter approximation works very well as long as we are interested only in the irradiance on a horizontal or nearly horizontal plane. For some other purposes it may be of interest to model more exactly the variations in sky radiance, and how this can be done (in a relative sense also for cloudy conditions and for ultraviolet-B radiation) in a simple way has been described by Grant et al. (1996a,b, 1997) and Grant & Heisler (1997) in a series of papers, with a summary presently available at http://shadow.agry.purdue.edu/research.model.skyrad. html

Skylight is elliptically polarised (i.e. partly plane polarised), which is important for some animals who are able to determine the direction of polarisation and use it for orientation (see, e.g., Labhart 1999). The degree of polarisation can be approximated by $p=p_o \sin\alpha/(1+\cos^2\alpha)$, where α is the angular distance from the sun. The value of p_o is never more than 94%, mostly lower, and depending on aerosol in the air, reflection from the ground etc. The direction of the major electrical vector is approximately along the circumference of "circles" on the sky with the sun in the centre (e.g., Schwind & Horváth 1993). A few comments should be added to this simplified description (again, a more exact mathematical description can be obtained using the radiative transfer theory). Thus the polarization is increased in the spectral bands where the terrestrial atmosphere absorbs strongly (Aben et al. 1999). When the sun is higher than about 20° above the horizon there are two points within 20° of the sun, one above and one below, where polarization is zero. When the sun is higher than about 20° above the horizon, there is one such point, located about 20° above the antisolar point (Bohren 1995).

4. CLOUD EFFECTS

Clouds usually decrease both the irradiance and the degree of polarization of daylight. However, under some circumstances clouds can cause the irradiance above the values it would have had without clouds. This effect is particularly pronounced when most of the sky is overcast, but the sun is not in cloud, and when the ground is snow-covered or otherwise highly reflecting.

5. EFFECTS OF GROUND AND VEGETATION

Reflection from the ground is particularly important in the ultraviolet, since ultraviolet reflected upward by the ground is partially scattered downward again by the atmosphere, and the ground cover thus affects also downwelling radiation. The effect of reflection from the ground is greatest when it is covered by snow.

Reflection from the ground can be quite important also for the visible spectrum, and be important for plant growth, as shown by Hunt et al. (1985) and Kasperbauer & Hunt (1987).

Penetration of light into the ground is important for the germination of seeds. Soil transmission generally increases with increasing wavelength, thus giving buried seeds a far-red biased environment (Kasperbauer & Hunt 1988).

Plant canopies absorb visible light and ultraviolet radiation, but reflect and transmit far-red light and near infrared radiation. Light in or under green vegetation is therefore strongly biased towards the longer wavelengths, a fact which is of paramount importance to the plants subjected to this regime. The plant-filtered light forces the phytochrome system to the Pr (inactive) state (Kasperbauer 1987, Smith 1986, Holmes & Smith 1977). This will be further dealt with in Chapter 14.

It is now possible to measure light also *inside* plants and animals (Vogelmann 1986, Marijnissen & Star 1987, Star et al. 1987).

6. UNDERWATER DAYLIGHT

Pure water transmits best in the UV-A to blue region of the spectrum. We showed the absorption spectrum of pure water already in Chapter 2 . The estimation of underwater light by computation is difficult and uncertain for several reasons. Natural waters do not consist of chemically pure water. Fresh water bodies and coastal sea water contain dissolved substances and particles that attenuate UV and blue light more strongly than green, and therefore the light penetrating such waters contains less of UV and blue than does light penetrating into clear ocean water. In clear ocean water it is water itself that is the main attenuator. N.G. Jerlov (1970) has described several types of natural waters with respect to the penetration of visible light (cf. Figs 3 and 4). As for the penetration of UV-B into different types of natural water, the reader is referred to Smith and Baker (1979) as well as Baker & Smith (1982).

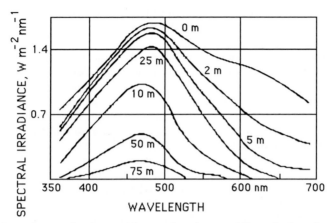

Figure 3. Spectral energy distribution of downward irradiance at different depths in the east Mediterranean (after Jerlow 1970, modified).

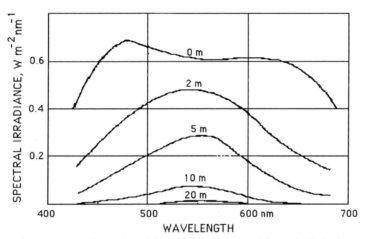

Figure 4. Spectral energy distribution of downward irradiance at different depths in the northern Baltic (after Jerlow 1970, modified).

Another problem is the water surface. Smith & Baker (1979) handle the direct sunlight using and Fresnel's law, but the water surface is seldom a flat horizontal interface as is required for this treatment. For the diffuse component of daylight, the transmittance coefficient can simply be set equal to 0.94 (see literature cited by Smith & Baker (1979).

7. THE ULTRAVIOLET-B DAYLIGHT SPECTRUM AND BIOLOGICAL ACTION OF UV-B

At the short-wavelength end of the daylight spectrum is the ultraviolet-B spectral band, 280 to 315 nm. This band is of particular interest, because it is highly biologically active (mostly inhibitory). It is more difficult to measure than visible light and ultraviolet-A radiation, because irradiance and fluence rate is lower. It is also more difficult to model than other daylight, because the spectral irradiance at ground surface is highly variable and dependent on other factors in addition to those influencing the longer wavelength components. The main factors influencing ultraviolet-B spectral irradiance at ground level are the elevation of the sun above the horizon, and the amount of ozone in the atmosphere. A program to study the effects of these and other factors on the ultraviolet-B spectral irradiance and estimate the biological action was designed by T.M. Murphy and myself (Björn & Murphy 1985), and is further described by Björn (1989) and Björn & Teramura (1993). A more accurate code, based on radiative transfer theory is that of S. Madronich, which is presently available on the Internet.

UV-B is more highly scattered than longer daylight components, and even under clear skies much of it reaches the ground as skylight rather than direct sunlight. Thus the fluence rate can be appreciable even in shadow. If the ground is snow-covered, and especially if the snow is fresh, much radiation can reach the observer from snow. Snow also increases the ultraviolet component of skylight, because radiation reflected from the ground is to an appreciable extent scattered down again by the atmosphere.

Also the underwater ultraviolet radiation has its special measuring and modelling problems. In freshwater bodies and coastal water the amount of UV-B absorbing dissolved substances is usually so high that UV-B radiation does not penetrate very far. Exceptions are some alpine lakes. But in clear ocean water, such as that in the Southern Ocean, UV-B radiation can be measured down to 60 m, and biological effects can be recorded at a depth of 20 m.

Ways of modelling underwater UV-B have been described by Smith & Baker () and Baker & Smith (1982), but these models describe only downward irradiance, not fluence rate. We shall return to ultraviolet-B radiation in Chapter 11.

LITERATURE

Aben, I., Helderman, F., Stam, D.M. & Stamnes, P. (1999) Spectral fine-structure in the polarisation of skylight. *Geophys. Res. Lett.*, *26*, 591-594.

Baker, K.S. & Smith, R.C. (1982). Bio-optical classification and model of natural waters. 2. *Limnol. Oceanogr.*, *27*, 500-509.

Bird, R.E. & Riordan, C. (1986). Simple solar spectral model for direct and diffuse irradiance on horizontal and tilted planes at the earth's surface for cloudless atmospheres. *J. Climate & Appl. Meteorology*, *25*, 87-97.

Björn, L.O. (1989). Computer programs for estimating ultraviolet radiation in daylight. *In* Diffey, B.L. (ed.) *Radiation measurement in photobiology, pp. 161-189.* London: Academic Press, London, ISBN 0-12-215840-7.

Björn, L.O. & Murphy T.M. (1985). Computer calculation of solar ultraviolet radiation at ground level. *Physiol. Vég.*, *23*, 555-561.

Bohren, C.F. (1995). Optics, atmospheric. *In* Trigg, G.L. (ed.) Encyclopedia of applied physics, vol. 12, pp.405-434. New York:VCH Publishers. ISBN 1-56081-071-8.

Chandrasekhar, S. (1950) Radiative transfer theory. Oxford University Press. Reprinted (1960) by Dover Publications, New York (pp. xiv + 393).

Grant. R.H. & Heisler, G.M. (1997) Obscured overcast sky radiance distributions for UV and PAR wavebands. J. Appl. Meteor. 36:1336-1345.

Grant, R.H., Gao, W. & Heisler, G.M. (1996a) Photosynthetically active radiation: sky radiance distributions under clear and overcast conditions. Agric. Forest Meteorol. 82:267-292.

Grant, R.H., Heisler, G.M. & Gao, W. (1996b) Clear sky radiance distributions in ultraviolet wavelength bands. Theor. Appl. Climatol. 56:123-135.

Grant, R.H., Gao, W. & Heisler, G.M. (1997) Ultraviolet sky radiance distributions of translucent overcast skies. Theor. Appl. Climatol.

Green, A.E.S. (1983). The penetration of ultraviolet radiation to the ground. *Physiol. Plant., 58,* 351-359.

Green, A.E.S. & Chai, S.-T. (1988). Solar spectral irradiance in the visible and infrared regions. *Photochem. Photobiol., 48,* 477-486.

Green, A.S. & Chai, S.-T. (1988). Solar spectral irradiance in the visible and infrared regions. *Photochem. Photobiol, 48,* 477-486.

Holmes, M.G. & Smith, H. (1977). Spectral distribution of light within plant canopies. *In* Smith, H. (ed.) *Plants and the daylight spectrum, pp. 147-158.* Academic Press, New York.

Hunt, P.G., Kasperbauer, M.J. & Matheny, T.A. (1985). Effect of soil surface color and Rhizobium japonicum strain on soybeen seeling growth and nodulation. *Agronomy Abstr., 85,* 157.

Jerlov, N.G. (1970). Light: General introduction. *In* Kinne, O. (ed.) *Marine Ecology,* vol. 1, part 1, *pp. 95-102,* 1970, London: Wiley- Interscience. ISBN 0-471-48001-0.

Kasperbauer, M.J. (1971). Spectral distribution of light in a tobacco canopy and effects of end-of-day light quality on growth and development. *Plant Physiol., 47,* 775-778.

Kasperbauer, M.J. (1987). Far red light reflection from green leaves and effects on phytochrome-mediated assimilate partitioning under field conditions. *Plant Physiol., 85,* 350-354.

Kasperbauer, M.J. & Hunt, P.G. (1987). Soil color and surface residue effects on seedling light environment. *Plant Soil, 97,* 295-298.

Kasperbauer, M.J. & Hunt, P.G. (1988). Biological and photometric measurement of light transmission through soils of various colors. *Bot. Gaz., 149,* 361-364.

Marijnissen, J.P.A. & Star, W.M. (1987). Quantitative light dosimetry in vitro and in vivo. *Lasers in Medical Science, 2,* 235-242.

Schwind, R. & Horváth, G. (1993) Reflection-polarization pattern at water surfaces and correction of a common representation of the polarization pattern of the sky. *Naturwiss., 80,* 82-83.

Smith, H. (1986). The perception of light quality. *In* Kendrick, R.E. & Kronenberg, G.M.H. (eds) *Photomorphogenesis in Plants, pp. 187-217.* Dordrecht: Martinus Nijhoff Publishers. ISBN 90-247-3317-0.

Smith, R.C. & Baker, K.S. (1979). Penetration of UV-B and biologically effective dose-rates in natural waters. *Photochem. Photobiol., 29,* 311-323.

Star, W.M., Marijnissen, H.P.A., Jansen, H., Keijzer, M and van Gemert, M.J.C. (1987). Light dosimetry for photodynamic therapy by whole bladder wall irradiation. *Photochem. Photobiol., 46,* 619-624.

Vogelmann, T.C. (1986). Light within the plant. *In* Kendrick, R.E. & Kronenberg, G.M.H. (eds) *Photomorphogenesis in Plants, pp. 307-337.* Dordrecht: Martinus Nijhoff Publishers. ISBN 90-247-3317-0.

LARS OLOF BJÖRN

6. ACTION SPECTROSCOPY IN BIOLOGY

1. INTRODUCTION

Action spectroscopy is a method for finding out what the initial step is in a photobiological or photochemical process. More exactly, the method serves to identify the kind of molecule absorbing the active light.

The basic principle of the method is the following: The more light that is absorbed, the greater its effect on the material systems under study. By comparing the *effects* of light having different wavelengths, a measure is obtained of the relative *absorption* at different wavelengths by the molecule *directly* affected by the light. This can then be compared to absorption spectra of various compounds. If everything works out one can identify the compound absorbing the active light in the photoprocess under study.

A hypothetical example: Molecule A, which is present in an organism, has the absorption spectrum shown in Fig. 1. Absorption of light by A causes a certain effect, say formation of anthocyanin in a plant. If a certain anthocyanin

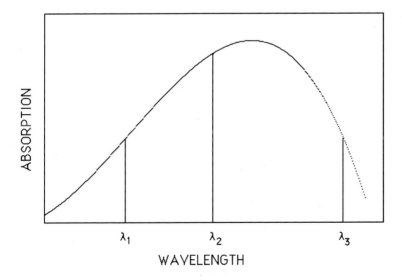

Figure 1. The concept of action spectrum (see text).

L.O. Björn (ed.), Photobiology, 97–114.
© 2002 *Kluwer Academic Publishers. Printed in the Netherlands.*

synthesis is obtained by irradiating for t minutes with N photons per m^2 and second of wavelength λ_1 (or λ_3), it ought to suffice with half as many photons of wavelength λ_2, since such light is absorbed twice as strongly. Or, conversely: If it is experimentally found that the two lights have the action described, this is an indication that molecule A is mediating the light effect. With only two wavelengths investigated the conclusion is still very uncertain. If the agreement between efficiency of the light and the absorptive power of A is extended over a wider spectral region, the conclusion will be more firmly founded.

The rest of this chapter will be more historical in character than other chapters. My experience is that action spectroscopy is better understood by studying several real examples of its use than theory alone. In addition, the papers cited here include some which can stand as good examples for young scientists. In some cases they reflect real scientific ingenuity, and despite the rapid development of science they have withstood the ravages of time remarkably well.

2. THE OLDEST HISTORY: INVESTIGATION OF PHOTOSYNTHESIS BY MEANS OF ACTION SPECTROSCOPY

Action spectroscopy may have roots in Young's and von Helmholtz' theories about colour vision. The first one, to my knowledge, to directly use action spectroscopy was T.W. Engelmann (1882a, b, 1884). He projected, under the microscope, spectra onto different algae, and assayed the amounts of oxygen formed by as a consequence of photosynthesis taking place in the algae. He estimated the relative amounts of oxygen by watching the accumulation of oxygen-loving (aerotactic) bacteria (Fig. 2).

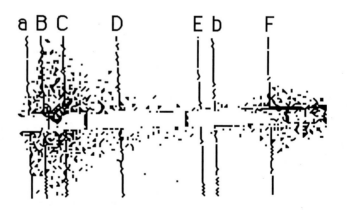

*Figure 2.. A piece of a filamentous green alga (*Cladophora *, of which two whole cells and parts of two more cells are seen) with swimming bacteria in the sunlight spectrum projected in a microscope. The letters indicate the Fraunhofer lines in the solar spectrum, which are used for wavelength calibration: a=718 nm, B=687 nm, C=656 nm, D=589 nm, b=518 nm, F=486 nm. The accumulation of bacteria is greatest in the red region around 680 nm, and in the blue region below 486 nm. These regions correspond to the main absorption bands of chlorophyll. From Engelmann (1882a).*

He compared the oxygen forming efficiency of different lights by reducing the light until the swimming movements of the algae stopped due to oxygen deficiency. In this way he could ascertain that in green algae it is chlorophyll which absorbs the light active in photosynthesis, while other pigments also participate in other kinds of algae. Fig. 3 shows some of his comparisons between absorption spectra and action spectra. The chlorophyll present in red algae (as evident from their absorption spectra) does not show up in the action spectrum for their oxygen production. The cause of this surprising fact was not revealed until after World War II when Duysens, Emerson and others discovered that two different photochemical systems cooperate in plant photosynthesis.

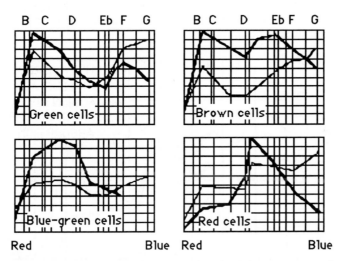

Figure 3. Absorption spectra for algal cells (per cent of incident light not penetrating the cells, thin lines) and action spectra for photosynthesis (thick lines). The high efficiency of light around 520 nm for photosynthesis by brown algae is due to light capture by the carotenoid fucoxanthol, the high efficiency around 620 nm in blue green algae to light capture by phycocyanin, and the high activity around 560 nm in red algae to light capture by phycoerythrin. From Engelmann 1884.

In addition to being an important step in the development of action spectroscopy and also in the history of photosynthesis research, Engelmann's experiments are important as early examples of a very sensitive "bioassay" of a chemical compound. The method of measuring oxygen by means of bacteria was so unconventional and the stated sensitivity in relation to other methods available at the time so remarkable that Engelmann was challenging the scientific authorities of his time. The algologist Pringsheim (1886) in Berlin as well as the Russian photosynthesis expert Timiriazeff (1885) found reason to criticize him using very harsh words.

Engelmann drew the correct conclusion that, in addition to chlorophyll, there are other pigments (coloured substances) able to absorb light and make it available to the photosynthesis process. He also understood (Engelmann 1882b) that there are other pigments in plant cells, which do not participate in photosynthesis but, on the contrary, "shadow" or "screen" the photosynthetically active pigments.

Engelmann's student Gaudikov studied chromatic adaptation (a designation in today's language, more consistent with the usual meanings of adaptation and acclimation, would be chromatic acclimation) in red algae and cyanobacteria, i.e. their acclimation to light of different colours. However, action spectra for this process was not determined until the 1960s by the Japanese Fujita & Hattori (1962) and, in the 1980s, with greater precision, by the Americans J. Scheibe, S. Diakoff, and T.C. Vogelmann (see Diakoff & Scheibe 1973, Vogelmann & Scheibe 1978).

As for action spectra of photosynthesis, a few investigations were carried out during the intervening years, but real progress beyond Engelmann's results did not take place until the 1940s. For details of this development the reader is referred to Haxo (1960). An early attempt was made also by Levring (1947) in Sweden to determine action spectra for photosynthesis in various algae, but he used wide spectral regions isolated with filters.

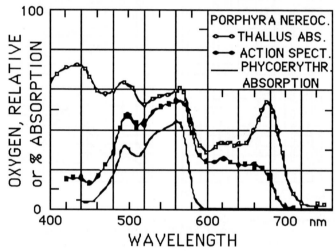

Figure 4. The action spectrum for photosynthetic oxygen production in the red alga Porphyra nereocystis *compared to the absorption spectrum of the same alga ("Thallus abs.") and to the absorption spectrum of extracted phycoerythrin. From Haxo and Blinks (1950).*

In this connection it is interesting to see how different scientists emphasized different aspects: Engelmann discussed in detail his method, spectral bandwidth etc. but lumped the algae together under the headings "green cells", "red cells" etc. Levring was less critical with regard to method but careful to state the species used, and published separate spectra for closely related species.

It was above all the spectra measured by the Americans Haxo and Blinks (1950) by a polarographic method for oxygen measurement which was to yield results valid to this day (Fig. 4). Due to them it became possible to do very careful comparisons between action spectra for photosynthesis and absorption spectra for various pigments in plants. Per Halldal brought this method to Sweden and improved it further.

3. INVESTIGATION OF RESPIRATION USING ACTION SPECTROSCOPY

The great Otto Warburg and his constant coworker Erwin Negelein (who, by the way, also determined action spectra for photosynthesis) over many years studied how the respiration of yeast cells is inhibited by carbon monoxide, and how this inhibition can be removed by light (Fig. 5). They developed action spectroscopy to an accurate quantitative method. As explicitly stated in one of their many papers, although they carried out the experiments together, it was Warburg who was the ingenious theoretician.

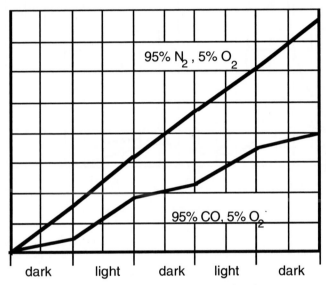

Figure 5. The effect of light on oxygen uptake (respiration, vertical axis) in yeast in an atmosphere of 95% nitrogen and 5% oxygen (straight line), and in a mixture of 95% carbon monoxide and 5% oxygen. The horizontal coordinate is time, with alternating light and dark periods. From Warburg (1926).

Their investigation led to the conclusion that the "Atmungsferment" (which we now call cytochrome c oxidase) is a protein to which iron- containing heme is bound, and that the inactive complex formed with carbon monoxide is dissociated by light. The conclusion rests on the observation that the action spectrum for removing the inhibition of respiration by carbon monoxide agrees very well with the absorption spectrum for a complex between carbon monoxide and heme (Fig. 6). There is only a small shift in wavelength which is explained by the binding to protein.

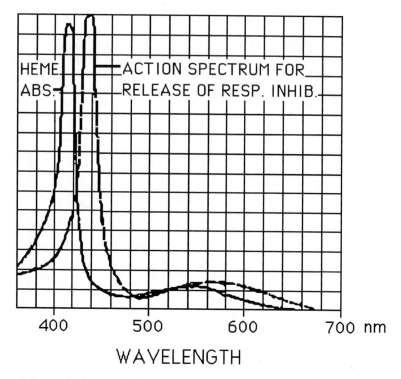

Figure 6. Comparison between the absorption spectrum for heme and the action spectrum for release from CO inhibition of yeast respiration. From Warburg & Negelein (1929b), cf. also Warburg & Negelein (1929a).

4. THE DNA THAT WAS FORGOTTEN

Already in the beginning of the last century Hertel (1905) in Jena had begun to study how microorganisms are affected by ultraviolet radiation. He managed to isolate 9 different spectral lines from 210 to 558 nm and quantify the radiation using a thermopile. Considering the time, this was no small feat. Unfortunately the evaluation of the biological effect was only semiquantitative. For constructing action spectra he determined the irradiance which gave a just noticeable effect on the organism, observable under the microscope. This effect could be stimulation of movement in *Paramaecium* or contraction in rotifers. Unfortunately he had no spectral line between 232 and 280 nm, and he therefore missed that region, which would later prove to be particularly interesting. Hertel's action spectra for ultraviolet damage to microorganisms are shown in Figure 7. His experiments gave rise to a long-lived opinion that the deleterious action of ultraviolet radiation rises at an even rate towards shorter wavelength.

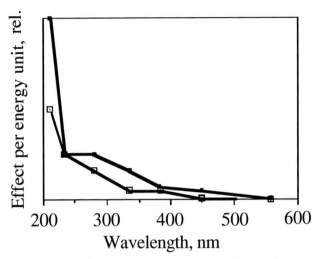

Figure 7. Action spectra for induction of swimming movements in Paramaecium *(top curve) and for contraction in* Rotatoria *(curve with squares). Redrawn after Hertel (1905).*

In the late 1920s, however, Gates (1928, 1930) found that the ability of ultraviolet radiation to kill bacteria varies with wavelength in the same way as does the ability of nucleic acid to absorb radiation (Fig. 8). This was the first indication of the fundamental importance of nucleic acid to life, and a key experiment at the entrance to molecular biology.

At the same time as Gates' first report, another one was published, which was also on the road leading to the great revolution. It was Griffith's discovery (1928) of bacterial transformation. But it was not until after 1944 that the biological role of DNA became generally accepted by the demonstration of bacterial transformation by DNA (Avery, MacLeod & MacCarty 1944).

One may wonder why Gates' experiments did not lead to a quicker development of DNA research. His work was in no way inferior to the transformation work. Perhaps an important reason was his way of publishing. In his first paper (1928) he does not show any convincing data. Despite this he clearly spelt out (although as we may think in retrospect, in a very cautious way) what was later proven to be essentially correct: "The close reciprocal correspondance between the curves of absorption of ultraviolet energy by these nuclear derivatives not only promotes the possibility that a single reaction is involved in the lethal action of ultra-violet light, but has a wider significance in pointing to these substances as essential elements in growth and reproduction". After discussing some experiments done by others, which strengthened his views he wrote: "Thus, while the relation of thymonucleic acid {i.e. DNA} to cell growth and reproduction remains a matter of conjecture, nevertheless the high concentration in the thymus gland and the coincidence of the evidence from these three independent series of experiments seem worthy of note, without further comment at present."

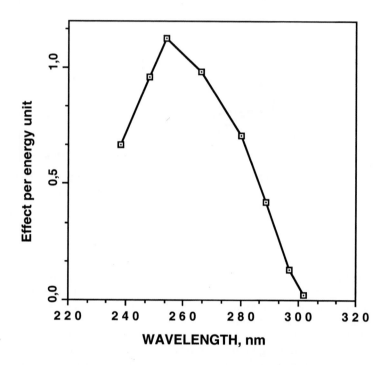

Figure 8. Action spectrum for "killing" (inactivation of cell division) in the bacterium Staphylococcus aureus. The graph is very similar to the absorption spectrum for DNA. Redrawn after Gates (1930).

In the next paper (1930) Gates includes data for a bacterium, but the diagram is not drawn in such a way that it is easy to see the similarity to the DNA absorption spectrum, and DNA is mentioned neither in the discussion nor in the summary. In one more paper from 1930 the deleterious action of ultraviolet radiation is treated from another point of view. In one of his last publications, from 1931, Gates compares the action spectra for two bacterial species to the absorption spectra of the bacteria. About the critical substance, the destruction of which causes the death of the bacterial cells, Gates writes in conclusion: "An examination of the evidence for its concentration in the cell nucleus, and the further search for evidence of its chemical character are reserved for the final paper of this series." Gates never got the opportunity to publish this final paper. He died on June 17, 1933. Even if one paper was published posthumously, his followers obviously did not consider his ideas about DNA important, or even correct.

Contributing reasons to the fact that Gates' ideas never got the attention that their importance deserved, were firstly that he was wrongly cited by later scientists (Hollaender & Claus 1936, for instance, "the maximum at 2499 A as reported by Gates"). Secondly, later scientists like Giese & Leighton (1935) went over to studying phenomena, e.g. swimming movements in *Paramaecium*, which were very protein-dependent. Action spectra for such processes have maxima around 280 nm. This diverted the interest from nucleic acids to protein.

5. PLANT VISION

One of the greatest triumphs of biological action spectroscopy is the discovery of the "vision pigment" of plants, phytochrome. However, at this point of the story action spectroscopy is getting more complicated.

When the phytochrome saga opens it was known since some time that some effects of red light on plants could be cancelled by exposing the plants to far-red light after the red. For instance, some lettuce seed do not germinate unless they are exposed to light after they have been allowed to take up water. Red light was found to be most efficient for this effect. Germination could be prevented by exposing the seeds to far-red light (720-740 nm wavelength) after the red.

Another example of red/far-red antagonism was the mode of growth of bean seedlings developing in darkness. The tip of such a seedling is curved to a "plumular hook", but if the seedling receives just a minute of red light, the hook straightens out during subsequent growth (Fig. 9). Withrow, Elstad & Klein (1957) tackled the problem of quantifying the straightening effect of different kinds of light. For each of various fluences of light of different wavelengths they measured by how many degrees the hooks of the bean plants were straightened out (Fig. 10).

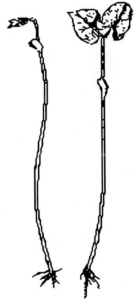

Figure 9. Bean plants grown in darkness (left) and in darkness except for a few minutes red light (right).

106

They also quantified how efficient different kinds of light were to counteract the straightening effect of a previously administered saturating fluence of red light (Fig. 11). Based on the results, they could postulate the existense of a light sensitive growth regulator, phytochrome. Phytochrome is formed in the plant in an inactive form (called P_r), which is transformable into the active form (P_{fr}) by red light.

The spectral curves that were obtained for the "straightening" and "bending" effects (Fig. 12) were postulated to correspond to absorption spectra for P_r and P_{fr}, respectively. It would never have been possible to isolate the phytochrome, had not its "spectral signature" been determined beforehand in this way.

In Figures 10 and 11 the lines for different wavelengths are not parallel. This means that the shape of the action spectra that are constructed from these lines will depend on the chosen level of action. That the curves are not parallel has to do with the fact that two photochemical reactions are involved to varying extents in all cases, i.e. the transformation of P_r to P_{fr}, and the transformation of P_{fr} to P_r. The "most correct" shapes, i.e. those most closely corresponding to the absorption spectra of P_r and P_{fr} (Fig. 13), are obtained by investigating the effects of very small fluences, resulting in small effects.

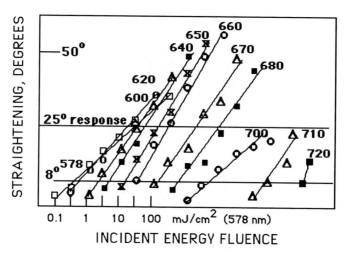

Figure 10. The straightening effect on bean hooks of light of various wavelengths as a function of the energy fluence. The scale on the horizontal axis is for the wavelength 578 nm, as an example to show that this scale is logarithmic. The scales for the other wavelengths have been moved in steps of 0.3 to make all the data points and regression lines visible. In the original publication (Withrow, Klein & Elstad 1957) there is one more data panel like this for the spectral range 385-578 nm.

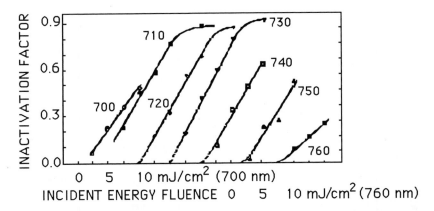

Figure 11. Inactivation of red-induced hook-straightening in bean plants by light of different wavelength as a function of fluence. In this case a linear fluence scale was found to give better linearity of regression than the logarithmic scale in the previous graph. To make all data points and regression lines distinguishable, the fluence scales were given different zero points for different wavelengths; only the extremes are shown here. Redrawn from Withrow, Klein & Elstad (1957).

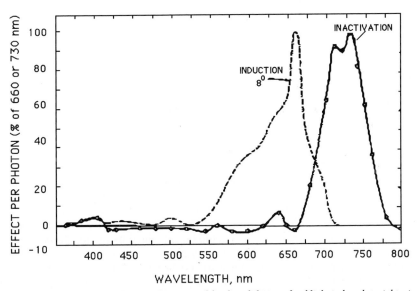

Figure 12. Action spectra for straightening and for the inhibition of red light induced straightening of the bean hook, constructed from regression lines of the kind shown in Figures 10 and 11. From Withrow, Klein & Elstad (1957).

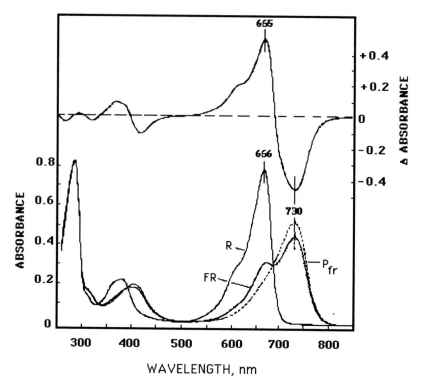

Figure 13. Absorption spectra of purified phytochrome from oats. The curve marked R is the measured absorption of far-red irradiated solution, containing almost only P_r. The curve marked FR is the measured absorption of red irradiated solution, containing mostly P_{fr} and some P_r. The dotted curve marked P_{fr} is the estimated spectrum of pure P_{fr}, which cannot be measured directly, since red light only partially converts P_r to P_{fr}. The top curve is a so-called difference spectrum showing the difference in absorption between fr-red irradiated and red-irradiated solutions. Since only the change due to irradiation shows up in this, almost the same difference spectrum can be obtained from plant tissue. After Vierstra & Quail (1983a,b).

The complications of this light sensitive system are well demonstrated by the action spectrum determined by Hartmann (1967) for the inhibiting effect of prolonged and relatively strong illumination on the extension growth of lettuce hypocotyls (Fig. 14). At first glance it does not seem to have anything to do with the shapes of the absorption spectra of the two phytochrome forms. But Hartmann could show that also this phenomenon could be explained by phytochrome being the mediator of the light action. However, in this case one has to take into account not only the photochemical reactions, but also the fact that the physiologically active form of phytochrome, P_{fr}, is unstable and disappears if there is no Pr present from which P_{fr} can be continually reformed. For this reason that light gives the greatest physiological effect which causes only a small part of the phytochrome to be continuously converted to P_{fr}. Light of longer wavelength has no effect because too little P_{fr} is formed. Light of too short a wavelength has no effect because P_{fr} is formed too quickly and all phytochrome disappears before it can act for sufficient time. The inhibition of the growth of the lettuce hypocotyls is an example of a so-called high intensity reaction (HIR).

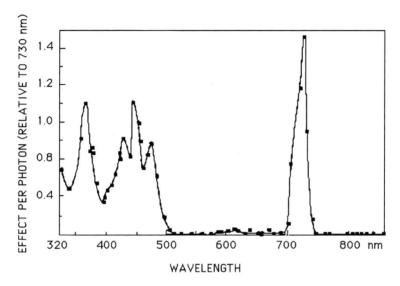

Figure 14. Action spectrum for the inhibition of extension growth of lettuce hypocotyls by prolonged irradiation with strong light (high intensity reaction). After Hartmann (1967).

6. PROTOCHLOROPHYLLIDE PHOTOREDUCTION TO CHLOROPHYLLIDE *a*

The present author and many other researchers have studied the action spectra for synthesis of chlorophyll and formation of chlorophyll. Important early contributions were made by J.H.C. Smith at the Carnegie Institution, T.N. Godnev and A.A. Shlyk in Belorussia, and H. Virgin in Sweden. Virgin's former students have continued this line of research in Göteborg.

Chlorophyll formation is governed by several light sensitive processes: Conversion of protochlorophyllide to chlorophyllide with enzyme-bound protochlorophyllide as the light absorber, conversion of phytochrome, and action on the so-called blue light receptor.

One specific question in this context was whether radiation absorbed in the aromatic amino acids of the enzyme NADPH-protochlorophyllide photooxidoreductase would be able to cause the conversion of protochlorophyllide to chlorophyllide in the same way as does radiation absorbed in the protochlorophyllide itself. In Fig, 15 the absorption of the NADPH-protochlorophyllide photoreductase complex with its substrates NADPH and protochlorophyllide is shown by the solid curve. In the visible region the absorption is due to the protochlorophyllide. The hight peak at 280 nm, on the other hand, is due to aromatic aminoacids in the protein. The action spectrum for protochlorophyllide photoreduction, shown by dots, has essentially the same features in the visible region, but lacks the high peak at 280 nm. The conclusion is that energy absorbed in the aromatic aminoacids cannot be used for photoreduction. There is also a small difference between absorption and action spectrum in the blue region (the so-called Soret peak of the spectra). This is probably because there are two fractions of protochlorophyll with slightly different spectra, of which only one can be converted by light (there may be no reductant, NADPH bound to the "inactive" complexes). There is also some blue-absorbing carotenoid contributing to the absorption spectrum.

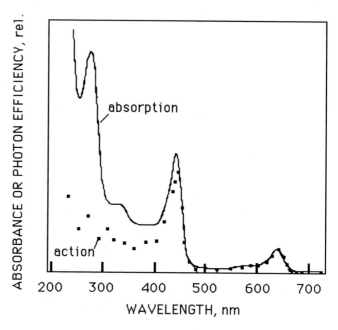

Figure 15. Solid line: The absorption spectrum of purified "protochlorophyll holochrome" (= complex between protochlorophyl- lide, NADPH and NADPH-protochlorophyllide photooxidoreductase) extracted from etiolated bean plants (Schopfer & Siegelman 1969) and the action spectrum for protochlorophyllide photoreduction to chlorophyllide a (Björn 1969a). From Björn (1969b), redrawn.

7. THE ELUSIVE BLUE-LIGHT RECEPTOR

Numerous "blue-light phenomena" have been studied in plants and fungi: phototropism, chloroplast rearrangements, plastid differentiation, nastic movements to mention a few. The discussion about the possible nature of the molecule absorbing the active light in these processes has centered mainly on carotenoids and flavoproteins, because the action spectra are similar to absorption spectra of these compounds (which are so similar to each other and so variable with conditions and molecular details that a conclusion on action spectra alone seems impossible). As described in Chapter 8, it has eventually been found that most blue-light phenomena are mediated by flavoproteins, but that some, e.g., stomatal movements, may be carotenoid-mediated.

One interesting feature of all "blue light phenomena" which have been studied in sufficient detail is that the fluence response regression lines are non-parallel, just in the case of phytochrome. This would indicate that the blue-light receptor is also a "photoreversibly photochromic pigment", which is changed by one kind of light in one direction and changed back by another kind of light to its original form. It has proved very difficult to demonstrate photoreversal of blue light reactions, and the reason for this may be that the absorption spectra of the two forms of the blue light receptor overlap much more than is the case with the two phytochrome forms.

As an example of this we show here an effect of blue light on stomatal opening. Stomata are regulated in a pluralistic way (not only by light, but also by water conditions and carbon dioxide concentration). Light exerts influences via photosynthesis, via phytochrome and via the blue light receptor. Using a background of red light to keep the effects on photosynthesis and phytochrome constant, the regulation via the blue light receptor can be isolated. Figures 16 and 17 show the fluence response regressions and the action spectrum derived therefrom.

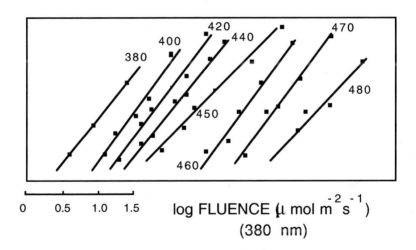

Figure 16. The blue-light effect on stomatal opening in wheat leaves as a function of fluence for different wavelengths. The log fluence scale is shown only for 380 nm; for other wavelengths the scales are consicutively transposed by 0.5 units in the order of the wavelengths. From Karlsson (1986).

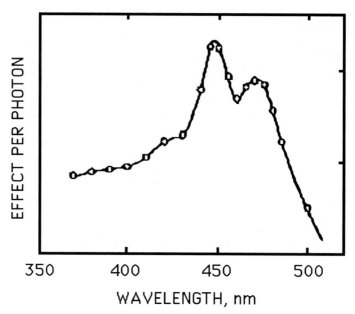

Figure 17. Action spectrum derived from data in Figure 15.

8. ANOTHER USE FOR ACTION SPECTRA

So far we have only discussed the use of action spectroscopy for identifying the molecules absorbing the light driving the various photoprocesses. At the end I would like to point out one more important reason for determining action spectra. For this we shall go back to near the beginning of this chapter, to the damaging effect of ultraviolet radiation, but see it from another angle.

In recent years there has been concern about the possible depletion of the stratospheric ozone layer. Such depletion would result in increased levels of ultraviolet radiation at the surface of the earth (unless other changes in the atmosphere would compensate for the depletion in stratospheric ozone). Experiments have been carried out to forecast the biological effects of such changes in radiation. Ozone depletion has been simulated by exposing the organisms to be studied to artificial ultraviolet radiation. One problem has been that the artificial radiation cannot be given the same spectral composition as the additional solar radiation that would leak through a depleted ozone layer. Therefore weighting functions have been needed to calculate how much artificial ultraviolet radiation that has to be administered to simulate a certain ozone depletion, and for this one has to determined action spectra for different ultraviolet effects. Initially one relied on ordinary, "monochromatic" action spectra, determined as the other spectra in this chapter. However, since so many different ultraviolet effects with different action spectra are involved, it has turned out to be more realistic to determine "polychromatic action spectra". For this one starts with a full spectrum, and for different samples cuts away more and more of the short-wavelength part.

LITERATURE

Avery, O.T., MacLeod, C.M. & McCarty, M. (1944). Studies on the chemical nature of the substance inducing transformation of pneumococcal types. *J. Exp. Med., 79*, 137-157.

Björn, L.O. (1969a). Action spectra for transformation and fluorescence of protochlorophyll holochrome from bean leaves. *Physiol. Plant., 22*, 1-17.

Björn, L.O. (1969b). Studies on the phototransformation and fluorescence of protochlorophyll holochrome in vitro . *In* Metzner, H. (ed.), *Progr. in photosynthesis research, Vol. II*, pp. 618-629. Tübingen.

Diakoff, S. & Scheibe, J. (1973). Action spectra for chromatic adaptation in *Tolypothrix tenuis* . *Plant Physiol. , 51*, 382-385.

Engelmann, Th. W. (1882a). Ueber Sauerstoffausscheidung von Pflanzenzellen im Mikrospektrum. *Bot. Ztg., 40*, 419-426.

Engelmann, Th. W. (1882b). Ueber Assimilation von *Haematococcus* . *Bot. Ztg., 40*, 663-669.

Engelmann, Th. W. (1884). Untersuchungen über die quantitativen Beziehungen zwischen Absorption des Lichtes und Assimilation in Pflanzenzellen. *Bot. Ztg. , 42*, 81-94, 97-106 & Tafel II.

Fujita, Y. & Hattori, A. (1962). Photochemical interconversion between precursors of phycobilin hromoproteids in *Tolypothrix tenuis* . *Plant Cell Physiol., 3*, 209-220.

Gates, F.L. (1928). On nuclear derivatives and the lethal action of ultra-violet light. *Science, 68*, 479-480.

Gates, F.L. (1930). A study of the bactericidal action of ultra violet light. I. The reaction to monochromatic radiations. *J. Gen. Physiol., 14*, 31-42.

Gates, F.L. (1931). A study of the action of ultra violet light III. The absorption of ultra violet light by bacteria. *J. Gen. Physiol., 14*, 31-42.

Griffith, F. (1928). The significance of pneumococcal types. *J. Hygiene, 27*, 113-159.

Hartmann, K.M. (1967). Ein Wirkungsspektrum der Photomorphogenese unter Hochenergiebedingungen und seine Interpretation auf der Basis des Phytochroms (Hypokotylwachstumshemmung bei *Lactuca sativa* L.). *Z. Naturforsch., 22b*, 266-275.

Haxo, F.T. (1960). The wavelength dependence of photosynthesis and the role of accessory pigments. - *In* Allen, M.B. (ed.). *Comparative biochemistry of photoreactive systems*, pp. 339-376. New York: Academic Press.

Haxo, F.T. & Blinks, L.R. (1950). Photosynthetic action spectra of marine algae. *J. Gen. Physiol. , 33*, 389-422.

Hertel, E. (1905). Ueber physiologische Wirkung von Strahlen verschiedener Wellenlänge. *Zschr. Allgem. Physiologie, 5*, 95-122.

Hollaender, A. & Claus, W.D. (1936). The bactericidal effect of ultraviolet radiation on *Escherichia coli* in liquid suspensions. *J. Gen. Physiol. , 19*, 753-765.

Giese, A.C. & Leighton, P.A. (1935). Quantitative studies on the photolethal effects of quartz ultra-violet radiation upon Paramecium . *J. Gen. Physiol. , 18*, 557-571.

Karlsson, P.E. (1986). Blue light regulation of stomata in wheat seedlings. II. Action spectrum and search for action dichroism. *Physiol. Plant., 66*, 207-210.

Levring, T. (1947). *Submarine daylight and the photosynthesis of marine algae*. Göteborgs Kgl. Vetenskaps- och Vitterhets-samhälles Handl., 6:e följden, ser. B, band 5, nr 6. 90 s.

Pringsheim, N. (1886). Zur Beurtheilung der Engelmann'schen Bakterienmethode in ihrer Brauchbarkeit zur quantitativen Bstimmung der Sauerstoffabgabe im Spektrum. *Ber. d. deutsch. bot. Ges., 4*, 40-46.

Schopfer, P. & Siegelman, H.W. (1969). Purification of protochlorophyllide holochrome. *In* Metzner, H., (ed.), *Progress in photosynthesis research Vol. II*, pp. 612-618, Tübingen.

Timiriazeff, C. (1885). État actuel de nos conaissances sur la fonction chlorophyllienne. *Ann. des sc. nat. Botanique (3)* T.II.

Vierstra, R.D. & Quail, P.H. (1983a). Purification and initial characterization of 124-kilodalton phytochrome from *Avena* . *Biochemistry, 22*, 2498-2505.

Vierstra, R.D. & Quail, P.H. (1983b). Photochemistry of 124-kilodalton *Avena* phytochrome *in vitro* . - *Plant Physiol., 72*, 264-267.

Vogelmann, T.C. & Scheibe, J. (1978). Action spectra for chromatic adaptation in the blue-green alga *Fremyella diplosiphon*. *Planta, 143*, 233-239.

Warburg, O. 1926. Über die Wirkung des Kohlenoxyds auf den Stoffwechsel der Hefe. *Biochem. Z., 177*, 471-486.

Warburg, O. & Negelein, E. (1929a). Über die photochemische Dissoziation bei intermittierender Belichtung und das absolute Absorptionsspektrum des Atmungsferments. *Biochem. Z., 202*, 202-228.

Warburg, O. & Negelein, E. (1929b). Absolutes Absorptionsspektrum des Atmungsferments. *Biochem. Z., 204*, 495-499.

Withrow, R.B., Klein, W.H. & Elstad, V. (1957). Action spectra of photomorphogenetic induction and its inactivation. *Plant Physiol. , 32*, 453-462.

LARS OLOF BJÖRN

7. SPECTRAL TUNING IN BIOLOGY

1. INTRODUCTION

The justification for a book about photobiology rests partly on combination of various specialities for interdisciplinary comparisons. One topic suitable for comparisons is "spectral tuning". By this I mean the principles for how spectra of pigments, and factors that can modify their spectral responses, are adjusted to the needs of the organisms that produce them. A number of examples will be found in this chapter.

Spectral tuning is relevant for vision (not only for colour vision), photosynthesis, bioluminescence, flower colours, and adaptive colouration of animals, and especially for animals that move around among green plants, to increase contrast of edges. These processes are not independent. Flower colours are adapted to the vision of pollinators and as a contrast to photosynthetic pigments. Bioluminescence and vision have evolved together. Phytochrome has evolved to discriminate between direct daylight and light modified by chlorophyll absorption. The basis is the spectrum of the Sun. To begin with, let us see what the relation is between spectrum of the most important of all pigments, chlorophyll a, and the spectrum of the Sun.

2. WHY ARE PLANTS GREEN?

Many people have discussed the spectrum of chlorophyll in relation to daylight. Some have come to the conclusion that they do not match well, as absorption "in the middle of the spectrum", i.e. the green band, is weak. A common idea is that an ideal pigment for photosynthetic energy conversion ought to be either black, absorbing all available radiation, or absorb most efficiently at the "peak" of daylight.

But what is the "peak" wavelength for daylight? The maximum of the daylight spectrum depends on how we plot it. For the present purpose, to simplify comparisons and calculations, we may represent the daylight spectrum by that of a 6000 K blackbody radiator, and thus apply Planck's radiation law (Chapter 1). We can then calculate that if we plot the spectrum as energy per uniform wavelength interval the maximum is at 480 nm. But if we instead plot it as photons per uniform wavelength interval, the maximum is at 600 nm, and if we, following the habits of physicists, plot the spectrum as energy per uniform frequency interval, or

L.O. Björn (ed.), Photobiology, 115–151.

photons per uniform frequency interval, the peak will be seen at frequencies corresponding to 800 nm and 1200 nm, respectively.

Thus the "maximum of the daylight spectrum" is an ambiguous concept, and we have to find another way of optimising our pigment. As for the idea that an ideal pigment should absorb everything, we should remember that the better a substance absorbs, the better it emits, and the transformation of radiant energy into other energy forms is just the balance between absorption and re-radiation. That total absorption is not an ideal is even more apparent in the case of colour vision. Vertebrate cones, which are cells receiving light signals for colour vision, are shorter than the rods involved in "non-colour night vision" (scotopic vision), and therefore absorb a smaller portion of the light and discriminate between wavelength bands better than they would if they were as long as rods.

Photosynthesis depends on photochemistry, and photochemistry works particle to particle, photon to molecule. The useful energy storage can be regarded as the product of the number of reacting photons and the free energy that each converted photon contributes. Björn (1976) following this principle comes to the conclusion that the long-wave absorption band of a pigment giving maximum energy conversion in direct sunlight should be rather narrow and have a maximum at 707 nm. Furthermore the pigment should be highly fluorescent (when not quenched by photochemistry). The maximum chemical potential difference that can be created by a one-step system is

$$\mu_o = kT \ \ln[\Phi r^2/4R^2)] + h\nu_o(1-T/T_s) - b^2[h^2/(2k)]T(1/T^2 - 1/T_s^2), \text{ where}$$

k=Boltzmann's constant, h Planck's constant, T ambient temperature, T_S the temperature of the radiating surface of the Sun, Φ the fluorescence yield, r the radius of the Sun, R the Earth-Sun distance, ν_o the frequency of the spectral peak of the absorption band, and b a parameter determining the width of the absorption band such that the half-band width is $2b\sqrt{(2\ln2)}=2.35b$. With numerical values inserted, this becomes

$$\mu_o = -0.342 \ eV + (19/20) \ h\nu_o - 6.6 \cdot 10^{-28}b^2 \ eV \ s^2 \ (eV \text{ stands for electron volts}).$$

The effective chemical potential difference under conditions of maximum energy conversion is 0.13 eV lower than that (just as the voltage of an electrical battery is lowered when power is drawn from it).

All this is for full sunlight. The optimum position of the absorption peak is lowered by 12 nm for every tenfold decrease of fluence rate, even if the spectrum of the daylight is not changed. Thus it appears that the long-wavelength band of chlorophyll a in vivo is rather well matched to the conditions of our planet. The "blue" absorption band of chlorophyll (the Soret band) does not contribute to chemical potential, but leads to increased photon absorption and thus increased energy conversion, as do various accessory pigments.

Various types of bacterial chlorophylls have absorption bands at longer wavelengths. They are not adapted to direct daylight, but to the light that penetrates down to the places (such as anoxic sediments below algae absorbing shorter wavelength light) where these bacteria live. The record in long wavelengths is held by bacteriochlorophyll b. Its spectrum peaks at 1020 nm, beyond two infrared absorption bands of water.

3. WHAT DETERMINES SPECTRA OF PIGMENTS?

Generally speaking, the absorption spectrum of a pigment is determined by (1) the structure of the chromophore(s) and (2) the environment of the chromophore.

The most important feature in chromophore structure is the arrangement of conjugated double bonds (alternating single and double bonds), i.e. the π electron clouds. The environment of the chromophore in many cases consists of, or is at least dominated by, a protein to which the chromophore is bound, but there are also important cases, such as vacuolar flower pigments, where the chromophore is not protein bound, and other factors are the main concern.

In general, the absorption peak with longest wavelength for a molecule with conjugated bonds increases with the length of the conjugated bond system. It depends, as we shall see, also very much on the shape of the conjugated system, whether it is straight or not and, in the case of macrocycles as in chlorophylls, on its symmetry. If conjugation is broken by single bonds and thus divided into two conjugated systems, the absorption spectrum is similar to the sum of the contributions from the two systems.

As a simple example of the effect of conjugated system length, let us consider a series of polyene hydrocarbons, $CH_2=CH-(CH=CH)_{n-2}-CH=CH_2$, with n conjugated double bonds. Such a simple and regular system is fairly well understood. Energy levels for the ground state and the first excited state can be computed, and the wavelength of the long-wavelength absorption maximum corresponding to their difference determined. Two common approaches are the free electron (FE) theory and the linear combination of atomic orbitals (LCAO) method. In the former the Schrödinger equation is applied to an "gas" of π-electrons in a "box potential", i.e. a potential which is constant over an interval corresponding to the length of the conjugated system and zero outside of it. In the simplest version of this the wavelength of the maximum would vary linearly with the length of the conjugated system, i.e. with n. However, one must consider, that the length of the bonds between carbon atoms is not constant (every second bond is longer, and there are "edge effects"), and therefore the potential is not constant in the "box", and Kuhn has developed a method to allow for this, the result of which is shown in Fig. 1 (another complication, as we shall see later, is that the chain is not straight and may be folded in various ways). In the LCAO method one starts by computing the orbitals of the individual atoms, and then sums up their wave functions. In a refined version due to Hückel one allows for the variation in bond length. Both methods give quite good results for low values of n, but underestimate the wavelength for large values of n. The reader interested in more information on these computational methods is referred to a textbook on electron spectra of organic molecules, such as that by Jensen & Bunker (2000), or the still very readable one by Murrell (1963).

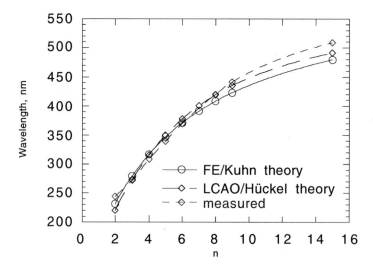

Figure 1. The wavelength of the absorption maximum with longest wavelength for a series of polyenes with formula $CH_2=CH-(CH=CH)_{n-2}-CH=CH_2$. The graph shows experimental values, as well as values calculated by two different methods, the Kuhn version of the free electron theory, and the Hückel version of the linear combination of atomic orbitals theory.

The carotenoids, and the retinals of visual pigments, are biologically important molecules resembling polyenes. Also the phycobiliproteins show a change in wavelength position related to the length of the conjugated system, but as we shall see also other factors are important for the tuning. A complication with simple polyenes and carotenoids is that the first excited state is "optically forbidden" or "dipole forbidden", i.e. cannot be reached from the ground state by light absorption (Schulten & Karplus 1972). The absorption spectrum in the daylight region is due to transition to the second excited state (or, for carotenoids, the third excited state). The first excited state can still, for some carotenoids, participate in energy transfer to chlorophyll and make this very efficient (Thrash et al. 1979, Ritz et al. 2000), while for other carotenoids it is too low (Polívka et al. 1999, but c.f. Frank et al. 2000).

4. RELATION BETWEEN THE ABSORPTION AND MOLECULAR STRUCTURE OF CHLOROPHYLLS

Chlorophylls can be classified into three main groups, depending on whether the nucleus is that of porphin with 11 conjugated double bonds, dihydroporphin with 10, or tetrahydroporphin with 9 conjugated double bonds. In Fig. 2 it is shown which chlorophylls belong to each group, as well as the positions in organic solvent (in most cases ethyl ether) of their main absorption bands.

As is evident from Fig. 2 the long-wavelength transition, called Q_y, corresponds to a smaller energy change in the tetrahydroporphin type bacteriochlorophylls than in the dihydroporphin type pigments, and to a smaller energy change in the dihydroporphin pigments than in the porphin pigments. Even in the porphin pigments there is a certain asymmetry (not shown in Fig. 2) due to the side chains, so even in these one can distinguish between Q_x and Q_y transitions. Q_x and Q_y transitions can be distinguished through measurements of polarisation of fluorescence excited by plane polarised light. Another way is to align the molecules in some way, for instance in thin films, and study the absorption dichroism.

But the effects due to the chromophore environment are sometimes even larger than the effects of the differences in the conjugated double bond system. Thus the in vivo environment (mainly the bonding to protein) changes the Q_y band as shown in table 1.

Table 1
Position of long-wavelength (Q_y) band (nm)

Pigment	in ethyl ether	in vivo
Protochlorophyll	624	630-650
Chlorophyll a	662	670-700
Chlorophyll b	644	650
Chlorophylls c	626-637	
Bacteriochlorophyll a	773	800-890
Bacteriochlorophyll b	795	1020
Bacteriochlorophyll c		725-750

Those chlorophylls having the smallest energy gaps and the absorption bands at longest wavelengths are the reaction centre special pairs. Their special absorption properties are due to formation of exciton complexes (see Chapter 1)

The reader who wishes to understand chlorophyll absorption spectra in more detail is referred to Linanto & Korppi-Tommola (2000). The theory of electronic spectra of organic molecules, and the computational methods used to understand the spectra, are treated in many books, e.g., Murrell (1963, 1967).

5. TUNING OF CHLOROPHYLL *a* AND *b* ABSORPTION PEAKS BY THE MOLECULAR ENVIRONMENT

Chlorophyll *a* is the key pigment in all organisms with oxygenic photosynthesis, i.e. cyanobacteria, algae and plants. It functions both in reaction centres I and II, and as antenna pigment in both photosystems. A great number of spectral forms can be distinguished *in vivo* and in thylakoid preparations.

In organic solvents the long-wavelength absorption peak of chlorophyll *a* lies at 662 nm (ethyl ether) to 663 nm (acetone), when the solution is dilute and the chlorophyll in a monomeric state. In vivo all chlorophyll a fractions are "red-shifted", i.e. shifted to longer wavelengths compared to this. Some of the antenna chlorophylls in photosystem II (PSII) have the smallest shift, while a few molecules in photosystem I have the largest shift and, perhaps surprisingly, have peaks at even longer wavelength than photosystem I (PSI) reaction centre chlorophyll (P700).

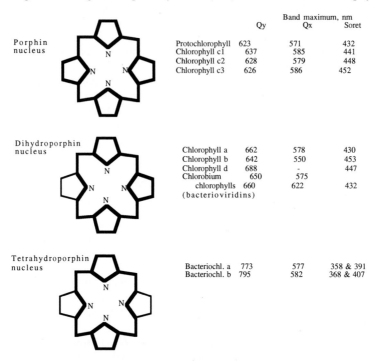

		Band maximum, nm		
		Qy	Qx	Soret
Porphin nucleus	Protochlorophyll	623	571	432
	Chlorophyll c1	637	585	441
	Chlorophyll c2	628	579	448
	Chlorophyll c3	626	586	452
Dihydroporphin nucleus	Chlorophyll a	662	578	430
	Chlorophyll b	642	550	453
	Chlorophyll d	688	-	447
	Chlorobium	650	575	
	chlorophylls (bacterioviridins)	660	622	432
Tetrahydroporphin nucleus	Bacteriochl. a	773	577	358 & 391
	Bacteriochl. b	795	582	368 & 407

Figure 2. The ring systems for the three main classes of chlorophylls. These ring systems are flat. The conjugated double bond systems are indicated by heavy lines. While the conjugated system is rather isodiametric (approximating circular form) for the porphins, it is more elongated for the other two groups. The long axis of this elongated system is often referred to as the y axis, and the transition moment for the so-called Qy transition lies along this axis. The x axis, and the direction of the Qx transition, is almost perpendicular to this, but also in the plane of the ring. The short-wavelength "Soret" band is complex with transitions along both axes. For instance, for bacteriochlorophyll a the 358 Soret component is in the y direction and the 391 component in the x direction.

Cinque et al. (2000) were able to specify the spectra of two particular antenna chlorophyll chromophores in maize by comparing recombinant chlorophyll proteins which lacked them with the wild type. They found that the long-wavelength (Q_y) peak of CP29 chlorophyll a is at 680 nm, and that of LHC II chlorophyll b at 652 nm. Thus the redshifts (bathochromic shifts) compared to solution in, e.g., ethyl ether are very different. However, the overall shapes of the two spectra are very similar in organic solution and protein environments.

Some chlorophyll species are extremely red-shifted *in vivo*. Halldal (1968) discovered that the green alga *Ostreobium*, growing inside a coral on the Great Barrier Reef, contained unusually large amounts of such long-wavelength chlorophyll, "C_a720". It seemed to be an adaptation to the unusual environment of this organism. Other algae growing outside *Ostreobium* filtered away much of the daylight, and left mostly far-red light for *Ostreobium* to use. Halldal asked his student Öquist to find out whether a more "ordinary" alga, *Chlorella*, would also be able to adapt to far-red light by forming more long-wavelength chlorophyll, and to some extent it could (Öquist 1969). But some species of *Ostreobium* are exceptional, and are still attracting the interest of scientists. Although most of the long-wavelength chlorophyll is associated to photosystem I, there seems also to be long-wavelength chlorophyll in photosystem II (Fuad et al. 1983, Zucchelli et al. 1990, Koehne et al. 1999). Investigators are puzzled by the fact that these pigment forms seem to be able to deliver energy to the reactions centres, which have absorption spectra peaking at shorter wavelengths, and were thought to require higher energy quanta. The explanation may be, as in the case of dragonfish vision, that thermal quanta deliver the extra energy needed. This need not violate the second law of thermodynamics if the whole photosynthetic system is considered.

Some of the recent and detailed investigations have been done on cyanobacteria. It has been shown that also in this case the long-wavelength forms, at physiological temperatures (but not at very low temperature), can transfer absorbed energy to P700 (Pålsson et al. 1998). The main long-wavelength absorption peaks of these long-wavelength forms are variously given as 708 and 719 nm (Pålsson et al. 1998), and 705, 714, and 723 nm (Kochubey & Samokhval 2000, both for *Synechococcus elongatus*. For *Synechocystis sp.* PCC 6803 only a single long-wavelength form peaking at 710 nm was identified (Gill & Wittmershaus 1999), possibly due to methodological differences. The 719 nm form seems to arise from the 708 form when monomers of PSI are combined to the trimers present in vivo (Palsson et al. 1998, Jordan et al. 2001). In a recent structural model (Jordan et al. 2001) for PSI the 710/719 nm chlorophyll is tentatively identified with the aC-A32/aC-B7 molecule pair, which seems to connect the monomers energetically. Thus the long-wavelength band may arise by excimer splitting, just like a long-wavelength spectral form of phycocyanobilin. However, Koehne et al. (1999) have obtained evidence that at least some long-wavelength forms in the eukaryotic alga *Oestrobium* has another origin.

To date the long-wavelength record for chlorophyll a tuning is held by the cyanobacteriumm *Spirulina platensis*, which has a form peaking at 738 nm (Shubin et al. 1991, Koehne & Trissl 1998, Karapetyan et al. 1997).

P700 of photosystem I as well as P680 of photosystem II are both dimers (within the photosystem monomer), with the tetrapyrrols closely stacked in a parallel manner and with tightly overlapping π orbitals, forming excimers. It is thus not difficult to understand that there is a large red-shift from the absorption peak of chlorophyll a in dilute organic solution.

Carotenoids may be important for the spectral fine-tuning of some other chlorophyll forms. Several of the carotenoids in PSI show extended overlap of their p-orbitals with those of chlorophyll molecules (Jordan et al. 2001).

For cyanobacterial PSII a structural model has been published by Zouni et al. (2001). According to this model, $Chlz_{D1}$ and $Chlz_{D2}$ are identified with chlorophyll a bound to histidine in polypeptides D1 and D2. These chlorophylls have absorption peaks at ...nm (Schelvis et al. 1994).

6. PHYCOBILIPROTEINS AND PHYCOBILISOMES

In no other case is spectral tuning more important and critical than in photosynthetic antenna pigments. Light is absorbed by one chromophore and transferred over a series of other chromophores to photosynthetic reaction centres. The most common method for energy transfer is the Foerster mechanism. For this to be efficient, the emission spectrum of the energy donor must in each step match the absorption spectrum of the receiver.

There are many kinds of antenna pigments in different organisms: chlorophylls, carotenoids, pteridines, phycobiliproteins. We shall concentrate here on phycobiliproteins, which occur in cyanobacteria and several groups of algae, most important of which are the red algae. The description below will primarily reflect the conditions in cyanobacteria.

Cyanobacteria use several principles for spectral tuning, the first of which is variation of the structure of the chromophores (bilins). Different lengths of the conjugated bond system results in different transition energies: the greater the length, the lower the energy. The following types of chromophores are known, although no single organism possesses all of them (Table 2):

Table 2
Number of conjugated double bonds and absorption maxima for the phycobilins

Phycobilin	Number of conjugated double bonds	Absorption maximum in free form, nm	Absorption maximum in phycobiliprotein
Phycourobilin (PUB)	5		495
Phycoerythro-bilin (PEB)	6	530	545-565
Phycoviolo-bilin (PVB)	7		510-570
Phycocyano-bilin (PCB)	8	600	610-671

Table 3

The occurrence of phycobilin chromophores among cyanobacterial phycobiliproteins and their α and β peptides, after MacColl (1998), slightly simplified. For corresponding information for rhodophycean phycobiliproteins, see Tab. 1 of Holzwarth (1991), and for cryptophycean (cryptomonad) phycobiliproteins Glazer & Wedemayer (1995).

	On α peptide	On β peptide
Allophycocyanin	1 PCB	1 PCB
C-Phycocyanin	1 PCB	2 PCB
Phycoerythrocyanin	1 PVB	2 PCB
R-Phycocyanin II	1 PEB	2 PCB
Phycocyanin WH8501	1 PUB	2 PCB
C-Phycoerythrin	2 PEB	2 PEB
CU-Phycoerythrin (1)	3 PUB	1 PEB+1 PUB
CU-Phycoerythrin (2)	1 PEB+ 2 PUB	2 PEB
CU-Phycoerythrin (3)	3 PUB	2 PEB
CU-Phycoerythrin (4)	2 PEB+ 1 PUB	2 PEB+ 1 PUB

When the chromophores attach to proteins (which takes place via sulfur bridges at one or two points) two changes are immediately apparent: The absorption bands become sharper, and the intensity of the long-wavelength band increases in comparison to those at lower wavelengths. The sharpening of the bands takes place because the conformation of the chromophore, which in the free form is very flexible and therefore distributed over a large number of conformational states, becomes restricted. The relative intensification of the long-wavelength band takes place because the chromophore becomes more straight, while in the free state it is, on average, more circular, as one turn of a helix (Scheer & Kufer 1977, Knipp et al. 1998), with overlapping ends. The spectrum therefore is in this case more "porphyrin-like", with a pronounced Soret type band. Another effect of the binding to protein is that the position of the long-wavelength band is shifted, to very different extents in different cases which will be described below.

The chromophore is not fixed in the protein with complete rigidity. Of particular interest are cases in which the conformation of the chromophore can change under the influence of light, causing photochromicity of the chromoprotein, analogous to the behaviour of phytochrome. The behaviour was first noticed by Scheibe (1972) in an extract containing phycocyanin. The phenomenon was further examined in a number of papers by G.S. Björn, summarised by G.S. Björn (1980, L.O. Björn (1979) and Björn & Björn (1980). The experiments were mostly carried out on extracts, but in one case (G.S. Björn 1980) photoreversible photochromism was shown to occur also in vivo. In this case irradiation with light of 505 nm results in a decrease of absorption at this wavelength and an increase at 570 nm, while irradiation with 570 nm light reverses this effect, and the reaction can be repeated over and over again.

This particular case has been further explored by other researchers, in particular the group around H. Scheer in Germany (Zhao et al. 1995, Zhao & Scheer 1995; see also Ohad et al. 1979 and Scharnagel & Fischer 1993). It has been found that the change in absorption spectrum is primarily due to rotation around a double bond beween carbon atoms 15 and 16 in a phycoviolobilin chromophore in the α subunit of phycoerythrocyanin.

How rigidly the chromophore is held by the protein depends partly on the covalent (thioether) bonds between chromophore and protein. The bonds may go from either the A ring or the D ring, or both, and this affects the spectral properties. The extent to which the chromophore is stretched and kept rigid also affects another property of great importance for the function, namely the excited state life-time. A phycobilin chromophore which is not fixed in a protein has great flexibility, which gives greater possibilities for thermal relaxation, i.e. shorter life-time and less efficient energy transfer and photochemical efficiency. This is in contrast to what is the case with chlorophylls, which are already in the free state rigid structures. The attachment to protein also favours a protonated state, which also lengthens the excited state life.

The ordered arrangement of chromophores in a protein matrix affects the chromophore spectrum by one more mechanism. It keeps certain chromophore in close vicinity of one another, which results in the formation of exciplexes. As described before, this splits the energy levels of the isolated chromophores in a higher and a lower level, in turn splitting the absorption bands in a corresponding manner.

Thus, even though cyanobacterial phycocyanins and allophycocyanins all contain only a single type of chromophore, namely phycocyanobilin, they exhibit a wide range of absorption bands, peaking from 620 nm in C-phycocyanin to 671 nm in allophycocyanin B and also in the "terminal linker polypeptide" in the centre of the phycobilisomes, close to the chlorophyll in the thylakoid membraner.

Phytochromes constitute a quite different type of phycobiliproteins, which have a light-sensing function (Chapters 8, 14 and 15). Plant phytochromes, which are those studied in most detail and contain phytochromobilin (closely related to phycocyanobilin) as chromophore, are interconvertible between two forms, of which one has evolved to absorb maximally near the absorption maximum of chlorophyll, while the other one absorbs maximally just outside the chlorophyll absorption. Therefore they are well suited for detecting the change in light spectrum caused by the presence of competing plants. Phytochromes of non-photosynthetic bacteria, on the other hand (whose exact biological function is yet to be explored) contain a different type of phycobilin chromophore, and absorb at longer wavelengths (Bhoo et al. 2001).

7. CHROMATIC ADAPTATION OF CYANOBACTERIAL PHYCOBILISOMES

Although also other photosynthetic organisms have some ability acclimatize to different spectral regimes, cyanobacteria are the real experts. Most of the so-called "chromatic adaptation" of cyanobacteria depends on their ability to change the amounts of phycoerythrin and phycocyanin depending on the balance between green and red light in their environment. "Chromatic adaptation" (or "Complementary chromatic adaptation") is since far back an established term, although "chromatic acclimation" would be more in line with the common use of "acclimation" for phenotypic adjustment of the individual, and "adaptation" for genotypic adjustment through evolution.

In direct daylight, or in red light, some cyanobacteria do not build any phycoerythrin into their phycobilisomes. These cyanobacteria sometimes form thick mats, or grow under other chlorophyll-containing organisms which filter out the red light but let some green light through. In the deeper layers of the mats, as well as under other organisms the cyanobacteria are able to utilize the green light efficiently by equipping the phycobilisomes with phycoerythrin at the same time as they decrease the amount of phycocyanin. Some phycocyanin must always be present to allow for energy transfer from phycocyanin to allophycocyanin and on to chlorophyll in the thylakoids.

Long ago it was found that a short pulse of light of particular wavelength is suffices to link the phycobilisome construction onto a phycoerythrin-rich or phycoerythrin-deficient path. The change in pigmentation itself can take place during a subsequent dark period. And just like with phytochrome responses of plants, in a series of alternate pulses of red or green light, it is just the wavelength composition of the final pulse that matters. From this it could be concluded that the regulation is mediated by a photochromic pigment of some kind, in analogy with phytochrome (Ohki & Fujita 1978).

It was also established through action spectroscopy, that this photochromic regulator must be a phycobiliprotein (Fujita & Hattori 1962, Diakoff & Scheibe 1973, Vogelmann & Scheibe 1978). Similar action spectra were also obtained for regulation of some other developmental processes in cyanobacteria (Lazaroff & Schiff 1962, Robinson & Miller 1970). Furthermore, the action spectra were also similar to those for conversions of "phycochrome b" (G.S. Björn), later identified as the β-subunit of phycoerythrocyanin. Phycochrome b itself could not, however, be the regulator, since chromatic adaptation takes place also in cyanobacteria lacking phycoerythrocyanin. Nevertheless, the spectral similarity with the photochromic β-subunit of phycoerythrocyanin made it plausible that the chromophore of the regulator is very similar to phycourobilin, the chromophore of the β-subunit.

Phytochrome in plants could be identified and characterized largely because plants lack other phycobiliproteins. In cyanobacteria, where phycobiliproteins make up a large part of the total protein, the situation is much less favourable, and the mechanism of chromatic adaptation remained elusive until recently.

Recent developments in the field are described by Grossman et al. (2001). As in some of the older investigations, *Fremyella diplosiphon* was used as experiemental organism. It appears that the chromoprotein involved as photoreceptor is the product of a gene called rcaE. This is a 74 kDa polypeptide (Kehoe & Grossman 1997). The C-terminal region has motifs typical of bacterial sensor kinases. The N-terminal half binds a linear tetrapyrrole covalently at a cysteine within a phytochrome-like domain. The central part of the polypeptide contains a PAS domain (Kehoe & Grossman 1997).

Grossman et al. 2001 outline the events during chromatic adaptation under red light as follows: Red light causes RcaE to undergo an autophosphorylation followed by transfer of the phosphate groups to the response regulator RcaF. The phosphate group is then transferred to still one polypeptide, RcaC, which acts on genes cpcB2A2 and cpeBA to repress the synthesis of phycoerythrin and stimulate that of phycocyanin. In green light the phosphorylation is prevented, and the effect on the genes is reversed.

8. VISUAL TUNING

Visual pigments of animals span a spectral range of 300 to 700 nm (Marshall & Oberwinkler 1999) i.e. more than an octave of the electromagnetic spectrum. They are proteins with, in most cases, either 11-cis-retinal or 11-cis-3-dehydroretinal (Fig. 3) as chromophores. Proteins with 11-cis-retinal alone cover a range of absorption

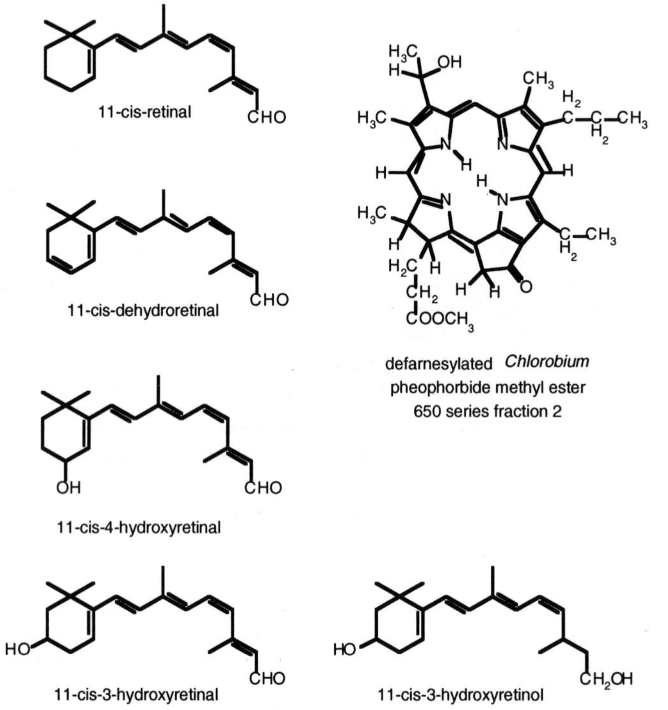

Figure 3. Structures of the five chromophores known from animal visual pigments. Several compounds related to the Chlorobium pheophorbide derivative also occur in dragonfish eyes.

spectra with maxima from 360 to 635 nm (Kleinschmidt & Harosi 1992). The term rhodopsin is somewhat ambiguous, and sometimes covers all chromoproteins related to the human visual pigments, including light-sensitive proteins in algae and archaebacteria, but sometimes even visual pigments containing 11-cis-dehydroretinal (which are then termed porphyropsins) are excluded. The spectra of visual pigments to some extent depends on which chromophore they contain; dehydroretinal (also called retinal$_2$) with its longer conjugated double bond system giving a red-shift of 10-50 nm compared to 11-cis-retinal. Two other "primary chromophores" involved in animal vision are also known: 11-cis-4-hydroxyretinal has been found (as well as retinal and 3-dehydroretinal) in the eyes of the bioluminescent squid *Watasenia scintillans* (Matsui et al. 1988) and give s blue-shift compared to retinal. 3-hydroxyretinal occurs in several insect orders (Vogt 1983, Vogt et al. 1987, Tanimura et al 1986). Many vertebrates, including man, have a differentiation of light-sensitive cells in the retina between rods specialised for "black-and-white vision" in weak light, and cones, specialised for colour vision in stronger light.

In addition to these "primary chromophores" there are, in certain cases, "sensitizing chromophores" attached to the same proteins. These chromophores act in analogy to photosynthetic antenna pigments: they absorb light and transfer the energy to the primary chromophores. Only two such sensitizing chromophores have been detected so far, i.e. 11-cis-3-hydroxyretinol in Diptera (Vogt 1989) and defarnesylated *Chlorobium* pheophorbide methyl ester in bioluminescent dragonfish (Douglas et al. 1998, 1999). More about the latter can be found in Chapter 16.

In addition to visual pigment structure, spectral filters in the form of coloured oil drops contribute to spectral tuning of photoreceptor sensitivity in some animals, especially birds (Maier & Bowmaker 1993, Bowmaker et al. 1997, Vorobyev et al. 1998, Hart et al. 2000) and reptiles (Schneeweis & Green 1995). We also have filters in our own eyes, namely in the yellow spot of the retina, macula lutea. This is the spot of highest visual acuity, devoid of blue-sensitive cones. Here the yellow pigment serves to prevent blue light to reach the green- and red-sensitive cones, for which it would degrade acuity by chromatic aberration.

However, most of the spectral tuning is achieved by variation of a few of the amino acids in the protein to which the chromophores are attached.

Humans use exclusively the 11-cis-retinal chromophore. In free form in methanol solution 11-cis-retinal absorbs maximally at 380 nm (in protonated Schiff's base form it absorbs maximally at 440 nm). Human rhodopsin, the protein-chromophore complex of rods, used in twilight vision, peaks at 493 nm (Wald & Brown 1958). The three human cone pigments, used in colour vision, peak at 426, 530 and 552 nm (or 557 nm; all persons do not have exactly the same type). The human cone pigments are often referred to as SW (for short wavelength), MW and LW pigments, respectively.

The effect of protein primary structure on spectral properties of visual pigments is studied (a) by comparing various naturally occurring pigments and (b) by site-directed mutagenesis experiments in which certain amino acids are changed.

An important determinant of the spectrum seems to be the negative charges on amino acids in proximity to the chromophore. We shall give two interesting examples of how this can works:

(1) Ultraviolet vision has been demonstrated in many vertebrate (fish, amphibian, reptilian, avian and mammal) species (Jacobs 1992). Yokoyama et al. (2000) have convincingly shown that ultraviolet-sensitive pigment in birds evolved from violet-sensitive pigments by a single amino acid substitution, namely by a

change of serine to C at position 84. This shifts the absorption spectrum of the wild type pigeon pigment with a maximum at 393 nm to one peaking at 358 nm. For the corresponding chicken pigment the shift was from 415 nm to 369 nm. Conversely, the zebrafinch UV pigment peaking at 359 nm could be shifted to 397 nm by a change at position 84 from C to serine. It should be noted, however, that ultraviolet absorbing pigments in other vertebrate groups have arisen independently, and by other substitutions.

(2) Human trichromatic colour vision has arisen recently during evolution; most animals have only dichromatic colour vision and some, such as whales, do not have more than one type of cone pigment. Some New World monkeys also possess trichromatic colour vision, but that has arisen independently. The human type of trichromacy has arisen by gene duplication of a long-wavelength pigment, and mutation of one of the gene copies to produce a middle-wavelength (MW) pigment. According to Neitz et al. (1991) the spectral differences between these pigments depend on 3 amino acid differences, and in addition there are several differences without spectral effect. The differences and their spectral effects are detailed in the table below (values vary somewhat between investigators using different methods) (Table 4):

Table 4. Spectral effects of amino acid substitutions in human cone pigments

Position	A.a. in MWP	A.a. in LWP	$\Delta\lambda$ on change
180	alanine	serine	+6
277	phenylalanine	tyrosine	+9
285	alanine	threonine	+15

In each case the change from a non-hydroxyl to a hydroxyl amino acid results in a "red-shift", i.e. absorption at longer wavelengths.

Those people who possess a 557 nm LW pigment have serine at position 180, while those with a 552 nm have alanine. This position is variable also in the MW pigment, but seems to produce a smaller spectral shift there, and investigations are not as thorough as for the LW pigment (see Sharpe et al. 1999 for details). The same person may, in fact, possess more than three different cone pigments (Neitz et al. 1993).

Nathans (1992) has discussed spectral tuning in visual pigments in more general terms. Retinal undergoes a large decrease in dipole moment in going from the ground state to the photoexcited state: In the ground state a positive charge is localised mainly to the Schiff base nitrogen, and this charge is distributed more evenly throughout the π-electron system upon photoexcitation. A negative charge, such as from glutamate or aspartate along the polyene chain of retinal would favour charge delocalisation in the ground state and thus a smaller energy gap, i.e. a red shift. Polar groups along the polyene chain would favour or disfavour charge delocalisation depending on orientation. Polarisable groups along the polyene chain would stabilise the excited state (produce a red-shift) through compensatory charge movement. Twisting around single or double bonds would decrease respectively increase charge delocalisation. Moving the Schiff base counterion further from the chromophore would decrease the effect of the ground state dipole moment and produce a red-shift. A record shift for a single amino acid change (which has not been found in nature), from 500 to 380 nm, was produced experimentally by changing E to Q at position 113 (references in Yokoyama 1997).

Returning to the human MW and LW pigments, one may wonder why their absorption spectra are not more different. Colour vision would appear to be more efficient if they were. The difference between the LW (552 or 557 nm) absorption peak and that of the MW (530 nm) pigment is much smaller than between MW and SW (426 nm) pigment maxima. The difference between the human LW and MW pigments is much smaller than the difference between corresponding pigments in, e.g., the fruitfly *Drosophila* (in this animal the two pigments absorbing at longest wavelengths peak at 420 nm and 480 nm, respectively; in addition the fly has two pigments peaking in the ultraviolet). One explanation that has been proposed is that the image-forming optics of the human eye has a large chromatic aberration and the effect of this is minimized if the spectra are not too different. The perception of shapes and position depends mainly on the LW and MW cones, and the focusing of an image on the retina is adjusted for the average of their wavelengths. The blue colour is mentally "painted" into the outlines formed by these receptors. In the part of the retina used for the sharpest vision, the luteum, has very few SW receptors, and contains a yellow pigment which absorbs blue light. The similarity of LW and MW spectra makes good focusing possible. On the other hand, the small difference in the human pigments may be just a consequence of the fact that the gene duplication has occured so recently, and evolution has not had time to result in a bigger difference yet.

The fruitfly has a completely different system for image generation, without the chromatic aberration problems, and the visual acuity of the fruitfly eye is much lower than that of the human eye.

What are the evolutionary pressures causing visual pigment spectra to be tuned? Generally speaking, of course, colour vision provides more information than monochromic vision. We prefer colour television to black and white. According to Osorio & Vorobyev (1996) and Regan et al. (1998) the main importance of the differentiation into LW and MW pigments in primates has been to aid our forefathers in detecting fruits agains a green background and judging the ripeness of fruits. This view has been questioned by Lucas et al. (1998) and Dominy & Lucas (2001), who provide evidence that trichromatic vision is important for the selection of leaves at an optimal developmental stage for consumption. As for ultraviolet vision in birds, one well documented advantage for birds of prey is that UV vision allows them to see urine of rodents, and thus to locate their whereabouts. Ultraviolet vision in birds is also important for recognition of plumage colouration of conspecifics and for detection and identification of edible berries (Siitari et al 1999).

For insects depositing eggs on leaves, it is important to find the leaves and to judge their age and health status. It turns out that for this task a *red-light* receptor can be very important. Most insects do not have red-sensitive receptors, but both sawflies (Peitsch et al. 1992) and moths (Kelber 1999) that oviposit (lay eggs) on leaves do. Excitation of green-sensitive photoreceptors give an attractive signal, and the red-sensitive receptors provide a contrasting, repelling signal. In the case of moths, probably also their ultraviolet-, violet-, and blue-sensitive receptors play a role in their orientation and choice of leaves (young leaves are preferred). On the other hand, in selecting green leaves, vision probably does not play an important role in the discrimination between plant species; for this chemical cues are more important.

The daylight penetrating deepest into the ocean is in the blue-violet region, and in consequence with this the vision of deep-water fish is tuned to this wavelength band (Lythgoe 1984), while surface-living fish and fish in shallow freshwater has a visual sensitivity peaking, like ours, in the green spectral region (although the span of pigments in fish is much wider than ours, with pigment absorption peaks spanning from the ultraviolet to the red). Although also several bioluminescent deep sea fishes have maximum sensitivity in the blue-violet (Fernandez 1978), others, who use their bioluminescence for environmental illumination, show remarkable deviations from this rule (Douglas et al. 1998, 1999). Other aspects of the connection between bioluminescence and the vision of deep sea fishes have been treated by Warrant (2000).

Deep-diving whales have rod pigments peaking at 485 nm, while rod pigments of aquatic animals foraging closer to the surface (seals, manatees) peak near 500 nm. To the surprise to some investigators none of six whale species and seven seal species possess no SW cones (nor any LW cones), only MW cones (Peichl et al. 2001), with a pigment absorbing maximally around 524 nm (Fasick et al. 1998). It has been claimed that this means that they do not possess colour vision (Peichl et al. 2001), but this may be jumping to a conclusion. Although colour discrimination in whales has not been established, rods are saturated in strong light and cones useless in weak light, there may be intermediate depth and light levels where signals are obtained from both rods and cones and give whales and seals a dichromatic colour vision (Fasick et al. 1998). A corresponding phenomenon in man was demonstrated many years ago through a very interesting experiment by John J. McCann and Jeanne L. Benton, described by Land (1964). They first illuminated a multicoloured display with "monochromatic" (narrowband) light of 550 nm (500 nm would probably have worked as well) which was so weak that only the rods of a human observer were stimulated. Of course no colours could be discriminated under such circumstances. They then added a second narrowband beam of 656 nm wavelength. The irradiance of this second light was adjusted so that only the LW cones were stimulated. Thus only the rods and the LW cones were operative. Nevertheless the obsever was able to give names to colours in the display almost as if it was illuminated by natural daylight and all three types of cones had been stimulated.

Colour vision is not restricted to di-, tri- and tetrachromatic versions. The record is probably held by the mantis shrimp (Osorio et al. 1997, Marshall & Oberwinkler 1999, Cronin et al. 2001), which may have up to 16 types of visual pigment. And not only that. It has the ability to further tune the sensitivity spectra of their receptors by colour filters as required by the light environment they are inhabiting. They may live close to the water surface (in full daylight spectrum), or as deep as 30 m (in a restricted blue-light environment). All 16 types of light-sensitive pigments may not correspond to separate sensory input channels, but no doubt they have a polychromatic vision.

9. TUNING OF ANTHOCYANINS

Anthocyanins are the most common vacuolar pigments, giving colour to many flowers, fruits and autumn leaves. We may think of the cell sap of plant vacuoles as structureless, and as interactions between anthocyanins and their environment as a dull subject, but if we do so we are in error.

The great pioneer in the elucidation of chemical structures of plant compounds, Richard Willstätter, got a surprise when he compared the structures of the blue pigment of the cornflower (*Centaurea cyanus*) with that of the red pigment of a rose. He found that the pigments were chemically identical (he was not completely right, but that does not destroy the story). He thought that the difference in colour came about from a difference in pH of the cell sap of the two plants. He was not right there either, but he got the main point, that anthocyanins can produce very different colours with practically identical chromophores.

The basic structure of an anthocyanin is shown in Fig. 4. The molecule consists of two fused six-membered rings connected to a third six-membered ring. A system of conjugated double bonds extends over all rings. The fused rings carry three hydroxy groups, of which one or two form glycoside bonds with sugar molecules, often glucose. The sugar free compound (the aglycon) is

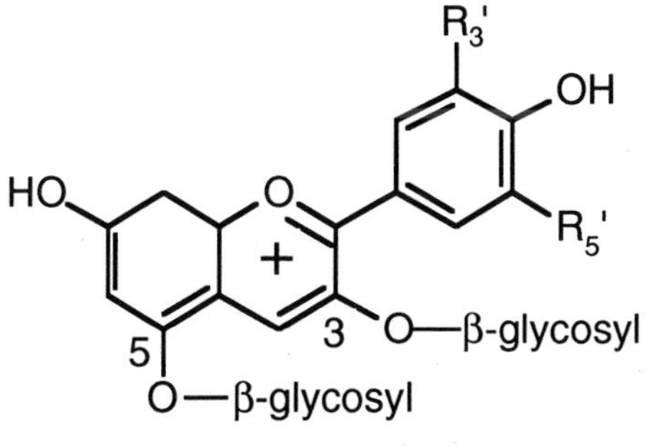

Fig. 4. The general structure of an anthocyanin in the flavylium cation form.

called anthocyanidin. The anthocyanidin of red rose flowers is cyanidin and that of cornflower succinyl-cyanidin (complexed , so they are indeed closely related, and the little difference does not explain the colour difference. Pure cyanin (the glycosylated cyanidin) is red in acid solution, and if it is made alkaline the colour changes towards blue, so it is understandable that Willstätter ascribed the colour difference between roses and cornflowers to a pH difference of the cell sap. However, the blue colour aquired upon alkalinisation does not last long, the colour fades away completely. This is because the molecule, already when the pH exceeds about 5, takes up water (Fig. 5). Furthermore, no plants are known with an alkaline cell sap, so the cornflower's blue colour must be explained in another way.

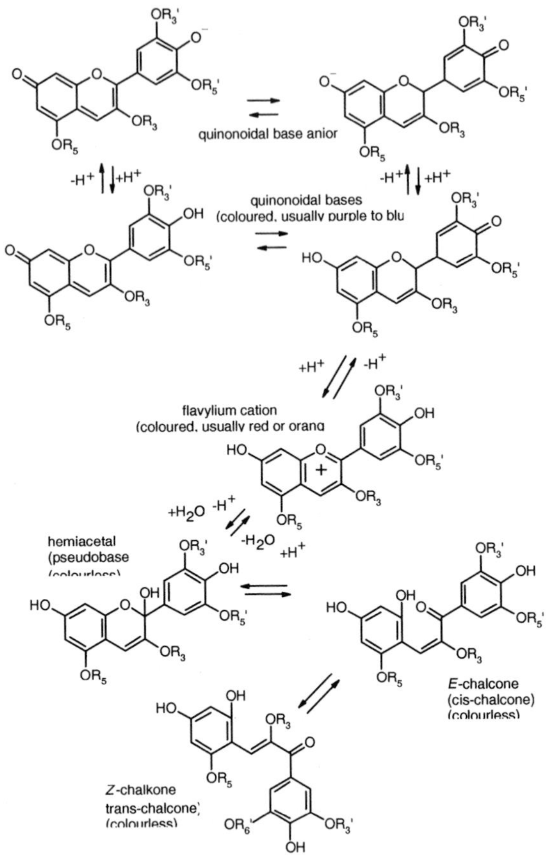

132

Figure 5. An anthocyanin can exist in many different interconvertible molecular forms, as shown in this diagram: as different quinoidal bases anions or unionized quinonoidal bases, as flavylium cation, as hemiacetal (pseudobase), or as chalcone (E or Z form). Although hemiacetals and chalcones are colourless, they absorb ultraviolet radiation, and can therefore appear coloured to some animals. Their presence can also modify the hue of the coloured forms. The R-groups in this diagram are numbered according to the conventional numbering of the carbon atoms to which they are bound. The groups can all be hydroxyl. R_3' and R_5' can also be hydrogen, methoxyl, a chelated metal ion, such as Fe^{3+} or Al^{3+}, a sugar or sugar derivative, while R_3 and R_5 can be a sugar or an acylated sugar. The equilibria between different forms depends on many things, such as the chelation with metal ions. In at least one case it is also light dependent (Figueiredo et al. 1994).

The explanation rests in the phenomena of copigmentation and self-association. Copigmentation means that the colour of the anthocyanin is influenced by other molecules in its environment. Self-association means that the anthocyanin molecule can associate with other anthocyanin molecules, and this also affects its colour. Both self-association and associations between anthocyanins and uncoloured phenolic compounds cause overlaps between the π electron clouds of the individual molecules, and hence changes in the electron levels. The concentration of anthocyanin in cell sap can exceed 20 mM, and this is more than sufficient for the molecules to associate to one another, and sometimes form helical stacks through a combination of hydrophobic bonds between the rings and hydrophilic bonds between the sugar residues. These associations also prevent the formation of pseudo-base, and hence the bleaching of colour that would otherwise take place at the pH prevalent in the cell sap.

Many different kinds of (uncoloured) molecules and ions produce copigmentation effects with anthocyanins. The most important ones are some metal ions and colourless flavonoids (absorbing in the ultraviolet spectral region) and other phenolic compounds. In the case of the blue cornflower pigment (Fig. 4), Fe^{3+} and Mg^{2+} have been reported as copigmenting ions, and a flavone as copigmenting phenol. Anthocyanins having two ortho-hydroxy groups at the B-ring (the leftmost

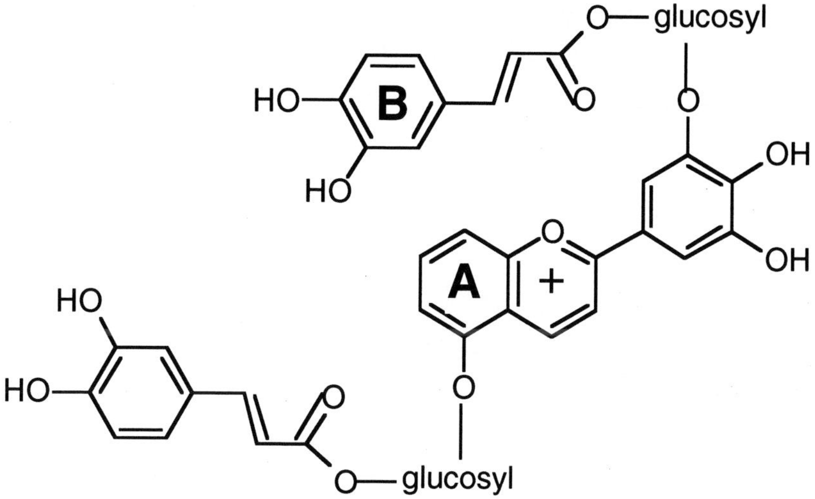

Fig. 6. The structure of gentiodelphin, schematic. In reality the molecule is more bent, so ring B closely overlaps the double ring A, and the p orbitals of the ring systems fuse.

ring in Fig. 6) form blue chelate complexes with trivalent metal ions such as iron(III) and aluminium(III), but not with magnesium. This latter ion is, however, very important in some other cases.

Goto and Kondo (1991) have written a very readable account of copigmentation, illustrated with colour pictures. They describe structures of several more complex anthocyanins, with aromatic groups attached to the sugar residues, and a particularly interesting case pigment complex, commelinin (rendering the blue colour to *Commelina communis*). This consists of six molecules each of malonylawobanin (a complex anthocyanin) and flavocommelinin (a flavone glycoside) arranged around two magnesium ions. The intense blue colour is partly due to exciton coupling between adjacent anthocyanin units. The molecular mass of this complex is nearly 1000.

134

The importance of the sugar groups in anthocyanins lies not in a direct effect on light absorption, but in their contribution to tho folding and ordering of the chromophoric groups such that overlaps of p-electron clouds can take place.

The colour of anthocyanins is affected not only by their chemical environment, but also by physical factors: temperature and — light! One summer day when I and my wife were having tea in our garden, she remarked that she had noticed how some flowers of a variety of *Phlox paniculata* had shifted colour when they were reached by sunlight. I gave her a lecture on the psychology of colour perception and the mis-

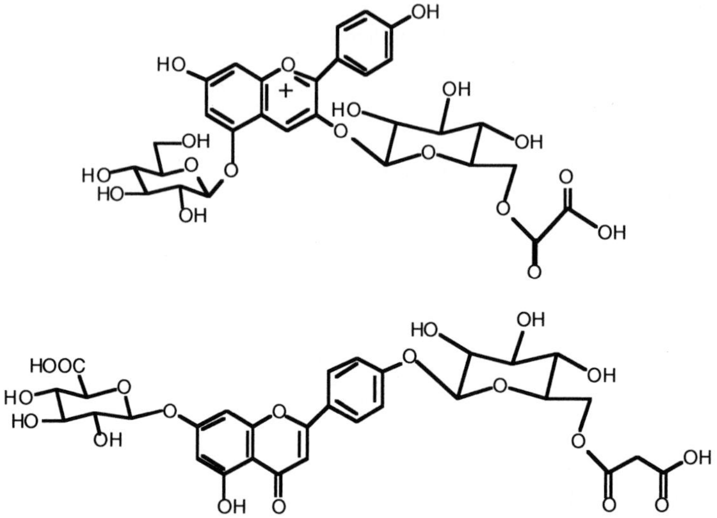

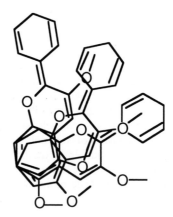

Figure 7. According to Goto et al. (1986) the blue pigment of cornflower, protocyanin, is built up of 6 molecules each of the malonylflavylium cation (top) and the succinylcyanin shown in the figure, and one Fe and one Mg ion.

Figure 8. An attempt to show the chiral stacking of delphin molecules. The sugar groups are omitted for clarity.

takes we can make, but she insisted. So we took the flowers to the lab, and measured the reflection spectra after dark adaptation, and after exposure to strong light for an hour (Björn et al. 1985). And indeed, the reflectance spectrum did change with light conditions. Colour photos of the phenomenon have been published in Swedish journals (Björn 1985 a,b). At the time we hypothesised that the colour shift was caused by a light-activated proton pump in the tonoplast (the membrane surrounding the vacuole) which would change the pH of the cell sap. We attempted to show such a pH change using a white *Phlox* variety and artificial pH indicator dyes, but did not succeed. About a decade later Figueiredo et al. (1994) demonstrated that the absorption spectrum of an artificial anthocyanin-like compound, 4´,7-dihydroxyflavylium chloride, could be reversibly affected, due to Z/E photoisomerisation, by ultraviolet radiation (changing under irradiation from pale yellow to intense yellow), so it is possible that the colour shift of our *Phlox* is due to an analogous phenomenon.

Thus a number of factors can affect the colour of anthocyanin-containing flower petals. Obviously the reason that plants have evolved coloured flowers is that they attract pollinators. Also the pollinators benefit from this, so a question naturally arises: To what extent have flowers adapted to pollinator vision, and to what extent has pollinator vision (and mental capacity, such as long-term and short-term membory) adapted to flower colours?

Chittka and coworkers (Chittka & Menzel 1992, Chittka 1996, Chittka et al. 1999, Chittka & Dornhous 1999) have studied this question. By comparison of colour vision in various arthropod groups they claim to be able to follow the evolution of colour vision in insect pollinators. Photoreceptors among most Crustacea and Insecta fall into three rather distinct spectral classes: Ultraviolet receptors with maxima around 350 nm, blue receptors with maxima at 400 to 460 nm and green receptors with maxima from 470 to 550 nm. All these types seem to be very ancient. In addition, a few groups have red receptors with maxima around 600 nm. Chittka and coworkers see little evidence of insect adaptation to flower colours, and believe that the main adjustment has been of flower colours to insect vision.

Only a minority of pollinators, such as hummingbirds and some butterflies, have red sensitive photoreceptors. However, it must be borne in mind that also pollinators without such receptors can see red flowers and distinguish them from other flowers and from green leaves.

The flowers of many plants vary in colour with stage of development (Lunau 1996), and in this way signal to pollinators when a visit will be rewarded. Of particular interest are flowers that change colour during the year in synchronisation with the availability of different pollinators.

What has been described here for flower colours has, to some extent, a counterpart in the colouration of fruits of plants which depend on animals for seed dispersal.

The ultraviolet receptors of insects makes it possible for them to see patterns invisible to us. Fig. 9 shows an example of such a flower pattern, due to ultraviolet-absorbing flavonoids.

136

Figure 9. A flower of Potentilla reptans *photographed in visible light (left) and in ultraviolet-A radiation (right).*

10. LIVING MIRRORS AND THE TUNING OF STRUCTURAL COLOUR

10.1. Introduction

In the living world we encounter many examples of metal-like specular reflection. This section deals with how such mirrors are constructed without the aid of metals. We shall, in fact, deal only with dielectric substances, i.e. electric non-conductors.

Examples of metal-like shiny reflection occur on wings of beetles (Parker et al. 1998) and butterflies (Berthier & Lafait 1999, Parker 1999, Vucusic et al. 1999), and birds, as well as on fish and other aquatic animals (Herring 1994). As those readers who have driven a car at a country-road at night know, many night-active wild animals have efficient reflectors in their eyes, as do our favourite pets, dogs and cats. These reflectors, located in the back of the eye behind the retina, double the sensitivity of the eyes to light. They do this by throwing those photons, which have escaped absorption by the light-sensitive pigments at their first passage through the retina, through the retina again to give them a second chance.

There are even eyes having mirrors as image-forming optical elements (Land 1966), as astronomical telescopes have. Some light emitting organs of marine animals have reflectors like the headlights of a car, that throw the light in a certain direction. Mirrors reflecting the daylight from the surface make it easier for many fish to avoid detection.

These mirrors are built up of alternating layers of high and low refractive index. We have previously (Chapter 1) treated reflection at a single boundary between phases of different refractive index. When several such boundaries are stacked upon one another, we *cannot* simply compute the reflectivity at each of them using Fresnel's formulas (Section 9 in Chapter 1) and add them (or compute the transmittivities and take their product), for two reasons: (1) There will be multiple reflections of photons bouncing back and forth between the boundaries, and (2) there will be interference effects.

To understand how such complex mirrors function we have to consider the phases of electromagnetic oscillations, which was not necessary for a single boundary. Phase and phase difference are usually expressed in angular measure (radians), where 2π radians correspond to a whole period (a whole wave) of the electromagnetic oscillation. Reflection of light travelling from a medium of lower refraction index differs from reflection of light travelling from a medium of higher refractive index towards one with lower refraction index in one important respect (Fig. 10): During reflection towards a denser medium the oscillation may undergo a phase change by π radians (180°).

The detailed general treatment of reflection in dielectric multilayers is complicated. We shall therefore here focus on some special cases, which are more easily understood and nevertheless demonstrate more general principles.

10.2. Reflection in a single thin layer

Everybody is familiar with the reflections from soap films, and knows that a range of colours can be produced by such reflection. In this case we have a thin film of higher refractive index (approximately that of water) with the same medium (air) on both sides of the film. We can generalise this to have a bulk medium of one refraction index, n_0, on one side, a thin film with a different refraction index, n_1, and on the other side a bulk medium with a third refraction index, n_2 (Fig. 11). For the soap film case $n_0=n_3<n_1$. A well-known example in which $n_0<n_1<n_2$ is a camera lens with an antireflective coating. As we shall see we can treat these and similar cases with the same set of equations.

We denote the incident angle by α_0. The angle that the light forms with the normal to the film within the film is α_1, and Snell's law gives the relationship $n_0*\sin \alpha_0=n_1*\sin \alpha_1$, or $\sin \alpha_1=\sin \alpha_0*n_0/n_1$.

The ordinary Fresnel reflection factors for a single surface (Chapter 1) are
$[(n_1*\cos \alpha_0-n_0*\cos \alpha_1)/(n_1*\cos \alpha_0+n_0*\cos \alpha_1)]^2$ and
$[(n_0*\cos \alpha_0-n_1*\cos \alpha_1)/(n_0*\cos \alpha_0+n_1*\cos \alpha_1)]^2$

for light having the electric vector parallel to the plane of incidence (p-polarisation), and perpendicular to it (s-polarisation), respectively. These are the reflection coefficients for power or intensity, and of course they are always represented by positive real numbers.

It turns out that for a thin film, due to the interference effects, not only these angle-dependent factors, but also another angle-dependent factor has to be taken into account. This is a function also of the layer thickness (d), the vacuum wavelength (λ, and the refraction index (n_1) of the layer. To compute this, we first calculate the phase difference between the rays reflected from the first and from the second surface (Fig. 11). This phase difference can be thought of as composed of three components. The first component is π, due to the phase reversal at the first surface (if $n_1>n_0$ as with the soap film and the camera lens, otherwise 0). The second component is due to the longer distance (AB+BC) travelled in the denser medium by the ray reflected from the second surface, while travelling through the film with refraction index n1, and therefore amounts to 2*n1*d*(2*π/λ)cos α1 radians. The factor 2 in the second term stems from the fact that the ray reflected from the second surface traverses the layer twice, and n1*d is the "optical thickness" of the layer, i.e. the geometrical thickness corrected for the shortening of the wavelength of light in proportion to the refraction index.

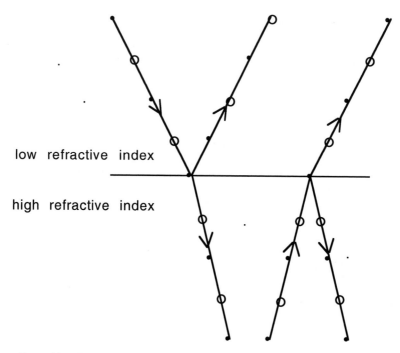

low   refractive   index

high   refractive   index

Figure 10. Reflections at an interface between a medium of higher and one of lower refractive index. Filled and empty circles denote opposite phases (i.e. phases differing by π radians) of the electromagnetic waves.

The third component of the phase difference is due to possible phase reversal during reflection at the second surface. It is 0 in the soap film case, because n1>n2, and π in the camera lens case, because then n1<n2.

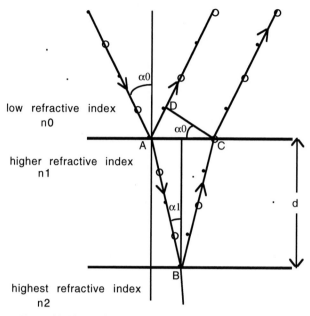

Figure 11. Phase relations during reflection at two consecutive phase boundaries. The drawing is fo the case n0<n1<n2, corresponding to, e.g., the antireflection layer on a camera lens. In this case phase reversal takes place at both reflective surfaces, while with a soap film reversal takes place only at the first surface.

The fourth component of the phase difference is due to the shorter distance, AD in Fig. 11, travelled in the n0 medium by the ray going through the film medium. The phase difference due to this is

$-AD*n0*(2*\pi/\lambda)=$
$-AC*\sin \alpha0*n0*(2*\pi/\lambda)=$
$-2*d*\tan \alpha1*\sin \alpha0*n0*(2*\pi/\lambda)=$
$-2*d*n1*(\sin \alpha1/\cos \alpha1)*n1*\sin \alpha1*(2*\pi/\lambda)=$
$-2*d*n1*\sin^2 \alpha1*(2*\pi/\lambda/\cos \alpha1$ radians.

If we combine this with the second component we get

$2*n1*d*(2*\pi/\lambda)/\cos \alpha1-2*d*n1*\sin^2 \alpha1*(2*\pi/\lambda)/\cos \alpha1=$
$2*n1*d*(1-\sin^2 \alpha1)*(2*\pi/\lambda)/\cos \alpha1=$
$2*n1*d*\cos^2 \alpha1*(2*\pi/\lambda)/\cos \alpha1=2*n1*d*(2*\pi/\lambda)/\cos \alpha1.$

The Fresnel formulae given above, with expressions in squared brackets, as already mentioned, represent intensity reflection coefficients. For many computations, however, it is advantageous to work with the corresponding unsquared expressions, the so called amplitude reflection coefficients, describing the change in amplitude of the electric wave motion upon reflection (the energy of the wavemotion is proportional to the square of the amplitude). These coefficients, in contrast to the intensity reflection coefficients, can be either positive or negative (if light absorbing materials are involved, the intensity reflection coefficient is represented by a complex number, but we shall not consider this case here). A negative amplitude reflection coefficient means that the electric field changes direction during reflection, in other words that we have a phase change of p radians. When we do calculations based upon the amplitude reflection coefficients we do not have to worry about the phase changes of π radians during reflection that were mentioned above, since they will come automatically with the sign of the amplitude reflection coefficient used.

The following QuickBasic programme will compute the reflectance over the range 0.3 to 1 μm for a thin film with refraction index n1 between media with refraction indices n0 (incidence side) and n2, respectively. In reality the refraction indices will vary with wavelength, but this has not been taken into account. If run on a Macintosh computer the result will appear in the clipboard and can be transferred to other programs from there. If you wish to have the output in another form, line 200 should be modified accordingly, and for other computers and Basic dialects modifications may have to be done.

```
INPUT "incident angle in degrees", a0:pi=3.14159: a0=a0*pi/180
INPUT "thickness of thin layer in micrometres", d
sina1=n0*SIN(a0)/n1: a1=ATN(sina1/SQR(1-sina1*sina1))
sina2=n1*sina1/n2: a2=ATN(sina2/SQR(1-sina2*sina2))
50 :PRINT "Enter p for parallel (p) polarisation, s for perpendicular (s)
polarisation."
PRINT "Polarisation directions are of electric vector relative to incidence plane."
INPUT "electric vector direction", polarisation$
IF polarisation$="p" THEN 150
IF polarisation$="s" THEN 100
PRINT "Mistake! Try again!": GOTO 50
100 : r1=(n0*COS(a0)-n1*COS(a1))/(n0*COS(a0)+n1*COS(a1))
r2=(n1*COS(a1)-n2*COS(a2))/(n1*COS(a1)+n2*COS(a2)):GOTO 170
150 :r1=(n1*COS(a0)-n0*COS(a1))/(n0*COS(a1)+n1*COS(a0))
r2=(n2*COS(a1)-n1*COS(a2))/(n1*COS(a2)+n2*COS(a1)):GOTO 170
170 :OPEN "O",1,"clip:"
FOR L=.3 TO 1 STEP .01
delta=4*pi*n1*d/L*COS(a1)
I=(r1*r1+r2*r2+2*r1*r2*COS(delta))
I=I/(1+r1*r1*r2*r2+2*r1*r2*COS(delta))
200: PRINT#1, I: NEXT L: CLOSE 1: END
```

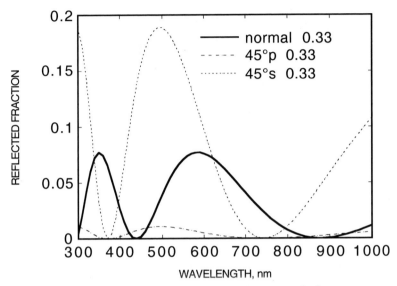

Figure 12. Reflection in a soap film of 0.33 μm thickness.

We show in Figure 12 the output of this programme for a soap film (n=1.33) of 0.33 μm thickness, both for normal incidence ($\alpha0$=0) and for 45° incident angle; in the latter case for both polarisations. This film would look yellow in a perpendicular direction, bluish green in a 45° direction. Note that the reflection spectrum shifts to shorter wavelength with increasing incidence angle, that there are several peaks in each spectrum corresponding to phase shifts of integer multiples of 2π, and that for oblique angles the s-polarised light is reflected better than the p-polarised light.

Then angles of maximum and minimum reflectance vary according to

$\lambda max=[(4*n1*d)/(2m+1)]*SQR[(1-(n1/n2)*sin2\alpha0]$
$\lambda min=[(4*n1*d)/(2m)*SQR[(1-(n1/n2)*sin2\alpha0],$

where m denotes any positive integer and SQR square root.

Next we run the programme for a typical camera lens antireflection coating with n0=0, n1=1.38 (corresponding to magnesium fluoride) and n2=1.89 (heaviest flintglass). The thickness of the layer is adjusted to correspond to one quarter wavelength of yellow light (589 nm) in the layer, i.e. (0.589/4)/1.38 μm (Figure 13).

142

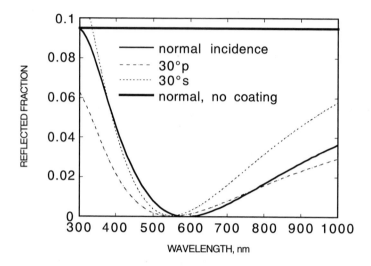

Figure 13. Reflectance of a lens of heaviest flint glass (n2=1.89) coated with a quarter-wave thickness (0.107 mm) of magnesium fluoride (MgF₂, n1=1.38). Reflected fraction of the incident light is shown for normal incidence and for 30° incidence angle with different polarisations. For comparison the reflected ratio at normal incidence without coating is also shown. For oblique angles even higher reflectance would be obtained without coating. The mixture of violet-blue and red reflected light gives such a lens a purple teint.

To obtain zero reflection with an antireflective coating on a lens two conditions must be fulfilled: (1) The thickness of the coating must be equal to one quarter of the wavelength of the light (or this plus an integer multiple of the wavelength), which can be achieved only for a certain wavelengths. (2) The coating material must have a refraction index which is the square root of the refraction index of the lens material (i.e. $n1^2=n2$). In practice compromises have to be made.

The reader might wonder what the construction of antireflection coatings on camera lenses has to do with photobiology. The fact is that even some eyes, notably in the insect orders Lepidoptera and Diptera, have antireflective sufaces (Miller et al. 1966, Parker et al. 1998, Parker 1999). In this case, however, the antireflective surface is not obtained by addition of a layer of material of a different refraction index, but by finishing off the bulk material not with a smooth surface, but with tapering "nipples" smaller than the wavelength of light, which gives the same effect as if the refraction index would decrease gradually. The main biological advantage might be to make the bearer of the eyes less conspicuous by avoiding a shiny surface, rather than to improve vision. This "biological antireflection coating" is more efficient than the man-made one, and has not yet been achieved technically on the same scale, but can be exploited on the microwave scale for military purposes to make surfaces invisible to radar.

10.3. Reflection by multilayer stacks

We shall now come to the more complicated case of multilayer stacks forming efficient biological mirrors with nearly 100% reflectivity. The most important case is that of normal incidence. Eyes, for instance, are constructed such that the light hits the retina almost perpendicularly, and therefore also the reflective backing of the retina. However, using the computer programme at the end of the section also oblique incidence and both s and p polarisation can be handled. The stacks have alternating layers of lower refraction index (n0) of thickness d0 and layers of higher refraction index (n1) and thickness d1. This computer programme is based on the treatise by Huxley (1968) and is in agreement with the example treated by Land (1966).

Fig. 14 shows examples of the output of this computer programme. Within a certain wavelength range, about 100 nm, the reflectance reaches nearly 100% for only 20 high-index layers, i.e. 40 interfaces, while it is lower around this region,

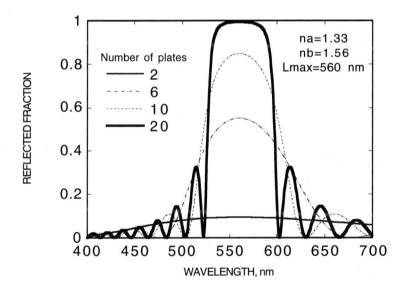

Figure 14. Reflectance at normal incidence of a stack of guanine plates (n=1.80) with spaces between the plates having refractive index as seawater (n=1.34). The thickness of both plates and spaces was chosen such that thickness times refraction index (optical pathlength) amounts to 140 nm. This leads to high reflection at 4x140 nm= 560 nm.

and oscillates with wavelength. The number of these oscillations increases with increasing number of layers. In reality this results in shiny reflectance only within a certain wavelength range, which gives the surface a coloured appearance, as can be seen in, for instance green and blue beetles. The hue of this metallic shine also changes with incident and viewing angles (Fig. 15, cf. Fig. 13). A very common animal in Europe where this effect can be seen is the head of the male mallard, which changes between green and blue depending on from where it is seen. However, all such angle-dependent colours do not result from the multilayer construction described here; there are also many examples where gratings (Chapter 3) produce the colourful effects (Pfaff & Reynders 1999, Srinivasarao 1999). Also very complicated structures, in which both mulitlayer and grating effects contribute to colours have been described, especially for the scales of butterfly wings (Ghiradella 1991, Parker 1966, Srinivasarao 1999, Vukusic et al. 2000), and also for such cases mathematical models have been constructed (Gralak et al. 2001).

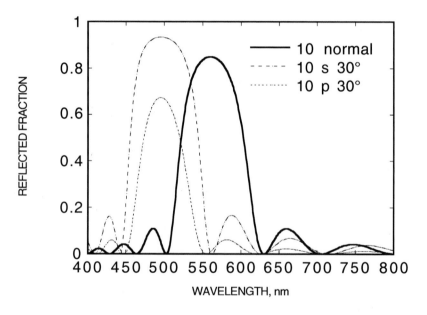

Figure 15. Reflectance of a stack of 10 plates similar to those in Fig.14, and with the same medium in between. Reflectance was computed for normal incidence, and for an incident angle of 30° and both polarisations.

In those cases when high reflectance takes place over a wider wavelength range, such as often from the tapetum of eyes, this can be achieved in a number of ways. One mirror with high reflectance in one wavelength band (i.e. with one spacing and thickness of high-index plates) can be positioned behind one mirror with another spectral tuning. Or the dimensions of the layers can be continuously varied from wide to narrow ("chirped layers"), or have "chaoticly" varying dimensions. An example of an animal in which layer thickness is varied in a way referred to as "doubly chirped" (thickest layers in the centre) is the silverfish (Large et al. 2001). In the heering, a neutral, silvery reflectance is obtained by a combination of different scales reflecting red, green, and blue wavebands (Denton & Land 1971).

Chemical compounds used by organisms for the layers of high refractive index include guanine (in eyes), chitin (in insect wings) and cellulose (in plants). The low refractive index layers are mostly aqueous solutions, or air.

The computer programme used for computations for Fig. 15 is:

```
pi=3.14159
INPUT "refraction index of medium and spaces", na
INPUT "thickness of spaces, micrometres", da
INPUT "number of plates=", p
INPUT "thickness of plates, micrometres", db
INPUT "refraction index of plates", nb
INPUT "incidence angle, degrees", aa:IF aa=0 THEN aa=.001:aa=pi*aa/180
sinab=na*SIN(aa)/nb:ab=ATN(sinab/SQR(1-sinab*sinab))
27 : INPUT "choose polarisation, p for parallel, s for perpendicular", pol$
IF pol$="s" THEN 29
IF pol$="p" THEN 30
PRINT "MISTAKE! Try again!"
28 : GOTO 27
29 :r=SIN(aa-ab)/SIN(aa+ab):GOTO 31
30 : r=-TAN(aa-ab)/TAN(aa+ab)
31 :OPEN "O",1,"clip:"
FOR L= .4 TO .85 STEP .001
fia=(2*pi/L)*da*na*COS(aa):fib=(2*pi/L)*db*nb*SQR(1-sinab*sinab)
kprim=-(COS(fia+fib)-r*r*COS(fia-fib))/(1-r*r)
k=(SIN(fia+fib)-r*r*SIN(fia-fib))/r/SIN(fib)/2
IF kprim*kprim<1 GOTO 100
REM In the following case mu is real
mu1=kprim+SQR(kprim*kprim-1)
mu2=kprim-SQR(kprim*kprim-1)
90 :murat=mu1/mu2:IF murat>1 THEN murat=1/murat
95 :m2=(murat)^p
RR=1/(1+4*m2*(1-k*k)/(1-m2)/(1-m2)):GOTO 200
100 :REM In the following case mu is complex
costheta=(COS(fia+fib)-r*r*COS(fia-fib))/(1-r*r)
theta=ATN(SQR(1-costheta*costheta)/costheta)
RR=1/(1+(k*k-1)/SIN(p*theta)/SIN(p*theta))
200 :PSET(1000*(L-.39),310-300*RR):PRINT# 1,RR
300 :NEXT L:CLOSE 1
400 :FOR L=.4 TO .85 STEP .05
CALL MOVETO(1000*(L-.39),310):CALL LINETO(1000*(L-.39),320)
IF 10*L<>INT(10*L) THEN 500
```

```
CALL MOVETO(1000*(L-.41),330): PRINT 1000*L;"nm"
500 :NEXT L
END
```

Many beetles with metallic lustre have another interesting optical property: The light reflected from them is circularly polarised, usually left circularly polarised (references given by Srinivasarao 1999). This effect is produced by liquid crystals oriented in a special way, so called cholesteric liquid crystals. Some beetles can change colour quickly by shifting between air and liquid in the space between the chitin layers (Hinton & Jarman 1972, 1973).

11. THE INTERPLAY OF SPECTRA IN THE LIVING WORLD

This chapter may seem very diverse to the reader, without a really common theme. However, spectral tuning in one pigment group is not independent of that in another one; in fact they all depend on one another. As mentioned already, it all starts with the spectrum of the Sun, and how that has determined the spectrum of the main photosynthetic pigment, chlorophyll a. The shielding effect of chlorophyll has lead to the evolution of accessory pigments, like the phycobiliproteins, which can pick up light not absorbed by chlorophyll and make it available to photosynthesis. Also the long-wavelength bacterial chlorophylls of photosynthetic bacteria deep in sediments save the light energy that has been wasted by chlorophyll a containing organisms above them.

The daylight available from the Sun, and that reflected from green plants, has also determined the spectral properties of visual pigments. The spectra of daylight, of chlorophyll and of visual pigments, in turn, have tuned the floral pigments so that pollinators could easily detect flowers against a background of photosynthesising leaves, and also discriminate one kind of flower from another, and to judge the stage of development of the flower

Likewise, the colour of fruits have evolved to make the fruits conspicuous to seed dispersers, and make it possible for them to determine the degree of ripeness.

Flavonoids, pigments produced by plants and having many functions in plant life (such as shown in Fig. 9), are used "secondhand" by some insects. Butterflies who ingest them with their food use them to enhance their wing patterns and make themselves more attractive to the opposite sex (Knuttel & Fiedler 2001).

For plants, it was important to see their competitors for light, and adapt their mode of growth thereafter. In phytochrome, first appearing already among prokaryotes, they got an excellent tool for detecting the modification of the daylight spectrum by chlorophyll. The short-wave form of phytochrome, Pr, has an absorption peak near 660 nm, near the centre of absorption of the combination of chlorophylls a and b. Its long-wavelength form, Pfr, has an absorption peak near 730 nm, just beyond the chlorophyll absorption bands. The balance between the two phytochrome forms is sensitively dependent on the presence of chlorophyll in the environment. Also for some insects, a red-light receptor plays an important role in the recognition of plant leaves, but in this case the green spectral region is used for contrast.

Pigments sending out bioluminescence have also evolved in an environment of other pigments. "Counterilluminating" fish with glowing belleys evade detection by mimicking the downwelling daylight. The dragonfish, in contrast, has evolved a bioluminescence that is undetectable for other animals, but visible to itself. Fireflies active in darkness send out green light, in the region of the spectrum where eyes are most sensitive. Those flying at dusk, in contrast, use yellow light, which makes them more visible against green leaves.

It all fits together.

REFERENCES

Berthier, S. & Lafait, J. (1999). Des tissues aux coleurs changeantes. Pour la Science 266, déc. http://www.pourlascience.com/numeros/framerevue4.htm

Bhoo, S.-H., Davis, S.D., Walker, J., Karniol, B. & Vierstra, R.D. (2001). Bacteriophytochromes are photochromic histidine kinasis using a biliverdin chromophore. *Nature, 414*, 776-779.

Björn, G.S. (1979). Action spectra for conversions of phycochrome b, a reversibly photochromic pigment in a blue-green alga, and its separation from other pigments. *Physiol. Plant., 46*, 281-286.

Björn, G.S. (1980). *Photoreversibly photochromic pigments from blue-green algae (cyanobacteria).* Diss. Lund University, CODEN LUNBDS/(NBFB-1009)/1-28/(1980).

Björn, L.O. (1979). Photoreversibly photochromic pigments in organisms: properties and roles in biological light perception. *Quart. Revs. Biophys., 12*, 1-23.

Björn, L.O. (1985a). Varför håller växterna inte färgen? *Forskning och Framsteg, 85 (6),* 40-46.

Björn, L.O. (1985b), Växternas ljusperception. *Svensk Bot. Tidskr. 79*, 249-264.

Björn, L.O. & Björn, G.S. (1980). Yearly review: Photochromic pigments and photoregulation in blue-green algae. *Photochem. Photobiol., 32*, 849-852.

Björn, G.S., Braune, W. & Björn, L.O. (1985). Light-induced, dark reversible colour shift in petals of *Phlox. Physiol. Plant., 64*, 445-448.

Bowmaker, J.K., Heath, L.A., Wilkie, S.E. & Hunt, D.M. (1997). Visual pigments and oil droplets from six classes of photoreceptor in the retinas of birds. *Vision Res., 37*, 2183-2194.

Britt, S.G., Feiler, R., Kirschfeld, K. &m Zuker, C.S. (1993). Spectral tuning of rhodopsin and metarhodopsin in vivo. *Neuron 11*, 29-39.

Chittka, L. 1996. Does bee color vision predate the evolution of flower color? Naturwiss. 83:136-138.

Chittka, L. & Menzel, R. (1992). The evolutionary adaptation of flower colors and the insect pollinators's color vision. *J. Comp. Physiol. A. 171*, 171-181.

Chittka, L. & Dornhaus, A. (1999), Comparisons in physiology and evolution, and why bees can do the things they do, http://www.ciencia.cl/CienciaAlDia/volumen2/numero2/articulos/ articulo5-eng.html.

Chittka, L., Thomson, J.D. & Waser, N.M. (1999). Flower constancy, insect psychology, and plant evolution. *Naturwiss., 86*, 361-377.

Cinque, G, Croce, R. & Bassi, R. (2000). Absorption spectra of chlorophyll a and b in Lhcb protein environment. *Photosyntheis Res. 64*, 233-242.

Cronin, T.W., Caldwell, R.L. & Marshall, J. (2001). Sensory adaptation — Tunable colour vision in a mantis shrimp. *Nature, 411*, 547-548.

Denton, E.J. & Land, M.F. (1971). Mechanism of reflecion in silvery layers of fish and cephalopods. *Proc. Roy. Soc. Lond. A, 178*, 43-61.

Diakoff, S. & Scheibe, J. (1973). Action spectra for chromatic adaptation in *Tolypothrix tenuis. Plant Physiol., 51*, 382-385.

Dominy, N.J. & Lucas, P.W. (2001). Ecological importance of trichromatic vision to primates. *Nature, 410*, 363-367.

Douglas, R.H., Partridge, J.C., Dulai, K., Hunt, D., Mullineaux, C.W., Tauber, A.Y. & Hynninen, P.H. (1998). *Science 393*, 423-424.

148

Douglas, R.H., Partridge, J.C., Dulai, K.S., Hunt, D.M, Mullineaux, C.W. & Hynninen, P.H. (1999). Enhanced retinal longwave sensitivity using a chlorophyll-derived photosensitiser in *Malacoseus niger*, a deep-sea dragon fish with far red bioluminescence. *Vision Res., 39*, 2817-2832.

Fasick, J.I. & Robinson, P.R. (1998). Mechanism of spectral tuning in the dolphin visual pigments. *Biochemistry, 37*, 433-438.

Fasick, J.I. & Robinson, P.R. (2000). Spectral-tuning mechanisms of marine mammal rhodopsins and correlations with foraging depth. *Visual Neurosci. 17*, 781-788.

Fasick, J.I., Cronin, T.W., Hunt, D.M. & Robinson, P.R. (1998). The visual pigments of the bottlenose dolphin (*Tursiops truncatus*). *Visual Neurosci. 15*, 643-651.

Fernandez, H.R.C. (1978). Visual pigments of bioluminescent and nonbioluminescent deep-sea fishes. *Vision Sci., 19*, 589-592.

Figueiredo, P., Lima, J.C., Santos, H., Wigand, M.-C., Brouillard, M., Macanita, A.L & Pina, F. (1994). Photochromism of the synthetic 4′,7-dihydroxyflavylium chloride. *J. Am. Chem. Soc., 116*, 1249-1254.

Frank, H.A., Bautista, J.A,, Josue, J.S. & Young, A.J. (2000). Mechanism of nonphotochemical quenching in green plants: Energies of the lowest excited singlet states of violaxanthin and zeaxanthin. *Biochemistry, 39*, 2831-2837.

Fuad, N., Day, D.A., Ryrie, I.J. & Thorne, S.W. (1983). *Photobiochem. Photobiophys. 5*, 255-262.

Fujita, Y. & Hattori, A. (1962). Photochemical interconversion between precursors of phycobilin chromoprotein in *Tolypothrix tenuis*. *Plant Cell Physiol. 3*, 209-220.

Ghiradella, H. (1991). Light and color on the wing: Structural colors in butterflies and moths. *Appl. Optics 30*, 3492-3500.

Gill, E.M. & Wittmershaus, B.P. (1999). Spectral resolution of low-energy chlorophylls in Photosystem I of *Synechocystis* sp. PCC 6803 through direct excitation. *Photosynthesis Res., 61*, 53-64.

Glazer, A.N. & Wedemayer, G.J. 1995. Cryptomonad biliproteins – an evolutionary perspective. *Photosynthesis Res., 46*, 93-105.

Goto, T. & Kondo, T. (1991). Structure and molecular stacking of anthocyanins - Flower color variation. *Angew. Chem. Int. Engl., 30*, 17-33.

Gralak, B., Tayeb, G. & Enoch, S. (2001). Morpho butterfly wings color modelled with lamellar grating theory. Optics express, 9, 567-578.

Grossman, A.R., Bhaya, D. & He, Q. (2001). Tracking the ligh environment by cyanobacteria and the dynamic nature of light harvesting. *J. Biol. Chem. 276*, 11449-11452.

Halldal. P. (1968). Photosyntheic capacities and photosynthetic action spectra of endozoic algae of the massive coral. *Favia, Biol. Bull., 134*, 411-424.

Hart, N.S., Partridge, J.C., Bennett, A.T.D. & Cuthill, I.C. (2000). Visual pigments, cone oil droplets and ocular media in four species of estrildid finch. *J. Comp. Physiol. A, 186*, 681-694.

Herring (1994). Reflective systems in aquatic animals. *Comp. Biochem. Physiol., 109A*, 513-546.

Hinton, H.E. & Jarman, G.M. (1972). Physiological color change in the Hercules beetle. *Nature, 238*, 160-161.

Hinton, H.E. & Jarman, G.M. (1973). Physilogical color change in the elytra of the Hercules beetle, *Dynastes hercules. J. Insect Physiol., 19*, 533-549.

Holzwarth, A.R. (1991). Structure-function relationships and energy transfer in phycobiliprotein antennae. *Physiol. Plant., 83*, 518-528.

Huxley (1968). A theoretical treatment of the reflexion of light by multilayer structures. J. Exp. Biol., 48, 227-245.

Jacobs, G.H. 1992. Ultraviolet vision in vertebrates. *Am. Zool., 32*, 544-554.

Jensen, P. & Bunker, P.R. (2000). *Computational molecular spectroscopy.* ISBN 0-471-48998-0. John Wiley & Sons.

Jordan, P., Fromme, P., Witt, H.T., Klukas, O., Saenger, W. & Krauß, N. (2001). Three-dimensional structure of cyanobacterial photosystem I at 2.5 nm resolution. *Nature, 411*, 909-917.

Kehoe, D.M. & Grossman, A.R. (1996). Similarity of a chromatic adaptation sensor to phytochrome and to ethylene receptors. *Science, 273*, 1409-1412.

Karapetyan, N.V., Dorra, D., Schweitzer, G., Beszmertnaya, I.N. & Holzwarth, A.R. (1997). Fluorescence spectroscopy of the longwave chlorophylls in trimeric and monomeric photosystem I core complexes from the cyanobacterium *Spirulina platensis. Biochemistry, 36*, 13830-13837.

Kelber, A. (1999). Ovipositing butterflies use a red receptor to see green *J. Exp. Biol., 202*, 2619-2630.

Kleinschmidt, J. & Harosi, F. (1992). Proc. Natl. Acad. Sci USA, 89, 9181-9185.

Knipp, B., Müller, M., Metzler-Nolte, N., Balaban, T.S., Braslavsky, S.E. & Schaffner, K. (1998). NMR verification of helical conformations of phycocyanobilin in organic solvents. *Helv. Chim. Acta*, 81, 881-888.

Knuttel, H. & Fiedler, K. (2001). Host-plant derived variation in ultraviolet wing-patterns influences mate selection by male butterflies. J. Exp. Biol., 204, 2447-2459.

Kochendoerfer, G.G., Lin, S.W., Sakmar, T.P. & Mathies, R.A. (1999). How color visual pigments are tuned. *Trends Biochem. Sci.*, 24, 300-305.

Kochubey, S.M. & Samokhval, E.G. (2000). Long-wavelength chlorophyll forms in Photosystem I from pea thylakoids. *Photosynthesis Res.* 63, 281-290.

Koehne, B., Elli, G., Jennings, R.C., Wilhelm, C. & Trissl, H.-W. (1999). Spectroscopic and molecular characterization of a long wavelength absorbing antenna of *Ostreobium sp. Biochim. Biophys. Acta*, 1412, 94-107.

Land, E. (1964). The retinex theory of color vision. *Sci. Am.*, 108-128.

Land, M.F. (1966). A multilayer interference reflector in the eye of the scallop, *Pecten maximus. J. Exp. Biol.*, 45, 433-447.

Large, M.C.J., McKenzie, D.R., Parker, A.R., Steel, B.C., Ho, K., Bosi, S.G., Nicorovici, N. & McPhedran, R.C. (2001). The mechanism of light reflectance in silverfish. *Proc. R. Soc. Lond. A*, 457, 511-518.

Lazaroff, N. & Schiff (1962). Action spectrum for developmental photoinduction of the blue-green alga *Nostoc muscorum. Science*, 137, 603-604.

Lucas, P.W., darvell, B., Lee. P.K.D., Yuen, T.D.B. & Choong, M.F. (1998). Colour cues for leaf food selection by long-tailed macaques (*Macaca fascicularis*) with a new suggestion for the evolution of trichomatic colour vision. *Folia Primatol.*, 69, 139-152.

Lythgoe, J.N. (1984). Visual pigments and environmental light. *Vision Sci.*, 24, 1539-1550.

MacColl, R. (1998). Cyanobacterial phycobilisomes. *J. Struct. Biol.*, 124, 311-334.

Maier, EJ. & Bowmaker, J.K. (1993). Color-vision in the passeriform bird, Leiothrix lutea — correlation of visual pigment absorbancy and oil droplet transmission with spectral sensitivity. *J. Comp. Physiol. A*, 172, 295-301.

Makino, C.L., Groesbeek, M., Lugtenburg, J. & Baylor, D.A. (1999). Spectral tuning in salamander visual pigments studied with dihydroretinal chromophores. *Biophys. J.*, 77, 1024-1035.

Marshall, J. & Oberwinkler, J. (1999). The colourful world of the mantis shrimp. *Nature*, 401, 873-874.

Matsui, S., Seidou, M., Uchiyama, I., Sekiya, N., Hiraki, K., Yoshihara, K. &Kito, Y. (1988). 4-hydroxyretinal, a new visual pigment chromophore found in the bioluminescent squid, *Watasenia scintillans.. Biochim. Biophys. Acta*, 966, 370-374.

Miller, W.H., Møller, A.R. & Bernhard, C.G. (1966). *The corneal nipple array. In The functional organization of the compound eye* (Bernhard, C.G., ed.), pp. 21-33. Oxford:Pergamon Press.

Murrell, J.N. (1963). *The theory of the electronic spectra of organic molecules.* Methuen, London. (German edition "*Elektronenspektren organischer Moleküle*", Bibliographisches Institut Mannheim 1967).

Nathans, J. (1990). Determinants of visual pigment absorbance: Identification of the retinylidene Schiff's base counterion in bovine rhodopsin. *Biochemistry*, 29, 9746-9752.

Nathans, J. (1992). Rhodopsin: Structure, function, and genetics. *Biochemistry*, 31, 4923-4931.

Neitz, M., Neitz, J. & Jacobs, G.H. (1991). Spectral tuning of pigments underlying red-green color vision. *Science*, 252, 971-974.

Neitz. J., Neitz, M. & Jacobs, G.H. (1993). More than three different cone pigments among people with normal color vision. *Vision Res.*, 33, 117-122.

Ohad, I., Clayton, R.K. & Bogorad, L. (1979). Photoreversible absorption changes in solutions of allophycocyanin purified from *Fremyella diplosiphon*: Temperature dependence and quantum efficiency. *Proc. Natl. Acad. Sci. USA*, 76, 5655-5659.

Ohki, K. & Fujita, Y. (1978). Photocontrol of phycoerythrin formation in the blue-green alga *Tolypothrix tenuis* growing in the dark. *Plant Cell Physiol.*, 19, 7-15.

Öquist, G. (1969). Adaptations in pigment composition and photosynthesis by far red radiation in *Chlorella pyrenoidosa. Physiol. Plant.*, 22, 516-528.

Osorio, D. & Vorobyev, M. (1996). Colour vision as an adaptation to frugivory in primates. *Proc. R. Soc. Lond. B*, 263, 593-599.

Osorio, D., Marshall, N.J. & Cronine, T.W. (1997). Stomatopod photoreceptor spectral tuning as an adaptation for colour constancy in water. *Vision Res., 37,* 3299-3309.

Pålsson, L.O., Flemming, C., Gobels, B., van Grondelle, R., Dekker, J.P. & Schlodder, E. (1998). Energy transfer and charge separation in photosystem I: P700 oxidation upon selective excitation of the long-wavelength antenna chlorophylls of *Synechococcus elongatus. Biophys. J., 74,* 2611-2622.

Parker, A.R. (1998). The diversity and implications of animal structural colours. *J. Exp. Biol., 201,* 2343-2347.

Parker, A.R. (1999). Light-reflection strategies. *American Scientist, 87,* 248-255.

Parker, A.R., McKenzie, D.R. & Large, M.C.J. (1998a). Multilayer reflectors in animals using green and gold beetles as contrasting examples. *J. Exp. Biol., 201,* 1307-1313.

Parker, A.R., Mckenzie, D.R. & Ahyung, S.T. (1998b). A unique form of light reflector and the evolution of signalling in *Ovalipes* (Crustacea:Decapoda: Portulidae). *Proc. R. Soc. Lond. B, 265,* 861-867.

Peichl. L., Behrmann, G. & Kröger, H.H. (2001). For whales and seals the ocean is not blue: a visual pigment loss in marine mammals. *Eur. J. Neurosci., 13,* 1520-1528.

Pfaff, G. & Reynders, P. (1999). Angle-dependent optical effects from submicron structures of films and pigments. *Chem. Rev., 99,* 1963-1981.

Regan, B.C., Julliot, C., Simmen, B., Viénot, F., Charles-Dominique, P. & Mollon, P. (1998). Frugivory and colour vision in *Alouatta seniculus,* a trichromatic platyrrhine monkey. *Vision Res., 38,* 3321-3327.

Ritz, T., Damjanovic´, A., Schulten, K, Zhang, J.P. & Koyama, Y. (2000). Efficient light harvesting through carotenoids. *Photosynthesis Res., 66,* 125-144.

Robinson, B.L. & Miller, J.N. (1970). Photomorphogenesis in the blue-green alga *Nostoc commune. Physiol. Plantarum, 23,* 461-472.

Scharnagel, C. & Fischer, S.F. (1993). Reversible photochemistry in the a-subunit of phycoerythrocyanin: Characterisation of chromophore and protein by molecular dynamics and quantum chemical calculations. *Photochem. Photobiol., 57,* 63-70.

Scheer, H. & Kufer, W. (1977). Studies on plant bile pigments, IV: Conformational studies on C-phycocyanin from *Spirulina platensis. Z. Naturforsch., 32c,* 513-519.

Scheibe, J. /1972). Photoreversible pigment: occurrence in a blue-green alga. *Science, 1976,* 1037-1039.

Schelvis, J.P.M, van Noort, P.I., Aartsma, P.I. & van Gorkom, H.J. (1994). Energy transfer, charge separation and pigment arrangement in the reaction center of Photosystem II. *Biochim. Biophys. Acta, 1184,* 242-250.

Schneeweis, D.M. & Green, D.G. (1995). Spectral properties of turtle cones. *Visual Neurosci., 12,* 333-344.

Schulten, K. & Karplus 1972. On the origin of a low-lying forbidden transition in polyenes and related molecules. *Chem. Phys. Lett., 14,* 305-309

Sharpe, L.T., Stockman, A., Jägle, H. & Nathans, J. (1999). *Opsin genes, cone photopigments, color vision, and color blindness. In* Gegenfurtner, K. & Sharpe, L.T. (eds) *Color vision: from genes to perception.* New York: Cambridge Univ. Press.

Shubin, V.V., Murthy, S.D.S, Karapetyan, N.V., & Mohanty, P.S. (1991). Origin of the 77-K variable fluorescence at 758 nm in the cyanobacterium *Spirulina platensis. Biochim. Biophys. Acta , 1060,* 28-36.

Siitari, H., Honkavaara, J. & Viitala, J. (1999). Ultraviolet reflection of berries attracts foraging birds. A laboratory study with redwings (*Turdus iliacus*) and bilberries (*Vaccinium myrtillus*). *Proc. R. Soc. Lond. B, 266,* 2125-2129.

Srinivasarao, M. (1999). Nano-optics in the biological world: Beetles, butterflies, birds, and moths. *Chem. Rev., 99,* 1935-1961.

Thrash, R.J., Fang, H.L.-B. & Leroi, G.E. (1979). On the role of forbidden low-lying excited states of light-harvesting carotenoids in energy transfer in photosynthesis. *Photochem. Photobiol., 29,* 1049-1050.

Vogelmann, T.C. & Scheibe, J. 1978. Action spectra for chromatic adaptation in the blue-green alga *Fremyella diplosiphon. Planta, 143,* 233-239.

Vucusic, P., Sambles, J.R., Lawrence, C.R. & Wootton, R.J. (1999). Quantified interference and diffraction in single *Morpho* butterfly scales. *Proc. R. Soc. Lond. B, 266,* 1403-1411.

Vucusic, P., Sambles, J.R. & Lawrence, C.R. (2000). Colour mixing in wing scales of a butterfly. *Nature, 404,* 457.

Wald, G. & Brown, P.K. (1958). Human rhodopsin. *Science, 127*, 222-226.

Warrant, E. (2000). The eyes of deep-sea fishes and the changing nature of visual scenes with depth. *Phil. Trans. R. Soc. Lond., 355*, 1155-1159.

Weiss, C., Jr. (1972). The pi electron structure and absorption spectra of chlorophylls in solution. *J. Mol. Spectrosc., 44*, 37-80.

Yokoyama, S. (1997). Molecular genetic basis of adaptive selection: Examples from color vision in vertebrates. *Annu. Rev. Genet., 31*, 315-336.

Yokoyama, S. (1997). Molecular genetic basis of adaptive selection: Examples from color vision in vertebrates. *Annu, Rev. Genet., 31*, 315-336.

Yokoyama, S., Radlwimmer, F.B. & Blow, N.S. (2000). Ultraviolet pigments in birds evolved from violet pigments by a single amino acid change. *Proc. Natl Acad. Sci. USA, 97*, 7366-7371.

Zhao, K.H. & Scheer, H. (1995). Type-I and Type-II reversible photochemistry of phycoerythrocyanobilin alpha-subunit from *Mastigocladus laminosus* both involve Z-isomerisation, E-isomerisation of phycoviolobilin chromophore and are controlled by sulphydryls in apoprotein. *Biochim. Biophys. Acta, 1228*, 244-253.

Zucchelli, G., Jennings, R.C. & Garlaschi, F.M. (1990). The presence of long-wavelength chlorophyll a spectral forms in the light-harvesting chlorophyll a/b protein complex II. *J. Photochem. Photobiol. B: Biol., 6*, 381-394.

Zhao, K.H., Haessner, R., Cmiel, E. & Scheer, H. (1995). Type I reversible photochemistry of phycoerythrocyanin involves Z/E-isomerisation of a-84 phycoviolobilin chromophore. *Biochim. Biophys. Acta, 1228*, 235-243.

Zoumi, A., Witt, H.T., Kern, J, Fromme, P, Krauß, N., Saenger, W. & Orth, P. (2001). Crystal structure of photosystem II from *Synechococcus elongatus* at 3.8 Å resolution. *Nature, 409*, 739-743.

LARS OLOF BJÖRN

8. THE PHOTOCHEMICAL REACTIONS IN BIOLOGICAL LIGHT PERCEPTION AND REGULATION

1. INTRODUCTION

Many photochemical reactions involved in the sensing of and regulation by light and ultraviolet radiation by organisms consist of cis-trans (and trans-cis) isomerisations. We shall start with this class of photosensors, and then go on to other mechanisms. There are many more known and unknown light-sensing molecular systems than those briefly described below, but (except for the first one) I have tried to concentrate on those more widespread. As examples of light-sensing pigments with a very limited distribution, one can mention stentorin and blepharismin of certain ciliates (Lenci et al. 2001).

The term "photoreceptor" means different things to different people. In zoology it means a cell which responds to light, such as rods and cones of our eyes, but to plant scientists it means a pigment molecule, such as rhodopsin or phytorchrome, which absorbs light at the start of a chain of events leading to light perception or regulation of a physiological process by light. We shall use the term here in this latter sense.

"Photoreceptor" in this sense is a concept related to "photoenzyme", i.e. enzyme active only in light. A class of photoreceptors, the cryptochromes, are thought to have evolved from certain photolyases. Another photoenzyme, NADPH-protochlorophyllide oxidoreductase, can also be regarded as a photoreceptor, helping the plant to regulate chlorophyll synthesis and chloroplast development. (Beale 1999). An example of a flavine enzyme whose activity is stimulatedby blue light is glycine oxidase of *Chlorella* (Schmid & Schwarze 1969, Schmid 1970).

2. CIS-TRANS AND TRANS-CIS ISOMERISATION

Double bonds, and conjugated double bond systems provide molecules with a certain rigidity. Molecular groups cannot rotate freely around double bonds, nor around single bonds in a continuous conjugation suite, as they can around isolated single bonds provided there is room enough. When a double bond is involved, there are two opposite torsion angles for which the energy has a minimum value, and which thus represent stable conformations. Often the carbon atoms at the double bonds carry one hydrogen atom and one larger atomic group. These atoms then lie

L.O. Björn (ed.), Photobiology, 153–179.

in the same plane, which is the same plane as the corresponding groups on the carbon atom at the other end of the double bond (in Fig. 1 this is the plane of the paper).

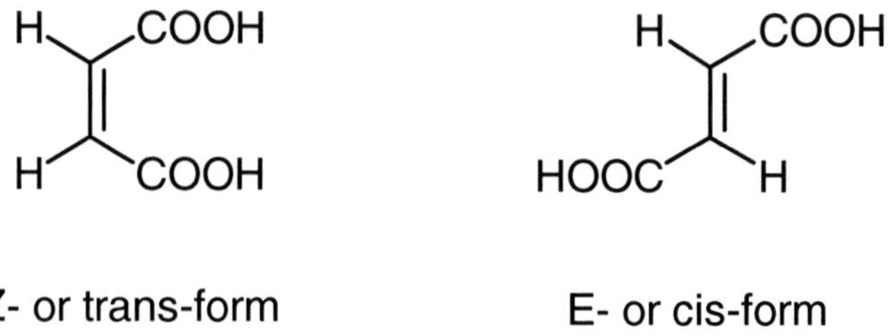

Z- or trans-form   E- or cis-form

Figure 1. A simple example of cis-trans isomerism.

If the larger atomic groups are on the same side of the line through the double bond and the carbon atoms at its ends, the molecule is said to be in cis-configuration; if they are on opposite sides the molecule has a trans-configuration. However, for larger molecules this designation may be difficult to apply, and another nomenclature has been introduced, i.e Z- (for German "zusammen", together) and E- (for German "entgegen", opposite) configurations. In this system a priority is assigned for the atoms immediately attached to the double bond, such that higher priority is assigned to atoms of higher atomic number. Thus carbon atoms in the exampel in Fig. 1 have first priority, hydrogen atoms second priority. When atoms of same priority are on the same side we have Z-configuration, otherwise E-configuration.

By a very rough consideration we can appreciate why cis-trans (E-Z) isomerisations are suitable for light sensing. A typical carbon-carbon single bond has a bond energy of 387 kJ per mole, while the typical double bond has a strength of about 610 kJ per mol. The difference is 263 kJ per mol. We can think of a rotation around a double bond to consist of the breaking of one of the bonds in the double bond, rotation around the remaining (single) bond, and reformation of a double bond. Thus one would need to add 263 kJ per mole (or $263,000/(6.02\ 10^{23}$ J per molecule) to achieve the rotation. The energy of a photon is $h\ c/$ (see Chapter 1), and by equating the two energies one obtains a typical wavelength for rotation of 455 nm, in the middle of the optical part of the electromagnetic spectrum to which the atmosphere is transparent. In rhodopsin of our rods the actual energy barrier for isomerisation is 238 kJ/mol (Okada et al. 2001), but in different other photoreceptors based on cis-trans (or trans-cis) isomerisation, the wavelength varies from the UV-B region (for urocanic acid) to the near infrared (for phytochrome). Some of the principles for this "tuning" are described in Chapter 7).

2.1. Urocanic Acid

Urocanic acid is present in human skin, and photoisomerisation from the trans to the cis form causes down-regulation of the immune system (Chapter 13). This radiation sensing reaction appears at first glance to be a very simple one: The pigment is structurally simple, of low molecular weight, and not protein bound. The isomerisation seems to be a very simple reaction (Fig. 2).

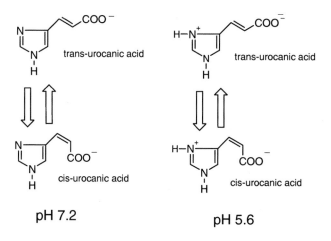

Figure 2. Photoisomerisations of anionic and zwitterionic forms of urocanic acid. The trans form is also called E-urocanic acid, the cis form Z-urocanic acid.

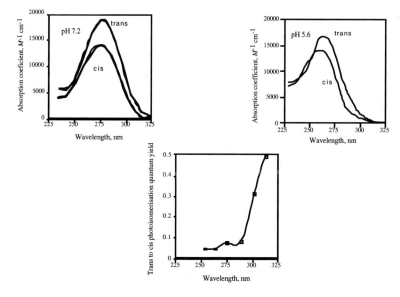

Figure 3. Absorption spectra for trans- and cis-urocanic acid at pH 7.2 (redrawn after Hanson et al. 1997) and at pH 5.6 (redrawn after Morrison et al. 1980), as well as the quantum yield for trans to cis photoisomerisation (from data of Morrison et al. 1984).

But the simplicity is only apparent. The first indication of this is the fact that the action spectrum for photoisomerisation is very different from the absorption spectrum of trans-urocanic acid. The absorption spectrum has a broad band peaking at about 280 nm, but radiation of this wavelength does not produce any photoisomerisation, i.e. the quantum yield is zero at this wavelength. The quantum yield is maximal, 0.5, at 310 nm, where absorption is much weaker (also for the cis to trans isomerisation the quantum yield is about 0.5 at long wavelengths). Various theoretical explantations have been given for this strange behaviour (Li et al. 1997, Hansson et al. 1997, Page et al. 2000, Ryan & Levy 2001), and the discussion is still going on. Urocanic acid is not the only substance which behaves in this way: also cinnamic acid and related compounds have a quantum yield for cis-trans-isomerisation which is wavelength dependent (cf. the section on yellow protein below).

Urocanic acid exists in several ionic forms depending on pH: at a low pH as a cation, at the slightly acidic pH in skin as a zwitterion, and at neutral or higher pH as an anion. Spectra for both trans and cis forms peak near 280 nm in neutral solution, but near 270 nm at ph 5.6 (Morrison 1980, Hansson et al. 1997, Li et al. 1997, Fig. 3).

Trans-urocanic acid is formed from histidine in the outermost layer of the skin, and after photoisomerisation the cis-urocanic acid diffuses inwards and acts on cells in deeper layers, but by what mechanism is so far not known.

2.2 Eukaryotic Rhodopsin

When we see the word rhodopsin our thoughts first go to the light-sensitive pigment of the rods in our own eyes. Very similar pigments, however, are present also in our cones and in the eyes of various insects, molluscs, and other animals. Recently it has been discovered that essentially the same type of pigment occurs also in various algae (Foster et al. 1984, Hegemann & Deininger 2001, Gualtieri 2001). Also several types of archaebacteria (archaea) contain a kind of rhodopsins, but the latter are sufficiently different, both with respect to the chromophore and the protein structure to warrant treatment in a separate section (2.3). Both eukaryotic and archaean rhodopsins, however, are membrane bound and have 7 membrane spanning helices in the molecule. It should also be noted that there are eukaryots which have proteins more similar to the archaean than to typically "eukaryotic" rhodopsin.

The chromophore of rhodopsin is retinal (in some animals dehydroretinal; see Chapter 7). We show it first in isolated form (Fig. 4) , to display in a simple way the phototransformation from the 11-cis form to the all-trans form, corresponding to the primary process of vision. The side-chain changes from a bent to a straight form. In rhodopsin the terminal carbon atom of retinal polyene chain is covalently tethetered to a lysine side chain (lysine296) of the protein (Fig. 5), forming a "Schiff base" in its protonated form. The counterion for the positive charge is formed by glutamine 113. The protein forms an antiparallel b-sheet in the vicinity of the chromophore, bringing the side chain of another aminoacid (position 181) in close contact to the polyene chain of retinal. In our rod rhodopsin this aminoacid is glutamine, in the red- and green-sensitive pigments of our cones it is histidine.

When light isomerises the rhodopsin-bound retinal from 11-cis to all-trans, the polyene chain cannot at once straighten out completely, because the straight chain does not fit in the protein pocket (see Okada et al. 2001 for a more detailed review of the events than the one given here). This restraint makes the retinal store energy, like a cocked spring. About two thirds of the photon energy is thus stored in the initial phase. Probably it is mainly the proximity of serine 186 that prevents the retinal from immediately reaching its trans equilibrium position. This intermediate stage is termed bathorhodopsin. Within a microseconds the proton on the Schiff base is translocated to glutamine 113, whereby the attractive force between the two parts of the protein disappears, and the protein helices can adjust their relative postitions to allow the retinal to straighten out completely. This brings the rhodopsin to a low energy state called metarhodopsin I. From this the rhodopsin, within a millisecond, changes to metarhodopsin II. This has higher energy than metarhodopsin I, and the transformation is made possible by a simultaneous increase in entropy: the forces between different parts of the protein are decreased, and the different parts can move more freely with respect to each other. The process has some similarities to the melting of ice, and we may recall that the free energy change (ΔG) of a system consists of the change in total energy (enthalphy, ΔH) minus an entropy term, $T \cdot \Delta S$. In the transformation from metarhodopsin I to metarhodopsin II the free energy is thus decreased even though the total energy increases. In this respect eukaryotic rhodopsin resembles photoactive yellow proteins

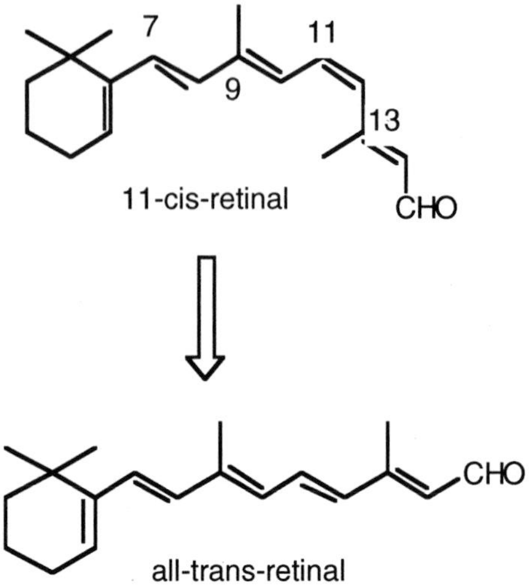

Figure 4. Photoisomerisation of the isolated retinal chromophore.

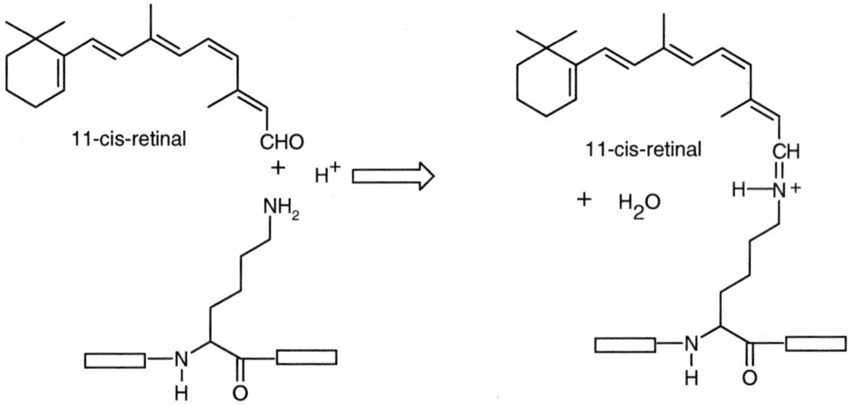

Figure 5. The binding of retinal to a lysine side chain in the rhodopsin protein, forming a protonated Schiff base. Rectangles symbolise aminoacids of the opsin.

(section 2.4), but differs from archaean rhodopsins (section 2.3). The transition also involves uptake of a proton.

Metarhodopsin II is the "signalling state" of rhodopsin. By its formation groups are exposed which can interact with a protein called transducin, a so-called G protein, and thus make the transition from a biophysical to a biochemical phase of the signal transduction. The activated transducin activates phophodiesterase which hydrolyzes cyclic guanosine-monophosphate (cGMP). When the concentrtion of cGMP has fallen sufficiently, sodium ion channels in the membrane closes, the electric membrane potential increases, and an electrical impulse is sent on to the nervous system.

Recently a new light-sensitive system has been discovered in vertebrates (Provencio et al. 1998, Barinaga 2002, Berson et al. 2002, Hattar et al. 2002). The start in this new development came with the study of how frogs can adjust their skin colour by changing the size, shape and position of the pigment-containing cells in their skin, the melanophores (Provencio et al 1998). It turned out that the skin cells contained a light-sensitive pigment which was named melanopsin. The same pigment was found also in the frog's retina, as well as in mouse retinas. However, it is not in the rods (nor in cones), but in some of the cells, retinal ganglion cells, inside (in front of) the visual receptors. Some of these cells have nerve connecions, not to the brain areas involved in vision, but to the suprachiasmic nuclei where the main clock of the body (see Chapter 15, section 6.2) is thought to reside. The logical conclusion is that melanopsin is involved in the resetting of the biological clock by light. However, some of the melanopsin-containg retinal ganglion cells have connections to the part of the brain regulating pupil size in response to light. Melanopsin-containing cells, in contrast to rods and cones,do not adjust their sensitivity in response to light level. They are therefore suited to record the light level, which is important, e.g., in photoperiodism and pupil size regulation. Although the melanopsins studied so far accur in vertebrates, their protein structure is more closely related to that of invertebrate opsins than to the vertebrate opsins of rods and cones. Their photochemical reactions have so far not been studied in detail, but are thought to be similar to those of rhodopsins.

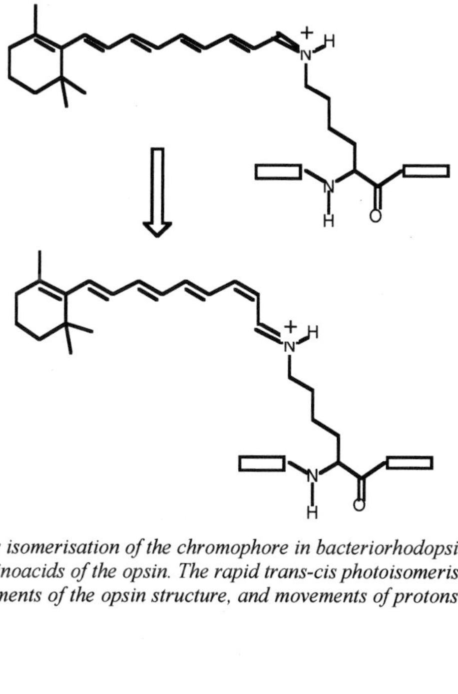

Figure 6. The trans-cis isomerisation of the chromophore in bacteriorhodopsin. As in Fig. 5 the rectangles symbolise aminoacids of the opsin. The rapid trans-cis photoisomerisation is followed by slower rearrangements of the opsin structure, and movements of protons (see Fig. 7).

160

2.3 Archaean Rhodopsins

Four types of archaean rhodopsins are known. In contrast to the eukarytic rhodopsins they all contain all-trans-retinal as the chromophore, and the photochemical step consists of its isomerisation to 13-cis retinal. In some cases also the reverse reaction has some importance.

Low oxygen tension

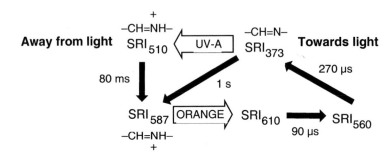

High oxygen tension

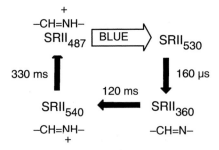

Away from light

Figure 7. The photocycles of sensory archaeal rhodopsins. Under low oxygen tension (top) Halobacterium and related organisms form bacteriorhodopsin a nd halorhodopsin, which pump ions, and sensory rhodopsin I (SRI). The latter is used to find a suitable light environment. As long as light is not very strong, the orange light sensing reaction causes accumulation of the long-lived intermediate SRI373, which is a singnalling state causing the organisms to move towards stronger light.. If the (UV-A) light becomes too strong, the UV-A sensing reaction causes conversion to SRI510, a signalling state which causes movement away from the strong light. In the presence of a high concentration of oxygen (bottom) only SRII is induced. The blue-light sensitive reaction causes movement away from light. As other non-oxygenic pigmented organisms these are probably much more light sensitive in the presence of oxygen, due to the possibility of formation of reactive oxygen species under illumination. As indicated in the figure, in some of the pigment forms the Schiff-base is protonated, in other forms not. After Hoff et al. (1997), modified.

Many species within the Haloarchaea (the subdivision of Archaea formerly referred to as halobacteria) have been investigated, but only four distinct types have been found, which are all present in species of the best investigated genus, *Halobacterium*. One of these rhodopsins called bacteriorhodopsin (BR) uses light energy to pump hydrogen ions out of the cells, and another one, called halorhodopsin (HR) pumps chloride ions into the cells. Both reactions contribute to making the inside of the cells negative, thus allowing the cells to accumulate cations at the expense of light energy. The light-driven export of protons also creates the proton motive force necessary for ATP synthesis and, indirectly the free energy necessary for swimming and biochemical syntheses. BR and HR are induced only under low oxygen tension, while under high oxygen tension the organisms can utilize oxygen for creation of the necessary free energy.

The two remaining archaerhodopsins, designated SRI and SRII (SR for sensory rhodopsin) are used by the halobacteria to orient with respect to light. SRI is induced only under low oxygen conditions, SRII only under high.

SRI, formed under low oxygen conditions, has two signalling states. One, SRI_{373}, formed by the orange component of weak daylight, causes the cells to move towards stronger light. This is not due to direct sensing of light direction as in the topophototaxis of eukaryotic flagellates, but by modulation of the frequencies of spontaneous reversals of swimming direction. If the light becomes very strong, another signalling state, SRI_{510}, is formed under the action of the UV-A component of daylight. This causes the cells to move towards weaker light.

Under high oxygen tension SRII, but not SRI (neither HR nor BR) is induced. SRI mediates only a light-avoiding signal.

SRI and SRII do not engage a signal-transmitting protein during only part of the photocycle (as eukaryotic rhodopsin engages transducin). Instead eqach one of them is permanently attached to its signal transmitting protein, HtrI or HtrII, respectively. Obviously the conformational change in the rhodopsins caused by the photoisomerisation of the chromophore is somehow transmitted to the signal transmitting protein, but the details of this are not known.

2.4 Photoactive Yellow Proteins (PYPs, xanthopsins)

Photoactive yellow proteins (PYPs) function as photoreceptors in purple bacteria, mediating negative phototaxis. One might think that this is too humble a function to warrant treatment in a book like this one, but PYP happens to be one of the best-known photoreceptor pigments, and we can learn some more general principles from it. PYPs are also referred to as xanthopsins, although this term is misleading, since the proteins are not opsins. Three photoreactions shuttles the pigment between several forms as shown in Fig. 8.

The PYP chromophore is trans-4-hydroxy cinnamic acid, and light causes photoisomerisation to the cis form (Fig. 9). This initial reaction is followed by rotation of one half of the molecule with respect to the other around a single bond. Genick et al. (1997, 1998) have succeeded in following in detail the changes in the protein structure associated with these changes (Fig. 10). They managed to crystallize the protein, and by time-resolved x-ray crystallography at low temperature capture the structure of the otherwise extremely short-lived (nanoseconds) intermediate.

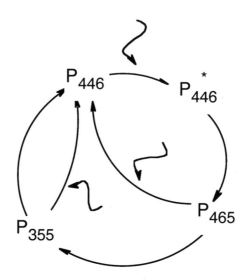

Figure 8. As shown in the diagram above, PYP is returned to the original P446 form either by dark conversion of the signalling state, P355, or by photochemical conversions from the intermediate or from the signalling state.

By conversion to the signalling state forces between different parts of the protein are weakened. It becomes more flexible, and the conversion can be likened to "melting", just as in the case of animal rhodopsin.

PYP is interesting also because it contains the prototype for a "PAS domain" (Pellequer et al. 1998). By this we mean a protein structure which occurs also many other signalling proteins (Taylor & Zhulin 1999), including some other photoreceptor proteins: phytochrome, phototropin (Salomon et al. 2000, Christie & Briggs 2001), and a blue-light receptor in the fungus *Neurospora* (Ballario & Macino 1997). PAS domains have been identified in proteins from all types of organisms: Archea, Bacteria and Eucarya. It comprises a region of 100-120 amino acids. They seem to occur almost exclusively in sensors of two-component "phosphorelay" regulatory systems. The activation of a PAS protein, either by the photoconversion of a chromophore, binding of an external activator, or by voltage sensing (as in some proteins regulating voltage-sensitive ion channeling) seems to involve the exposure of the PAS domain and initiation of protein kinase activity (either histidine kinaseor, as in the case of phytochrome, serine/threonine kinase).

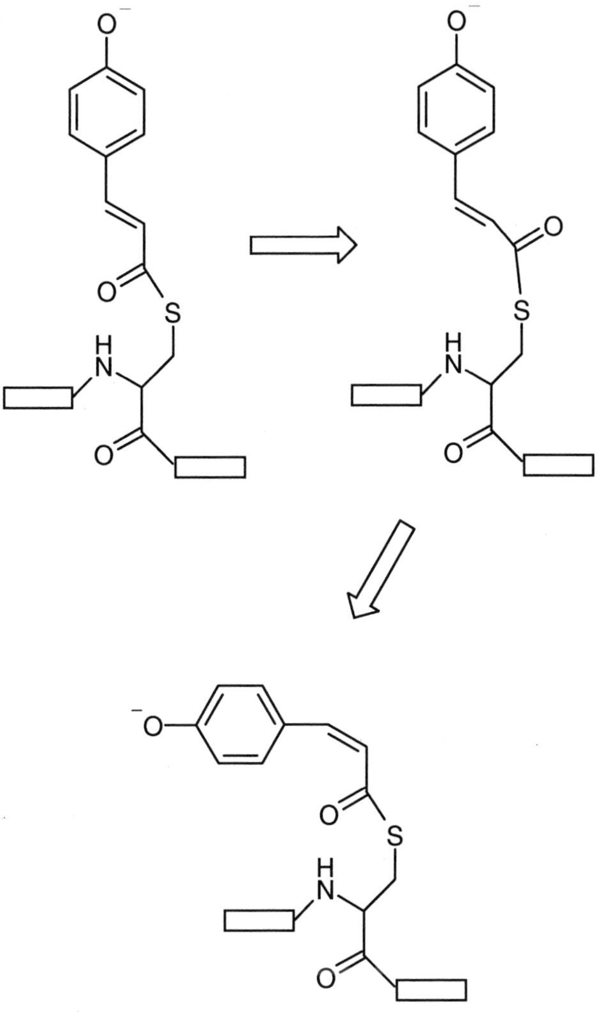

Figure 9. A sketch of change of the PYP chromophore structure in two steps. The rectangles symbolize part of the protein. The cystein residue which forms a thioester linkage with the cinnamic acid is outlined. Only the first step requires photon absorption and at physiological temperature is complete in a few nanoseconds. The second step takes several milliseconds.

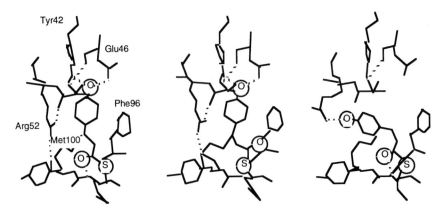

Figure 10. Changes in PYP induced by light. To the left the region around the chromophore before light absorption. The chromophore skeleton is outlined in bold, with oxygen and sulphur atoms indicated. Amino acid skeletons are outlined with thin lines. Dotted lines stand for non-covalent interactions (hydrogen bonds). In the centre is the intermediate structure a few nanoseconds after light absorption. The trans- cis isomerisation of the chromophore has taken place and its tail has flipped over, but the ring is still in its original position, with hydrogen bonds to the upper oxygen still intact. To the right is the "signalling state" reached after several milliseconds. The chromophore is still in its cis-state, but the whole chromophore has changed position, broken the original hydrogen bonds to the ring-attached oxygen atom and formed a new hydrogen bond. By these rearrangements both the chromophore and the PAS domain of the protein become accessible from the outside.

2.5. Phytochrome

The discovery of phytochrome is one of the classical detective stories of plant science (Butler 1980, 1982, Björn 1980, Sage 1992). It started with the discovery that some effects of red light, such as the germination of seeds, photomorphogenesis of etiolated plants, and the inhibition of flowering in short-day plants, could be reversed by irradiation with light of longer wavelength, so called far-red light (700-740 nm). By accurate action spectroscopy (see Chapter 6) the spectral properties of two different pigment forms were defined, and this made possible the detection in plants by absorption spectrophotometry and the subsequent purification of phytochrome.

It is now known that plants contain several phytochromes with different properties and regulatory roles. This is not the place to describe this in detail, and the reader is referred to, e.g., chapters in this volume on photomorphogenesis and photoperiodism in plants (Chapter 14) and on the biological clock and its resetting by light (Chapter 15), Kendrick & Kronenberg (1994), Sineshchekov (1995), Quail (2000) and the short updates with references by Genick & Chory (2001) and Parks et al. (2001).

Phytochrome or phytochrome-like proteins has also been found in various algae, a myxomycete, cyanobacteria, and other photosynthetic and non-photosynthetic bacteria (Schneider-Poetsch et al 1998, Davis et al. 1999, Jiang et al. 1999, Herdman et al. 2000, Lamparter & Marwan 2001, Hubschmann et al. 2001, Bhoo et al. 2001).

Phytochrome is synthesized by the plant in the red-absorbing form, called Pr, and can be converted, via several intermediates, to the far-red absorbing form by red light or direct daylight. It is the far-red absorbing form which is considered to be the active (signalling) state, but in some cases one or several intermediates may be active. The reverse conversion (via another set of intermediate states) can take place under far-red light, daylight filtered through vegetation or soil, or (with some phytochrome types and more slowly) in darkness. A pigment which changes its absorption spectrum in light (without being destroyed) is called photochromic; phytochrome is said to be photoreversibly photochromic, since the original state can be restored by another kind of light.

The chromophore in phytochrome has generally been believed to be an open chain tetrapyrrole (see Fankhauser 2001), phytochromobilin (Fig. 11), but recently it has been proposed (M. Furya as cited by Campbell & Liscum 2001) that the type of higher plant phytochrome called phytochrome B contains the related phycocyanobilin, the same chromophore as is present in the photosynthetic antenna pigments phycocyanin and allophycocyanin of cyanobacteria and red algae. Phycocyanobilin is also the chromophore in cyanobacterial phytochrome, while those phytochromes of nonphotosynthetic bacteria which have been investigated so far have biliverdin as chromophore (Bhoo et al. 2001).

An interesting optical property of phytochrome is that conversion from Pr to Pfr or vice versa results in rotation of the transition moment corresponding to the long-wavelength absorption band with respect to the bulk of the protein. This was first shown by Etzold (1965, Haupt (1970) by *in vivo* linear action dichroism, and has later been confirmed by various methods (Sarkar & Song 1981, Kadota et al. (1982), Sundquist & Björn (1983a,b), Tokutomi & Mimuro (1989). At first it was believed that the rotation amounts to 90°, but the newer experiments and reinterpretation of the old *in vivo* experiments (Björn 1984) Based on this and other evidence, Rospendowski et al. (1989) have drawn a cartoon of how the chromophore moves in the protein during conversion.

Phytochrome in solution is a dimer (Jones & Quail 1986), and there is evidence that it is dimeric also in vivo. Like many other sensors it has a PAS domain, (see 2.4 above) and is an autophosphorylating protein kinase (Boylan & Quail 1996, Watson 2000).

2.6. Photosensor for Chromatic Adaptation of Cyanobacteria

Many cyanobacteria have the ability to adjust the amounts of the photosynthetic antenna pigments phycocyanin (red-absorbing) and phycoeythrin (green-absorbing) according to the spectrum of ambient light. This regulation process is known as chromatic adaptation, although with present-day definitions it would more appropriately be called chromatic acclimation. Long ago action spectroscopy revealed that the photoreceptor for chromatic adaptation in

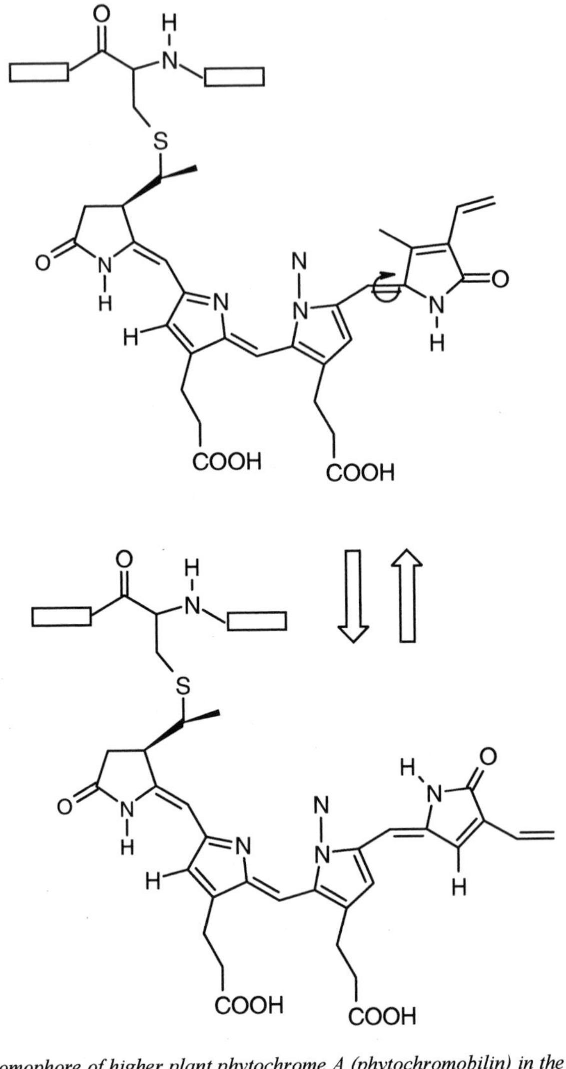

Figure 11. The chromophore of higher plant phytochrome A (phytochromobilin) in the Pr (top) and Pfr (bottom) forms. The rectangles symbolize aminoacid residues in the protein. The circular arrow in the upper formula indicates the double bond at which the pyrrol group to the right rotates during photoisomerisation. Phytochrome B may contain the very similar phycocyanobilin as chromophore, and phytochromes in non-photosynthetic bacteria contain biliverdin.

cyanobacteria must be a phycobiliprotein (Fujita & Hattori 1962, Diakoff & Scheibe 1973, Vogelmann & Scheibe 1978), just as phytochrome. It was also found (Lazaroff & Schiff 1962, Lazaroff 1973, Robinson & Miller 1970) that a similar system regulates also other processes in cyanobacteria. Scheibe (1962) was the first to show that a photoreversibly photochromic pigment could be obtained from a cyanobacterium, and this was further explored in a series of investigations by G.S. Björn (1980). Most of the early work is reviewed by Björn (1979). A photoreversibly photochromic preparation, phycochrome c, with properties similar to those postulated for the regulator of chromatic adaptation could be obtained from C-phycocyanin (Fig. 12).

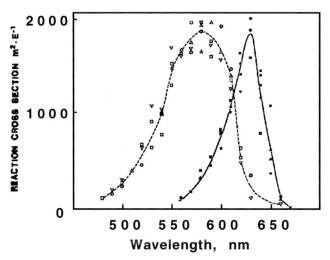

Figure 12. Action spectra for conversions of phycochrome c to the short-wavelength form (solid line and symbols) and to the long-wavelength form (dashed line, empty symbols). From Björn & Björn 1978. Somewhat similar action spectra were also obtained for phycochrome a, another phycocyanin containing polypeptide (Björn 1980).

Recently a new start on an old problem has been made from the other end at the Department of Plant Biology of the Carnegie Institution of Washington. The recent work has been reviewed by Grossman et al. (2001). Kehoe and Grossman (1996) found a gene, rcaE, coding for the protein RcaE, which is necessary for chromatic adaptation. RcaE binds a tetrapyrrole chromophore covalently in a domain similar to that of phytochromes, and also has a PAS domain typical for signal sensing proteins. It is believed that the chromophore is phycocyanobilin (D.M. Kehoe as cited by Campbell & Liscum 2001), which would fit well with the old spectral observations. According to the theory proposed by Grossman et al. (2001) there are, in addition to RcaE, two other proteins, RcaF and RcaC, involved in the signalling for chromatic adaptation. Under red light, RcaE autophosphorylates and transfers a phosphate group, via RcaF to RcaC. In RcaC there are two sites which can be phosphorylated, and the one active in this context is near the N-terminal (the role of the other site is unclear). This chain of events results in incrased phycocyanin production. Green light, on the other hand, causes RcaE to change to the non-phosphorylating conformation, resulting in phycoerythrin synthesis.

2.7. Violaxanthin as a Blue-light Sensor in Stomatal Regulation

Stomata are adjustable valves in the outer layer (epidermis) of leaves and other photosynthetic plant organs. They are designed to let suffient carbon dioxide in from the external air without causing the plant to dry out due to outward diffusion of water vapour. Their regulatory system senses the water status , both directly in the leaves, and indirectly in the rest of the plant body via the hormone abscisic acid. It also senses the internal carbon dioxide concentration. It senses light in several ways. One indirect way is via photosynthesis, since this causes the internal carbon dioxide concentration to fall. But the fastest and most dramatic light effect is blue-light specific, and there seems to be another light sensing molecule involved than the cryptochromes and phototropins dealt with in the next section: the xanthophyll zeaxanthin (see Zeiger 2000 and Assman & Wang 2001 for recent reviews).

The strongest evidence for participation of zeaxanthin as a blue-light sensor of stomata is the fact that stomata of an *Arabidopsis* mutant, npq1, which lacks a functional violaxanthin deepoxidase and therefore cannot accumulate zeaxanthin, does not show a blue-light specific response (Frechilla et al. 1999). On the other hand, mutants defective in cryptochromes 1 or 2 or phototropin 1 have a normal response (). Howevere, there is also evidence that there are several independent blue-light channels for stomatal regulation (Lasceve et al. 1999), and that violaxanthin is only one of the sensors.

Violaxanthin has nine double bonds, so there there are many possibilities for cis-trans isomerisations. Such a photoisomerisation has not been directly shown, but postulated from kinetic experiments and the fact that the blue-light effect can be reversed by green light (Iino et al. 1985, Frechilla et al. 2000). The reversal spectrum has peaks at 490, 540, and 580 nm, similar to a wavelength-shifted zeaxanthin spectrum. We would like in this context to remind of several experiments in the 1960's and 70's in which blue-light effects in algae were reversed by light of longer wavelength. For instance, inhibition of cell division in *Protosiphon botryoides* by 430 nm light could be most efficiently reversed by light of 589 nm wavelength (Carroll et al. 1970). At that time the effect was ascribed to an interaction between a flavoprotein and the copper-containing chloroplast pigment plastocyanin (see Björn 1979 for a review).

3. OTHER TYPES OF PHOTOSENSORS

3.1. Cryptochrome

The term cryptochrome has been in use for a long time in plant physiology, as a name for the unknown blue-light photoreceptor. The name derives from the fact that it was hiding for such a long time. Now it is known that there are at least two quite different types of blue-light receptors in plants, called cryptochromes (cry1 and cry2) and phototropin. Thus the term has acquired a more restricted meaning than it used to have. On the other hand it has recently been discovered that chromoproteins similar to the plant cryptochromes are present also in other organisms than plants, including humans. These proteins are also called cryptochromes, but may have arisen independently during evolution (Todo et al. 1996). Recent reviews covering both cryptochromes and phototropins are provided by Lin (2000) and Christie and Briggs (2001).

In plants cry2 represses phytochrome B action and plays a role in photoperiodism. cry1 regulates the period of the biological clock and is involved in the entrainment of the circadian oscillator (see Christie & Briggs 2001 for references).

Cryptochromes have two chromophores: 5,10-methenyltetrahydrofolic acid (Fig. 13) and flavin-adenin-dinucleotide (FAD). The latter is non-covalently bound to the protein. The role of the former is not known, perhaps it acts as an antenna pigment (in analogy with antenna pigments in photosynthesis) and transfers absorbed light energy to FAD, as the case is for the related photolyases.

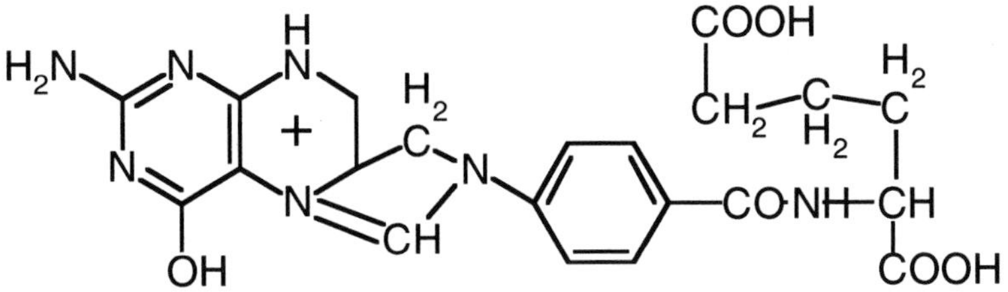

Figure 13. N5,N10-methenyl-5,6,7,8-tetrahydrofolic acid, one of the chromophores in cryptochrome, acting as an antenna pigment for the other chromophore, FAD (see Fig. 14).

The FAD part undergoes at least partial reduction upon illumination. The semiquinone formed by light action on cry1 has a long lifetime, and seems to be able to act as a chromophore, too, giving the cryptochrome a sensitivity to green light under some circumstances. It has long been known that for various blue-light effects in plants and fungi, in experiments designed to determine action spectra, are not parallel in log fluence vs effect diagrams (see, e.g., Shropshire and Withrow (1958). The explanation may be this light sensitivity of the semiquinone, or the participation of several photoreceptors (such a the two cryptochromes, phototropin, and phytochrome) in the effects studied.

The signal transduction chains associated with cryptochromes have been difficult to elucidate, not only because is more than one cryptochrome which probably act in diffferent ways, but because plants have also another blue-light receptor (phototropin), and because there are interactions with phytochrome. However, recently one signalling pathway proved surprisingly simple (Wang et al. 2001). In the dark, a protein called COP1 is present in the nucleus prevents the activity of several genes by preventing the action of their transcription factors. After photoactivation of cryptochromes their conformation is changed so they can bind to COP1 and prevent its action, thereby activating the genes.

Apart from their roles in the cell nucleus related to rhytmicity and gene regulation, cryptochromes seem also to have direct effects on membranes. Thus cry1 activates an anion channel in the cell membrane, and thereby influences the membrane potential. As for the mechanism of action there is, so far, hardly more than speculation. Flavins are known to mediate light driven electron transfer in other cases, but this has not been shown for cryptochromes. One indication for a role of electron transfer is the similarity between cryptochromes and photolyases. Merrow and Roenneberg (2001) speculate about relations between redox potential, cryptochromes and the mechanism of the circadian oscillator.

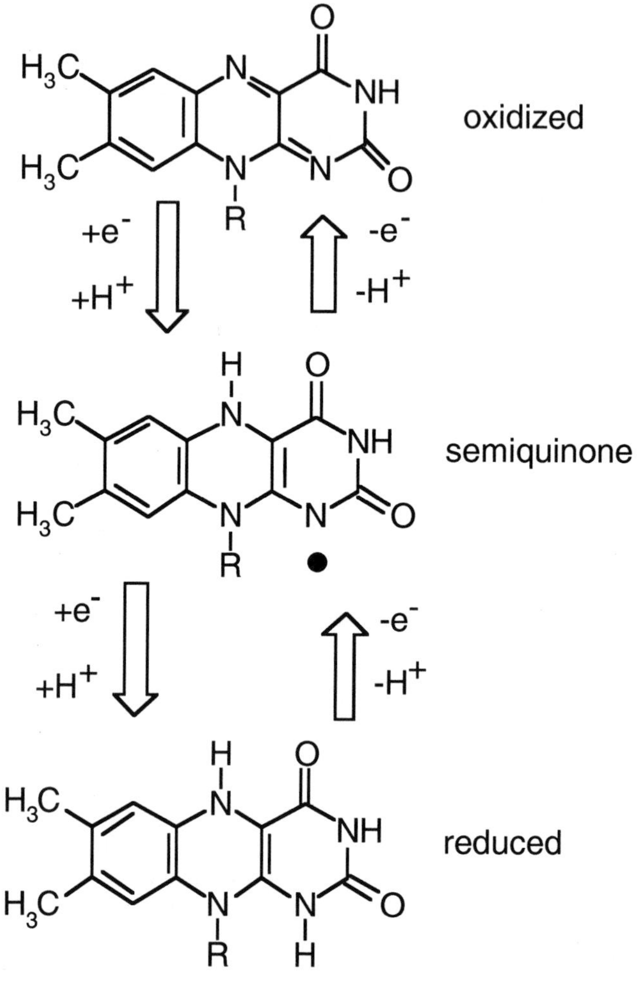

R= ribose-phosphate (FMN)
R=ribose-phosphate-phosphate-ribose-adenine (FAD)

Figure 14. The structures of FMN and FAD in oxidised and reduced forms, and in the half-reduced (semiquinone) form .

3.2. Phototropin

Phototropin is the photoreceptor primarily involved in plant phototropism, the phenomenon which, beginning with Darwin, has meant so much for stimulating interest research in plant photobiology and about plant hormones. However, also cryptochromes and phytochrome are involved in the very complex phenomenon of phototropism (Galland 2001, Iino 2001). On the other hand, phototropin is involved also in other blue-light reactions, such as high- and low-light induced chloroplast movements (Sakai et al. 2001, Jarillo et al. 2001, Kagawa et al. 2001) and inhibition of hypocotyl extension growth (Folta & Spalding 2001). So far two main types of phototropin, phot1 and phot2 (Briggs et al. 2001) have been identified, of which phot1 acts primarily in phototropism, phot2 primarily in chloroplast photomovement.

Phototropin, like the cryptochromes, is a flavoprotein, and it also has two chromophores per molecule. In the phototropin, however, both chromophores consist of covalently bound flavine mononucleotide (FMN, Fig. 14). The protein part is quite different from that of the cryptochromes. The chromophore binding regions are so-called PAS domains, designated LOV1 and LOV2 (Salomon et al. 2000, Christie & Briggs 2001). LOV stands for **L**ight, **O**xygen, or **V**oltage regulated. The properties of these two domains can be investigated separately using molecular biology techniques. The absorption spectra of both show a striking (and expected) similarity to the action spectra for phototropism determined long ago (Fig. 15).

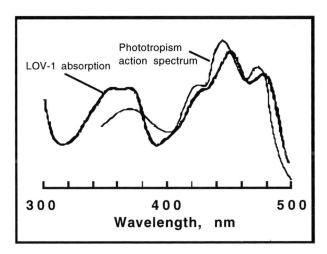

Figure 15. Comparison of the action spectrum for phototropism of oat coleoptiles determined by Thimann & Curry (1961) with the absorption spectrum of the LOV-1 protein domain of phototropin of an oat mutant determined by Salomon et al. (2000). The spectra for both LOV1 and LOV2 of the native oat are very similar, although not identical.

Upon illumination the chromophores, both in the LOV1 and LOV2 domains, undergo a spectral shift (Fig. 16); the absorbance in the blue region decreases, while that in the UV-A part of the spectrum changes only little. Illumination also causes quenching of the fluorescence. In darkness both LOV1 and LOV2 return to the original state with half-times at rom temperature of 11.5 s and 27 s, respectively. By aminoacidsubstitution it has been made likely that the spectral change is caused by the formation of a bond between cystein (Cys39) and the FMN chromophore, with simultaneous reduction of the flavin. Later this has been confirmed by other methods (Crosson & Moffat 2001). The C-terminal end of phototropin is a serine-threonine kinase, which supposedly is activated by a conformational change resulting from the light-induced change in the chromophore region. The detailed structure of the LOV2 domain has now been determined and compared to other PAS domains (Crosson & Moffat 2001).

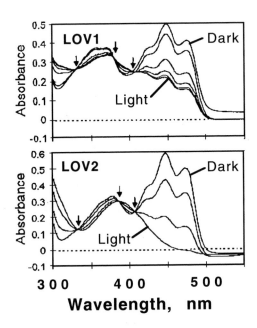

Figure 16. Light induced spectral changes taking place in the LOV1 and LOV2 domains of phototropin upon illumination. The arrows point to isosbestic points (wavelength positions with unchanged absorbance).

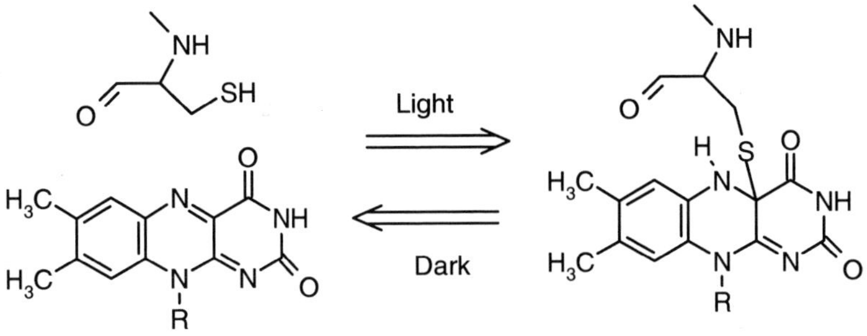

Figure 17. The proposed light induced, dark reversible bonding between FMN and the cysteinyl residue in phototropin. Redrawn and simplified from Crosson & Moffat (2001).

Red light effects and blue light effects in plants have traditionally been investigated by different sets of researchers, and there have been red light meetings and blue light meetings. It has become more and more difficult to uphold such a segregation, as more and more interactions between the signalling channels have been discovered (Mohr 1994, Neff & Chory 1998, Parks et al. 2001). Nozue et al. (1998) have even found a protein in a fern, which possesses both phytochrome and phototropin properties.

3.3. The Plant UV-B Receptor

Ultraviolet-B radiation (280-315 nm) affects plants and other organisms in many destructive and inhibitory ways (Chapter 11). Plants, however, exhibit also regulatory effects of UV-B. The most studied effect is the stimulation of UV-B protecting pigments, or processes which are related to this stimulation. In many cases this stimulation (or induction) is specifically achieved by UV-B radiation, and radiation of longer wavelength is ineffective. This indicates an ability of plants to perceive and react to UV-B radiation. The topic has been reviewed recently by Björn (1999), Brosché (2001) and Kalbin (2001).

Although pigment induction is the most thoroughly studied aspect, there are also UV-B specific photomorphogenetic effects, in particular inhibition of extension growth. Action spectroscopy reveals that the various ultraviolet effects on plants fall into two main categories. Either the efficiency per photon of the radiation increases monotonically towards shorter wavelength, into the UV-C region, or the effectiveness peaks at about 295 nm. It is the latter type of effect that is ascribed to a specific UV-B receptor. The effects increasing into the UV-C region, on the other hand, are regarded as stress effects caused by damage to DNA and other cell components.

Fig. 13 shows examples of action spectra for regulatory UV-B effects, including pigment formation, gene activation, and inhibition of extension growth.

Although much effort has been expended, the nature of the UV-B receptor is not yet known. Flavoproteins and phytochrome have been suggested, but in those cases when the effects are strictly UV specific, these pigments can hardly be active.

Another proposition is pterins, and certainly pterins can be found having absorption spectra fitting the UV-B action spectra well (Fig. 18). A fact increasing the plausibility of a pterin as UV-B receptor chromophore is the fact that they occur in cryptochromes, and also in the related photolyases (Chapter 11). But here they do not have a photochemical role of their own; they act only as antenna pigments absorbing light and transferring the energy to another chromophore. In any case, there is no hard evidence for the participation of pterins as UV-B receptor chromophores.

Another proposition is that an aromatic amino acid residue in a protein gives the spectral signature typical of UV-B regulatory effects. Several amino acids (tyrosine, phenylalanine, tryptophane, histidine) have absorption bands close enough that one can not exclude that spectral tuning (Chapter 7) could bring the peak to 295 nm.

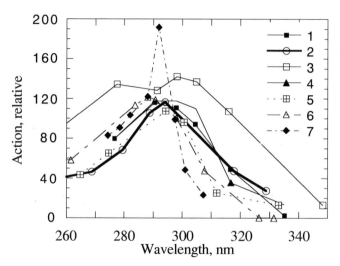

Figure 18. Action spectra for various processes thought to be mediated by a special UV-B receptor: 1. Induction of synthesis of flavonoid glycosides in cell cultures of parsley (Wellmann 1975 as cited by Wellmann 1983). 2. The same as 1 but according to Wellmann (1994). 3. Induction of anthocyanin synthesis in maize coleoptiles (Beggs and Wellmann 1985). 4. The sama as 3 but according to Wellmann (1983). 5. Inhibition of hypocotyl elongation in tomato (Ballaré et al. 1995). 6. Indution of anthocyanin formation in Sorghum bicolor in a background of red light (Yatsuhashi et al. 1982).

A third possible type of UV-B receptor chromophore is provitamin D (D_2 or D_3) or related compound. These compounds have absorption peaks at the right position, but also at shorter wavelength. It is possible that in vivo the short-wavelength absorption bands are hidden by other absorbing compounds, as is the case for the conversion of provitamin D_3 to previtamin D_3 in human skin (Fig. 19). It should also be noted, that other compounds with the same ring and pi-electron structure, and the same absorption spectra and photochemical properties are present as intermediates in plants, for instance in the biosynthetic pathway leading to the brassinosteroid group of plant hormones.

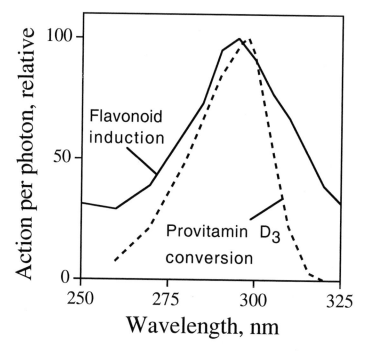

Figure 19. Action spectra for photochemical conversion of provitamin D₃ to previtamin D₃ in human skin (redrawn after MacLaughlin et al. 1982) and for induction of flavonoid biosynthesis in plant cell cultures (redrawn after Beggs and Wellmann 1994).

Since it is known that the UV-B receptor regulates genes, there have been attempts to trace UV-B reception from this end. The UV-B signalling pathway seems to depend on species, and on what gene or other endpoint is studied (see references in Brosché 2001, Frohnmeyer et al. 1997, 1998, 1999). For upregulation of the gene for chalcone synthase in *Arabidopsis* and parsley the following are components of the signal transduction chain: Increase in cytosolic calcium ion concentration, calmodulin, serine/threonine kinase activity (as for phytochrome) and synthesis of new protein. A single millisecond flash of UV-B is sufficient to increase calcium ion concentration in the cytosol (Frohnmeyer et al. 1999). In soybean there is no indication of serine/threonine kinase involvement. For regulation of another gene (for "early light inducible protein") by UV-B none of the above seem to be involved.

REFERENCES

Assmann, S.M. & Wang, X.-Q. (2001). From milliseconds to millions of years: guard cells and environmental responses. *Curr. Opinion Plant Biol., 4,*421-428.

Ballaré, C.L., Barnes, P.W. & Flint, S.D. (1995). Inhibition of hypocotyl elongation by ultraviolet-B radiation in de-etiolating tomato seedlings. I. The photoreceptor. *Physiol. Plant., 93,*584-592.

Ballario, P. & Macino, G. (1997). White-collar proteins: PASsing the light signal in *Neurospora crassa*. *Trends Microbiol., 5,* 458-462.

Barinaga, M. (2002). How the brain's clock gets daily enlightenment. *Science, 295,* 955-957.

Beale, S.I. (1999). Enzymes of chlorophyll biosynthesis. *Photosynthesis Res. 60,* 43-73.

Beggs, C.J. & Wellmann, E. (1985). Analysis of light-controlled anthocyanin formation in coleoptiles of *Zea mays* L.: The role of UV-B, blue, red and far-red light. *Photochem. Photobiol., 41,* 481-486.

Beggs, C.J. & Wellmann, E. (1994). Photocontrol of flavonoid synthesis. In Kendrick, R.E. & Kronenberg, G.H.M. (eds) *Photomorphogenesis in plants*, 2nd ed., pp. 733-751. Dordrecht: Kluwer Academic Publishers.

Berson, D.M., Dunn, F.A. & Takao, M. (2002). Phototransduction by retinal ganglion cells that set the circadian clock. *Science 295,* 1070-1073.

Bhoo, S.-H., Davis, S.D., Walker, J., Karniol, B. & Vierstra, R.D. (2001). Bacteriophytochromes are photochromic histidine kinasis using a biliverdin chromophore. *Nature, 414,* 776-779.

Björn, G.S. (1980). *Photoreversibly photochromic pigments from blue-green algae.* Diss. Lund University. LUNBDS/(NBFB-1009/1-28/(1980).

Björn, G.S. & Björn, L.O. (1978). Action spectra for conversions of phytochrome c from *Nostoc muscorum*. *Physiol. Plant., 43,* 195-200.

Björn, L.O. (1970). Photoreversibly photochromic pigments in organisms: properties and role in biological light perception. *Quart. Revs Biophys., 12,* 1-23.

Björn, L.O. (1984). Light-induced linear dichroism in photoreversibly photochromic sensor pigments. - V. Reinterpretation of the experiments on *in vivo* action dichroism of phytochrome. *Physiol. Plant., 60,* 369-372.

Björn, L.O. (1999). UV-B effects: Receptors and targets. In Singhal, G.S., Renger, G., Sopory, S.K., Irrgang, K.-D., & Govindjee (eds) *Concepts in photobiology: Photosynthesis and photomorphogenesis* (pp.821-832). New Delhi: Narosa Publishing House.

Boylan, M.T. & Quail, P. H. (1996). Are the phytochromes protein kinases? *Protoplasma 195,* 12-17.

Briggs, W.R., Beck, C.F., Cashmore, A.R. et al . (17 authors) (2001). The phototropin family of photoreceptors. *Plant Cell, 13,* 993-997.

Brosché, M. 2001. *Deconstruction of a plant UV-B stress response.* Diss. Göteborg University. pp. 60 . ISBN 91-628-4606-X.

Christie, J.M. & Briggs, W.R. (2001). Blue light sensing in higher plants. *J. Biol. Chem., 276,* 11457-11460.

Crosson, S. & Moffat, K. (2001). structure of a flavin-binding plant photoreceptor domain: Insights into light mediated signal transduction. *Proc. Natl Acad. Sci. USA, 98,* 2995-3000.

Davis, S.J., Vener, A.V. & Vierstra, R.D. (1999). Bacteriophytochromes: Photochrome-like photoreceptors from nonphotosynthetic eubacteria. *Science, 286,* 2517-2520.

Diakoff, S. & Scheibe, J. (1973). Action spectra for chromatic adaptation in *Tolypothrix tenuis*. *Plant Physiol., 51,* 382-385.

Etzold, H. (1965). Der Polarotropismus und Phototropismus der Chloronemen von *Dryopteris filix-mas* (L.) Schott. *Planta, 64,* 254-280.

Folta, K.M. & Spalding, E.P. (2001). Unexpected roles for cryptochrome2 and phototropin revealed by high-resolution analysis of blue ligh-mediated hypocotyl growth inhibition. *Plant J., 26,* 471-478.

Foster, K.W., Saranak, J., Patel, N., Zarilli, G., Okabe, M., Kline, T. & Nakashini, K. (1984). A rhodopsin is the functioning photoreceptor for phototaxis in the unicellular eukaryote *Chlamydomonas*. *Nature, 311,* 756-759.

Frohnmeyer, H., Bowler, C. & Schäfer, E. (1997). Evidence for some signal transduction elements involved in UV-light-dependent responses in parsley protoplasts. *J. Exp. Bot. 48,* 739-750.

Frohnmeyer, H., Bowler, C., Zhu, J.-K., Yamagata, H., Schäfer, E. & Chua, N.-H. (1998). different roles for calcium and calmodulin in phytochrome- and UV-regulated expression of chalcone synthase. *Plant J., 13,* 763-77.

Frohnmeyer, H., Loyall, L., Blatt, M.R. & Grabov, A. (1999). Millisecond UV-B irradiation evokes prolonged elevation of cytosolic-free Ca^{2+} and stimulates gene expression in transgenic parsley cell cultures. *Plant J., 20*, 109-117.

Fujita, Y. & Hattori, A. (1962). Photochemical interconversion between precursors of phycobilin chromoprotein in *Tolypothrix tenuis*. *Plant Cell Physiol. 3*, 209-220.

Galland, P. (2001). Phototropism in *Phycomyces*.. In Häder, D.-P. & Lebert, M. (eds) *Photomovement* (pp. 621-657). Amsterdam: Elsevier. ISBN 0-444-50706-X.

Genick, U.K., Borgstrahl, G.E.O., Kingman, N., Ren, Z., Pradervand, C., Burke, P.M., Srajer, V., Teng, T.-Y., Schildkamp, W., McRee, D.E., Moffat, K. & Getzoff, E.D. (1997). Structure of a protein cycle intermediate by millisecond time-resolved crystallography. *Science, 275*, 1471-1475.

Genick, U.K., Soltis, S.M., Kuhn, P., Canestrelli, I.L. & Getzoff, E.D. (1998). Structure at 0.85 Å resolution of an early protein photocycle intermediate. *Nature, 392*, 206-209.

Gilroy, S. & Trewavas, A. (2001). Signal processing and transduction in plant cells: The end of a beginning? Nature Revs Molec., *Cell. Biol. 2*, 307-314.

Grossman, A.R. , Bhaya, D. 6 He, Q. (2001). Tracking the light environment by cyanobacteria and the dynamic nature of light harvesting. *J. Biol. Chem., 276*, 11449-11452.

Gualtieri, P. 2001. Rhodopsin-like proteins: Light detection pigments in *Leptolyngbya, Euglena, Ochromonas, Pelvetia*. In Häder, D.-P. & Lebert, M. (eds) *Photomovement* (pp. 281-295). Amsterdam: Elsevier. ISBN 0-444-50706-X.

Hansson, K.M., Li, B. & Simon, J.D. (1997). A spectroscopic study of the epidermal ultraviolet chromophore *trans*-urocanic acid. *J. Am. Chem. Soc., 119*, 2715-2721.

Hartmann, U., Valentine, W.J., Christie, J.M., Hays, J., Jenkins, G.I. & Weisshaar, B. (1998). Identification of UV/blue light-responsive elements in the *Arabidopsis thaliana* chalcone synthasepromoter using a homologous protoplast transient expression system. *Plant Molecul. Biol., 36*, 741-754.

Hattar, S., Liao, H.-W., Takao, M., Berson, D.M. & Yau, K.-W. (2002). Melanopsin-containing retinal ganglion cells: Architecture, projections, and intrinsic photosensitivity. *Science 295*, 1065-1070..

Haupt, W. (1970). Localization of phytochrome in the cell. *Physiol. Vég., 8*, 551-563.

Hashimoto, T., Shichijo, C. & Yatsuhashi, H. (1991). Ultraviolet action spectra for the induction and inhibition of anthocyanin synthesis in broom sorghum seedlings. *J. Photochem. Photobiol. B.: Biol., 11*, 353-363.

Hegemann, P. & Deininger, W. (2001). Algal eyes and their rhodopsin photoreceptors. In Häder, D.-P. & Lebert, M. (eds) *Photomovement* (pp. 475-503). Amsterdam: Elsevier. ISBN 0-444-50706-X.

Herdman, M., Coursin, T., Rippka, R., Houmard, J. & Tandeau de Marsac, N. (2000). A new appraisal of the prokaryotic origin on eukaryotic phytochromes. *J. Mol. Evol., 51*, 205-213.

Hoff, W.D., Jung, K.-H. and Spudich (1997). Molecular mechanism of photosignaling by archaeal sensory rhodopsins. *Annu. Rev. Biophys. Biomol. Struct., 26*, 223-258.

Hubschmann, T., Borner, T., Hartmann, E. & Lamparter, T. (2001). Characterization of the Cph1 holo-phytochrome from Synechocystis sp. PCC 6803. *Eur. J. Biochem., 268*, 2055-2063.

Iino, M. (2001). Phototropism in higher plants. In Häder, D.-P. & Lebert, M. (eds) *Photomovement* (pp. 659-812). Amsterdam: Elsevier. ISBN 0-444-50706-X.

Jiang, Z.Y., Swem, L.R., Rushing, B.G., Devanathan, S., Tollin, G. & Bauer, C.E. (1999). Bacterial photoreceptor with similarity to photoactive yellow protein and plant phytochromes. *Science, 285*, 406-409.

Jones, A.M. & Quail, P. (1986). Quaternary structure of 124-kilodalton phytochrome from *Avena sativa* L. *Biochemistry, 25*, 2987-2995.

Kagawa, T., Sakai, T., Suetsugu, N., Oikawa, K., Ishiguru, S., Kato, T., Tabata, S., Okada, K. & Wada, M. (2001). *Arabidopsis* NPL1: A phototropin homolog controlling the chloroplast high-light avoidance response. *Science, 291*, 2138-2141.

Kalbin, G. (2001). *Towards the understanding of biochemical plant responses to UV-B*. Diss. Göteborg University. ISBN 91-628-4627-2.

Kehoe, D.M. & Grossman, A.R. (1996). Similarity of a chromatic adaptation sensor to phytochrome and ethylene receptors. *Science, 273*, 1409-1412.

Lamparter, T. & Marwan, W. (2001). Spectroscopic detection of a phytochrome-like photoreceptor in the myxomycete *Physarum polycephalum* and the kinetic mechanism for the photocontrol of sporulation by Pfr. *Photochem. Photobiol., 73*, 697-702.

Laudet, V. (1997). Evolution of the nuclear receptor superfamily: early diversification from an ancestral orphan receptor. *J. Molec. Biol., 19*, 207-226.

Lazaroff, N. (1973). Photomorphogenesis and Nostocacean development. *In* Carr, N.G. & Whitton, B.A. (eds) *Biology of blue-green algae* (Botanical Monographs, vol. 9). Oxford: Blackwell Scientific Publications.

Lazaroff, N. & Schiff, J. (1962). Action spectrum for developmental photoinduction of the blue-green alga *Nostoc muscorum. Science, 137*, 603-604.

Lenci, F. Ghetti, F. & Song, P.-S. (2001). Photomovement in ciliates. In Häder, D.-P. & Lebert, M. (eds) *Photomovement* (pp. 281-295). Amsterdam: Elsevier. ISBN 0-444-50706-X.

Li, B., Hanson, K.M. & Simon, J.D. (1997). Primary processes of the electronic excited states of trans-urocanic acid. *Phys. Chem. A., 101*, 969-972.

Lin, C., Robertson, D.E., Ahmad, M., Raibekas, A.A., Schuman Jorns, M., Dutton, P.L. & Cashmore, A.R. (1995). Association of flavin adenine dinucleotide with the *Arabidopsis* blue light receptor CRY1. *Science, 269*, 968-970.

Lin, C. (2000). Plant blue-light receptors. *Trends Plant Sci., 5*, 337-342.

Merrow. M. & Roenneberg, T. (2001). Circadian clocks: Running on redox. *Cell, 106*, 141-143.

Mohr, H. (1994). Coaction between pigment systems. In Kendrick, R.E. & Kronenberg, G.H.M. (eds) *Photomorphogenesis in plants*, 2nd ed. Dordrecht: Kluwer Acad. Publ. ISBN 0-7923-2551-6.

Morrison, H., Avnir, D., Bernasconi, C. & Fagan, G. (1980). Z/E photoisomerisation of urocanic acid. *Photochem. Photobiol., 32*, 711-714.

Morrison, H., Bernasconi, C. & Pandey, G. (1984). A wavelength effect on urocanic acid E/Z photoisomerisation. *Photochem. Photobiol., 40*, 549-550.

Neff, M.M. & Chory, J. (1998). Genetic interactions between phytochrome A, phytochrome B, and cryptochrome 1 during *Arabidopsis* development. *Plant Physiol., 118*, 27-35.

Nozue, K., Kanegae, T., I(maizumi, T., Fukuda,S., Okamoto, H., Yeah, K.-C., Lagarias, J.C. & Wada, M. (1998). A phytochrome from the fern *Adiantum* with features of the putative photoreceptor NPH1. *Proc. Natl Acad. Sci. USA, 95*, 15826-15830.

Okada, T., Ernst, O.P., Palczewski, K. & Hofmann, K.P. (2001). Activation of rhodopsin: new insights from structural and biochemical studies. *Trends Bioch. Sci., 26*, 318-324.

Page, C.S., Merchán, M. & Serrano-Andrés, L. (1999). A theoretical study of the low-lying excited states of trans- and cis-urocanic acid. *J. Phys. Chem. A, 103*, 9864-9871.

Parks, B.M., Folta, K.M. 6 Spalding, E.P. (2001). Photocontrol of stem growth. *Curr. Opinion Plant Biol., 2001*, 436-440.

Pellequeler, J.-L., Wagner-Smith, K.A., Kay, S.A. & Getzoff, E.D. (1998). Photoactive yellow protein: A structural prototype for the three-dimensional fold o the PAS domain superfamily. *Proc. Natl. Acad. Sci. USA, 95*, 5884-5890.

Portwich, A. & Garcia-Pichel, F. (2000). A novel prokaryotic UVB photoreceptor in the cyanobacterium *Chlorogloeopsis* PCC 6912. *Photochem. Photobiol., 71*, 493-498.

Provencio, I., Jiang, G., De Grip, W.J., Hayes, W.P. & Rollag, M.D. (1998). Melanopsin: An opsin in melanophores, brain, and eye. *Proc. Natl. Acad. Sci. USA, 95*, 340-345

Rospendowski, B.N., Farrens, D.L., Cotton, T.M. & Song, P.-S. (1989). Surface enhanced resonance Raman scattering (SERRS) as a probe of the structural differences between the Pr and Pfr forms of phytochrome. *FEBS Lett., 258*, 1-4.

Ryan, W. & Levy, D.H. (2001). Electronic spectroscopy and photoisomerisation of *trans*-urocanic acid in a supersonic jet. *J. Am. Chem. Soc., 123*, 961-966.

Sage, L.C. (1992). *Pigment of the imagination: a history of phytochrome research*. San Diego: Academic Press. ISBN 0126144451.

Salomon, M., Christie, J.M., Knieb, E., Lempert, U. & Briggs, W.R. (2000). Photochemical and mutational analysis of the FMN-binding domains of the plant blue light receptor, phototropin. *Biochemistry, 39*, 9401-9410.

Scheibe, J. (1962). Photoreversible pigment: occurence in a blue-green alga. *Science, 176*, 1037-1039.

Schmid, G.H. (1970). The effect of blue light on some flavine enzymes, Hoppe Seylers Z. *Physiol. Chem., 351*, 575-578.

Schmid, G.H. & Schwarze, P. (1969). Blue light enhanced respiration in a colorless *Chlorella* mutant, *Hoppe Seylers Z. Physiol. Chem., 350*, 1513-1520.

Schneider-Poetsch, H.A.W., Kolukisaoglu, U., Clapham, D.H., Hughes, J. & Lamparter, T. (1998). Non-angiosperm phytochromes and the evolution of vascular plants. *Physiol. Plant., 102*, 612-622.

Shropshire, W. & Withrow, R.B. (1958). Action spectrum of phototropic tip-curvature of *Avena*. *Plant Physiol., 33*, 360-366.

Sineshchekov, V.A. (1995). Photobiophysics and photobiochemistry of the heterogeneous phytochrome system. *Biochim. Biophys. Acta, 1228*, 125-164.

Spudich, J.L. (2001). Color-sensitive vision by halobacteria. In Häder, D.-P. & Lebert, M. (eds*) Photomovement* (pp. 151-178). Amsterdam: Elsevier. ISBN 0-444-50706-X.

Sundqvist, D. & Björn, L.O. (1983a). Light-induced linear dichroism in photoreversibly photochromic sensor pigments. -II. Chromophore rotation in immobilized phytochrome. *Photochem. Photobiol. 37*, 69-75.

Sundqvist, D. & Björn, L.O. (1983b). Light-induced linear dichroism in photoreversibly photochromic sensor pigments. -III. Chromophore rotation estimated by polarized light reversal of dichroism. *Physiol. Plant., 59*, 263-269.

Takeda, J., Ozeki, Y. & Yoshida, K. (1997). An action spectrum for induction of promoter activity of phenylammonia lyase gene by UV in carrot suspension cells. *Photochem. Photobiol., 66*, 464-470.

Taylor, R.R. & Zhulin, I.B. (1999). PAS domains: Internal sensors of oxygen, redox potential, and light. Microbiol. *Molecular Biol. Revs, 63*, 479-506.

Thimann, K.V. & Curry, G.M. (1961). Phototropism. In McElroy, W.D. & Glass, B. (eds) *Light and life* (pp. 646-672). Baltimore: Johns Hopkins Press. Library of Congress Catalog Card Number 60-16544.

Todo, T., Ryo, H., Yamamoto, K., Toh, H., Inui, T., Ayaki, H., Nomura, T. & Ikenaga, M. (1996). *Science, 272*, 109-112.

Tokutomi, S. & Mimuro, M. (1989). Orientation of the chromophore transition moment in the 4-leaved shape model for pea phytochrome molecule in red-light absorbing form and its rotation induced by the phototransformation to the far-red-light absorbing form. *FEBS Lett., 255*, 350-353.

Wade, H.K., Bibikova, T.N., Valentine, W.J. & Jenkins, G.I. (2001). Interactions within a network of phytochrome, cryptochrome and UV-B transduction pathways regulate chalcone synthase gene expression in *Arabidopsis* leaf tissue. *Plant. J., 25*, 675-685.

Wang, H., Ma, L.G., Li, J.M., Zhao, H.Y. & Deng, W.W. (2001). Direct interaction of *Arabidopsis* cryptochromes with COP1 in mediation of photomorphogenic development. *Science, 294*, 154-158.

Watson, J.C. (2000). Light and protein kinases. Adv. Botanical Res. Incorporating Adv. *Plant Pathol., 32*, 149-184.

Wellmann, E. (1975). Der Einfluss physiologischer UV-Dosen auf Wachstum und Pigmentierung von Umbelliferenkeimlingen. In Bacher, E. (ed.) *Industrieller Pflanzenbau* (pp. 229-239). Tech. Univ. Wien Selbstverlag.

Wellmann, E. (1983). UV radiation in Photomorphogenesis. In Shropshire Jr, W. & Mohr, H. (eds) *Enc. Plant Physiol., New Series 16B* (pp. 745-756). Springer Verlag.

Yatsuhashi, H., Hashimoto, T. & Shimizu, S. (1982). Ultraviolet action spectrum for anthocyanin formation in broom *Sorghum* first internodes. *Plant Physiol., 70*, 735-741.

Zeiger, E. (2000). Sensory transduction of blue light in guard cells. *Trends Plant Sci., 5*, 183-185.

G. ADRIAN HORRIDGE

9. THE DESIGN OF THE COMPOUND EYE
DEPENDS ON THE PHYSICS OF LIGHT

1. INTRODUCTION

This chapter considers the design of the insect compound eye as an optical device, taking advantage of techniques and ideas from the physical sciences. As equipment for optical, electrical and chemical analysis was developed in the 20th century, it became possible to discover how eyes actually work. The nervous system was revealed on the one hand as a vast assortment of neurons with individual specificities and responses, and on the other as an integrated system of circuits in parallel that can be understood only as a whole in context during normal behaviour. Even so, advance is limited by technique; for example, little is known about the responses of receptor cells that move under the influence of light.

In the visual system, all the components can actually be listed, and simple physics such as the properties of light can be most easily applied. Mechanistic analysis is possible where the response is closely related to the input. The best examples are the sensitivities of receptors. Only later it was realized that progress is difficult where numerous receptors are simultaneously active in different ways in parallel, but the interest lies in their joint action. This parallel activity in the retina has three consequences for analysis. First, analysis of one part is *valid only when the effects of other parts in parallel are excluded or measured.* Secondly, at each successive layer, *the signals of relevance to the next stage can be found only by direct demonstration that they are effective.* Thirdly, to understand the design of the parallel structures in an eye, the analysis must be of *the action of the whole and of the components at the same time.*

As every engineer knows, there is no discrepancy between the holistic approach towards the design of a complete functional structure that will operate in a certain way and the exact analysis of the properties of its components. The two approaches can be fitted together when the components and the working system are studied from every point of view. So you will find in this account, on the one hand, that minute details are related to ecological tasks. For example, the orientation of the visual pigment molecules enables a flying aquatic insect to locate a lake that is reflected in skylight. On the other hand, a property of the whole organ, eye size, is related to the wavelength of light and the luminance of the sun.

The compound eye is composed of ommatidia, each of which consists of a cuticular cornea, a transparent cone and receptor cells (usually 8). The actual receptor is a column of membranes called the rhabdom, composed of rhabdomeres formed by these cells. It is folded in the form of microvilli and stuffed with

L.O. Björn (ed.), Photobiology, 181–218.

rhodopsin molecules that absorb the light. Like camera-type eyes, insect compound eyes are organs for *sampling the visual world in angular co-ordinates* (Fig. 1). *Compromises between resolution and sensitivity exist at every level in the visual system,* relating to sensitivity, range of responsiveness, field size, field overlap, and the responses of the neurons at the next stage. The receptors that lie side by side in the retina and map the outside world are like arrays of neurons that map any sensory input and pass on the excitation centrally. Vision, being the best known sense, is therefore a convenient model of how neurons function in parallel arrays, and we can truly say that the eye is the window on the brain. To understand how information is processed in parallel and passed to the next level, we examine the insect retina in detail.

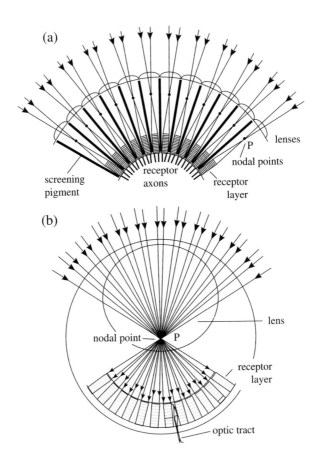

Figure 1. An eye is an array of photoreceptors with optical axes that diverge in angular space. (a) Compound eye. (b) Camera-type eye. The nodal point is the point through which light rays pass when they are not bent by the optics.

2. SINGLE PHOTORECEPTORS

2,.1. The anatomy of light capture

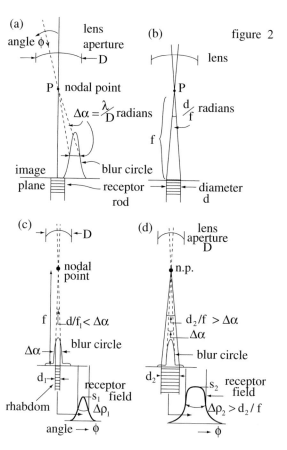

Figure 2. The match between optics and anatomy in the formation of the receptor field. (a) With a point source, a lens of aperture D forms a diffraction image (called a blur circle) of angular width $\Delta\alpha = \lambda /D$ radians at the 50% level of intensity in the focal plane of the lens, subtended at the nodal point of the lens P. As the point source is moved outside the eye, the blur circle moves across the receptor profile. (b) The receptor of width d μ m in the focal plane of the lens subtends an angle of d/f radians at the nodal point. (c) The convolution of the blur circle with the absorption profile of the receptor generates the receptor field, of angular width $\Delta\rho$ at the 50% contour. For narrow receptors d_1, this reduces to $\Delta\rho_1 = \lambda /D$ when d_1/f is negligible, and the sensitivity s_1 is sub-optimal. (d) For wide receptors $\Delta\rho_2 \approx d_2/f$ when $\Delta\alpha$ is negligible.

Let us strip down to its essentials the action of a single lens focused upon a single receptor. In a single ommatidium of the insect eye, the image of a distant point of light on axis is a bright patch called a *blur circle* (Fig. 2a). The spherical and chromatic aberrations are negligible with a small lens, and the blur circle (sometimes called an *Airy disc* when well focused) has an intensity distribution determined by diffraction theory and found accurately in tables of Bessel functions. To avoid this complication, the distribution of intensity in the blur circle is approximated by a Gaussian function, which simplifies the algebra but omits the weak bright halo around the edge. A convenient approximation in the theory below is that the angular diameter of the blur circle at its 50% intensity contour is $\Delta\alpha = \lambda/D$ radians (Fig. 2a). The reciprocal of λ/D is the *lens resolution* D/λ.

The optimum receptor that catches the blur circle is as narrow as possible to make use of the lens resolution but wide enough to catch as much as possible of the light focused on it. To achieve the best compromise in the match between the blur circle (Fig. 2a) and the receptor size (Fig. 2b), *the cross-sectional area of the receptor for photon capture must match the distribution of photons in the blur circle.* Larger receptors (Fig. 2d) subtend a larger angle in the outside world and therefore catch more light, but throw away lens resolution by making the image grainy, exactly as happens with a fast film, however good the camera lens. Receptors subtending less than λ/D radians in diameter (Fig. 2c), like extra-fine film in a cheap camera, simply throw away sensitivity with no extra gain in lens resolution, and are rare in compound eyes, although they occur in the crowded foveas of camera-type eyes.

Because light is absorbed by the visual pigment at a rate of only about 0.5% per μm along its length, the rhabdom is frequently a long rod with the incoming light focused on its distal tip. This is the reason that we have rods in our own eyes. To catch rays optimally, the rod must point directly at the nodal point of the lens, which explains the geometry of elongated photoreceptors in most eyes.

The rhabdom rods act as separate light guides because they are isolated from each other by a medium of lower refractive index. When light guides are about 1-2 μm (see below) their capture-cross-section for light can be approximated by a Gaussian distribution of diameter d at the 50% level of sensitivity, where d is the diameter of the receptor rod. Therefore, *to match the resolution of the lens to the capture cross-section of the receptor*, we have

$$\lambda/D = d/f \quad \text{............} \quad (1)$$

where f is the focal length, measured from the tip of the receptor (at the focal plane) to the nodal point of the lens (Fig. 2).

The *nodal point* is defined as the point through which rays pass as straight lines through the lens (Figs 1 and 2). The power of the lens of most non-aquatic insect eyes lies in the curvature of the outside (cornea) surface, all the internal surfaces having much less power, so that the nodal point in an insect ommatidium usually lies near the centre of curvature of the cornea. Aquatic eyes, on the other hand, must also generate lens power within the eye, and consequently they have internal optical structures with strongly non-homogeneous refractive index. Animals that see well in water as well as in air, such as water beetles, often have eyes with a flat cornea like a diver's face mask, and the lens power is all internal.

Modern technology cannot approach the adaptive refinements of natural photon capturing systems. The small size of the anatomical components makes the best use of the wavelength of light. The optical systems in insects have a variety of devices to optimize sensitivity and also to detect the directions of origin of the rays. The molecules of rhodopsin in the photoreceptor membranes in insects capture and convert about half of the absorbed photons, and the photoreceptor rods are sufficiently long to absorb most of the light. Cones of vertebrates are different, because they are short and absorb little in order to retain colour vision (see below).

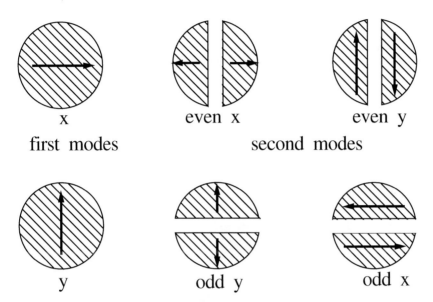

Figure 3. The 1st and 2nd modes of vibration of light in a thin light-guide. These are the commonest modes in diurnal insects with thin rod-shaped rhabdoms, and can be seen by examination of some butterflies with oil on the eye. The arrows show the direction of vibration of the electric vector (polarization). Outside the guide the energy falls to zero within about 1 micron (after Snyder 1975).

2.2. Ray optics and mode optics

If a solid rod is transparent at the wavelength used, and is surrounded by a material with a lower refractive index, electromagnetic radiation can be piped along it. Light is transmitted along a glass or plastic fibre, and radar waves can be piped along a rod of polystyrene. When the rod is thick, one can think of the process as total internal reflection of the waves at the inside of the rod surface. *Classical optics based upon ray tracing and absorption within the rod explains most of the properties of photoreceptors.*

When the diameter of the light pipe approaches the wavelength of the light, the rays bouncing from wall to wall can fit into the pipe in only a limited number of ways, called *modes* (Fig. 3), and the rod is properly called a wave guide. The fraction of the power carried in each mode inside a wave guide is governed by the value of the *mode parameter* V,

$$V = \pi d/\lambda \, (n_1^2 - n_2^2) \dots\dots\dots\dots \quad (2)$$

where d is the rhabdom diameter, λ is the wavelength, and n_1 and n_2 are the refractive indices inside and outside the light guide respectively (Fig. 4). In fact, n_1 and n_2 are difficult to measure and the value of V is very sensitive to their difference. Instead of calculating V from this equation, the first few modes have been observed in the light-adapted state in a few insects, from which the value of V has been inferred to lie between 1.5 and 4.0. The refractive indices are then calculated and it is assumed that rhabdoms of other insects have similar values, so that d and λ become the only variables. The transition between mode optics and ray optics is illustrated in Fig 5. Very approximately, the number of modes in an insect rhabdom turns out to be equal to the diameter in microns up to 3 μm, from which we can calculate that $n_1 - n_2 = 0.05$, so the refractive index of the rhabdom, n_1 is approximately 1.39.

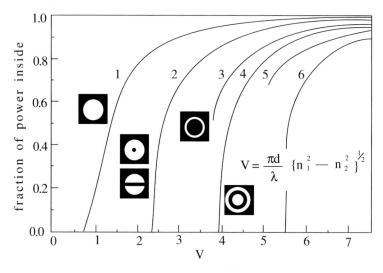

$$V = \frac{\pi d}{\lambda} \{ n_1^2 - n_2^2 \}^{1/2}$$

Figure 4. The fraction (n_j) of light power within the light guide for each of the first six mode types, with an indication of the light power in the cross section of the light guide (after Snyder 1975).

Each mode has a characteristic distribution of light energy in the cross section of the wave guide (Fig. 3). As the mode parameter V is reduced, the fewer the number of modes accepted, until only the first mode remains (Fig. 4). In the extreme situation the receptor rod is so thin that light will travel only longitudinally along its axis. This is called the *1st mode* of vibration. An additional property arises from the fraction of the light energy which lies *outside the surface* of the wave guide. This external energy can leak into neighbouring about 20° wide. This is significant because F numbers range as low as F = f/D = 2. The polarization of the light can also have an effect because the way the second mode fits into the light guide depends on the polarization plane. Therefore the limiting angle of entry may be influenced by the plane of polarization of the light, and by neighbouring screening pigment, as well as by the optics of the lens. The combined effect of these factors can be measured only by direct recording from the individual receptors.

In favourable circumstances, the modes can be directly observed in butterflies' eyes with an epi-illumination microscope, when the corneal curvature is neutralised with a little oil. The rays going down their rhabdom are reflected by the tapetum at the base of the retina and pass a second time through the rhabdom and out of the eye. Under a microscope, when the cornea is optically neutralized with a little oil, the first few modes can be seen at the distal end of the rhabdom. The effect of increase in light intensity can be directly observed as a progressive loss of the third and then the second modes, as screening pigment grains migrate closer to the rhabdom.

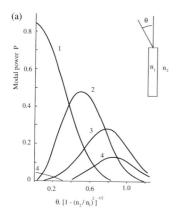 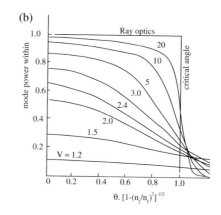

Figure 5. The distribution of power in the different modes. (a) Modal power excited as a function of the angle of incidence θ on the end of the light guide when the mode parameter $V = 5$. The abscissa is scaled as a function of θ so that the critical angle is unity. The numbers on the curves refer to the mode types. Note that axial light excites the first mode but not the second. (b) The total power transmitted within the fibre due to rays that are incident at an angle θ, for different values of the mode parameter V, after summing the contributions of all modes. As V increases, mode optics turn into ray optics, by addition of the curves in (a). (After Snyder 1975).

The dimensions of the rhabdoms in eyes of many large diurnal insects are exactly in the range where the light is controlled inside the retinula cells by radial movements of pigment grains that act like a shutter around the outside of the rhabdom. Migration of these pigment grains is towards the rhabdom in the light, away from it in the dark. As pigment moves to within 1 μm of the rhabdom surface, it bleeds off some of the light energy within. Similarly, a very small constriction of the diaphragm formed by the grains in the principal pigment cells around the cone tip reduces the entry of focused light from the cone to the rhabdom at the focal plane. The resulting closure can be observed like a diaphragm from the outside when the corneal curvature is neutralised with a little oil. The movements of screening pigment are locally controlled by the illumination, not necessarily acting through the receptors,

When mode theory was applied to photoreceptors in the early 1970's, rapid advances led to the generalisation that long narrow photoreceptors of 1 to 2 μm diameter are in the range of mode optics but larger ones obey classical ray optics. The thinnest in small insects, about 1 μm diameter, are in the range where they carry only the 1st mode for green light. More modes can be carried if the receptor is broader, but that throws away some lens resolution (Fig. 2d). In a camera-type eye, photoreceptors must also be as narrow as possible so that as many as possible can be packed in to optimize the spatial resolution. Even in bright light, the photoreceptors of some animals would carry insufficient light energy down their insides if they were any thinner.

A remarkable coincidence is revealed when we calculate the diameter of the receptor rod from two sets of assumptions. On the one hand, from the optics of the eye (Fig. 2), d/f is approximately equal to λ/D, and *f/D is the F number of the lens*, so d = $F\lambda$. Since F ranges from 2 to 4 for insect eyes, and λ is 0.5 μm for green light, *the receptor diameter d is predicted to be between 1 and 2 μm for all of the insect receptors which operate in bright light*. So a photoreceptor between about 1 and 2 μm wide makes full use of the resolution of light. On the other hand, observations reveal, in butterflies for example, that the first and second modes carry most of the power in the photoreceptor rod, therefore V must be between 1 and 3 from the theory of light guides. If n_1 = 1.39 and n_2 = 1.34, then d must be between 2 λ and 4λ, and λ = 0.5 μm. So a photoreceptor between about 1 and 2μm wide is also exactly the right size to receive the light power in such a way that it is controlled by pigment migration outside the rhabdom. In fact, photoreceptor rods that operate in bright daylight are commonly about 2 μm in diameter. The acceptance angle of the rhabdom tip must be large enough for the cone of light reaching it. There are numerous refinements of these principles, for example, muscid flies commonly have the receptor rods thinner at the tips to allow them to be more closely packed, but widening out below to retain the light power. The receptor rods are separate rhabdomeres in flies, but 6-8 fused rhabdomeres in most diurnal large flying insects.

In insects that are active in dim light, when photons are less plentiful, we always find larger cross-sectional areas of individual receptors, so that modes can be ignored and ray optics apply. The receptor field of view, which is the solid angle of diffuse light caught by the receptor, is then the projection of the receptor tip through the nodal point, and is d/f radians wide (Fig. 2d). Large receptor fields (or neural pooling of the outputs of small fields) are consistent features of all eyes that see in dim light. In insects, there are few signs of peripheral receptor pooling, as occurs in vertebrate rods; instead, summation to reduce noise is done optically or by high-level neurons which have large fields, and the local spatial resolution of colour and polarization plane is not sacrificed.

2.3. The field of the receptor

The concept of "field" is fundamental to the analysis of nervous systems. *The field is the plot of all the effective inputs,* in all the dimensions in which they exist. The field must be obtained by tedious exploration, which is why micro-electrode recording is the bottle-neck in the advance of knowledge about receptors and nervous systems. The spatial field of a photoreceptor is defined as *the angular distribution of sensitivity* when a point source is moved around the eye. Sensitivity in this case is defined as the reciprocal of the stimulus intensity required to give a constant response. So, the field may depend on the size of this arbitrary constant response, and on other factors such as polarization plane or wavelength, so that even for a primary receptor the spatial field is dependent on the kind of stimulus. The *optical axis of a receptor* is defined as the axis of symmetry of the spatial field. The responses of the receptors also have important temporal properties which limit the resolution of motion.

The field is measured with a flashing point source at intervals (of 0.1° in practice) in front of the eye (Fig. 6) and recording the heights of the graded electrical responses with a microelectrode. Then the point source is kept stationary near the receptor axis and its intensity at successive flashes is controlled with a series of neutral density filters. The graph of *response versus intensity on axis* is called the

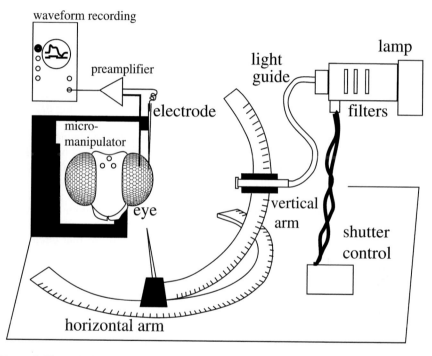

Figure 6. *The method of measuring field sizes of single retinula (receptor) cells. A small source (the end of a quartz light guide) is moved around the eye on a calibrated cardan arm that measures to an accuracy of 0.1° in two coordinates. Flashes of the source at each angle cause responses that are picked up by a microelectrode from a single cell, amplified and recorded. Angular sensitivity is the reciprocal of the number of photons required to give a constant response at each angle on the eye. The light guide can be replaced by a telescope to check from the shape and position of the pseudopupil that the optics of the eye are not damaged and to identify which eye region is stimulated.*

V/log I curve, and this curve acts as a calibration whereby the response measured at each angle is converted to a sensitivity relative to the maximum response on axis. A more accurate way to measure the field is to adjust the intensity of the point source at each angle relative to the axis until a constant response is obtained at each flash, and then plot the reciprocals of these intensities. The two methods would be equivalent as long as the intensity response curve has a constant shape, but this constancy is only approximately true for changes of average intensity, colour, angle, polarization plane and rate of repetition of the flashes. A problem arises because the control of screening pigments may be via a separate pathway with different properties. *The useful measure of the field size is the angular diameter of the angular sensitivity curve at the 50% contour, called the acceptance angle,* Δ ρ (Fig. 2c, d).

2.4. The theoretical acceptance angle.

The structure which catches the light is the cross sectional area of the receptor tip. As the blur circle moves over it, the peripheral end of the receptor accepts progressively more light, up to a maximum when the point source stimulus is on axis, then progressively less again as the blur circle moves off the receptor (Fig. 2a). In this way, the angular sensitivity of the receptor cell is a two dimensional bell-shaped surface which is formed by the *convolution of the blur circle with the absorption by the end of the receptor* (Figs 2c, d). This is only the first of many convolutions on the visual pathway because the same process occurs in the sliding of the stimulus over the field of every neuron during the visual processing of moving images.

At the receptor tip, geometrical optics can be assumed if the tip diameter is more than about 2 μm. The receptor absorption function then approximated by a cylinder of diameter d, and the blur circle by a Gaussian function. The convolution yielding the field shape is then accurately given by standard tables of the offset circle probabilities for the normal distribution function. These are the tables showing the distribution of shots hitting a circular target by a gun with a random error which aims at a point off the centre of the target. As the receptor subtense d/f increases in the range $\Delta\alpha/2 < d/f < 2\Delta\alpha/2$ radians, where $\Delta\alpha$ is the width of the blur circle at the 50% level, the field width increases only slowly, but the sensitivity (the height of the peak, *s* in Figs 2c, d) increases with increasing area of the receptor tip. When d/f > 2 $\Delta\alpha$, the value of the acceptance angle $\Delta\rho$ becomes independent of λ or D and approximates to d/f, while the angular sensitivity curves become progressively more square and flat topped (Fig. 2d).

The tips of receptors less than 2 μm in diameter act as wave guides. The absorption at the distal tip is approximated by a Gaussian surface of angular width Δ σ = d/f radians, subtended at the posterior nodal point. The *field size* is then given by the convolution of two Gaussian functions which gives a third Gaussian of width $\Delta\rho$ at 50% sensitivity, according to the relation

$$\Delta\rho^2 = \Delta\alpha^2 + \Delta\sigma^2 = (\lambda/D)^2 + (d/f)^2 \text{ radians} \dots\dots\dots \quad (3)$$

This neat equation allows us to calculate the value of $\Delta\rho$ entirely from anatomical measurements. There is a reasonable match with the values of $\Delta\rho$ measured electrophysiologically in butterflies, locust by day, dragonflies and flies. The diffraction component (λ/D) and the anatomical factor (d/f) both contribute to $\Delta\rho$ when they are nearly equal. When the receptor is narrow, the angular sensitivity curve is a little wider than the blur circle. When the receptor is comparatively broad, its angular subtense (d/f) dominates the angular sensitivity (Fig. 2d), in which case: $\Delta\rho$ = d/f radians. The relation between the facet aperture D and the receptor diameter d depends on *the compromise between resolution and sensitivity* reached by natural selection. To make full use of the resolution of the lens the rhabdom must be small. To optimize the sensitivity and also the resolution towards a small black spot, the tip of the receptor must catch exactly the light in the blur circle. The best match is then approximately

$$d/f \approx 2\Delta\alpha \approx \Delta\rho \text{ radians} \qquad \dots\dots\dots\dots \quad (4)$$

As the receptor subtense d/f is increased, there is increasing sensitivity to a diffuse source, but only by sacrifice of lens resolution. Some insects which are active both by day and night, notably locusts and praying mantids, increase the rhabdom diameter at night by a factor of at least 10 in cross-sectional area, greatly increasing their sensitivity to diffuse sources at night but retaining lens resolution by day.

Motion also causes blurring of the image, and the impulse response of the receptor to a brief flash can also be represented as a Gaussian, of angular width v.Δt where Δt is the duration of the impulse response and v is the angular velocity. Then, for a moving image,

$$\Delta\rho^2 = (\lambda/D)^2 + (d/f)^2 + (v\Delta t)^2 \qquad \dots\dots\dots\dots \quad (5)$$

The above discussion assumes that only the convolution of λ/D and d/f generates the static acceptance function of the receptor, but if mode optics apply at the tip of the rhabdom, the direction of incidence of the light in the cone has an effect. The first mode accepts mainly axial light (Fig. 5a). The second mode accepts mainly light that is off-axis by an amount that depends on the mode parameter V of the wave guide (the rhabdom). The more modes are accepted, the nearer the situation approaches ray optics and the field becomes square (Figs 2d and 5b). In practice, what actually happens at the cone-rhabdom interface cannot be observed so the acceptance angle $\Delta\rho$ must be measured electrophysiologically and the optics inferred from it.

The best example is the fly, for which Kuiper predicted in 1966 that $\Delta\rho$ should be independent of wavelength. Horridge *et al.* (1976) confirmed this from numerous measurements of fields of the fly *Eristalis*. This result contradicted Snyder's simple model based on the convolution of two Gaussians. As an explanation, it was suggested that "the second mode may be admitted at the short but not at the long wavelength end of the spectrum". Following theoretical work by Pask and others, Smakman *et al.* plotted the fields very carefully and calculated what modes are admitted. References will be found in Snyder (1979), Smakman *et al.* (1984). In one cell, at 588 nm only the first mode is admitted (80%); at 494 nm the second mode is partially admitted (37%), at 355 nm more of the 2nd mode (42%) and some of the 3rd mode (20%). *Shortening the wavelength is equivalent to increasing the diameter of the receptor*, and makes the field increasingly square.

2. 5. Absorption by the rhabdomere

The rhabdomere stands out as a discrete anatomical organelle because it consists of numerous tubules (microvilli) of lipid-rich cell membrane, 80 to 90 μm diameter, on the inside walls of which are attached molecules of the visual pigment rhodopsin. There are about 1000 rhodopsin molecules per micron of microvillus. The rhodopsin molecules (12 nm apart when packed in the microvilli) consist of a protein, called opsin, combined with a carotenoid side chain related to vitamin A. The alternating double bonds of this side chain are excited by passing photons and capture them, so *the receptor as a whole acts as a photon counter*, and in calibrations the light must be measured as a photon flux rather than in energy units. The properties of each photoreceptor cell depend on those of the molecules of visual pigment. The position of the peak absorption depends on the particular protein, which is the origin of the different spectral sensitivity curves of different rhodopsins. The complexity of the vibration patterns of the rhodopsin molecule makes a broad absorption spectrum, 70-110 nm wide at the 50% level. This property of the molecule is essential for catching natural light. The actual absorption in the rhabdom is about 0.5% per micron (see Equ. 12).

Polarization sensitivity arises because the rhodopsin side chains are asymmetrical and elongated, with the result that the ratio of absorption along the preferred plane of polarisation to that in the plane at right angles is up to 10:1. In addition, if the molecules lie at random but in the plane of the membrane in the walls of the microvilli, they will appear to be oriented because the incoming light strikes some of the membrane edge-on. This effect of this microvillus geometry yields an absorption ratio of 2:1 to the plane of polarization. Between these two effects, directed and passive orientation of the rhodopsin molecules, the ratio of the maximum to the minimum absorption (called the *dichroic ratio*) cannot be predicted, but must be measured. There are two ways to do this, by optical measurement or by counting individual photon captures by electrical recording.

In contrast to the kind of light, the photoreceptor takes no account of the direction of the light which excites it. *All of the directional sense of the receptor is determined by the optics and the distribution of screening pigment.*

The capture of a photon by a rhodopsin molecule causes a molecular change which initiates a chain of amplifying reactions which take place within the cell and finally result in the opening of ionic channels in the receptor cell membrane and an electrical response. The unit of action appears to be the microvillus. *One photon capture causes one small positive-going miniature potential (called a bump)* caused by the entry of a pulse of mainly Ca^{++} ions. This entry appears to be at the base of the microvillus. Photon arrivals are distributed randomly in time, so as the light intensity is increased, the bumps come closer together in an irregular noisy summation and eventually fuse into a receptor potential (see Fig. 19).

Insect rhodopsins have the interesting property that when bleached to metarhodopsin by the normal absorption of a photon, they are reconverted back to rhodopsin by photons of a different (usually longer) wavelength. This reaction, called photoregeneration, leads to an equilibrium concentration of available rhodopsin, depending on the wavelength content of the ambient light. One obvious effect is that screening pigments are often red or yellow, and allow entry to long wavelength solar power for the regeneration of the rhodopsin, but without loss of resolution in vision.

2.6. Properties of fused rhabdomeres

If single photoreceptors absorb all the light that falls on them, they will be black and therefore unable to respond differently to light of different colours or different polarization planes, so that discrimination of colour or polarization will be impossible. To get around this problem, vertebrate cones are shorter than the rods and absorb only a small fraction of the incident light and sacrifice sensitivity.

Most insect retinula cells have rhabdomeres that are fused to form a central rod along the axis of the ommatidium, so that light rays passing down the composite rhabdom are absorbed approximately in proportion to the contribution of each cell. The rhabdomeres, however, differ in their absorption properties; *each rhabdomere absorbs its own preferred kind of light*. The most abundant rhabdomeres, with a peak absorption for green light, remove green photons from the incident light but absorb less blue or UV photons. Because rhabdomeres share the light path in this way, the major consequence is that each can absorb its own fraction of the light and each retains its own differential sensitivity to colour and polarization plane but *all the light can be used*. Therefore a fused rhabdom with different absorption peaks can be longer and more sensitive than a separate rhabdomere of a single cell. In most insects, the commonest type of retinula cell has a peak which matches the most abundant environmental background colour. Frequently four of the seven or eight retinula cells are green sensitive, except in the dorsal part of the eye, where they are blue sensitive to match the sky.

Theoretically, absorption of green photons by the rhabdomere with a peak in the green removes photons before they can be absorbed off-peak by rhabdomeres with a peak in the blue or UV, so that the spectral sensitivity curves of these other receptors are sharpened, and vice-versa. References will be found in Snyder (1979).

A related result follows when the retinula cells form a *tiered retina,* i.e., one before another along the light path, as in many insects. Frequently one cell, No 8 or 9, at the base of the rhabdom, receives light that has passed through all the other rhabdomeres. The effect can be a marked sharpening of its spectral or polarization sensitivity. Another factor that is important for absolute and polarization sensitivity is the twist of the orientation of the microvilli, as in the worker honeybee. With the twist all of the light can be absorbed, not only that polarized in the appropriate plane, and the polarization no longer interferes with the spectral sensitivity.

These effects in the optics are small and difficult to measure because there is also an unknown amount of electrical coupling between neighbouring retinula cells. The electrical coupling can be purely resistive leakage, which makes the sensitivity curves more similar, or it may be an antagonistic current flow caused by generation of extracellular current within a region surrounded by a resistive membrane, so that, for example, the activity of large numbers of neighbouring green-sensitive cells can hyperpolarize the blue-sensitive cells and modify their spectral and polarization sensitivity curves. These effects are particularly strong in butterflies and moths.

The important point is that so many physiological factors interact on retinula cells and so few of them are measurable, that one has to be content with a list of possible factors, such as the outline of the optics, the anatomy and the absorption curves of rhodopsins with different peaks. The consequences of all interactions at the retinula cell are bundled together and conveyed as a single variable in its electrophysiological response, which after all is the output that acts as input for the next stage. In the nervous system in general, on account of convergence of this kind that is unknown in detail, this *redefining of the signal at each stage can only be done by microelectrode recording*. This is a fundamental lesson for the study of any nervous system.

2.7. Modulation at the photoreceptor

Eyes are adapted to see extended images rather than point sources, and a different treatment of lens resolution considers the blurring of an extended pattern that is passed through the optics although it is focused. When a regular striped pattern, of period $\Delta\theta$ radians subtended at the eye, forms an image in the focal plane of the lens, *the contrast in the image is demodulated or blurred by diffraction at the lens* by a factor which depends on the spatial frequency of the pattern and the aperture of the lens. The total light reaching the photoreceptor, however, depends on the angular sensitivity function, which depends on receptor size, screening pigment and the diffraction by the lens. The modulation is the useful light signal as a fraction of the total stimulus, and can be calculated by convolution of the moving pattern with the (light adapted) angular sensitivity field of the receptor (Fig. 7).

Assuming that the receptive field is Gaussian in shape, the absolute modulation of light intensity is given by

$$M = (I_{max}-I_{min})/(I_{max}+I_{min}) = m \cdot I \cdot \exp[-3.56 (\Delta\rho_{LA}/\Delta\theta)2] \dots\dots(6)$$

where m is the relative intensity modulation in the stimulus, I is a measure of the luminance of the stimulus and $\Delta\rho_{LA}$ is the width of the field of the receptor when it is light-adapted to the mean intensity of the stimulus, not of the dark-adapted receptor, which is the measurement usually available.

A receptor of known angular sensitivity therefore experiences a predictable temporal modulation of light above and below a mean intensity as the striped pattern moves relative to the eye. *This temporal modulation is the only indication of the image that the individual receptor receives,* and the electrical response that it generates is passed on to the second-order neurons. This is why the acceptance angle is a useful measure of the properties of the receptors in visual processing.

There is a range of values of the light modulation for a limited range of stripe periods from a minimum of $\Delta\theta = \Delta\rho$ to about $\Delta\theta = 8\Delta\rho$, beyond which there is no further increase in modulation. There is a sharp cut-off as $\Delta\theta$ approaches $\Delta\rho$ when the modulation of the signal disappears into noise but the illumination continues to generate noise. There is a definite *tuning to a limited range of spatial frequencies* in the range $\Delta\rho < \Delta\theta < 6\Delta\rho$.

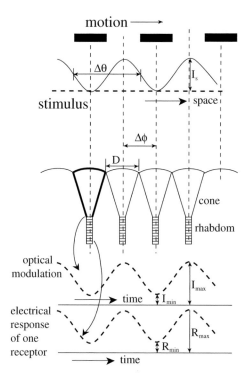

Figure 7. A regular striped pattern laid out in angular space in front of each facet can be represented by its sine-wave fundamental of period Δ θ . As the eye moves relative to the pattern, each receptor generates a response which is an oscillation __in time__. In the example shown here, there is still modulation in the receptors when the array of receptors is at the limit for sampling the pattern.

These linear results based on light intensity modulation will predict the modulation of the electrical response of the retinula cells to different spatial frequencies over a small range of intensity or contrast for slowly moving patterns. On account of the poor lens resolution and strong divergence of the visual axes in insect eyes, the detail in all patterns rapidly disappears as they recede from the eye. For a cell with an acceptance angle $\Delta\rho$ of 3°, a pattern of stripes of period 1 cm and 100% contrast in bright daylight generates no modulation when further away than 20 cm. Also, as the temporal frequency is increased by faster motion of the pattern, the electrical response falls off at a rate that depends on the species. Some insects respond up to 200 Hz, others fail at less than 30 Hz.

3. ASSEMBLING THE COMPOUND EYE

So far, everything said relates to the optics of a single ommatidium behind a single facet. Next we consider how the units, usually each consisting of one visual axis, are assembled to make an eye. Some of the oldest crustaceans, of the Burgess Shale, had compound eyes with 360° vision, which is an advantage for the free-swimming pelagic habit. Some of the earliest arthropods, notably trilobites of the Devonian period about 400 million years ago, had eyes with numerous sub-units, with a small retina in each facet. A few modern insects and several types of larvae also have this structure. On balance, it looks as though the fused rhabdom type evolved from the type with a small unit retina beneath each facet, and local processing behind the individual facets was an early rather than a late arrangement.

3.1. Spatial sampling and overlap of fields

The next step is to examine the way that ommatidia are arranged side by side and interact with their neighbours to make the array of the compound eye (Figs 1a and 8). The *sampling resolution* $(1/\Delta\phi)$ is the reciprocal of the angle between adjacent axes, $\Delta\phi$. In brief, the field of each ommatidium is matched to the spatial sampling resolution of the whole array. The field of each ommatidium of width $\Delta\rho$ at the 50% level is determined by the optics of each individual lens plus its receptor. This is matched to the spatial sampling resolution of the whole array for a repeated pattern, which depends only on the angle between axes, $\Delta\phi$. These two angles, $\Delta\rho$ and $\Delta\phi$, together with eye size, determine the performance. The point to watch is how the optimum design turns out to have overlapping rather than adjacent fields.

With a regular grating pattern in front of an eye, there is a minimum practical sampling density, because *along a line the angle between receptors* $\Delta\phi$ *has to be half of the period of the pattern* $\Delta\theta$ (Figs 7 and 8). If there is a greater density of receptors, some are superfluous, and if the angle $\Delta\phi$ is greater than $\Delta\theta/2$, the pattern cannot be reconstructed within the visual system. This is known as the *sampling criterion.* The sampling resolution is the highest resolvable spatial frequency of the stimulus; it is $1/2\Delta\phi$ periods per radian for a line of ommatidia, but is $1/\sqrt{3}\Delta\phi$ for a hexagonal array.

With an infinitely extended sinusoidal pattern there is a minimum period which an aperture D can resolve, given by $\Delta\theta = \lambda/D$ radians. When $\Delta\theta < \lambda/D$, a sinusoidal pattern generates no modulation of light at all in the receptors. For the same eye in water, if focus is retained, $\Delta\theta_W = \lambda/nD$, where n is the refractive index of water. Lens resolving power is therefore better under-water than in air but the focal length must be longer. In practice, aquatic insects have a flat corneal surface with all the lens power within, and performance is independent of the medium.

Let us put the required sampling criterion for a line of facets and a grating pattern ($2\Delta\phi = \Delta\theta$ radians), together with the diffraction limit along one line of facets ($\Delta\theta = \lambda/D$ radians), so that $D\Delta\phi = \lambda/2 = 0.25$ μm, when $\lambda = 0.5$ μm. The product $D\Delta\phi$ is called the *eye parameter;* it contains *only variables that are measurable from the anatomy of the eye*. The eye parameter should therefore, on this argument, be 0.25 μm for a line of receptors which normally function in bright light. This theory assumes that $\Delta\rho \approx \lambda/D$, a situation that is never reached on account of noise, but which sets the limit. There is actually a disadvantage of a mismatch between the sampling density and the lens resolution because an excess in one adds to the noise but not to the signal in the other.

The eye parameter calculated by this diffraction cut-off criterion along a line of facets is in practice too small to predict real values of $D\Delta\phi$ because it is the limit at which the modulation of light at the receptor is zero. Some modulation is essential.

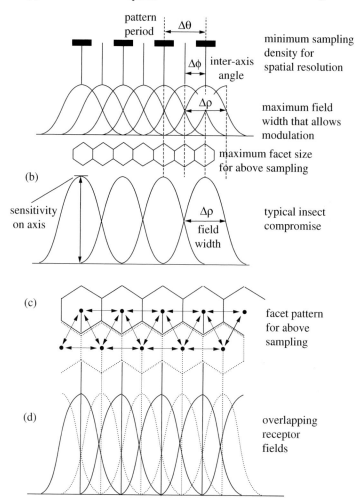

Figure 8. Matching the sampling density of visual axes to the width of the fields of individual receptors. (a) With an ideal match along a line of facets, the spacing of the receptor axes just enables the pattern to be resolved spatially, and at the same time the width of the receptor field allows just sufficient modulation from the same pattern to reach the receptor. (b) Insects frequently have fields which overlap near the 50% level of sensitivity (along a line of facets). As a result the facets can be larger than in (a) and sensitivity on axis is increased. (c) In a two-dimensional eye, the interactions for detection of orientation and motion extend between rows of facets. (d) The partial overlap between rows of facets brings back the full sampling of vertical edges.

198

Because photon arrivals are random with a Poisson distribution in time, the signal-to-noise ratio decreases at lower ambient light intensity (Equ. 6). Therefore, as the light level is lowered, the signal also has to be greater so that the modulation exceeds the noise.

Taking into account these compromises in the optics, the photon noise and the transduction noise, optimum values for the eye parameter were calculated for the range of common ambient intensities (Fig. 9). Some insects specialize with fixed large facets for life in dim light, but the increased interommatidial angle $\Delta\phi$ reduces the sampling density. A dynamic alternative is to increase the angular receptor size d/f, so that receptors catch more light with wider fields. However, $\Delta\phi$ is usually fixed, so that the sampling ratio $\Delta\rho/\Delta\phi$ then changes drastically. In bright light the limit is set by the transduction and synaptic noise (see below).

When the actual fields for insects that fly in daylight are plotted on the pattern of ommatidia in two dimensions rather than along a single line, the resulting eye maps show that adjacent axes are commonly separated by the width of the field $\Delta\rho$, so that eye parameters are about 0.5 rather than 0.25. It is probable that interactions in two dimensions compensate for the undersampling along a single row (Fig. 8d).

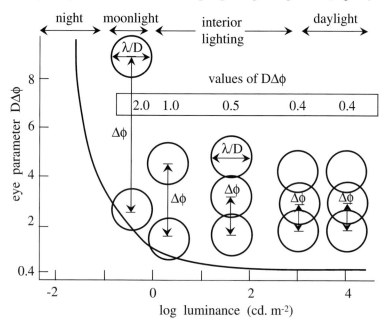

Figure 9. The theoretical effect of average ambient illumination upon the design of an eye. Greater sensitivity is given by greater aperture D and greater spatial resolution is given by smaller interommatidial angle $\Delta\phi$. Therefore the brighter the light, the greater can be the spatial resolution. A value of the product $D\Delta\phi$ is predicted for each luminance, with a minimum near 0.4 (μ m), taking transduction noise into account. This product is represented geometrically by the separation of the minimum theoretical fields of adjacent receptors measured at their 50% intensity contour. These fields touch or overlap in insects that are active in normal daylight luminances (modified after Howard & Snyder 1983).

3.2. The mapping of the eye parameter $D\Delta\phi$ for a compound eye

We can make a map of a compound eye to show the distribution of both lens and sampling resolutions (see Fig. 14a). A circle of diameter λ/D radians is drawn with its centre at each ommatidial optical axis on a map of the axes of the eye in angular coordinates. Larger circles imply smaller facets and poorer lens resolution. For most insect eyes, clearer maps result when circles of diameter $5\lambda/D$ radians are drawn at every fifth ommatidial axis, as in the maps illustrated. If the circles exactly touch, we have $\lambda/D = \Delta\phi$ radians, so that the eye parameter $D\Delta\phi = \lambda = 0.5$ μm. If the centre of a circle falls on the neighbouring circle, we have $\lambda/D = 2\Delta\phi$ so that $D\Delta\phi = \lambda/2 = 0.25$ μm. The separation of the circles on *eye maps* can therefore be read as the local value of the eye parameter $D\Delta\phi$ (Fig. 9).

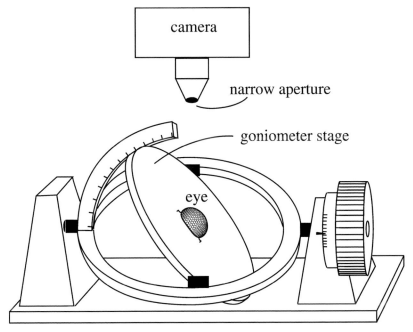

Figure 10. Equipment for measuring D and $\Delta\phi$ and making maps of the eye parameter $D\Delta\phi$.
The centre of the pseudopupil is the visual axis looking at the centre of the camera. Dust grains are used as markers on the surface of the eye, which is photographed every 5° or 10° around the eye in two dimensions. The angular co-ordinates of the pseudopupil are then marked on a map of the facets, from which a map of the visual axes is made in angular co-ordinates (Fig. 14a).

These eye maps are easily but tediously made as follows from living insects which have a *pseudo-pupil*, which is the black spot which does not scatter back light when the compound eye is observed with a lens (Horridge 1978). The visual axis of each ommatidium is found from the direction from which the pseudopupil is centred on that facet (Fig. 10). The virtue of this method is that the map is independent of the axes of the eye, facet lines or head symmetry, and is applicable irrespective of facet shape and the regularity of facet rows. There are the usual difficulties of mapping a spheroid on a plane. The key to understanding maps of the eye parameter is to remember that the circles of diameter λ/D radians indicate the theoretical minimum fields of the receptors. One should imagine a striped pattern of period 2Δφ superimposed on the map, with alternate circles coinciding with peaks and troughs in the pattern. It is then clear from the separation of circles in Fig. 8 that eyes designed for dim light can re-assemble only large spatial frequencies. In all insect eyes the actual field widths (Δρ) are larger than the circles of λ/D because the receptors are not of negligible size, but eye maps of Δρ (see Fig. 14b) are exceedingly tedious to prepare, whereas maps of DΔφ (Fig. 14a) can be made in a few hours from the pseudopupil.

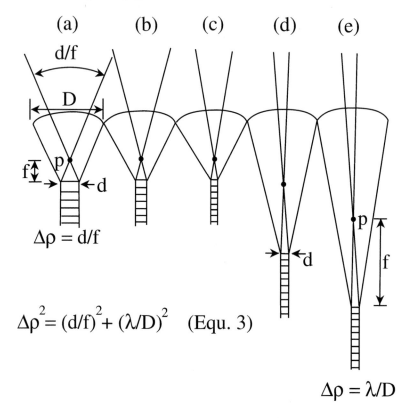

Figure 11. The effect of focal length f and rhabdom width d for a constant facet diameter D. (a) -(c) the decrease in rhabdom (receptor) width d causes a corresponding decrease in field width. For a large d/f the field width Δρ is approximately d/f radians. (d) - (e) Increase in the focal length f causes reduction in the field width Δρ which now depends on the diffraction factor λ /D radians. For intermediate values of d/f, see Equ. (3).

3.3. Regional differences in spatial resolution

The different regions of the eye usually have distinct functions. The part looking forwards needs more resolution for specific targets in chasing behaviour or colour discrimination, and sometimes to triangulate by binocular overlap; the parts looking sideways are more concerned with large landmarks and the flowfield caused by locomotion. In some groups there is a dorsal strip of specialized polarization-sensitive UV receptors looking upwards. The geometry of the compound eye lends itself conveniently to regional differences in spectral sensitivity, interommatidial angle $\Delta\phi$, facet diameter D, receptor capture cross-section d and focal length f to meet a range of task-oriented needs. Therefore in the equation of the match between lens optics and receptor subtense $\lambda/D \approx d/f$, which has occupied so much of the previous discussion, each of the variables can be regionally varied to suit the visual tasks (Fig. 11). Two problems arise; the eye must function as a whole, and the various compromises expressed in the number and size of ommatidia, size of receptors and interommatidial angle must all be met simultaneously in each eye region. There is commonly a local concentration of visual axes called a fovea or *acute zone* looking forwards, upwards, or horizontally, with implications for behaviour (Fig. 12). The acute zone excels in spatial sampling, but the greater area of each facet means that other regions of the eye are sampling less, so that complete coverage is lost, except perhaps at the acute zone.

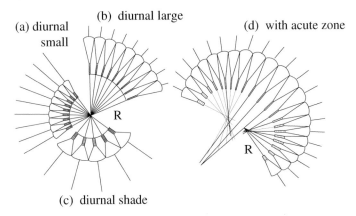

Figure 12. Matching rhabdom diameter, facet size and interommatidial angle in relation to eye size and habits. (a) Small numerous facets for a diurnal eye. (b) Larger facets, longer focal length and narrower receptors achieve the same spatial resolution as in (a) but with greater lens resolution for small dark objects against the sky. (c) In dimmer light the facets and the receptors must both be wide for greater sensitivity, not for lens or spatial resolution. (d) To make an acute zone in a diurnal eye, the geometry of (b) is graded into that of (a) giving a gradient of spatial resolution with uniform sensitivity to a diffuse source. Note the gradients of eye radius R, facet size and cone length in (d).

Many diurnal insects have a band looking out horizontally around the eye, where the interommatidial angle measured in the horizontal plane is greater than that in the vertical plane. The locust and honeybee are typical examples, whereas mantids do not have it. One possible explanation is that turning in the horizontal (yaw) plane during flight smears the resolution of each ommatidium in the horizontal direction but not in the vertical direction, so that the optimum array is astigmatic. Another explanation is that the greater vertical resolution helps flying insects to discriminate landmarks or potential food supplies along the horizon.

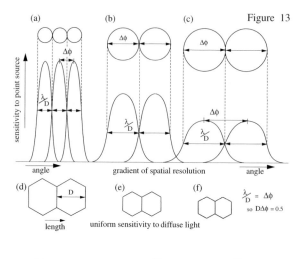

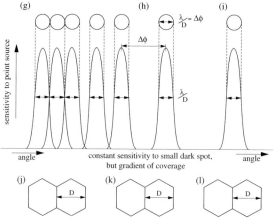

Figure 13. Two naturally occurring ways to build an acute zone in a compound eye. (a) -(f) In dragonflies, the acute zone is a flatter part of the eye with (a) smaller interommatidial angle and (d) larger facets. At the side of the eye (c), the interommatidial angles and receptor sizes are greater and (f) facets are smaller. The sensitivity to an extended source, which is the volume under the Gaussian surface, could be constant across the eye. (g) - (l) In mantids the acute zone is formed mainly by greater focal lengths and smaller interommatidial angles. The facets are similar in size (j-l) so that there is uniform sensitivity to a small black spot, but there is a gradient of sampling coverage.

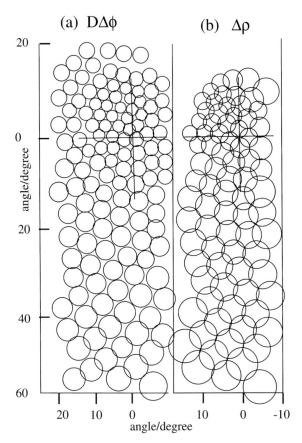

Figure 14. Maps of the front of the eye of the dragonfly Austrogomphus. (a) A map with circles of diameter 5λ /D with centres at the axes of every fifth facet, plotted in angular coordinates. This is a map of the regional distribution of maximum lens resolution and of spatial resolution 1/Δ φ . (b) The same region, with circles of diameter of 5Δ ρ , plotted on the same axes as before, showing the measured dark-adapted acceptance angles and their overlaps in each part of the eye. From such a map it is possible to calculate the pattern of light modulation in all receptors as caused by the motion of any pattern of illumination moving across the visual fields. (from Horridge 1980).

There are two extremes in the design of an acute zone: the common way, found in some male flies and dragonflies, is to *increase the facet diameter in step with the decrease in the angle between axes, so that the F number and therefore the sensitivity to a diffuse source remain constant* (Fig. 13a-f). As an example, the dragonfly *Austrogomphus* has an acute zone looking forwards above the mid-line, and circles of DΔφ just touch both inside and outside this zone despite the large differences in D (Fig. 14a). The way that this compromise is achieved across the eye is shown in Fig. 15. There is a gradient of λ/D and also of d/f in step with it (mainly due to a gradient of rhabdom diameter), so that the gradient of Δρ matches the gradient of Δφ and none of the resolution is wasted.

Another way to make an acute zone is to reduce $\Delta\phi$ locally with little change in facet size simply by increasing eye radius and focal length locally, illustrated by some grasshoppers, dragonflies, butterflies and especially mantids (Fig. 13g-i). There is a local decrease in interommatidial angle and rhabdom diameter, and a large increase in the length of the cones (Fig. 16). This compromise involves less difference in facet size so that a point source or a small black spot is detected equally well everywhere on the eye, although there is a gradient of sampling coverage (Fig. 13g-l). Whether these regional differences in spatial resolution have an effect on motion perception, depends on whether motion detectors are tuned differently in different parts of the eye.

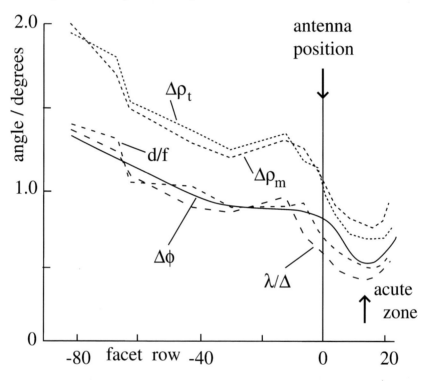

Figure 15. The anatomical width of the rhabdom is matched to the width of the blur circle behind facets of different sizes in the eye of the dragonfly Austrogomphus. The angles subtended by the blur circle, λ /D, and by the rhabdom tip, d/f, agree over a wide range from the ventral part of the eye (left) to the centre of the acute zone (right). Because λ /D is approximately equal to d/f, the theoretical field width, Δ ρ $_t$, is √ 2 times either of them. The measured values of the field widths, Δ ρ $_m$, agree with theory. The values of interommatidial angle Δ φ closely follow the values of λ /D, so that DΔ φ is constant over the eye and equal to 0.5 μ m. Note the small angles at the acute zone (from Horridge 1980).

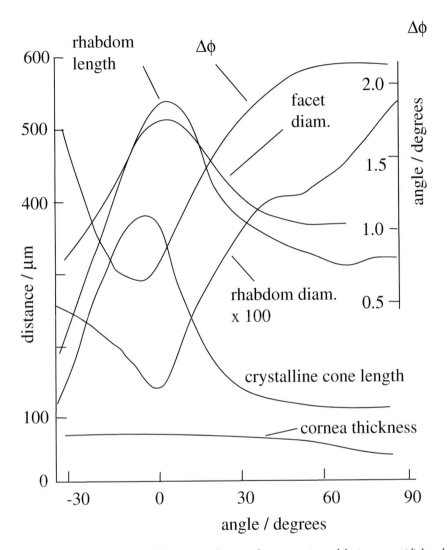

Figure 16. Measurements of the anatomy that are relevant to optics and the interommatidial angle, Δφ ; across the front of the eye of the mantid Tenodera. This is quite a different design from that of the dragonfly in Fig. 15. (from Horridge 1980)

Some insects which detect their females or prey in the air from underneath, for example drone honeybees, male mayflies (Ephemeroptera), male owl flies (Ascalephidae) and biting flies (Simuliidae), have a separate upward-looking eye region with greatly increased resolution. In owl flies and mayflies, astonishingly, this region is sensitive only to ultraviolet light and has very numerous small facets down to 15μm diameter, presumably to achieve high spatial resolution by using only short wavelengths as well as by reducing the interommatidial angle.

To obtain good modulation from a sinusoidal pattern of period $\Delta\theta$ and from larger periods, the acceptance angle must approximate to the relation $\Delta\rho = \Delta\phi = \Delta\theta$ /2 (all in radians), so that adjacent fields cross at the 50% contour. The amount of overlap of circles of λ/D radians on an eye map represents the selection pressure for increased spatial resolution at the expense of sensitivity at that point on the eye. *The ratio of $\Delta\rho$ to λ/D represents the extra size of the receptors to obtain sensitivity.* These ideas, however, do not take into account the many variables which must contribute to the diversity of eyes, such as different fields within one ommatidium, eye tremor, peering, binocular overlap, or regional sensitivity to short wavelengths. Some eyes are adapted to see specialized features (such as a flash of light by a firefly) or a mate of predictable size (such as a queen bee against the background of the sky) rather than generalized features (such as maximum spatial resolution). Also. the effects of large changes in field size that commonly occur with adaptation to light intensity are difficult to predict.

In brief, the eye parameter is half of the ratio of (the minimum angular period which passes the lens) to (the minimum angular period which the eye can reconstruct). This match of the input to the output in the fields of these sensory neurons is the best example we have of the neural equivalent of a physical impedance match when energy is transferred across an interface. A similar matching of the channel capacity, channel overlaps and number of channels must apply to the division of information among neurons at every subsequent stage in visual processing, but we rarely have sufficient data or understand, or even list, the compromises. *The retina is the best model we have of essential compromises in parallel processing.*

3.4. Eye size

Larger eyes are more sensitive for two reasons; they catch more light and have room for larger or more numerous receptors. These two factors generate an interesting compromise between insects of different sizes with similar diurnal apposition eyes. We have seen that the eye parameter $D\Delta\phi$ is a constant depending on the ambient intensity at which the eye functions. Therefore, because $\Delta\phi = D/R$ where R is the radius of the eye, D^2/R is a constant that is equal to the eye parameter. Therefore *facet diameter D is proportional to the square root of eye radius* (Fig. 17), so that larger eyes can have larger facets and also more of them without change in eye parameter. The constant of proportionality depends on the luminance in which the eye functions, with larger facets but fewer of them in dimmer light.

The usual way to vary the field size at different light levels, given that the facet diameter is fixed, is to adjust the receptor width and the focal length, so that the overlap of adjacent fields can be small or large. Some insects, notably Hemiptera and some beetles, can change the focal lengths quickly, but in most of them the focal length is fixed and dark-adaption effectively means a migration of screening pigment, or an increase in rhabdom diameter. An enormous diversity of insect eyes is generated by these degrees of freedom coupled with selection pressures from different lifestyles. Insects wear their visual habits on their eyes, in varied proportions in different parts of the retina (Fig. 18).

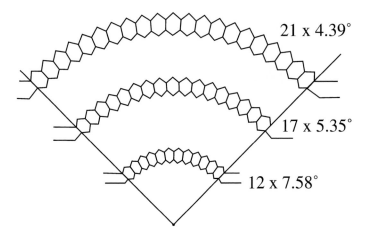

Figure 17. Theory suggests that facet diameter in an apposition compound eye should be proportional to the square root of the eye radius and the proportionality factor depends on habitat. Larger eyes have larger facets and also greater spatial resolution.

3.5. The effect of transduction noise

The theory so far may have conveyed the message that the individual ommatidia of large insects, particularly those that fly in strong sunlight, is limited in lens resolution by the diffraction of light in each single lens. A few insects which chase prey in bright sunlight with facet diameters less than 30 μm have acceptance angles in the range 1.0° to 1.5°, implying that their receptor size contributes little to the field. On the other hand, most insects sacrifice some spatial sampling for greater receptor size and therefore sacrifice both lens resolution and sampling resolution for sensitivity at low light levels.

There is a variety of evidence suggesting that even *the eyes of diurnal insects are not at the diffraction limit* although they function in bright sunlight. First, direct measurements on most insects reveals that the rhabdoms and the *field sizes are larger than calculated from the aperture*. Secondly, measurements of interommatidial angles reveal that *all eyes undersample*. Thirdly many insects with apposition eyes are active in sunlight but have mechanisms of dark-adaption that remove screening pigment and increase sensitivity in dim light at the expense of lens resolution but without change in interommatidial angle. These indirect observations are supported by direct measurements of the noise. Recent work will be found in Anderson & Laughlin (2000).

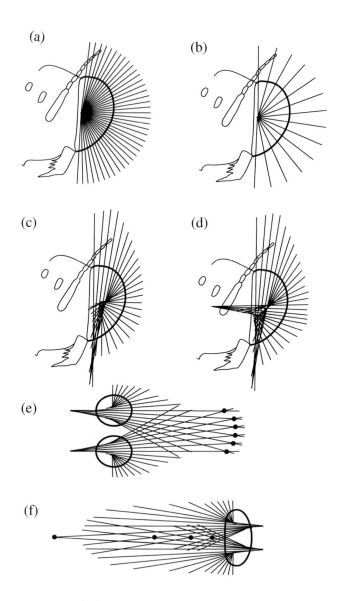

Figure 18. Diversity of geometry for different ambient intensities and visual tasks. (a) A diurnal eye that functions in bright light with small facets and high spatial sampling frequency. (b) An eye for dim light has fewer, larger facets. (c) An upward-looking acute zone is made by local increase in eye radius, as in dragonflies that catch their prey from below. (d) Two acute zones are possible in a large eye. (e) Coincidences of certain visual axes, as in the praying mantis, for prey capture at a fixed range. (f) Coincidences of visual axes for discrimination of range along the head axis, almost always associated with the grabbing action of the mouthparts, as in dragonfly larvae.

The arrivals of photons are Poisson-distributed in time, and therefore the *photon or shot noise* is proportional to the square root of the mean number of photons. Therefore

signal/noise ratio is proportional to $N/\sqrt{N} = \sqrt{N}$ .... (6)

To see better in dim light, the signal must be increased relative to the noise, so this implies that the photon flux must be increased. Each photon gives rise to a bump, and bumps fuse, so another way to reduce noise is to increase the integration time for the electrical response, as in "slow" eyes, but this means a reduction in the speed of response (Equ. 5). The ways to gather more photons are to increase the aperture D, the receptor size d, reduce the F number f/D, or combine the signal from several ommatidia. The last possibility implies optical pooling with a superposition eye (see Fig. 22) or summation of neurons below the retina. Examples of all these ways to increase sensitivity occur.

Finally, *direct measurements of the signal/noise ratio show that the diffraction-limited noise-free eye is unattainable even in the brightest light.* At lower luminances, the noise comes from the shot noise, but *in sunlight the dominant noise is transduction noise,* mainly from the variety of sizes of miniature potentials produced by single photon captures. To this is later added the *synaptic noise* of the neurons below. The spontaneous activation of rhodopsin molecules in the dark, so-called *dark-noise,* is negligible in insects.

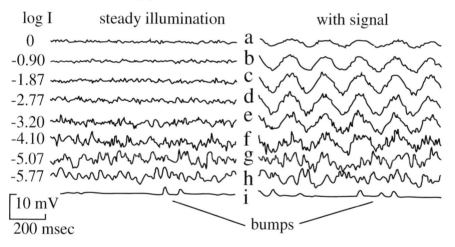

Figure 19. A modulated signal and accompanying noise recorded in a locust retinula cell. Traces on the left are to steady illumination, those on the right to sine-wave modulation of 0.17 at 5 Hz at different mean light intensities as shown. (a) - (b) baseline calibrations. (c) - (i) at the background intensities $\log_{10} I$, as shown on the left. Increasing background intensity reduces the noise and also the modulation (after Howard and Snyder 1983).

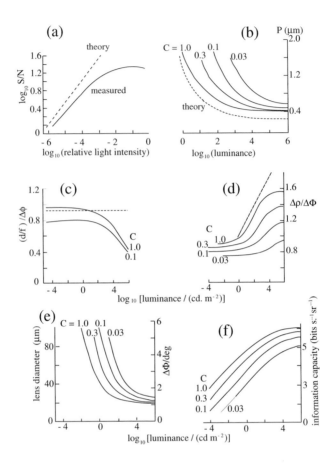

Figure 20. Optimum values of compound eye parameters for different contrasts and luminances (i.e. visual tasks and habitats). (a) The measured signal/noise ratio for locust retinula cells plotted against the log intensity. The signal/noise ratio saturates in the normal range for daylight. The dashed line with a slope of _ predicts the curve if shot noise were the only source of noise. (b) Optimum eye parameter P = DΔφ for each luminance and contrast C, according to the strategy in the text. The dashed line corresponds to the diffraction limited eye. (c) Optimum ratio (near unity) of rhabdom subtense (d/f) and interommatidial angle (Δφ). (d) Optimum ratio (around unity) of acceptance angle (Δρ) to interommatidial angle (Δφ) for a number of contrasts. (e) Optimum lens diameter and interommatidial angle Δφ for an eye of radius 1000 μm suited to various luminances for a number of contrasts. (f) Maximum information capacity attainable by a compound eye of radius 1000 μm suited to different environmental luminances and various mean contrasts.(after Howard and Snyder 1983).

The receptor capacity of the retinula cell to give a further increment of electrical response ultimately depends on the number of simultaneously active channels per receptor. In white-eyed *Drosophila* the photon capture rate saturates at about 2.5 x 10^4 events which overlap, which is about three orders of magnitude less than the number of rhodopsin molecules. Recordings from single locust receptors show that the signal/noise ratio follows the square root rule that is the consequence of shot noise up to moderate intensities, but then saturates at about S/N = 40 (Figs 19, 20a). Saturation of the signal/noise ratio implies that Weber's Law holds ($\Delta I/I$ = constant). At luminances above 10^4 cd m^{-2} (daylight), photon shot noise is negligible compared to the *photoreceptor transduction noise*.

When the information capacity is maximised in a compound eye taking account of the intrinsic noise as measured (Fig. 20a), the optimum dimensions of a standard isotropic eye of radius 1 mm can be calculated for a variety of contrasts (Fig. 20 b-f). For bright light and high contrast, the optimum eye parameter is not less than 0.4 and for a typical value of useful contrast C = 0.3, the optimum is P = $D\Delta\phi$ = 0.5, which is that commonly found. The values of receptor size, acceptance angle, lens diameter, and interommatidial angle all saturate at the intensity of bright daylight due to intrinsic receptor noise. Insect eyes of this size can detect small dark objects subtending 0.2° against the bright sky (i.e. another fly at a range of 0.5 - 1.0 m). Eyes suited to bright environments have no way to increase sensitivity in the short-term except by standing still, increasing receptor subtense, or by having superposition optics of some kind.

4. EYES FOR LOW LIGHT LEVELS

4.1. The need for more light

Having dealt at length with the resolution of the single retinula cell and of the whole retinal array, we return we return to the sensitivity towards reception of an extended image. One lesson to be learned from the signal-to-noise ratio and from the modulation of the receptor is the need to collect light. There are two ways to use summation to reduce noise; optical summation to increase the intensity at the receptor, and summation of neuron fields at a deeper level.

The word sensitivity can mean several different kinds of measurement, e.g., the height s (for axial light) of the angular sensitivity curve (Fig. 2), as used above. For diffuse light, the sensitivity of a receptor is the volume under the whole angular sensitivity curve, as used below, mainly. There are other measures of sensitivity, e.g., the slope of the curve of the stimulus plotted against the response, the minimum detectable increment of stimulus above noise, or the absolute threshold stimulus that is detectable in a given time. When the word is used, the meaning must be defined.

4.2. Diffuse sources and extended receptors

For axial rays from a point source, the light flux F_a entering an eye is proportional to the intensity I of the source in candelas, the area of the lens S_a and the inverse square of the distance (Fig. 21)

$$F_a = I. S_a / X^2 \text{-----------------(7)}$$

When a receptor behind a lens of area S_a views an extended object of area S_e and luminance L candela / m^2 at a distance X, then the solid angle that the lens subtends at the object is S_a / X^2 (from the definition of a solid angle) and the total flux passing through the lens (Fig. 21) is

$$F_a = L. S_e. S_a / X^2 \text{ --------------------(8)}$$

This light forms an image of area S_i and the retinal illumination is $E_i = F_a / S_i$.

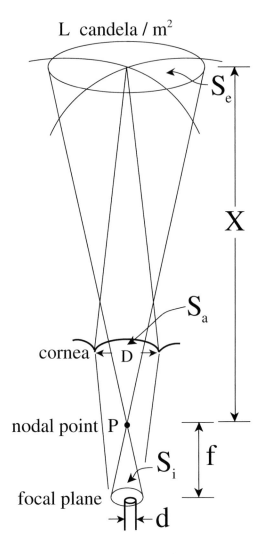

Figure 21. An object of area S_e and luminance L candela / m^2 forms an image of area S_i thro ugh a lens of area S_a and some of the light falls on a receptor of diameter d.

Considering the rays that pass through the nodal point, the angle within the eye subtended by the image is the same as the angle subtended by the object (from the definition of nodal point). So,

$$S_e / X^2 = S_i / f^2 \text{----------------------}(9)$$

From these equations and from the relation $S_a = \pi D^2 / 4$, (where D is the diameter of the aperture of the lens), the image illuminance is

$$E_i = L. \pi D^2 / 4f^2 \text{----------------------}(10)$$

This equation, well known for camera lenses, tells us that the brightness of the image is proportional to that of the source, and is inversely proportional to the square of the F number (f/D) of the eye, and nothing else, so *surfaces do not change in apparent brightness as they move nearer or further away.*

The light flux entering a receptor is proportional to the image illuminance and the area of the receptor surface, so

$$F_p = E_i \pi.d^2 / 4 \text{--------------------}(11)$$

where d is the receptor diameter.

The light absorbed by a length x of the receptor gets exponentially less as it passes through the receptor, so

$$\text{Flux (absorbed) / Flux (incident)} = 1 - e^{-kx} \text{-------------}(12)$$

where k is the coefficient of absorption, which is usually taken to be the fraction absorbed in 1 micron. If the value of k is 0.005, then absorption is 40% in 100 μm, 64% in 200 μm, and 99.5% in 1000 μm. These numbers show that there are diminishing returns in having longer rhabdom rods, and in practice we find receptor lengths from 20-200 μm, except for a few freaks like large dragonflies with rhabdoms 1 mm long (but tiered).

Putting all together, we have

$$\text{the absorbed flux} = L. (\pi / 4)^2. (D / f)^2.d^2. (1 - e^{-k x}) \text{--------}(13)$$

This equation ignores the polarization of the light, screening pigment, and the relation between the spectral sensitivity of the receptors and the colour of the scene. It can be applied to any eye as a test of maximum possible sensitivity to ordinary scenes. Because they are adapted to the same visual world, most eyes turn out to have similar values of absorbed flux and also F number f/D. Bees and humans (and cheap cameras) have similar F numbers and similar sensitivity although humans have 100 times more spatial resolution than bees.

The absorbed flux is proportional to $(D.d/f)^2$ which is the square of the eye parameter if d/f is approximately equal to $\Delta\phi$, as in many eyes, showing again that the eye anatomy is directly related to the ecological ambient intensity.

Normal bright sunlight contains about 10^{14} photons $/cm^2/s$ and moonlight has similar white light which is less intense by a factor of 10^7. Insects flying by day necessarily have the sun in view, but somehow they ignore it by saturation and by regeneration of visual pigment by red light. They function with light that is reflected from natural surfaces with not more than 10% reflectance. So a facet of 500 μm^2 collects approximately 5 x 10^7 photons/s or 5 x 10^5 photons in one integration period of 10 ms from natural surfaces in sunlight. These photons are divided among the receptors. Eyes with optical superposition (Fig. 22) are designed to increase the light at the receptor given the historical situation of being evolved from a compound eye.

Eyes suited to low luminances are all dominated by the lower signal-to-noise ratio caused by photon shot noise. Increase in facet size to increase sensitivity implies reducing the number of facets and therefore undersampling. With a moving stimulus, this matters less for the non-directional motion detection system than for the classical directional unit motion detectors. There is a lower limit to eye size in fast flying insects.

4.3. Eyes with neural summation

Special conditions apply in the eyes of many Diptera, Hemiptera, and Coleoptera which have 6 to 8 separate rhabdomeres in each ommatidium. For the fly *Musca,* the interommatidial angle is about 4 times that expected from the maximum spatial resolution at the diffraction limit, and does not follow the sampling theorem. The limit of spatial resolution is set by the angular separation between receptor axes in the same ommatidium, but rhabdomeres have to be separated by a minimum distance to avoid optical cross-talk between the receptors by bleeding of light energy from one rhabdomere to the next. To avoid serious cross-talk between rhabdomeres of 2 μm in diameter, the rhabdomeres must be separated by a space of 0.5 μm. So, the angle between receptor axes is determined by rhabdomere diameter and focal length in a way that does not apply with a fused rhabdom. In addition, if there is exact summation of the terminals of receptors looking in the same direction, the interommatidial angle must be equal to the angle between the rhabdomere axes. The problem is illustrated in the housefly, which lives out of direct sunlight. The rhabdomeres are about 2 μm in diameter, so with a focal length of 30 μm and rhabdomeres spaced at 3 μm to avoid cross-talk, the angular separation of axes and therefore the interommatidial angle would be 6°. To alleviate the problem, the rhabdomeres are narrowed at their tips, where they also touch, so their centres are separated by 2.5°- 4°, but below this level they widen and separate. In flies, two of the receptors behind each facet, cell 7 and 8, have rhabdomeres one above the other, which are only about 1 μm in diameter and carry mainly the 1st mode. The geometry of packing in the axes is further assisted by extending the pattern of summation in one direction. In this way the interommatidial angle can be reduced to about 3°.

As an alternative way to solve the packing problem, in bibionid flies the receptor axes coincide with those of the next-adjacent facets, and in the aquatic bug *Lethocerus* (Belastomatidae) the optical axes coincide with those of facets that are even further separated, to allow for even larger rhabdomeres. In the forward-looking part of the male fly's eye, the so-called love spot for chasing females, rhabdomeres 7 and 8 are the same size as 1-6, so that in this region sensitivity is further enhanced for the sake of getting sufficient absolute contrast to be able to follow a small spot during aerial acrobatics.

4.4. Superposition eyes

In a superposition eye, a bundle of rays from a distant point entering by a patch of neighbouring facets (Fig. 22) are bent by the optics of the cornea and cone so that they all focus on a blur circle at the rhabdom tips. The nodal point P now lies, by definition, at the centre of the eye, so that $\Delta\phi = d/f$. Rays arriving at an angle α at one side of a radius of the eye cross the clear zone *on the same side of the radius* at an angle β to it (Fig. 22b). The paths of oblique rays in the cone and cornea are similar to those generated by two convex lenses separated by the sum of their focal lengths (Fig. 22c). Then α / β is approximately f_2 / f_1. The actual focusing power of the superposition eye lies partly in the curvature of the cornea (except in aquatic insects) and in the non-homogeneous refractive index of the cornea and cone, which each form a lens cylinder. This is a longitudinal lens which bends rays towards the axis by having a greater refractive index along the longitudinal axis (Fig. 22d). To achieve the appropriate optical path, even for oblique rays, the refractive index η must vary along the radius of the cylinder as a parabolic relation

$$\eta = \eta_0 - kr^2 \quad (14).$$

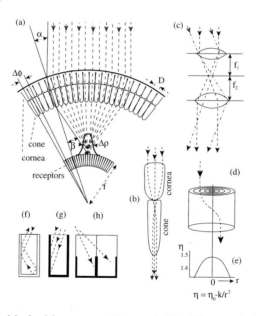

Figure 22. The optics of the fixed-focus superposition eye. (a) Parallel rays strike the cornea at an angle α on one side of the eye's radius and emerge through the clear zone at an angle β on the same side of the radius. They are focused to a blur circle on the receptor layer. P, nodal point; f, focal length. (b) Three oblique rays in the cornea and cone of one ommatidium. (c) The ideal optics equivalent to (b) with two lenses with a common focal plane. (d) A lens cylinder with higher refractive index along the axis bends the rays appropriately. (e) The gradient of refractive index η. (f)-(h) Rhabdoms with (f) tracheal sleeve and tapetum that reflect the rays, (g) fully sheathed by screening pigment, (h) partially surrounded by screening pigment.

In a few superposition eyes, namely in some advanced moths such as Bombycoidea, Sphingoidea and Agaristidae among the Noctuoidea, and in the skipper butterflies, a sharp optical separation of the elongated columnar rhabdoms is achieved by a tracheal sleeve around each one, even though the rhabdoms within also act as separate light guides. This separation ensures that oblique rays do not cause blurring at receptor level. Many superposition eyes e.g., most night flying beetles including the fireflies, do not have this sharp separation. In such an eye of very low F number, oblique rays will pass laterally and be absorbed by a rhabdom of another ommatidium, broadening the receptive fields (Fig. 22h). A convolution of the blur circle intensity with the width of the rhabdom is then not an appropriate model. As the image is degraded by this spread, correspondingly the sensitivity to large objects that fill the field is increased. Large fields are tolerable if the main task is only to see trees at night against the sky, or the flash of another firefly. Superposition eyes present various solutions to this compromise. With good optics it is possible to have closely packed ommatidia and complete separation by sleeves of trachea (Fig. 22f), or less effectively by screening pigment (Fig. 22g). Partial separation by short sleeves (Fig. 22h) is common and is appropriate for moderately well-focused eyes, while no separation at all is appropriate for poorly focused eyes. References will be found in Warrant and McIntyre (1991). Note that the firefly (and some others) have an additional mechanism in parallel. There is no optical separation in the thick main rhabdom layer, but there is a thin layer of smaller rhabdoms that retain resolution at a lower sensitivity, maybe by day. To see the timing and the vague direction of a firefly flash at night requires maximum sensitivity but no resolution at all. Peculiar fluted rhabdom shapes in some nocturnal eyes are probably also designed to catch or retain more light.

During an extended study of the development of the eye of the dung beetle *Onitis aygulus* (Scarabaeidae) it was discovered that the optics changes rapidly over the first three days of adult life and then more slowly for as much as three weeks. All the main optical components grow, change shape and increase in refractive index. The acceptance angle ($\Delta\rho$) decreases from $22°$ down to $4°$ while the young adult remains buried, so the improvement in the focus is not controlled by a feedback from the quality of the image.

4.5. Measurement of sensitivity and optical gain

Educated guesses from models are instructive but the measurement of sensitivity by recording photon captures is definitive. Sensitivity can be measured as the photon flux per facet required to give a threshold or a 50% of maximum response, or as the slope of the curve of the response in mV plotted against intensity, or better, as the fraction of photons usefully captured and recorded as miniature potentials or bumps. At low light levels, the individual photon captures are recorded as brief potentials of about 1 mV. In the locust, 50% of the photons arriving on the facet surface are captured by the rhabdom. In the gyrinid water beetle *Macrogyrus* there are two photon captures for every photon that falls on the facet belonging to the receptor that is recorded from (with green light at 552 nm). Photons from the same distant point source are also going to many other receptors in parallel, and the optical gain achieved by the superposition optics is 32-48 times (Horridge *et al.* 1983).

The retinula cells can be calibrated by counting the effective photon captures (bumps) by intracellular recording at fluxes less than 10 per receptor per second. It is then observed that a reasonable signal/noise ratio is reached at a flux of approximately 100 photons per receptor per second, which is approximately 1000 photons per facet per second, or 10 photons/per facet per 10 ms period. Referring above, we found that full sunlight provides 5×10^5 photons per facet per 10 ms period, which is more than the transduction can use. In shadow, if intensities are down by a factor of 100, an eye with a $500\mu m^2$ aperture will still cope quite well, but, as we go towards deeper shadow or sunset, we reach a flux of 10^8 useful photons / cm/s at the cornea, which is approximately 5 per receptor per 10 ms period, i.e. the lowest limit for useful vision. Plenty of insects, however, find it necessary to fly in even lower luminances, e.g. in moonlight which is a factor of 10^7 less than sunlight.

Sensitivity is increased by extending the integration time, which is determined by the properties of membranes, channels and synapses. *There are "fast" and "slow" eyes*; the latter are characteristic of species that are active in dim light. The temporal properties of the membranes and synapses of the receptors and lamina cells, and presumably the whole system of processing motion and small objects, depends on a mix of ion channels of different types. Even deep in the brain level, the preferred target size of dragonfly object detector neurons increases as the light intensity is lowered. We can expect that the control of integration times by the signal strength and the temporal frequency extends through the whole visual system.

Most of the diversity of insect eyes, including optical superposition, neural superposition in flies and bugs, flexible cones and large rhabdoms, represents efforts to obtain greater sensitivity to overcome photon noise. By being active in dim light, they escape enemies such as swallows and carnivorous insects. Better to sacrifice spatiotemporal resolution than not to see at all. Insects also solve the problem of sensitivity by having higher-order neurons with large spatial fields, which can retain the local spatial resolution of the retina and smooth out the noise.

5. RELEVANCE FOR ARTIFICIAL VISION

Insects illustrate the effects of Darwinian selection for the best retina for each life style. Currently we use many of the principles of the eye appropriately embodied in the size and distribution of grains in film, which, like the photoreceptor array, records the picture but does not see it. Then we have the video camera which puts the sequence of pixels into a time series in a raster sequence, and this can be stored in an array of memory units called a frame-grabber. The timing and size of the pixels in a video camera is aimed at the need to reproduce the picture, and the human brain does the processing. All these devices use the sun as a passive light source. *To recognise pictures in real time is beyond the limit of the fastest sequential light-weight computer* if much computation is done on each frame. The alternative is an array of photodetectors as in an eye and parallel processing to relieve the haste.

Hardware that is based on the principles of design of natural retinas has only just become available. Absorption of photons takes time. The photoreceptors have to be small to achieve the spatial resolution that the wavelength of light makes available. Being small, the receptors take time to catch sufficient photons to overcome photon noise at normal light levels. Human beings manage with a sampling rate of about 30 Hz with an F4 lens; large fast-flying insects need a higher sampling rate of about 100 Hz with an F2 lens because they have less time to avoid collision when objects suddenly come within the range of their coarsely diverging sampling array. The limiting factor in each case is probably the signal-to-noise ratio of transduction (which is improved by having more photopigment) and the limited time available to collect the light. The photoreceptor rods of insects are long to increase light absorption to 50-60%. None of this, especially the small receptors with overlapping fields, or local circuits for a 2-dimensional array, can be manufactured at present.

As said, photon capture is slow, limiting the flicker fusion to (say) 100 Hz, but digital processing is fast (say 10 MHz), so it would be possible for 500 photocells in a line to time-share one processing chip. The chip could be on a wafer and the 500 photocells would be along one edge of the wafer. If the heat could be removed, 500 wafers could then be stacked to make a square retina of 500 x 500 pixels, but processing only along one dimension. The first stages of task-oriented processing circuitry would have to be within this block otherwise it would have too many parallel output wires. This is only the beginning of selection of relevant features from the image because the correlations to detect features would be in one dimension. Trying to design artificial visual processing reveals how subtle, appropriate and versatile are the receptors and neurons in eyes 1992).

REFERENCES

Anderson, J.C. & Laughlin, S.B. (2000). Photoreceptor performance and the co-ordination of achromatic and chromatic inputs in the fly visual system. *Vision Research*, 40: 13-31.

Horridge, G.A. (1978). The separation of visual axes in apposition compound eyes. *Philosophical Transac. Royal Soc. London, B 285*, 1-59.

Horridge, G.A. (1980). Apposition eyes of large diurnal insects as organs adapted to seeing. *Proc. Royal Soc. London, B 207*, 287-309.

Horridge, G.A. (1992). What can engineers learn from insect vision? *Philosophical Transac. Royal Soc. London, B 337*, 271-282.

Horridge, G.A., Marcelja, L., Jahnke, R. & McIntyre, P. (1983). Daily changes in the compound eye of a beetle (*Macrogyrus*). *Proc. Royal Soc. London, B 217*, 265-285.

Horridge, G.A., Mimura, K. & Hardie, R.C. (1976). Fly photoreceptors, III, Angular sensitivity as a function of wavelength and the limits of resolution. *Proc. Royal Soc. London, B 194*, 151-177.

Howard, J., Blakeslee, B. & Laughlin, S.B. (1987). The intracellular pupil mechanism and photoreceptor signal: Noise ratios in the fly *Lucilia cuprina*. *Proc. Royal Soc. London, B 231*, 415-435.

Howard, J. & Snyder, A.W. (1983). Transduction as a limitation on compound eye function and design. *Proc. Royal Soc. London, B 217*, 287-307.

Smakman, J.G.J., van Hateren, J.H. & Stavenga, D.G. (1984). Angular sensitivity of blowfly photoreceptors: intracellular predictions and wave-optical predictions. *Journal of Comparative Physiology, A 155*, 239-247.

Snyder, A.W. (1979). The physics of vision in compound eyes. In *Handbook of Sensory Physiology*, Vol VII/6A: *Vision in Invertebrates*, ed. H. Autrum, 255-314. Berlin: Springer.

Warrant E.J. and McIntyre, P.D. (1991). Strategies for retinal design in arthropod eyes of low F-number. *Journal of Comparative Physiology, A 168*, 499-512.

LARS OLOF BJÖRN AND PIRJO HUOVINEN

10. PHOTOTOXICITY

1. INTRODUCTION

Phototoxicity means that something which is not toxic in itself is converted into a toxin or produces a toxin by the action of light. We can divide phototoxicity into several classes:

Type I phototoxicity arises when a pigment, after absorption of light and acquiring an excited state, either combines directly with an important cell constituent (Fig. 1), or transfers electrons or hydrogen atoms. The transfer may take place from or to another molecule, which then becomes a toxic radical or radical ion, or produces toxins in subsequent reactions. As an exampel of the action of a type I phototoxin we show in Fig. 1 how 8-methoxypsoralen (called "MOPS" in medical jargon) combines with thymine residues in DNA,

Type II phototoxicity arises when a pigment (photosensitizer) after absorption of light goes from the excited singlet state to a triplet state, and then reacts with molecular oxygen and produces singlet excited oxygen (see Chapter 1), which is highly toxic.

In some cases a pigment molecule excited by light absorption transfers an electron to molecular oxygen, thereby producing superoxide anion (see Fig. 1). According to the above definitions this is type I phototoxicity, but in the literature it has also been designated type II phototoxicity, because in practice it is easier to distinguish between oxygen independent and oxygen dependent phototoxicity. The main cellular targets of both types I and II phototoxins are DNA, membrane lipids, and membrane proteins. A wide variety of organisms (except those having special protection systems) can be poisoned by most of the substances; i.e. they are rather unspecific with regard to poisoned organism.

As a third type of phototoxicity we can categorise those cases when a substance is converted into a toxin by a photochemical reaction which does not fall into any of the above categories.

As an example in which several mechanisms contribute to the photodestructive action, we show in Fig. 2 a schematic description of how membrane lipids are peroxidised by a photoexcited pigment (cf. Samadi et al. 2001).

An interesting consequence of lipid peroxidation is that a weak light (ultraweak luminescence) is emitted during the reaction. Lipid peroxidative chain reactions can be initiated also in ways other than through phototoxic action.

In our disposition of the topic "phototoxicity" we shall not follow the categorisation into types I and II, but rather subdivide into the different contexts in which phototoxicity has been observed.

L.O. Björn (ed.), Photobiology, 219–238.

220

We shall not include photoallergic reactions here, which, as they involve the immune system, are of a different character. Photoallergy will be treated in the chapter on the photobiology of skin.

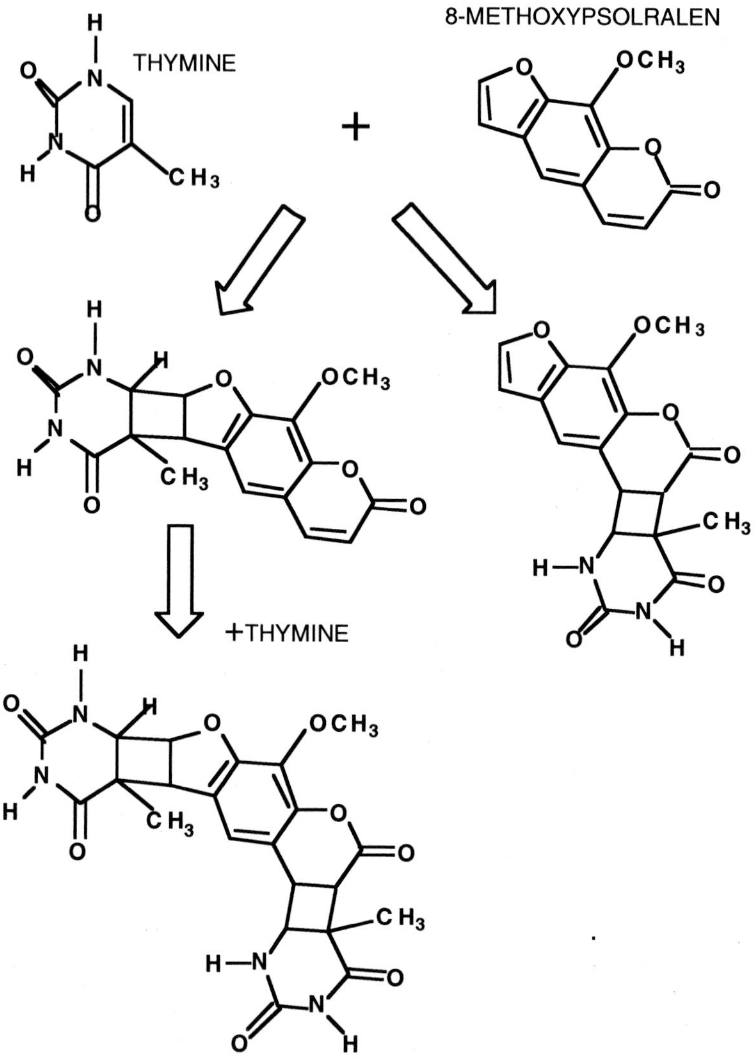

Figure 1. Formation of photoadduct between 8-methoxypsoralen and thymine residues in DNA. The thymine is shown for simplicity as free molecules, but is in reality part of a DNA molecule. One 8-methoxypsoralen molecule can combine with two thymine residues, and if they are bound to opposite DNA strands cross-bridges can form between the strands. Although "phototoxicity" sounds dangerous, this and other similar reactions are also exploited in phototherapy of certain diseases.

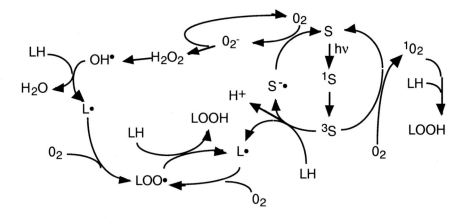

Figure 2. Diagram showing how a pigment (S), excited by light (hν) via an excited singlet state (¹S) to an excited triplet state (³S) can damage a membrane lipid (LH) in several ways (note that the lipid molecules enter the reactions at four points in the diagram): (1) Type II reaction (right part of the diagram): The triplet pigment may react with triplet (ground state) oxygen (O₂) to form singlet oxygen (¹O₂), which can directly convert the lipid to a lipid peroxide (LOOH). (2) Classical type I reaction (lower part of the diagram): The triplet pigment abstracts a hydrogen atom from the lipid, creating a lipid radical (L•), which combines with triplet oxygen to form a lipid peroxy radical (LOO•). The latter abstracts a hydrogen atom from another lipid molecule to form a lipid peroxide. In this way a new lipid radical is formed, and a chain reaction is created. (3) Oxygen-dependent hydrogen abstraction (upper part of the diagram: An electron is donated to triplet oxygen, creating superoxide anion, which via formation of hydrogen peroxide and hydroxide radical abstracts hydrogen from the lipid. Also in this case the lipid is degraded to lipid peroxide and a chain reaction is initiated.

2. PHOTOTOXICITY IN PLANT DEFENCE

The most important defences of plants against parasites and grazers are of a chemical nature, and among chemical defences phototoxicity plays an important role, especially among flowering plants. The phototoxic substances employed by plants can also affect people when they appear in food, perfumes and other cosmetic products, and even if we just touch certain plants.

Downum (1992) estimates that 75-100 different phototoxic molecules have been isolated from flowering plants. Phototoxins or phototoxic activity has been reported for about 40 of more than 100 angiosperm families, representing all subclasses except Alismatidae and Arecidae. Many plants have several phototoxic substances. From *Ammi majus* as well as from *Angelica archangelica* the following ones are reported: angelicin, bergapten, 8-methoxypsoralen, and pimpinellin; from the former one in addition furocoumarin and from the latter one psoralen. The plant family Apiace (former name Umbelliferae) dominates the cases of phototoxicity of most importance to man.

The phototoxins affect bacteria, fungi, nematodes, insects, and other organisms. This wide spectrum is due to the fact that the toxins attack cellular constituents common to all cells. DNA is a major target for type I acting chemicals, such as acetophenones, coumarins, furanochromones, furanoquinolines, pterocarpans, and sesquiterpenes. Examples of type II acting compounds are isoquinolines and thiophenes.

Photosensitisers generally have many double bonds, i.e. many π-electrons, and most of them are polycyclic. The most common types in plants are acetylens and furanocoumarins, but many other types also occur.

Since absorption of ultraviolet radiation is a common feature of organic compounds, and absorption for polycyclic systems and acetylenes with conjugated triple bonds (compounds with many π-electrons) extends into the UV-A region, it is not surprising that UV-A (of which there is much more in daylight than of UV-B) in most cases is the most important spectral region for inflicting phototoxicity. However there are exceptions, and hypericin (present in *Hypericum*, St John's wort) with its many fused phenyl rings absorbs and is excited to phototoxicity even by yellow and orange light, while some other substances require UV-B radiation. Detailed information on action spectra is still lacking in most cases. Guesses made based on absorption spectra are not reliable, since cases are known in which the phototoxic action takes place with radiation of longer wavelength than that absorbed by the pure substance. The reason for this is probably that the spectrum is shifted when the substance binds to cellular components.

The mode of action of hypericin has been debated, but it has now been established (Delaey et al. 2000) that it required oxygen for phototoxicity. As several other phototoxic compounds from plants (e.g., psoralen, 8-methoxypsoralen), it has been used in the phototherapy of diseases.

Specific plant species causing problems for man and domestic animals naturally vary among countries, but the following are worth mentioning here:

Fig, *Ficus carica*. Fig leaves can be troublesome only for those involved in picking and handling them professionally. One source speculates that fig leaves could have caused trouble for Adam and Eve!

Angelica, *Angelica archangelica*. This and other *Angelica* species are used as traditional medicine from Korea to Lapland, and also in drinks. Have caused problems in growers and collectors.

Buckwheat, *Fagopyrum esculentum*. Causes trouble mainly in grazing cattle.

Celery, *Apium graveolens*. Has caused burns when ingested before visiting suntan parlor. Contains 5-methoxypsoralen, 8-methoxypsoralen (xanthotoxin), and 4,5',8-trimethylpsoralen. Of special interest is that this plant can contain tenfold increased contents of psoralen derivatives after infection with a fungus, *Sclerotinia sclerotium* (pink rot disease). Persons handling celery professionally are at risk. Disease-resistant celery contains increased levels of furocoumarins.

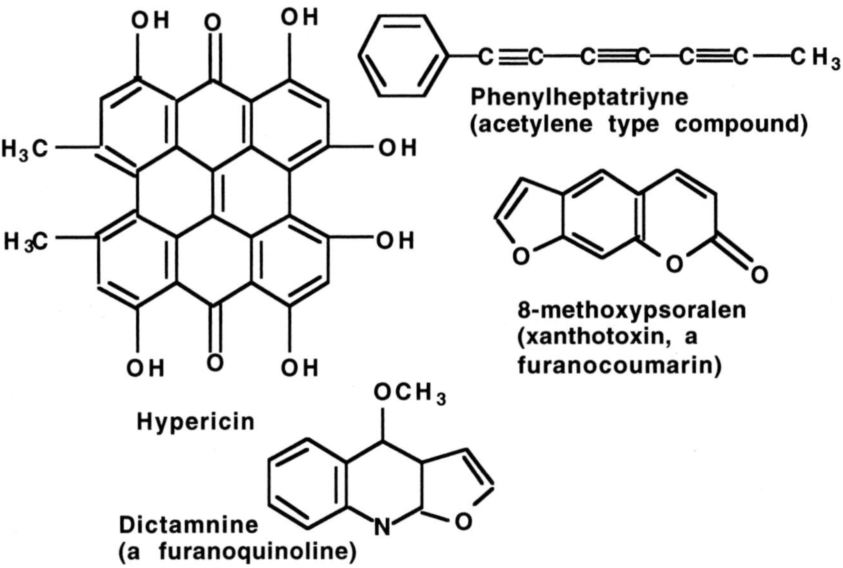

Figure 3. Examples of phototoxic substances from plants.

Hogweed (*Heracleum*), especially Russion hogweed (*Heracleum mantegazzianum*), light produces severe blisters in skin that has been in touch with the plant. the plant has spread over large areas of Europe and North America. Heracleum species contain angelicin, bergapten, pimpinrllin and 5-methoxypsoralen.

Spring parsley (also erroneously called wild carrot), *Cymopteris watsonii*, growing in Oregon, Nevada and Western Utah. Problems with grazing sheep and cattle. Newborn lambs and calves die because mothers become so touch-sensitive that they refuse nursing. The plant contains furocoumarins, 8-methoxypsoralen (xanthotoxin) and bergapten.

Lei flowers, especially *Pelea anisata*. Leis are the greeting wreaths that visitors receive on their arrival to Hawaii.

Burning bush of Moses (also called gas plant), *Dictamnus albus*. The plant grows wild in Europe and Asia, and is used as a garden plant also in other parts of the world. It belongs to the family Rutaceae, which harbours also other plants with some phototoxicity, among them *Citrus* species and *Ruta graveolens*, garden rue.

St John's worth, *Hypericum* species contain hypericin and can cause trouble both to grazing animals and to persons who consume drinks based on *Hypericum* extracts and are exposed to light afterwards.

Some of the phytophototoxins are used for medical treatments. The most noteworthy example is treatment of vitiligo and psoriasis with 8-methoxypsoralen and related substances. In fact, the juice of the Egyptian plant *Ammi majus* has been used for this purpose since 2000 B.C. (Pathak & Fitzpatrick 1992).

We can only give some examples of detailed mechanisms of action in phototoxicity.

For further information on phototoxic plants and plant phototoxins we refer to Pathak, M.A. (1986), Downum (1992), Lovell (1993) and to the following Internet sites:

http://telemedicine.org/Botanica/Bot5.htm
http://www.ars-grin.gov/cgi-bin/duke/chemical_activity.pl

3. PHOTOTOXINS OF FUNGAL PLANT PARASITES

Phototoxins are not used only for plant defence, but also for attack on plants by parasitic fungi. So far only one case has been thoroughly researched, but a number of plant pathogenic fungi produce photosensitising substances. A review of the subject (Daub & Ehrenshaft 2000) has recently appeared.

The best known example of a plant parasite using a phototoxin to weaken its host is the genus *Cercospora*. About 500 parasititc *Cercospora* species are known and cause, e.g., leaf spot of sugar beet, grey leaf spot of corn, purple seed stain of soybean, frogeye leaf spot of tobacco and brown eye spot of coffee. For sugar beet the active pigment, cercosporin (Fig. 4), has been isolated from 34 *Cercospora* species grown in culture, while other species do not produce cercosporin, and still can parasitise plants.

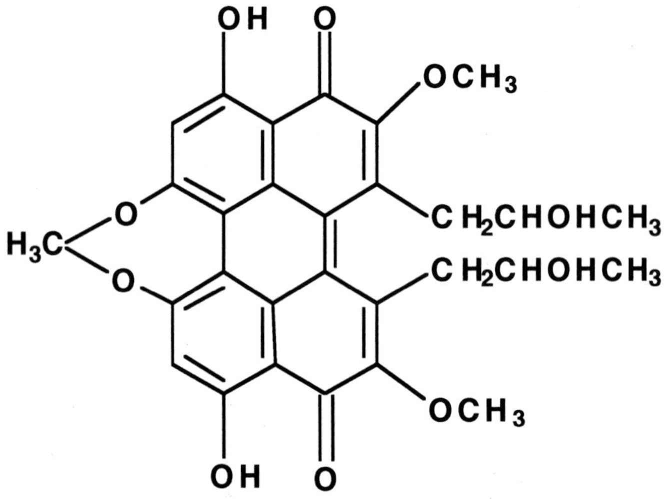

Figure 4. Cercosporin, the phototoxin of the parasitic fungus Cercospora.

Cercosporin is a type II phototoxin. After reaching its triplet state during illumination it reacts with oxygen to form singlet oxygen. The singlet oxygen destroys the cell membrane of host cells, which leads to leakage of nutrients to the fungus.

Of course, also the fungus has cell membranes which could be damaged by cercosporin, so it must have some defence against its own toxin. In culture they can accumulate up to millimolar toxin in the medium without observable toxic effects. In fact, it defends itself in two ways: (1) As long as the cercosporin is inside the hyphae, it is kept in a reduced form, which in light produces only a small amount of singlet oxygen. After secretion to the environment it is oxidised to the highly active form. The two forms can be easily distinguished under the microscope, since the reduced form has a green, the oxidised one a red fluorescence. (2) In addition the fungus is extraordinarily well equipped with a set of triplet and singlet oxygen quenchers. That they are efficient is shown by the fact that *Cercospora* is resistant also to the effects of other singlet oxygen producing phototoxins. Among the quenchers of singlet oxygen pyridoxine is thought to be particularly important for *Cercospora*.

Interestingly, *Cercospora* does not produce cercosporin in darkness (when it would be of no use); its synthesis is triggered by light.

Pigments having structures related to cercosporin (perylenequinones, cf. Fig. 4), and presumably having a corresponding function, are produced by a number of other fungi: by *Cladosporium* species, by the bamboo pathogens *Shiraia bambusicola*, and *Hypocrella bambusae*, and by *Stempylium botryosum*, and some *Alternaria* and *Elsinoe* species. Also light-requiring fungal toxins of other types are known, produced by *Cercospora species* and *Dothistroma pini* (Jalal et al. 1992, Stoessl et al. 1990).

4. PHOTOTOXIC DRUGS AND COSMETICS

Many phototoxic drugs are either antibiotics or medications for blood pressure and heart disease, but there are also many others. In combination with light they may cause extreme sunburn, vesicles, hives and edema. Among antibiotics photosensitivity reactions have more commonly been noted after administration of the following: Doxycycline ("Vibramycine" etc.), demeclocyline, tetracycline (Fig. 5), nalidixic acid and lomefloxacin. For blood pressure and heart medications a similar "short list" is hydrochlorothiazide (occurs as an ingredient in a large number of formulations), chlorothiazide, furosemide and amiodarone. Amiodarone is responsible for an unusually high number of cases. Among other drugs causing photosensitivity reactions chlorpromazine and other phenothiazines, and birth control pills containing estrogens may be mentioned.

Somewhat surprising is the fact that also sun lotions containing para-amino benzoic acid (PABA) or esters of it, which are sold to protect from the sun, are also a common cause of photosensititivity. These substances were selected for their ability to absorb UV-B radiation (daylight with wavelength below 315 nm), since formerly this radiation was supposed to be the only threat from sunlight. At the long-wavelength edge of their absorption band they let radiation through to depths where they can cause the photosensitivity reactions.

It is well known that use of perfumes in combination with sunlight is unwise, because many perfumes are phototoxic or at least discolour the skin when exposed to sunlight. This is, of course, because many, if not most, of them are based on plant extracts and often contain substances mentioned in the section on phototoxins in plant defence. Already Freund (1916) described skin discolourations which he attributed to Eau de Cologne containing bergamot oil, although he did not clearly understand the role of sunlight. Bergamot orange, *Citrus bergamia*, like many other *Citrus* species, has later been found to contain photosensitising substances.

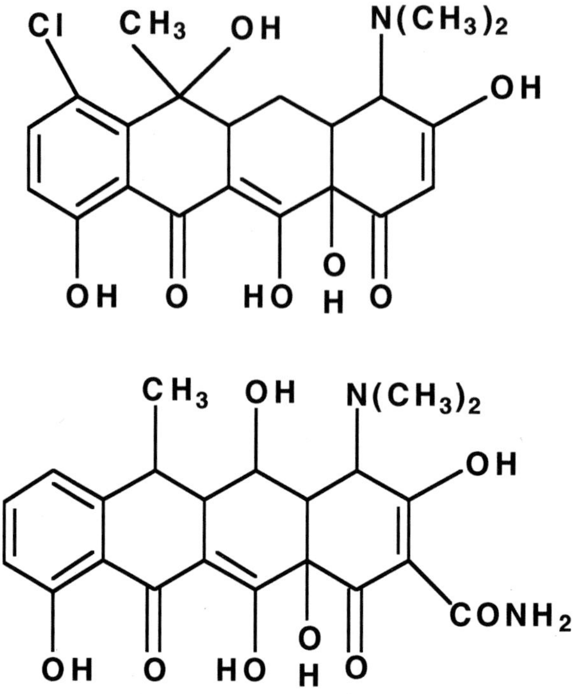

Figure 5. Top: Chlortetracycline, the first tetracycline, introduced in 1948. Tetracycline itself (introduced in 1952) has the same structure with hydrogen in place of chlorine. Bottom: Doxycycline, introduced in 1968, is one of the most potent photosensitisers among the tetracyclines.

A book (Miranda 2001) has been assembled from 28 articles in the journal Photochemistry & Photobiology, and is currently freely downloadable from the journal's homepage.

5. METABOLIC DISTURBANCES LEADING TO PHOTOTOXIC EFFECTS OF PORPHYRINS OR RELATED COMPOUNDS

A number of different disturbances in both men and animals lead to the appearance in the skin of phototoxic compounds such as uro- and coproporphyrinogens (porphyrin precursors), protoporphyrin IX (the immediate precursor of heme. Fig. 6), and phylloerythrin (a breakdown product of chlorophyll, Fig. 7). These substances are phototoxins of type II, generating singlet oxygen in light.

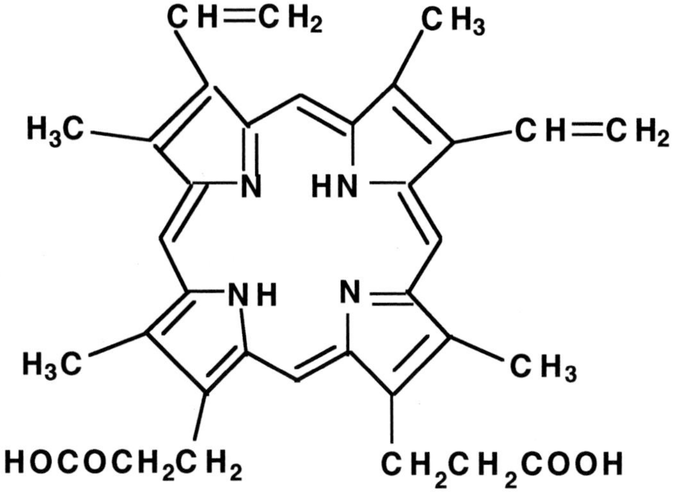

Figure 6. Protoporphyrin IX, the immediate precursor of heme, which accumulates in protoporphyria due to lack of ferrochelatase (or inhibition of the enzyme due to, e.g., lead poisoning).

In human patients a variety of diseases have been described which go under the common designation of porphyria. With the exception of a type called acute intermittent porphyria they lead to photosensitivity of the skin: variegate porphyria (Frank & Cristiano 1998) and hereditary coproporphyria (acute porphyrias with increased levels of both porphyrin precursors and porphyrins) and porphyria cutanea tarda, erythropoietic protoporphyria and congenital porphyria (non-acute porphyrias with increase levels of porphyrins). Porphyria is either due to a disturbance in the liver (hepatic porphyria or protoporphyria) or in the red blood cells (erythropoietic porphyria or protoporphyria). To complicate things further, an erythrohepatic porphyria has recently been described (Gauer et al. 1995), and erythropoietic porphyria may lead to secondary damage to the liver.

Porphyria may be inherited or acquired, and even in cases when it is caused by environment, life-style, alcohol (Doss et al. 1999), lead poisoning, liver transplantation (Sheth et al. 1994) etc., inherited predisposition may play a role. Gross et al. (2000) remarks in a recent review: "The molecular genetics of the porphyrias is very heterogenous. Nearly every family has its own mutation." Correct treatment of porphyria is therefore not easy, and requires very careful examination. Porphyria cannot be cured, but symptoms can often be ameliorated, also in other

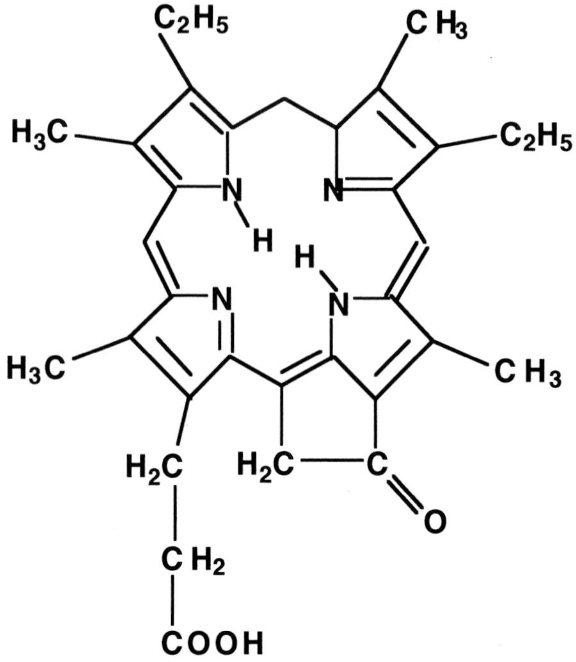

Figure 7. Phylloerythrin, which causes phototoxicity in animals due to malfunctioning of the liver. In healthy animals the substance is broken down in the liver.

ways than avoidance of light. There are prospects for a future cure of the erythropoietic protoporphyria. In this the enzyme ferrochelatase is lacking in the red blood cells, causing accumulation of protoporphyrin IX. It may become possible to cure this by retro-viral mediated gene transfer to the bone marrow (Todd 1994). At present the symptoms can be alleviated using β-carotene; interestingly the same compound as used by plants to quench triplet chlorophyll.

In ruminants another group of diseases with names as geeldikkop ("yellow head") and alveld is important. Geeldikkop, affecting sheep in South Africa, is the best studies of these. It is caused by saponins in the plant *Tribulus terrestris* (puncture vine or calthrops of the family Zygophyllacea) grazed by the sheep (Miles et al. 1994, Wilkins et al. 1996). Liver damage caused by these saponins prevents breakdown of phylloerythrin, a substance produced from chlorophyll by acid in the stomach and rumen bacteria. The phylloerythrin is circulated to skin capillaries, where it can be exposed to light. In other parts of the world *Panicum* species such as kleingrass or bambatsi grass, *P. coloratum* (Muchiri et al. 1980, Bridges et al. 1987, Regnault 1990) and switchgrass, *P. virgatum* (Puoli et al. 1992) cause the same disease in sheep and also in horses (Cornick et al 1988).

Similar symptoms were induced by *Myoporum laetum* in calves (Raposo et al. 1998), and also buttercup (*Ranunculus bulbosus*) has been suspected as a cause in cattle (Kelch et al. 1992). Mold fungi in hay and fungi in pasture can cause similar problems (Scruggs & Blue 1994, Casteel et al. 1995). Finally, it has been known for a long time that cyanobacterial toxins in drinking water can cause liver damage with associated photosensitivity in cattle. In the case of the fungus *Pithomyces cartarum* in lamb pasture (Hansen et al. 1994) it is not clear whether the photosensitivity is due to primary photosensitization or liver damage.

6. POLYCYCLIC AROMATIC HYDROCARBONS (PAHs) AS PHOTOTOXIC CONTAMINANTS IN AQUATIC ENVIRONMENTS

6.1. Nature and occurrence of PAHs

Some compounds have a potential to become toxic or acquire increased toxicity when they interact with natural or simulated sunlight. Such compounds with a possible environmental relevance include e.g. photoactive insecticides, such as naturally occurring α-terthienyl (Kagan et al. 1984, 1987) and some photodynamic dyes (Larson & Berenbaum 1988), a carbamate insecticide (Zaga et al. 1998), trinitrotoluene (TNT, an explosive) and some related compounds (Davenport et al. 1994), and polycyclic aromatic hydrocarbons (PAHs) (Newsted & Giesy 1987, Arfsten et al. 1996). PAHs, composed of multiple aromatic rings (Fig. 8) and present in coal and petroleum products, are widespread organic environmental contaminants, some having carcinogenic potential (Neff 1979, 1985). PAHs can be introduced into the environment, e.g., through incomplete combustion of organic matter. In aquatic environments oil spills, surface runoff from land, and industrial and domestic wastewaters are among the possible sources of PAH contamination, as well as airborne PAHs entering aquatic systems through dry fallout and rainfall (Neff 1979, 1985). Photoenhanced toxicity of some petroleum products (Pelletier et al. 1997) and creosote (Schirmer et al. 1999) has been related to phototoxicity of PAHs present. Furthermore, liquid-phase elutriates of petroleum containing sediments (Davenport & Spacie 1991), urban stormwater runoff (Ireland et al. 1996), as well as PAH-contaminated sediments (Ankley et al. 1994, Monson et al. 1995) contain phototoxic components, suggesting the role of PAHs. Interaction with solar radiation has also increases the toxicity of weathered oil, but possibly other constituents than PAHs contributed to this photoenhanced toxicity (Cleveland et al. 2000, Little et al. 2000).

Although generally considered relatively acutely nontoxic under normal laboratory lightning, numerous PAHs, such as anthracene, benzo[a]pyrene, fluoranthene, and pyrene have a potential to become highly toxic in the presence of UV radiation, and a risk that PAHs constitute through this photoenhanced toxicity, especially to aquatic organisms, has been recognized (Landrum et al. 1987, Arfsten et al. 1996, Schaeffer & Larson 1999). Since bioassays used to test the toxicity of chemicals are commonly carried out in the laboratory under artificial lightning without UV radiation, the risk related to photoactive compounds in natural conditions can be underestimated with traditional toxicity testing.

6.2. Mechanisms of PAH phototoxicity

Because of their chemical structure many PAHs absorb energy in the UV waveband (Newsted & Giesy 1987, Huang et al. 1993, Diamond et al. 2000, Huovinen et al. 2001) (Fig. 8). According to the quantitative structure/activity relationship (QSAR) model the phototoxicity of PAHs can be related to the HOMO-LUMO gap (i.e. energy difference between the highest occupied molecular orbital and the lowest unoccupied molecular orbital), which has been suggested as a suitable ground state index of the electronic structure relating to absorbed energy and molecular stability (Mekenyan et al. 1994). However, the comparison of the phototoxic potency of PAHs is complicated because it is also related to the bioaccumulation potential of each compound (Boese et al. 1998). Contaminated environments generally contain a mixture of numerous PAHs, and phototoxicity of PAH mixtures is regarded as somewhat additive (Swartz et al. 1997, Boese et al. 1999, Erickson et al. 1999). Also substituted PAHs can contribute to phototoxicity (Boese et al. 1998, Kosian et al. 1998). With some exceptions, phototoxicity is likely in a substituted PAH only if the aromatic structure of its parent compound is phototoxic (Veith et al. 1995).

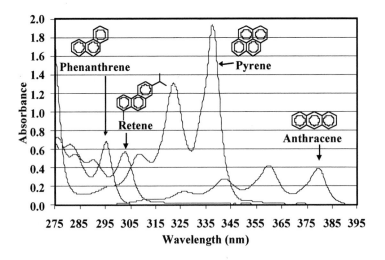

Figure 8. Absorption spectra of anthracene, pyrene, phenanthrene and retene (7-isopropyl-1-methylphenanthrene) (10 mg l^{-1} in dimethylsulfoxide). Modified from Chemosphere, Vol. 45, 2001. Pages 683-691, Huovinen et al, Photoinduced toxicity of retene to Daphnia magna under enhanced UV-B radiation), with permission from Elsevier Science.

Phototoxicity of PAHs is reported to occur mainly via photosensitization and/or photomodification reactions. The role of PAHs as active photosensitizers has been related to their capability of forming triplet states (Newsted & Giesy 1987, Larson & Berenbaum 1988) and transfer their triplet energy to oxygen, potentially resulting in the formation of biologically damaging singlet oxygen (Larson & Berenbaum 1988,

Chapter 1 of this volume). Photosensitization reactions of bioaccumulated PAHs inside organisms are regarded as important mechanisms for phototoxicity, which is supported by studies demonstrating enhanced toxicity when bioaccumulation of PAHs in aquatic organisms is followed by exposure to UV radiation in clean water (Bowling et al. 1983, Allred & Giesy 1985, Ankley et al. 1994, 1997, Boese et al. 1997, Monson et al. 1999, Huovinen et al. 2001). Phototoxicity is considered a function of both PAH dose in tissue and UV intensity (Ankley et al. 1995).

In addition to photodegradation (Neff 1979, 1985), PAHs may be photomodified into more toxic form e.g. via photooxidation (McConkey et al. 1997, Mallakin et al. 1999). Photomodification of PAH can result in a complex mixture of products (Mallakin et al. 1999). The enhanced toxicity of many photoproducts can probably be attributed to increased aqueous solubility and thus potentially increased bioavailability, as well as increased bioactivity (Duxbury et al. 1997, McConkey et al. 1997). Although many photomodified PAHs are toxic as such, they can also be phototoxic as well (Huang et al. 1993, Mallakin et al. 1999). According to model predictions, photosensitization and photomodification contribute additively to phototoxicity (Huang et al. 1997a, Krylov et al. 1997, Mezey et al. 1998).

6.3. Factors affecting exposure to phototoxicity of PAHs in aquatic systems

Due to their hydrophobic nature, PAHs tend to accumulate in sediments and organic particles (Neff 1979, 1985) resulting in a decrease in their availability to organisms. However, disturbance of contaminated sediment, e.g., during a hurricane or dredging may result in mobilization and resuspension of PAHs in the water, increasing the risk of phototoxicity (Davenport & Spacie 1991, Ireland et al. 1996). On the other hand, because of their lipophilic nature, PAHs also tend to bioaccumulate in organisms (Neff 1979, 1985). In addition to waterborne PAHs, a possible cause of exposure to PAHs is their bioaccumulation from contaminated sediments (Ankley et al. 1994, Boese et al. 1998), and potentially also from food.

The potential for UV exposure varies in different types of waters. UV-B penetration depths can range from a few centimeters in highly humic lakes (Kirk 1994, Lean 1998, Huovinen et al. 2000) to dozens of meters in clear oceanic waters (Smith et al. 1992, Kirk 1994). The spectra of underwater UV irradiance change with depth, as penetration decreases with decreasing wavelength (Kirk 1994, Scully & Lean 1994, Lean 1998, Huovinen et al. 2000). This spectral change and variation among natural waters affect the potential for phototoxicity (Barron et al. 2000), since the phototoxic response is related to the UV absorption characteristics of a compound (Newsted & Giesy 1987, Diamond et al. 2000, Huovinen et al. 2001) (Fig. 8). Aquatic biota in PAH contaminated areas (particularly in clear, shallow waters and littoral areas, which often provide habitats for various aquatic organisms during reproduction and early development) may be at risk. UV exposure and thus phototoxicity can also be increased e.g. during low flow (Ireland et al. 1996), or potentially when organisms move up in the water column. Also other factors, such as increased turbidity, which reduce the penetration of UV radiation in the water column, can attenuate phototoxicity as well (Ireland et al. 1996).

In addition to strongly contributing to attenuation of UV radiation (Kirk 1994, Scully & Lean 1994, Morris et al. 1995, Williamson et al. 1996, Lean 1998,

Huovinen et al. 2000), humic substances have a complex role in aquatic systems in potentially affecting the phototoxicity of PAHs. Dissolved humic material may mitigate the potential for photoinduced toxicity (Gensemer et al. 1998, 1999) by reducing the bioaccumulation of PAHs to organisms (Oris et al. 1990, Weinstein & Oris 1999). However, UV radiation may also induce photochemical reactions, such as photochemical degradation in dissolved organic carbon, which may increase the UV transparency (Morris & Hargreaves 1997) and thus the risk for phototoxicity. Humic substances are also potential photosensitizers (Larson & Berenbaum 1988).

6.4. Phototoxicity of PAHs to aquatic biota

Phototoxicity of PAHs has been demonstrated in a variety of aquatic organisms, including bacteria, phytoplankton, aquatic higher plants, zooplankton, benthic invertebrates, insect larvae, as well as in early life stages of amphibians, bivalves and fish. Responses in biota to PAH phototoxicity range from acute lethality to chronic effects, such as reproductive impairment.

The major photooxidation product of phenanthrene is more toxic to marine bacteria *Photobacterium phosphoreum* than phenanthrene (McConkey et al. 1997). Anthracene can inhibit algal growth and primary production (Gala & Giesy 1992) and reduce cell viability (Gala & Giesy 1994) in the green alga *Selenastrum capricornutum* in the presence of UV radiation. Growth inhibition due to phototoxicity of benzo[a]pyrene has also been shown (Cody et al. 1984). Exposure of natural phytoplankton assemblages to intact or photomodified anthracene in sunlight diminishes PS II photosynthetic efficiency and photosynthetic quantum yield, intact anthracene being more phototoxic than its photomodified product (Marwood et al. 1999). Photomodification of PAHs into more toxic form has been suggested as an important mechanism for phototoxicity to the aquatic higher plant the duckweed *Lemna gibba*, causing, e.g., inhibition of growth (Huang et al. 1993, Mallakin et al. 1999). Anthracene inhibits photosynthesis after its photomodification, with the primary site of action possibly being the electron transport at or near PSI, followed by inhibition of PSII (Huang et al. 1997b). Intact and photooxidized PAHs can accumulate in the thylakoids and microsomes of *L. gibba*, which possibly are the most susceptible subcellular compartments (Duxbury et al. 1997). Models using 16 different PAHs have indicated an additive role of photomodification and photosensitization in toxicity to *L. gibba* (Huang et al. 1997a, Krylov et al. 1997, Mezey et al. 1998).

Acute phototoxicity of several PAHs has been demonstrated in the aquatic crustacean *Daphnia magna* (Newsted & Giesy 1987). In *D. pulex*, activation of anthracene on or within organisms was the cause of acute phototoxicity observed in the presence of solar radiation (Allred & Giesy 1985). Photosensitization of bioaccumulated anthracene and pyrene appeared to be the primary mechanism for acute photoinduced toxicity in *D. magna* (Huovinen et al. 2001). Also sublethal effects, such as reduced feeding efficiency due to fluoranthene phototoxicity have been shown in *D. magna* (Hatch & Burton 1999a). Chronic effects due to anthracene phototoxicity, like reduced fecundity, were by Holst & Giesy (1989 Genetic and ecological fitness were less severely impacted than clutch size and survivorship (Foran et al. 1991.

Photoactivation of PAHs bioaccumulated in the benthic invertebrates *Lumbriculus variegatus* (an oligochaete) and *Hyalella azteca* (an amphipod) from contaminated field sediments i thought to cause increased mortality (Ankley et al. 1994, Monson et al. 1995). UV exposure can increase the toxicity of PAH contaminated sediments to the infaunal amphipods *Rhepoxynius abronius* and *Leptocheirus plumulosus*, decreasing survival and ability to rebury (Boese et al. 2000). Exposure via water to fluoranthene and subsequently to UV radiation demonstrated increased mortality in *L. variegatus* as a function of both PAH dose in tissue and UV intensity (Ankley et al. 1995). Furthermore, the reported effects of phototoxicity of fluoranthene include reduced feeding efficiency of *H. azteca* (Hatch & Burton 1999a). PAH phototoxicity can reduce survival and adult emergence in larvae of the mosquito *Aedes aegypti*, and quite similar phototoxic potency has been observed in highly carcinogenic benzo[a]pyrene and non-carcinogenic pyrene (Kagan & Kagan 1986).

Increased mortality due to phototoxicity of fluoranthene has been demonstrated in glochidial larvae of the freshwater mussel *Utterbackia imbecillis* exposed to waterborne PAH (Weinstein 2001) and in embryos of the marine bivalve *Mulinia lateralis* with body burden of PAH through maternal transfer from benthic adults (Pelletier et al. 2000). In addition to fluoranthene, anthracene and pyrene as well as some petroleum products containing PAHs have displayed phototoxicity to larvae and juveniles of *M. lateralis* (Pelletier et al. 1997).

Juvenile bluegill sunfish (*Lepomis macrochirus*) kept in anthracene contaminated water died upon exposure to sunlight in clean water, indicating photoactivation of PAH inside organism rather than formation of toxic photoproducts in the water (Bowling et al. 1983). Structural changes in gills and dorsal epidermis have been detected to result from anthracene phototoxicity (Oris & Giesy 1985), and the general disruption of cell membrane integrity and function including gills, blood, and other tissues appear to be an important mode of acute phototoxicity in fish (Oris & Giesy 1985, McCloskey & Oris 1993). In juvenile fathead minnow (*Pimephales promelas*) respiratory stress has been indicated as the cause of death resulting from phototoxicity of fluoranthene, a disruption of mucosal cell membrane function and integrity being the mode of phototoxic action (Weinstein et al. 1997). Furthermore, studies using a cell line from fish gill indicated a specific action on lysosomes contributing to photocytotoxicity of PAHs (Schirmer et al. 1998). Studies with fish liver microsomes (Choi & Oris 2000a) and fish hepatoma cell line (Choi & Oris 2000b) suggest that lipid peroxidation induced by reactive oxygen species is an important factor in the phototoxicity of anthracene to fish. Although reducing the reproductive output in fathead minnow as such, maternal transfer of anthracene in eggs with subsequent UV exposure can further decrease reproductive potential by decreasing hatching rate of eggs and causing teratogenic effects in fry (Hall & Oris 1991).

Phototoxicity of fluoranthene reduces survival of the northern leopard frog larvae (*Rana pipiens*) (Monson et al. 1999), and causes sublethal effects on locomotor behavior, signs of necrosis and structural alterations in the skin of the bullfrog larvae (*Rana catesbeiana*) (Walker et al. 1998), and enhances teratogenic effects in the African clawed frog (*Xenopus laevis*) (Hatch & Burton 1998). Reduced survival of late embryonic stages of *R. pipiens* results from phototoxicity of anthracene (Kagan et al. 1984).

Species vary in their sensitivity to the phototoxicity of PAHs (Boese et. al. 1997, Hatch & Burton 1998, Spehar et al. 1999), which could be related to e.g. to behavioral (Hatch & Burton 1999b) and potentially to metabolic and morphological differences. Previous exposure of organisms to UV radiation can lead to development of protective mechanisms reducing their sensitivity (Boese et al. 1997). In algae carotenoid pigments may mitigate the effects of PAH phototoxicity by quenching singlet oxygen generated from PAH photosensitization (Gala & Giesy 1993). Furthermore, a possibility for repair of phototoxic effects has been demonstrated (Oris & Giesy 1986).

References

Allred, P.M. & Giesy, J.P. (1985). Solar radiation-induced toxicity of anthracene to *Daphnia pulex*. *Environ. Toxicol. Chem.*, *4*, 219-226.

Ankley, G.T., Collyard, S.A., Monson, P.D. & Kosian, P.A. (1994). Influence of ultraviolet light on the toxicity of sediments contaminated with polycyclic aromatic hydrocarbons. Environ. *Toxicol. Chem.*, *13*, 1791-1796.

Ankley, G.T., Erickson, R.J., Sheedy, B.R., Kosian, P.A., Mattson, V.R. & Cox, J.S. (1997). Evaluation of models for predicting the phototoxic potency of polycyclic aromatic hydrocarbons. *Aquat. Toxicol.*, *37*, 37-50.

Ankley, G.T., Tietge, J.E., DeFoe, D.L., Jensen, K.M., Holcombe, G.W., Durhan, E.J. & Diamond, S.A. (1998). Effects of ultraviolet light and methoprene on survival and development of *Rana pipiens*. *Environ. Toxicol., Chem. 17*, 2530-2542.

Arfsten, D.P., Schaeffer, D.J. & Mulveny, D.C. (1996). The effects of near ultraviolet radiation on the toxic effects of polycyclic aromatic hydrocarbons in animals and plants: a review. *Ecotox. Environ. Safety 33*, 1-24.

Barron, M.G., Little, E.E., Calfee, R. & Diamond, S. (2000). Quantifying solar spectral irradiance in aquatic habitats for the assessment of photoenhanced toxicity. *Environ. Toxicol. Chem., 19*, 920-925.

Bjellerup, M. (1986). *Tetracycline phototoxicity. An experimental and clinical study.* Diss. Lund University, CODEN LUMEDW/(MEHM-1002)/1-59(1986).

Boese, B.L., Lamberson, J.O., Swartz, R.C. & Ozretich, R.J. (1997). Photoinduced toxicity of fluoranthene to seven marine benthic crustaceans. *Arch. Environ. Contam. Toxicol., 32*, 389-393.

Boese, B.L., Lamberson, J.O., Swartz, R.C., Ozretich, R. & Cole, F. (1998). Photoinduced toxicity of PAHs and alkylated PAHs to a marine infaunal amphipod (*Rhepoxynius abronius*). *Arch. Environ. Contam. Toxicol., 34*, 235-240.

Boese, B.L., Ozretich, R.J., Lamberson, J.O., Swartz, R.C., Cole, F.A., Pelletier, J. & Jones, J. (1999). Toxicity and phototoxicity of mixtures of highly lipophilic PAH compounds in marine sediment: Can the ∑PAH model be extrapolated? *Arch. Environ. Contam. Toxicol., 36*, 270-280.

Boese, B.L., Ozretich, R.J., Lamberson, J.O., Cole, F.A., Swartz, R.C. & Ferraro, S.P. (2000). Phototoxic evaluation of marine sediments collected from a PAH-contaminated site. *Arch. Environ. Contam. Toxicol., 38*, 274-282.

Bowling, J.W., Leversee, G.J., Landrum, P.F. & Giesy, J.P. (1983). Acute mortality of anthracene-contaminated fish exposed to sunlight. *Aquat. Toxicol., 3*, 79-90.

Bridges, D..H., Camp, B.J., Liningston, C.W. and Bailey, E.M. 1988. Kleingrass (*Panicum coloratum* L.) poisoning in sheep. *Vet. Path., 24*, 525-531.

Casteel, S.W., Rottinghaus, G.E., Hohnson, G.C. & Wicklow, D.T. (1995). Liver disease in cattle induced by consumption of moldy hay. *Vet. Hum. Toxicol., 37*, 248-251.

Choi, J. & Oris, J.T. (2000a). Evidence of oxidative stress in bluegill sunfish (*Lepomis macrochirus*) liver microsomes simultaneously exposed to solar ultraviolet radiation and anthracene. *Environ. Toxicol. Chem., 19*, 1795-1799.

Choi, J. & Oris, J.T.(2000b). Anthracene photoinduced toxicity to PLHC-1 cell line (*Poeciliopsis lucida*) and the role of lipid peroxidation in toxicity. *Environ. Toxicol. Chem., 19*, 2699-2706.

Cleveland, L., Little, E.E., Calfee, R.D. & Barron, M.G. (2000). Photoenhanced toxicity of weathered oil to *Mysidopsis bahia*. *Aquat. Toxicol., 49,* 63-76.

Cody, T.E., Radike, M.J. & Warshawsky, D. (1984). The phototoxicity of benzo[a]pyrene in the green alga *Selenastrum capricornutum*. *Environ. Res., 35,* 122-132.

Cornick, J.L., Carter, G.K. & Bridges, C.H.(1988). Kleingrass-associated hepatoxicosis in horses. *J. Am. Vet. Med. Assoc., 193,* 932-935.

Daub, M.E & Ehrenshaft, M. (2000). The photoactivated *Cercospora* toxin cercosporin: Contributions to plant disease and fundamental biology. *Annu. Rev. Phytopathol., 38,* 461-490.

Davenport, R. & Spacie, A. (1991). Acute phototoxicity of harbor and tributary sediments from lower Lake Michigan. *J. Great Lakes Res., 17,* 51-56.

Davenport, R., Johnson, L.R., Schaeffer, D.J. & Balbach, H. (1994). Phototoxicology. 1. Light-enhanced toxicity of TNT and some related compounds to *Daphnia magna* and *Lytechinus variagatus* embryos. *Ecotoxicol. Environ. Safety, 27,* 14-22.

Delaey, E., Vandenbogaerde, A., Merlevede, W. & de Witte, P. (2000). Photocytotoxicity of hypericin in nomroxic and hypoxic conditions. *J. Photochem. Photobiol. B: Biology, 56,* 19-24.

Diamond, S.A., Mount, D.R., Burkhard, L.P., Ankley, G.T., Makynen, E.A. & Leonard, E.N. (2000). Effect of irradiance spectra on the photoinduced toxicity of three polycyclic aromatic hydrocarbons. *Environ. Toxicol. Chem., 19,* 1389-1396.

Doss, M.O., Kuhnel, A., Gross, U & Sieg, I. (1999). Hepatic porphyrias and alcohol. *Medizinische Klinik, 94,* 314-328.

Downum, K.R. 1992. Light-activated plant defence. *New Phytol., 122,* 401-420.

Duxbury, C.L., Dixon, D.G. & Greenberg, B.M. (1997). Effects of simulated solar radiation on the bioaccumulation of polycyclic aromatic hydrocarbons by the duckweed *Lemna gibba*. *Environ. Toxicol. Chem., 16,* 1739-1748.

Erickson, R.J., Ankley, G.T., DeFoe, D.L., Kosian, P.A. & Makynen, E.A. (1999). Additive toxicity of binary mixtures of phototoxic polycyclic aromatic hydrocarbons to the oligochaete *Lumbriculus variegatus*. *Toxicol. Appl. Pharmacol., 154,* 97-105.

Foran, J.A., Holst, L.L. & Giesy, J.P. (1991). Effects of photoenhanced toxicity of anthracene on ecological and genetic fitness of *Daphnia magna*: a reappraisal. *Environ. Toxicol. Chem., 10,* 425-427.

Frank, J. & Christiano, A.M. (1998). Variegate porphyria: past, present and future. Skin Pharmacol. *Appl. Skin Physiol., 11,* 310-320.

Freund, E. (1916). Über bisher noch nicht beschriebene kunstliche Hautverfarbungen. *Dermatol. Wochenschrift, 63,* 931-933.

Gala, W.R. & Giesy, J.P. (1992). Photo-induced toxicity of anthracene to the green alga, *Selenastrum capricornutum*. *Arch. Environ. Contam. Toxicol., 23,* 316-323.

Gala, W.R. & Giesy, J.P. (1993). Using the carotenoid biosynthesis inhibiting herbicide, Fluridone, to investigate the ability of carotenoid pigments to protect algae from the photoinduced toxicity of anthracene. *Aquat. Toxicol., 27,* 61-70.

Gala, W.R. & Giesy, J.P. (1994). Flow cytometric determination of the photoinduced toxicity of anthracene to the green alga *Selenastrum capricornutum*. *Environ. Toxicol., Chem. 13,* 831-840.

Gensemer, R.W., Dixon, D.G. & Greenberg, B.M. (1998). Amelioration of the photo-induced toxicity of polycyclic aromatic hydrocarbons by a commercial humic acid. *Ecotoxicol. Environ. Safety, 39,* 57-64.

Gensemer, R.W., Dixon, D.G. & Greenberg, B.M. (1999). Using chlorophyll *a* fluorescence to detect the onset of anthracene photoinduced toxicity in *Lemna gibba*, and the mitigating effects of a commercial humic acid. *Limnol. Oceanogr., 44,* 878-888.

Gonzales, E. & Gonzales, S. (1996). Drug photosensitivity, idiopathic photodermatoses, and sunscreens. *J. Am. Acad. Dermatol., 35,* 871885.

Hansen, D.E., McCoy, R.D., Hedstrom, O.R., Snyder, S.P. & Ballerstedt, P.B. (1994). Photosensitization associated with exposure to *Pithomyces cartarum* in lambs. *J. Am. Vet. Med. Assoc., 204,* 1668-1671.

Hatch, A.C. & Burton, G.A., Jr. (1998). Effects of photoinduced toxicity of fluoranthene on amphibian embryos and larvae. Environ. Toxicol. Chem. 17:1777-1785.

Hatch, A.C. & Burton, G.A., Jr. (1999a). Phototoxicity of fluoranthene to two freshwater crustaceans, *Hyalella azteca* and *Daphnia magna*: measures of feeding inhibition as a toxicological endpoint. *Hydrobiol., 400,* 243-248.

Hatch, A.C. & Burton, G.A., Jr. (1999b). Photo-induced toxicity of PAHs to *Hyalella azteca* and *Chironomus tentans*: effects of mixtures and behavior. *Environ. Pollut., 106,* 157-167.

Hall, A.T. & Oris, J.T. (1991). Anthracene reduces reproductive potential and is maternally transferred

236

during long-term exposure in fathead minnows. *Aquat. Toxicol., 19*, 249-264.

Holst, L.L. & Giesy, J.P. (1989). Chronic effects of photoenhanced toxicity of anthracene on *Daphnia magna* reproduction. *Environ. Toxicol. Chem., 8*, 933-942.

Huang, X.-D., Dixon, D.G. & Greenberg, B.M. (1993). Impacts of UV radiation and photomodification on the toxicity of PAHs to the higher plant *Lemna gibba* (duckweed). *Environ. Toxicol. Chem., 12*, 1067-1077.

Huang, X.-D., Krylov, S.N., Ren, L., McConkey, B.J., Dixon, D.G. & Greenberg, B.M. (1997a). Mechanistic quantitative structure-activity relationship model for the photoinduced toxicity of polycyclic aromatic hydrocarbons. II. An empirical model for the toxicity of 16 polycyclic aromatic hydrocarbons to the duckweed *Lemna gibba* L. G-3. *Environ. Toxicol. Chem., 16*, 2296-2303.

Huang, X.-D., McConkey, B.J., Babu, T.S. & Greenberg, B.M. (1997b). Mechanisms of photoinduced toxicity of photomodified anthracene to plants: inhibition of photosynthesis in the aquatic higher plant *Lemna gibba* (duckweed). *Environ. Toxicol. Chem., 16*, 1707-1715.

Huovinen, P.S., Penttilä, H. & Soimasuo, M.R. (2000). Penetration of UV radiation into Finnish lakes with different characteristics. *Int. J. Circumpolar Health, 59*, 15-21.

Huovinen P.S., Soimasuo M.R. & Oikari, A.O.J. (2001). Photoinduced toxicity of retene to *Daphnia magna* under enhanced UV-B radiation. *Chemosphere 45*, 683-691.

Ireland, D.S., Burton, G.A., Jr. & Hess, G.G. (1996). *In situ* toxicity evaluations of turbidity and photoinduction of polycyclic aromatic hydrocarbons. *Environ. Toxicol. Chem., 15*, 574-581.

Jalal, M.A.F., Hossain, M.B., Robeson, D.I. & van der Helm, D. (1992). *Cercospora beticola* phytotoxins: cebetins that are photoactive, Mg2+-binding, chlorinated anthraquinone-xanthone conjugates. *J. Am. Chem. Soc., 114*, 5967-5971.

Kagan, J. & Kagan, E.D. (1986). The toxicity of benzo[a]pyrene and pyrene in the mosquito *Aedes aegypti*, in the dark and in the presence of ultraviolet light. *Chemosphere, 15*, 243-251.

Kagan, J., Kagan, P.A. & Buhse, H.E., Jr. (1984). Light-dependent toxicity of a-terthienyl and anthracene toward late embryonic stages of *Rana pipiens. J. Chem. Ecol, 10*, 1115-1122.

Kagan, J., Bennett, W.J., Kagan, E.D., Maas, J.L., Sweeney, S.A., Kagan, I.A., Seigneurie, E. & Bindokas, V. (1987). a-Terthienyl as a photoactive insecticide: toxic effects on nontarget organisms. *In* Heitz, J.R. & Downum, K.R. (eds), *Light-activated pesticides*, pp. 176-191. ACS Symposium Series 339. Washington, D.C.: American Chemical Society.

Kelch, W.J., Kerr, L.A., Adair, H.S. & Boyd, G.D. (1992). Suspected buttercup (*Ranunculus bulbosus*) toxicosis with secondary photosensitization in Charolais heifer. *Vet. Hum. Toxicol., 34*, 238-239.

Kirk J.T.O. (1994). Optics of UV-B radiation in natural waters. *Arch. Hydrobiol. Beih. Ergebn. Limnol., 43*, 1-16.

Kosian, P.A., Makynen, E.A., Monson, P.D., Mount, D.R., Spacie, A., Mekenyan, O.G. & Ankley, G.T. (1998). Application of toxicity-based fractionation techniques and structure-activity relationship models for the identification of phototoxic polycyclic aromatic hydrocarbons in sediment pore water. *Environ. Toxicol. Chem., 17*, 1021-1033.

Krylov, S.N., Huang, X.-D., Zeiler, L.F., Dixon, D.G. & Greenberg, B.M. (1997). Mechanistic quantitative structure-activity relationship model for the photoinduced toxicity of polycyclic aromatic hydrocarbons: I. Physical model based on chemical kinetics in a two-compartment system. *Environ. Toxicol. Chem., 16*, 2283-2295.

Landrum, P.F., Giesy, J.T. & Allred, P.M. (1987). Photoinduced toxicity of polycyclic aromatic hydrocarbons to aquatic organisms. *In* Vandermeulen, J.H. & Hrudey, S.E. (eds), *Oil in freshwater: chemistry, biology, countermeasure technology*, pp.304-318. Proc. Symp. Oil Pollution in freshwater, Edmonton, Alberta, Canada. New York: Pergmon Press.

Larson, R.A. & Berenbaum, M.R. (1988). Environmental phototoxicity. Solar ultraviolet radiation affects the toxicity of natural and man-made chemicals. *Environ. Sci. Technol., 22*, 354-360.

Lean, D. (1998). Attenuation of solar radiation in humic waters. *In* Hessen, D.O. & Tranvik, L.J. (eds), *Aquatic humic substances. Ecology and biogeochemistry. Ecol. Studies 133*, 109-124. Berlin: Springer-Verlag.

Little, E.E., Cleveland, L., Calfee, R. & Barron, M.G. (2000). Assessment of the photoenhanced toxicity of a weathered oil to the tidewater silverside. *Environ. Toxicol. Chem., 19*, 926-932.

Ljunggren, B. (1978). *Drug phototoxicity. An experimental study on phototoxic inflammation with special reference to phenothiazines.* Diss. Lund University, pp. 37.

Ljunggren, B. & Bjellerup, M. (1986). Systemic drug photosensitivity. *Photodermatology, 3*, 26-35.

Lovell, C.R. (1993). *Plants and the skin.* Oxford: Blackwell Scientific Publ.

Mallakin, A., McConkey, B.J., Miao, G., McKibben, B., Snieckus, V., Dickson, D.G. & Greenberg, B.M.

(1999). Impacts of structural photomodification on the toxicity of environmental contaminants: anthracene photooxidation products. *Ecotoxicol. Environ. Safety, 43,* 204-212.

Marwood, C.A., Smith, R.E.H., Solomon, K.R., Charlton, M.N. & Greenberg, B.M. (1999). Intact and photomodified polycyclic aromatic hydrocarbons inhibit photosynthesis in natural assemblages of Lake Erie phytoplankton exposed to solar radiation. *Ecotoxicol. Environ. Safety, 44,* 322-327.

McCloskey, J.T. & Oris, J.T. (1991). Effect of water temperature and dissolved oxygen concentration on the photo-induced toxicity of anthracene to juvenile bluegill sunfish (*Lepomis macrochirus*). *Aquat. Toxicol., 21,* 145-156.

McCloskey, J.T. & Oris, J.T. (1993). Effect of anthracene and solar ultraviolet radiation exposure on gill ATPase and selected hematologic measurements in the bluegill sunfish (*Lepomis macrochirus*). *Aquat. Toxicol., 24,* 207-218.

McConkey, B.J., Duxbury, C.L., Dixon, D.G. & Greenberg, B.M. (1997). Toxicity of a PAH photooxidation product to the bacteria *Photobacterium phosphoreum* and the duckweed *Lemna gibba*: Effects of phenanthrene and its primary photoproduct, phenanthrenequinone. *Environ. Toxicol. Chem., 16,* 892-899.

Mekenyan, O.G., Ankley, G.T., Veith, G.D. & Call, D.J. (1994). QSARs for photoinduced toxicity: I. Acute lethality of polycyclic aromatic hydrocarbons to *Daphnia magna*. *Chemosphere, 28,* 567-582.

Mezey, P.G., Zimpel, Z., Warburton, P., Walker, P.D., Irvine, D.G., Huang, X.-D., Dixon, D.G. & Greenberg, B.M. (1998). Use of quantitative shape-activity relationships to model the photoinduced toxicity of polycyclic aromatic hydrocarbons: Electron density shape features accurately predict toxicity. *Environ. Toxicol. Chem., 17,* 1207-1215.

Miles, C.O., Wilkins, A.L., Erasmus, G.L., Kellerman, T.S. & Coetzer, J. (1994). Photosensitivity in South Africa. 7. Chemical composition of biliary chrystals from a sheep with experimentally induced geeldikkop. *Onderstepoort J. Vet. Res., 61,* 215-222.

Miranda, M.A. (ed.) (2001). Phototoxicity of drugs: A decade of nonsteroidal anti-inflammatory 2-arylpropionic acids research. Downloadable from the Photochemistry & Photobiology homepage.

Monson, P.D., Ankley, G.T. & Kosian, P.A. (1995). Phototoxic response of *Lumbriculus variegatus* to sediments contaminated by polycyclic aromatic hydrocarbons. *Environ. Toxicol. Chem., 14,* 891-894.

Monson, P.D., Call, D.J., Cox, D.A., Liber, K. & Ankley, G.T. (1999). Photoinduced toxicity of fluoranthene to northern leopard frogs (*Rana pipiens*). *Environ. Toxicol. Chem., 18,* 308-312.

Morris, D.P., Zagarese, H., Williamson, C.E., Balseiro, E.G., Hargreaves, B.R., Modenutti, B., Moeller, R. & Queimalinos, C. (1995). The attenuation of solar UV radiation in lakes and the role of dissolved organic carbon. *Limnol. Oceanogr., 40,* 1381-1391.

Morris, D.P. & Hargreaves, B.R. (1997). The role of photochemical degradation of dissolved organic carbon in regulating the UV transparency of three lakes on the Pocono Plateau. *Limnol. Oceanogr., 42,* 239-249.

Muchiri, D.J., Bridges, C.H., Ueckert, D.N. & Bailey, E.M. (1980). Photosensitization of seheep on kleingrass pasture. *J. Am. Vet. Med. Assoc., 177,* 353-354.

Neff, J.M. (1979). *Polycyclic aromatic hydrocarbons in the aquatic environment. Sources, fates and biological effects*, pp. 262. London: Applied Science Publishers Ltd.

Neff, J.M. (1985). Polycyclic aromatic hydrocarbons. *In* Rand, G.M. & Petrocelli, S.R. (eds), *Fundamentals of aquatic toxicology. Methods and applications*, pp. 416-454. New York: Hemisphere Publishing Corporation.

Newsted, J.L. & Giesy, J.P. (1987). Predictive models for photoinduced acute toxicity of polycyclic aromatic hydrocarbons to *Daphnia magna*, Strauss (Cladocera, Crustacea). *Environ. Toxicol. Chem., 6,* 445-461.

Oris, J.T. & Giesy, J.P., Jr. (1985). The photoenhanced toxicity of anthracene to juvenile sunfish (*Lepomis* spp.). *Aquat. Toxicol., 6,* 133-146.

Oris, J.T. & Giesy, J.P., Jr. (1986). Photoinduced toxicity of anthracene to juvenile bluegill sunfish (*Lepomis macrochirus* Rafinesque): photoperiod effects and predictive hazard evaluation. *Environ. Toxicol. Chem., 5,* 761-768.

Oris, J.T., Hall, A.T. & Tylka, J.D. (1990). Humic acids reduce the photo-induced toxicity of anthracene to fish and daphnia. *Environ. Toxicol. Chem., 9,* 575-583.

Pathak, M.A. (1986). Phytodermatitis. *Clin. Dermatol, 4,* 102-121.

Pathak, M.A. & Fitzpatrick, T.B. (1992). The evolution of photochemotherapy with psoralens and UVA (PUVA): 2000 BC to 1992 AD. *J. Photochem. Photobiol. B: Biology, 14,* 3-22.

Pelletier, M.C., Burgess, R.M., Ho, K.T., Kuhn, A., McKinney, R.A. & Ryba, S.A. (1997). Phototoxicity of individual polycyclic aromatic hydrocarbons and petroleum to marine invertebrate larvae and

238

juveniles. *Environ. Toxicol. Chem., 16,* 2190-2199.

Pelletier, M.C., Burgess, R.M., Cantwell, M.G., Serbst, J.R., Ho, K.T. & Ryba, S.A. (2000). Importance of maternal transfer of the photoreactive polycyclic aromatic hydrocarbon fluoranthene from benthic adult bivalves to their pelagic larvae. *Environ. Toxicol. Chem., 19,* 2691-2698.

Puoli, J.R., Reid, R.L. & Belesky, D.P. (1992). Photosensitization in lambs grazing switchgrass. *Agron. J., 84,* 1077-1080.

Raposo, J.B., Mendez, M.C., de adrade, B.B. & Riet-Correa, F. (1998). Experimental intoxication by *Myoporum laetum* in cattle. *Vet. Hum. Toxicol., 40,* 275-277.

Regnault, T.R.H. (1990). Secondary photosentization of sheep grazing bambatsi grass (*Panicum coloratum* var makarikariense). *Aust. Vet., J. 67,* 419.

Schaeffer, D.J. & Larson, R.A. (1999). Phototoxicology. *The Chemist, July-Aug.,* 18-24.

Schirmer, K., Chan, A.G.J., Greenberg, B.M., Dixon, D.G. & Bols, N.C. (1998). Ability of 16 priority PAHs to be photocytotoxic to a cell line from the rainbow trout gill. *Toxicology, 127,* 143-155.

Schirmer, K., Herbrick, J.-A.S., Greenberg, B.M., Dixon, D.G. & Bols, N.C. (1999). Use of fish gill cells in culture to evaluate the cytotoxicity and photocytotoxicity of intact and photomodified creosote. *Environ. Toxicol. Chem., 18,* 1277-1288.

Scruggs, D.W. & Blue, G.K. (1994). Toxic hepatopathy and photosensitization in cattle fed moldy alfalfa hay. *J. Am. Vet. Med. Assoc., 204,* 264-266.

Scully, N.M. & Lean, D.R.S. (1994). The attenuation of ultraviolet radiation in temperate lake. Arch. Hydrobiol. *Beih. Ergebn. Limnol., 43,* 135-144.

Sheth, A.P., Esterly, N.B., Rabinowitz, L.G. & Poh-Fitzpatick, M.B. (1994). Cutaneous porphyria-like photosensitivity after liver transplantation. *Arch. Dermatol., 130,* 614-617.

Smith, R.C., Prézelin, B.B., Baker, K.S., Bidigare, R.R., Boucher, N.P., Coley, T., Karentz, D., MacIntyre, S., Matlick, H.A., Menzies, D., Ondrusek, M., Wan, Z. & Waters, K.J. (1992). Ozone depletion: ultraviolet radiation and phytoplankton biology in Antarctic waters. *Science, 255,* 952-959.

Spehar, R.L., Poucher, S., Brooke, L.T., Hansen, D.J., Champlin, D. & Cox, D.A. (1999). Comparative toxicity of fluoranthene to freshwater and saltwater species under fluorescent and ultraviolet light. *Arch. Environ. Contam. Toxicol., 37,* 496-502.

Stoessl, A., Abramowski, Z., Lester, H.H., rock, G.L. & Towers, G.H.N. (1990). Further toxic properties of the fungal metabolite dothistromin. *Mycopathologia, 112,* 179-186.

Swartz, R.C., Ferraro, S.P., Lamberson, J.O., Cole, F.A., Ozretich, R.J., Boese, B.L., Schults, D.W., Behrenfeld, M. & Ankley, G.T. (1997). Photoactivation and toxicity of mixtures of polycyclic aromatic hydrocarbon compounds in marine sediment. *Environ. Toxicol. Chem., 16,* 2151-2157.

Todd, D.J. (1994). Erythropoietic protoporphyria. *Br. J. Dermatol., 131,* 751-766.

Veith, G.D., Mekenyan, O.G., Ankley, G.T. & Call, D.J. (1995). A QSAR analysis of substituent effects on the photoinduced acute toxicity of PAHs. *Chemosphere, 30,* 2129-2142.

Walker, S.E., Taylor, D.H. & Oris, J.T (1998). Behavioral and histopathological effects of fluoranthene on bullfrog larvae (*Rana catesbeiana*). *Environ. Toxicol. Chem., 17,* 734-739.

Weinstein, J.E. & Oris, J.T. (1999). Humic acids reduce the bioaccumulation and photoinduced toxicity of fluoranthene to fish. *Environ. Toxicol. Chem., 18,* 2087-2094.

Weinstein, J.E., Oris, J.T. & Taylor, D.H. (1997). An ultrastructural examination of the mode of UV-induced toxic action of fluoranthene in the fathead minnow, *Pimephales promelas. Aquat. Toxicol., 39,* 1-22.

Weinstein, J.E. (2001). Characterization of the acute toxicity of photoactivated fluoranthene to glochidia of the freshwater mussel, *Utterbackia imbecillis. Environ. Toxicol. Chem., 20,* 412-419.

Wilkins, A.L.Miles, C.O., DeKock, W.T., Erasmus, G.L., Basson, A.T. & Kellerman, T.S. (1996). Photosensitivity in South Africa. 9. Structure elucidation of a b-glucosidase-treated saponin from *Tribulus terrestris*, and the identification of saponin chemotypes of South African *T. terrestris. Onderstepoort J. Vet. Res., 63,* 327-334.

Williamson, C.E., Stemberger, R.S., Morris, D.P., Frost, T.M. & Paulsen, S.G. (1996). Ultraviolet radiation in North American lakes: Attenuation estimates from DOC measurements and implications for plankton communities. *Limnol. Oceanogr., 41,* 1024-1034.

Zaga, A., Little, E.E., Rabeni, C.F. & Ellersieck, M.R. (1998). Photoenhanced toxicity of a carbamate insecticide to early life stage anuran amphibians. *Environ. Toxicol. Chem., 17,* 2543-2553.

LARS OLOF BJÖRN AND RICHARD L. McKENZIE

11. OZONE DEPLETION AND EFFECTS OF ULTRAVIOLET RADIATION

1. INTRODUCTION

The fear of increased ultraviolet radiation at earth surface in connection with depletion of stratospheric ozone caused by human activities has spurred not only diplomatic activity and political action, but also the investigation of ultraviolet radiation effects on biological systems, from cells to the biosphere as a whole. What has emerged is also a greatly increased understanding of daylight ultraviolet radiation as an ecological factor and health factor under natural conditions.

A great number of reviews and special volumes of books and journals on this subject have seen the light in recent years, such as Young et al. (1993), Lumsden (1997), Rozema (1997), Rozema et al. (1997, 2001). In fulfilment of the Montreal Protocol the United Nations Environmental Protection Programme (UNEP) regularly evaluates the biological consequences of ozone depletion, and some of the reports from these evaluations are now available as special journal issues (van der Leun et al. 1995, 1999). Several popular-science books deal with the subject, e.g., Roan (1989) and Nilsson (1996). The field has even been considered interesting enough for scientists dealing with the sociology of the scientific process (Nolin 1995), and has resulted in an account of the negotiations involved in saving the ozone layer by one of the main actors (Benedick 1991, 1998).

After a few words about ozone and ozone depletion and how it affects ultraviolet radiation, I shall start with a short account of some effects of ultraviolet radiation at the molecular level, and go on to higher levels of biological organization.

2. THE OZONE LAYER

Ozone, O_3, is formed from oxygen, O_2, by the action of ultraviolet-C radiation in the 175-242 nm band in the stratosphere. The high energy UV-C photons first split some of the oxygen molecules into free oxygen atoms, O, which then combine with oxygen to form ozone. Most of the ozone is present at an altitude of 20-30 km, but some of it extends down to ground level. The ozone molecules very efficiently absorb ultraviolet-B radiation (of much higher fluence rate than the UV-C) and are thereby split again into free oxygen molecules. The strong absorption of UV-B radiation in ozone molecules causes heating of the air, and layers higher up in the stratosphere, where the radiation is stronger, are heated more than the lower layers.

L.O. Björn (ed.), Photobiology, 239–263.

240

This results in a lower density of the upper layers, and is the reason that the stratosphere, in contrast to the lower atmosphere (troposphere) is stratified, with very little vertical movement of air.

The absorption maximum of ozone is at practically the same wavelength as that of DNA, and ozone protects DNA very efficiently from the radiation from the sun (Fig. 1).

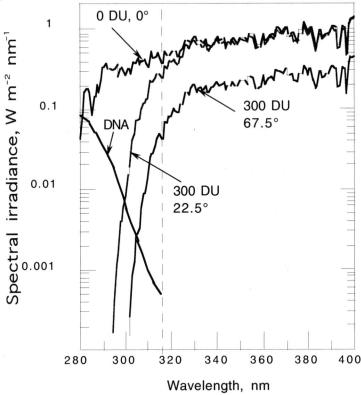

Figure 1. Spectrum of UV spectral irradiance for several observing conditions. Note the logarithmic scale. 300 DU (Dobson units) of ozone is an amount corresponding to a 3 mm thick layer of pure oxone at a prressure of 1 bar and a temperature of 0°C. The dashed vertical line marks the limit between UV-A radiation (315-400 nm) and UV-B radiation (280-315 nm). Ozone has appreciable absorption in the UV-B band, but only a small effect on UV-A radiation. Therefore the difference between the curves in the UV-A band corresponds mainly to the cos(solar zeinith angle) factor.

It is only at the long-wavelength edge of the ozone absorption band that sufficient radiation leaks through to interact with the long-wavelength edge of the DNA absorption band, as well as on other cell constitutents. There is more ozone at high latitudes than near the equator (Fig. 2). The average solar zenith angle is also lower at high latitudes (i.e. the average solar elevation above the horizon during the day lower), and both these circumstances contribute to a much lower yearly UV-B exposure at high latitudes compared to the tropics (Fig. 3).

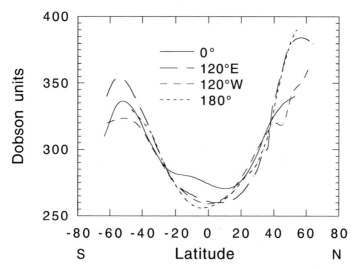

Figure 2. The latitudinal distribution of average ozone at three different longitudes. (data from Labitzke & van Loon 1997). The values are averaged for all seasons over the years 1979-1993. Ozone formed near the equator is transported to higher latitudes, resulting in high values there, but at the highest latitudes anthropogenic depletion lowers the values. The ozone layer is subject to natural annual variations, as well as to anthropogenic and natural long-term changes.

The ozone column (the total amount of ozone from ground level to space) varies over the year, and both amplitude and phase are latitude dependent. In the northern hemisphere the variation can be described by (ozone column - yearly average) = $0.07*(La+10)/90*\cos((Dn-90-(44-La)*3.1)*2*\pi/365.25)$ dobson units for latitudes lower than 44°N, and (ozone column - yearly average) = $0.07*(La+10)/90*\cos((Dn-90)*2*\pi/365.25)$ for La > 44°N, where La is the latitude in degrees and Dn the day number (January 1=1).

The ozone colum is also affected by the rhythms of the Sun. The changes due to the 11-year cycle are rather small, about 1.5-2% (Reid 1999). Although the sun radiates more ultraviolet radiation at a sunspot maximum than when the sun is "quiet", the change is much larger in the ozone-forming UV-C band than in the UV-B band, and therefore the net effect is to make the UV-B fluence rate slightly less at a "solar maximum". The effect due to changes in ozone transport are larger than the direct effect (Labitzke & van Loon 1997). Larger changes are likely to take place over longer solar cycles. (Lean & Rind 1999). From the viewpoint of UV-B conditions at ground level, the variations in cloudiness brought about by solar influence on the terrestrial magnetosphere and thus on cosmic ray flux are probably more important than the ozone variations in both 11-year and long term perspectives, but as yet they are largely unexplored.

242

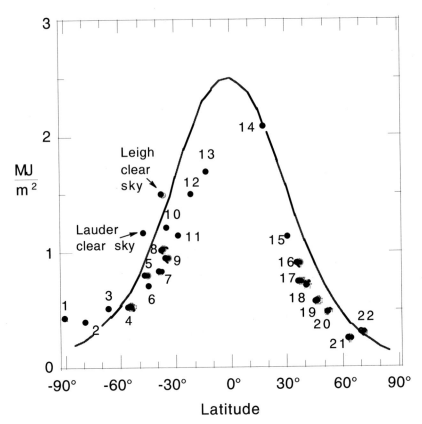

Figure 3. Latitude and cloud dependence of daylight ultraviolet-B radiation. The diagram shows (solid line) the annual erythemally active radiation (time-integrated irradiance) for various latitudes calculated for cloud free conditions and (numbered symbols) the actually measured values at different locations. The difference is mainly due to clouds. Note that clouds can increase the irradiance under certain conditions, which is particularly evident at high southern latitudes. For Lauder and Leigh clear sky values were also estimated from measurements on cloud-free days. The numbers represent the following locations: 1 South Pole, 2 McMurdo, 3 Palmer, 4 Ushuaia (Tierra del Fuego), 5 Lauder (New Zealand), 6 Hobart (Tasmania), 7 Melbourne, 8 Leigh (New Zealand), 9 Sydney, 10 Perth, 11 Brisbane, 12 Alice Springs, 13 Darwin, 14 Mauna Loa (Hawaii), 15 San Diego, 16 Kos (Greece), 17 Athens (Greece), 18 Thessaloniki, 19 Garmisch-Partenkirchen, 20 Oxford (England), 21 Reykjavik, 22 Barrow (Alaska).

3. OZONE DEPLETION

Thus ozone is continuously formed by UV-C radiation and decomposed again by UV-B radiation. There are also other processes involved in the natural dynamics of the ozone layer. But this situation has been disturbed by emission of artificially generated substances, mainly organic halogen compounds and nitrogen oxides.

Many organic halogen compounds, such as the chlorofluorocarbons (CFCs) are very inert substances as long as they stay in the troposphere, but when they eventually reach sufficient altitude in the stratosphere they are decomposed by the ultraviolet-C radiation there, and produce free halogen atoms. Of these fluorine atoms are of minor interest, but both chlorine and bromine react with "odd oxygen" (O as well as O_3) and form monoxides: $Cl+O ==> ClO$ and $Cl+O_3 ==> ClO+O_2$. Nitrogen monoxide functions in an analogous way: $NO+O ==> NO_2$; $NO+O_3 ==> NO_2+O_2$. These substances thus both destroy ozone already present, and prevent formation of new ozone by sweeping up the free oxygen atoms. Still, this would not be catastrophic in itself, if the ozone depleting substances would themselves be depleted by the process. However, the following reactions also take place: $ClO+O ==> Cl+O_2$; $NO_2+O ==> NO+O_2$. In other words, through reactions which further deplete the stratosphere of ozone-forming free oxygen, the ozone-destroying substances are regenerated; a chain reaction results. It has been estimated that a single halogen atom may destroy thousands of ozone molecules before it finally undergoes a chain-ending reaction of some kind. When we remember that the amount of ozone is so small, corresponding to about 3 mm of the whole atmosphere (or, 6 g out of the 10 tons of air on a square meter), it is understandable that the effect is serious. The ozone hole over Antarctica, which started to develop in the early 1970's and appears every antarctic spring is well-known. but it is less well-known that depletion has taken place everywhere in the world except near the equator, and that it is quite significant even in some populated parts of the world (Fig, 4).

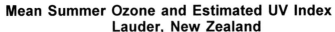

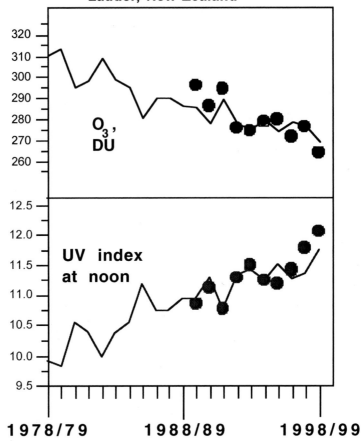

Figure 4. The decline in summertime ozone over the South Island of New Zealand over two decades (upper panel) and the calculated change in biologically active ultraviolet radiation over the same time period (lower panel).

4. MOLECULAR EFFECTS OF ULTRAVIOLET-B RADIATION

We shall describe the most important effects on the molecular level. As an example of the large number of effects, Fig. 5 gives an overview of UV-B effects on plants. Some of these effects are common to all organisms, while others are specific to plants. Some of the effects of special relevance to man are dealt with in Chapter 13.

4.1. Effects of ultraviolet radiation on DNA

Ultraviolet radiation directly absorbed by DNA (which may be UV-B under natural conditions, but also UV-C under experimental or otherwise artificial circumstances) can elicit a number of different lesions, but the great majority are either dimerisation or adduct formation affecting only adjacent pyrimidine bases (thymine and cytosine) on one DNA strand. The dimerisation products are referred to as cyclobutane pyrimidine dimers (CPDs) and the adducts, properly termed 6-4[pyrimidine-2′-one] pyrimidines, are often called (6-4) photoproducts (Fig. 5). Ultraviolet radiation transforms the (6-4) photoproducts to "Dewar photoisomers". This transformation can take place with UV-A (Takeuchi et al. (1998). UV-B, or UV-C radiations (Ravanat et al. 2001). As shown in Figs 5 and 6 several diastereoisomers of CPDs are theoretically possible, but only the cis-syn type is, in fact, formed in double-stranded DNA. In single-stranded or denatured DNA also the trans-syn form can be generated, but in low yield. Pyrimidine bases may react to form CPDs as well as (6-4) photoproducts in the combinations TT, TC and CT; for CPDs also CC is possible. The biological effects are very variable. In some cases replication is stopped at a lesion, in other cases replication continues, resulting either in a mutation (e.g., a change from thymine to cytosine, Jiang et al. 1993), or a normal DNA strand. Only specialists in the field can keep track of this, since also the same type of lesion may have different consequences in different organisms (e.g., Gibbs et al. 1993). The reader is referred to Ravanat et al. (2001) as an entrance to the literature.

CPDs are the most common UV-induced lesions in DNA. Under UV-C (254 nm) radiation 2-10 CPDs are formed per million bases per J m^{-2}. As DNA is more weakly absorbed by UV-B, it is not surprising that the same result requires about two orders of magnitude more UV-B. CPDs occur at higher frequencies than (6-4) photoproducts. Base order is important for UV susceptibility. Pyrimidine pairs with T (thymine) on the 5'-side, i.e. TT and TC, are about ten-fold more reactive than those with C (cytosine) on the 5'-side (CT and CC). C at the 3'-side favours (6-4) photoproduct formation. For more details see Ravanat et al. (2001) and Yoon et al. (2000). In diatoms the frequency of (6-4) photoproducts was, on average, 85% of that of CPDs (Karentz et al. 1991), but in most other organisms it seems to be lower.

Another kind of lesion affecting cytosine is photohydration followed by deamination to yield uracil hydrate.

Also the purine bases in DNA may be altered by ultraviolet radiation. Thus adenine may combine with either an adjacent adenine or with a thymine residue. This occurs with very low yield, but may be biologically important since it is not as easily repaired as the CPDs and (6-4) photoproducts. At least the adenine-thymine adduct is also highly mutagenic (Zhao & Taylor 1996).

DNA bases may also suffer photooxidative damage, and in this respect guanine is particularly sensitive. This photooxidation may take in various ways, and the

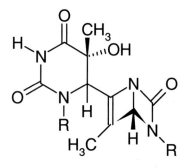

Unreacted pair of thymine residues in DNA

Cis-syn TT dimer

Trans-syn TT dimer

(6-4) TT photoproduct

Dewar TT photoisomer

Figure 5. Thymine residues in DNA, and the most common types of lesion due to ultraviolet radiation. Dimers and (6-4) photoproducts arise from thymine by the action of UV-B or UV-C radiation, while Dewar photoisomers are formed from (6-4) photoproducts under the influence of UV-A radiation. Since cytosine (C) reacts in a similar way, and also can react with thymine, forming TT, TC and CT dimers, there are in fact a multitude of possible lesion types, all with differently serious consequences for the organism.

frequency of various pathways has not yet been established. A direct photon hit in a base may expel an electron, and the electron hole can then migrate along the DNA chain until it encounters a guanine residue, via interaction with water, resulting in 2,6-diamino-4-hydroxy-5-formamidopyrimidine, Deprotonation instead of hydration may also occur, and gives another product. Another possibility is that a pyrimidine or purine base after photoexcitation enters a triplet excited state and reacts with oxygen, resulting in generation of singlet oxygen (see Chapter 1). The singlet oxygen then attacks a DNA base, preferably guanine. In this way 8-oxo-7,8-dihydro-2′-deoxyguanosine is formed. Finally singlet oxygen and other reactive oxygen species (see Chapter 1 and below) formed after photoexcitation of other chromophores may also attack DNA in various ways. It is not known with certainty which cellular chromophores that are most important for such processes, but flavines are among the likely canditates. In any case UV-A is more important than UV-B, partly because its fluence rate in daylight is higher, partly because it penetrates further into organisms and natural waters.

Lesions in addition to those mentioned can be induced by direct action of the radiation on DNA. They occur at lower frequency, but some of them may have some importance because they cannot be repaired by the rapid action of photolyases, only by the slower acting light-independent, "dark" repair systems

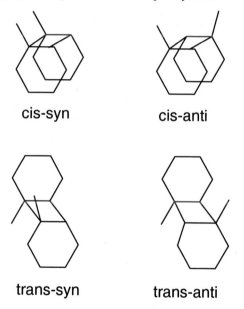

cis-syn   cis-anti

trans-syn   trans-anti

Figure6. Perspective drawings of the four different "diastereoisomers" that are possible for a cyclobutane dimer.Due to steric constraints only the syn forms can be generated in DNA, and of these the trans-syn only in single-stranded DNA. Redrawn after Ravanat et al. (2001).

4.2. Photolyases and photoreactivation

Most prokaryotic and eukaryotic organisms are equipped with two kinds of photolyases, enzymes which under the influence of light repair these lesions. In older literature they are called photoreactivating enzymes. The physiological effects of photoreactivation were observed by Haussner & Oehmcke (1932, cf. Fig. 7) long before the role of DNA was known. One type of photolyase repairs the CPDs, the other type repairs (6-4) photoproducts. If a (6-4) photoproduct has been converted to a Dewar photoisomer, it can no longer be repaired by photolyase action.

The photolyases are of interest not only from the viewpoint of DNA repair, but also because they have played a role in the evolution of other photobiological systems (see section 3.1 in Chapter 8). Since photolyases are so widespread among organisms, and related types occur in distantly related organismal groups, and also because the need for DNA repair probably was very great before the emergence of the ozone layer, they are thought to have evolved very early. In addition to amino acid sequence and the specificity for different lesion types, photolyases differ also in the chromophore constitution, and therefore also in the action spectra (Figs 8-9) for photorepair. All photolyases contain a flavin as the main chromophore, but most of them also an accessory chromophore.

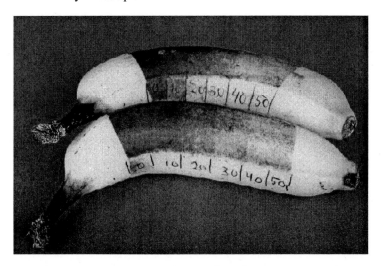

Figure 7. These bananas, while still unripe and green, were first irradiated for 2 minutes with ultraviolet-C radiation (254 nm, W m^{-2}), except for the ends, which were wrapped in aluminium foil during irradiation. They were then irradiated with white light for the number of minutes indicated on the different sections. After this they were allowed to ripen in darkness. Where the epidermal cells have been damaged by ultraviolet radiation they have turned brown, while in other areas they have become yellow during ripening. White light after the ultraviolet has an ameliorating effect. This experiment, which is recommended as a student experiment (see Chapter 17) is, in principle the same as that by which photoreactivation was discovered by Hausser & von Oehmcke (1933).

Based on aminoacid sequences the evolution of the photolyase/cryptochrome family of proteins is thought to have proceeded as follows (Todo 1999):

An ancestral gene encoded a CPD photolyase and duplicated very early to form several copies. One of the copies evolved to become class II CPD photolyase gene, which now occus only in eukaryotes. Another copy evolved to become class I CPD photolyase gene. One copy gave rise to several related genes coding for (6-4) photolyase and plant and animal cryptochromes.

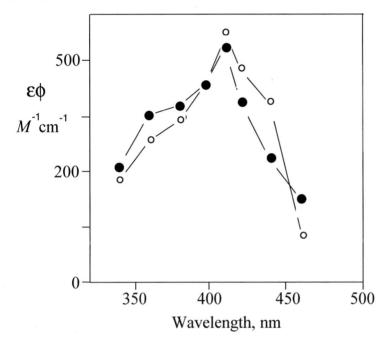

Figure 8. Absolute action spectrum for photoreactivation of (6-4) photoproducts by Drosophila melanogaster *photolyase: filled circles, repair of T(6-4)T; empty circles, repair of T(6-4)C. Redrawn after Zhao et al . (1997).*

4.3. Formation and effects of reactive oxygen species

Under the common name of reactive oxygen species (ROS) we lump together a number of oxygen-containing chemical species which arise in organisms by a variety of reactions. Experts dealing with effects of ionizing radiation, such as radiation arising from radioactive decay of atomic nuclei have been familiar with them for a long time. In photobiology, somewhat surprisingly, they are more important when we are dealing with effects of the less energetic UV-A photons than in the UVB and UV-C fields, where direct radiation effects on DNA and other macromolecules dominate the picture. This is because the cell contains a number of chromophores which, when excited by UV-A photons can give rise to ROS. Murphy (1990) and Murphy & Huerta (1990) found that UV-C irradiation of plant cells can give rise to hydrogen peroxide, probably by another mechanism as that described below for UV-B.

According to A.-H,-Mackerness et al. (2001) the main mechanism by which ultraviolet radiation increases the generation of ROS is indirect, by increasing the activity of NADPH oxidase and peroxidases. The generation of ROS by NADPH oxidase is well-known from the vertebrate defence against bacterial infection, and an "oxidative burst" similar to that generated in vertebrate neutrophils (a kind of blood corpuscles) takes place also in the plant's defence agains infection (review by Lamb & Dixon 1997). NADPH oxidase functions according to the reaction formula $O_2 + NADPH ==> O_2^- + NADP^+ + H^+$. The superoxide anion, O_2^- so formed is very reactive, and may give rise to hydrogen peroxide:

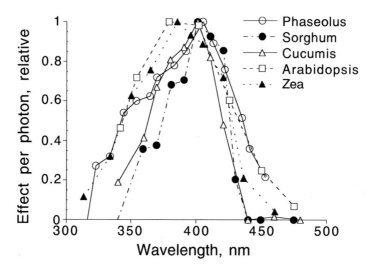

Figure 9. Collection of action spectra for photoreactivation of CPDs in plants by Class II photolyase, to show the variability; Phaseolus *(pinto bean, Saito & Werbin 1969),* Sorghum *and* Cucumis *(cucumber) (Hada et al. 2000),* Arabidopsis *(Pang & Hays 1991),* Zea *(maize, Ikenaga et al. 1974).*

$H^+ + O_2^- ==> {}^{\bullet}HO_2$
${}^{\bullet}HO_2 + {}^{\bullet}HO_2 ==> H_2O_2 + O_2$
${}^{\bullet}HO_2 + O_2 + H_2O ==> H_2O_2 + O_2 + OH^-$

In the presence of iron as a catalyst, the most reactive of all ROS, the hydroxyl radical, ${}^{\bullet}OH$, may be generated from superoxide anion and hydrogen peroxide:
$O_2^- + Fe^{3+} ==> O_2 + Fe^{2+}$
$Fe^{2+} + H_2O_2 ==> Fe^{3+} + OH^- + {}^{\bullet}OH$

Peroxidase can contribute to the formation of O_2^- by first oxidizing NADH with H_2O_2 as an oxidant. This results in formation of a ${}^{\bullet}NAD$ radical (two mols per mol of hydrogen peroxide), which reduces O_2 to O_2^-, a reaction which is stimulated by monophenols and Mn^{2+}. If the superoxide, after protonation, dismutases to form hydrogen peroxide as shown above, more hydrogen peroxide is generated than was originally consumed.

Enzyme systems and antioxidants protecting against effects of ROS are often increased in response to ultraviolet irradiation.

4.4. Effects of ultraviolet radiation on lipids

Lipids containing unsaturted fatty acids are particularly susceptible by photooxidation. Such attack may be initiated in two different ways (see Girotti 2001 for a review):

Type I photooxidation is started by the formation of a radical when photon absorption in a sensitizer (S) results either in expulsion of an electron resulting in a radical (which can abstract an electron or a hydrogen atom from a lipid molecule), or in formation of a triplet excited state ($^3S^*$). In the latter case the following reactions can take place (LH symbolizing a lipid molecule and RH another reductant such as ascorbate, glutathione, or a membrane component other than the lipid):

$$(1) \; ^3S^* + RH ==> S^{-\bullet} + R^\bullet + H^+$$
$$(2) \; S^{-\bullet} + O_2 ==> S + O_2^-$$
$$(3) \; S^{-\bullet} + O_2^- + 2H^+ ==> H_2O_2$$
$$(4) \; H_2O_2 + Fe^{2+} ==> OH^- + HO^\bullet + Fe^{3+}$$
$$(5) \; LH + HO^\bullet ==> L^\bullet + H_2O$$
$$(6) \; L^\bullet + O_2 ==> LOO^\bullet$$
$$(7) \; LOO^\bullet + LH ==> LOOH + L^\bullet$$

Reaction (1) above results in two radicals, $S^{-\bullet}$ and $R^\bullet$. Either of these could initiate a peroxidative chain reaction (6-7) creating new radicals via a two-step reduction of oxygen (2-3) and a Fenton reaction (4) generating the highly reactive hydroxy radical ($HO^\bullet$) that readily attacs the lipid and eventually generating lipid peroxide (LOOH). A product also generated in such peroxidations is malonyl aldehyde, which is often used as an indicator to monitor peroxidative lipid breakdown. Another characteristic is generation of light ("ultraweak radiation", see Chapter 16), which has been used to construct an action spectrum for lipid breakdown in plant leaves by ultraviolet radiation (Cen & Björn 1994). I should be pointed out that if suitable sensitizers are present in the appropriate compartments, this type of photooxidation can be mediated also by visible light. In addition to unsaturated fatty acids, other lipids, like cholesterol, can be broken down by essentially the same type of process.

The oxygen appearing in the above schemes is ordinary (triplet) oxygen. In a Type II photooxidation singlet excited oxygen ($^1O_2$, see Chapter 1) is first generated by energy transfer from a sensitizer which, after photoexitation to a singlet excited state has reached the triplet state. The singlet oxygen then directly reacts with the double bond of an unsaturated fatty acid, and oxygen is added to form a peroxide: $LH + {}^1O_2 ==> LOOH$.

4.5. Photodestruction of proteins

Proteins may be affected by ultraviolet radiation in various ways. Enzymes in which sulfhydryl groups are essential for the function are sensitive to photooxidative damage. One of the most sensitive proteins is photosystem II, a protein complex having a key role in oxygenic photosynthesis. In short-term experiments on photosynthesis this was frequently found to be the most ultraviolet-sensitive link. However, it seems that long-term, ecologically relevant inhibition of photosynthesis by UV-B radiation has other causes (see below).

Proteins can also be photochemically cross-linked to adjacent nucleic acids, which is one more mechanism by which ultravilet radiation can affect DNA.

Enzymes may be inactivated also by visible light if they absorb it (e.g., Björn 1969), or via sensitizers which absorb visible light.

4.6. UV absorption affecting regulative processes

Many organisms have special receptors for ultraviolet radiation, which enable them to regulate various processes in accordance with ambient radiation levels, or even to locate environments of higher or lower radiation, or to locate sources of radiation. Most of these, including receptor cells in the eyes of many animals, respond only to UV-A radiation. However, at least some plants and some algae possess receptors for UV-B, and in man trans-urocanic acid (cf. Chapter 8) can probably be regarded as a molecular UV-B photoreceptor, even if we presently do not understand the significance. However, many processes are regulated by UV-B seemingly without any photoreceptors specifically evolved for regulation. Thus, many genes are regulated in both plants and animals via photon absorption in DNA. It is too early to say whether this is, in general, to be regarded as a stress response to DNA damage inflicted by the radiation.

In other cases regulation proceeds via formation of reactive oxygen species (A.-H. Mackerness et al. 1999, 2000 and literature cited therein). This seems to be the case with the down-regulation by increased UV-B of several genes of importance for photosynthesis. Reactive oxygen species also affect a number of hormone systems (Fig. 8).

The specific UV-B receptor in plants, with an action maximum at 295 nm has not yet been identified. It regulates growth processes and also the synthesis of flavonoids and other substances. Flavonoids serve as ultraviolet-B absorbing filters, and also have other functions in the plants.

It may be wrong, however, to regard the regulation by reactive oxygen species and regulation by an UV-B receptor as two different pathways. As described in section 4.3 a major way for generation of ROS may be the increase in activities of NAD oxidase and of peroxidase. This increase may be achieved via the UV-B receptor.

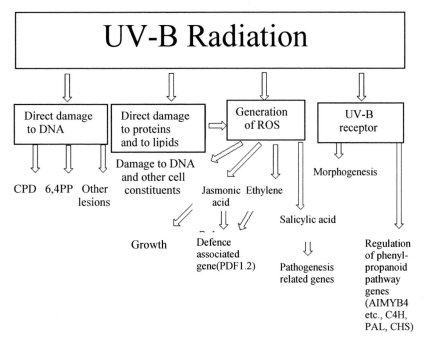

Figure 8. Overview of molecular effects of UV-B radiation on plants. ROS stands for reactive oxygen species (such as HO·, H_2O_2, O_2^- and $^1O_2$). Examples of morphogenesis are tendril and cotyledon curling and hypotyl growth.

5. ULTRAVIOLET EFFECTS ON INANIMATE MATTER THAT ARE OF BIOLOGICAL RELEVANCE

We have already described how ultraviolet radiation affects the composition of the upper stratosphere. It also has effects on the troposphere. Among other processes it contributes to the formation of smog. Increased UV-B in the troposphere leads to increased concentration of hydroxyl radical, which in turn results in increased removal of organic compounds and oxides of sulfur and nitrogen. These effects are probably more beneficial than negative for life in general. On the other hand, concentrations of hydrogen peroxide, organic peroxides and ozone in the troposphere increase when UV-B radiation increases. The overall effect on air quality is probably negative, and various organisms are affected to very different degrees. Tropospheric ozone in particular is clearly a problem for vegetation in many parts of the world.

In the aquatic UV-B decomposes dissolved organic matter, thus making the water more transparent and increasing penetration of ultraviolet radiation, and also making the organic substances more available for bacterial growth.

Photochemical reduction of iron and manganese is an important process in some aquatic ecosystems (reviewed by Wu and Deng 2000). For this UV-A, and to some extent visible light, is more important than UV-B. UV-A and UV-B can also decrease availabilityh of iron by chelating agents, which is a considerable problem in use of artificial iron fertilization in some countries.

In the terrestrial environment ultraviolet radiation affects dead plant matter (litter) in a complex way. The chemical composition of the litter is affected already in the living state: More secondary substances are usually formed in plant parts under increased radiation, making the derived therefrom more resistant. After litter has formed, organic compounds in it are photochemically broken down by ultraviolet radiation, provided they are sufficiently exposed. The activity of some litter-decomposing fungi, on the other hand, is hampered by ultraviolet radiation. In some investigations the overall effect was that organic carbon as well as nitrogen disappeared more rapidly under increased UV-B radiation (Gehrke et al. 1995, Paul et al. 1999), while in other cases decomposition was decreased by UV-B radiation (Moody et al. 2001 and references therein). The rate of litter breakdown is very important in some environments, since it may be the most important process making plant nutrients available for new growth.

8. UV-INDUCED APOPTOSIS

Induction of apoptosis (programmed cell death), because of its medical importance, has been studied most thoroughly in mammals, but is known to occur in other eukaryotes as well. Apoptosis is a normal process in plant development, for instance in xylem differentiation, and also induction of apoptosis in plants by UV-C has been observed (Danon & Gallois 1998). In the flagellate *Euglena* induction of an apoptosis-like process was observed by Scheuerlein et al. (1995).

According to Godar (1999a,b) several mechanisms for apaoptosis caused by UV-A radiation in the wavelength range 340-400 nm can be distinguished. Two of these depend on the generation of ROS. Formation of singlet oxygen results in "immediate pre-programmed cell death", which takes place in less than 20 minututes, does not require protein synthesis, and depends on the opening of a "megapore" in the mitochondrial membrane, which releases chemical apoptosis signals into the cytosol. Another pathway for "immadiate" apoptosis depends on the formation of superoxide anion, which probalby is transformed to hydroxyl ion before interacting with another part of the mitochondrial megapore, which releases cytochrome c. The presence of cytochrome c in the cytosol then causes apoptosis. A third, slower ("intermediate"), kind of apoptosis is caused by absorption of radiation in the plasma membrane, causing cross-links in the "Fas receptor" there, which via formation of "caspases" (apoptosis-signalling proteins) lead to opening of the mitochondrial megapore. Finally, there are several forms of "delayed" apoptosis, caused by lesions in DNA. At least some of these signalling channels also go via opening of the mitochondrial megapore. For UV-B-induced apoptosis, CPD formation in DNA can be involved, since Nishigaki et al. (1998) have shown that photorepair of CPDs can prevent UV-induced apoptosis.

7. UV-B RADIATION IN AN ECOLOGICAL CONTEXT

As we have seen above, ultraviolet-B radiation can act destructively on DNA and other cellular constituents, but it can also act in a regulating way, and it modifies the physical environment. In the early days of research regarding UV-B effects on organisms and ecosystems there was an overemphasis on the direct, destructive effects of UV-B on organisms. Nowadays most researchers in the field are of the opinion that the important effects of UV-B change resulting from changes in the ozone layer and cloudiness will not be these effects, but modulations in the way organisms influence one another. The literature on this is voluminous. We shall here only briefly give some examples, and refer the interested reader to recent reviews, books, and special journal issues devoted to the subject (e.g., Björn 1996, Helbling et al. 2001a, Rozema et al. 1997a,b, 2001, 2002, Hessen 2002).

7.1. Aquatic life

Few biological effects of the increase in UV-B that has resulted from ozone depletion have been directly recorded, so most effects are inferred from experimental manipulation of the radiation level. Direct detection of ozone depletion effects is hampered by the lack of baseline data, i.e. lack of information of the sitution prior to ozone depletion. In the Southern Ocean one can clearly monitor how photosynthesis decreases when the ozone hole sweeps over the monitored area (Smith et al. 1992). Different phytoplankton species are affected to very different extents.

Even if the impacts of ozone depletion are difficult to monitor in most water bodies, the impact of UV-B in situ can be inferred from the vertical distribution of photosynthesis, as photosynthesis is partly inhibited in the upper layers where the UV-B radition is strongest (e.g., Helbling & Villafañe 2002). The degree of vertical mixing of water, which varies with location and season, is important in this context. During the seasons most important for UV-B effects, the upper mixed layer is shallow in antarctic waters (mean depth 50 m), while in the Arctic it varies between 100 and 300 m. When arctic phytoplankton become exposed to UV-B, they are therefore to a higher degree "dark adapted" than antarctic phytoplankton, and therefore, on average, more sensitive (Helbling & Villafañe 2002).

In coastal waters, and in fresh water bodies in particular, UV-B does not penetrate as far as in the open sea, and therefore affects only a shallow layer. This can still be of importance, as larvae of many organisms important for the food web live near the surface in these waters. The food web in aquatic ecosystems is generally more complex than in terrestrial ecosystems, and organisms at each trophic level, with the possible exception of mammals and birds, can be affected by UV-B at some stage of development.

One might suspect that the smaller an unicellular organism is, the more susceptible it would be to UV-B radiation. A small organism has less possibility than a larger one to protect the most sensitive parts (especially its DNA) by shielding layers containing radiation-absorbing substances. Some evidence in this direction was presented by Karentz et al. (1991), who compared DNA damage and survival for a number of planktonic diatom species of different size exposed to ultraviolet-B radiation. Although the data were scattered, there was a clear tendency for fewer DNA lesions in species with a small surface area: volume ratio (as a large cell has in comparison with a small one, although there is also a shape factor involved). They also (but comparing only four species) showed a negative correlation between survival and frequency of DNA lesions. Such a size-selective effect of UV-B could have serious consequences for consumer organisms, who may simply find the surviving prey too big for consumption. In fact, van Donk et al. (2001) found UV-B irradiation to decrease the fraction of phytoplankton which was considered "*Daphnia*-edible".

Several other authors have also mentioned a higher tolerance among larger species. But this relationship cannot be upheld when distantly related organisms are compared, and its generality is questionable even when organisms within the same group are compared. Thus Peletier et al. (1996) found no relationship between cell size and UV-B sensitivity among benthic diatoms. Laurion and Vincent (1998) conclude that cell size is not a good predictor of UV-B sensitivity, and that, in particular, the cyanobacteria-dominated picophytoplankton is less sensitive than what would be assumed on the basis of a size-sensitivity relation. Also Helbling et al. (2001b) find that picoplankton is more resistant than a cell size relationship would predict.

In the case of marine Prymnesiophyceae it was found (Mostajir et al. 1999b) that prolonged exposure to UV-B caused an increase in cell size, which possibly diminished the risk of genetic damage, but above all change the availability and quality of these organisms as food for grazers.

Most organisms (including ourselves) have pigments which protect against ultraviolet radiation (the human skin pigment melanin protects in both optical and chemical ways, as do also some plant pigments). In cyanobacteria and many (but not all types of) algae so-called mycosporine-like amino acids (MAAs) constitute an important group of UV-protecting pigments. In general, however, they protect mainly against UV-A, rather than UV-B radiation (Sinha et al. 1998, 2001, Bishof et al. 2002). Although they are produced by cyanobacteria and some algae and fungi, they are taken up by grazing animals and used also by then as protecting pigments (Sinha et al. 1998). For many aquatic animals, carotenoids serve as important protectants (Hessen, 2002).

Despite the many negative effects of ultraviolet-B radiation, it is frequently found that concentrations of various aquatic organisms are increased when UV-B levels are increased. For both bacteria and phytoplankton species this can be due to a greater sensitivity of the predators, and therefore a decline in predation (Mostajir et al. 1999a). For bacteria it can also be due to increased availability of nutrients when dissolved organic matte is decomposed by the radiation (Herndl et al. 1997).

7.2. Terrestrial life

As in the case of aquatic life, few biological effects of the increase in UV-B that has resulted from ozone depletion have been directly recorded, so the majority of effects are inferred from experimental manipulation of the radiation level. Only in Antarctica, and in Tierra del Fuego, where the ozone hole provides a great increase in radiation level, can a ´direct biological impact of ozone depletion be recorded. Thus Rousseaux et al. (1999) found that CPD frequency in DNA in plant leaves tracked variations in ambient UV-B in Tierra del Fuego. Several researchers have been inspired to do research on the two higher plant species in Antarctica. Only the 4-year study of Day et al. (2001) provides data to allow the conclusion that the ozone hole has had an impact on plant life there. These researchers compared plants under UV-B transmitting and UV-B excluding filters. Since the "natural" (pre-hole) UV-B level is so low, the plants under UV-B excluding filters were probably developed in a way very similar to pre-hole plants. Many morphological differences between plants exposed to the two treatments were noted. There was some evidence that effects accumulated over the years.

Experiments with exclusion of ambient UV-B from natural ecosystems have also been conducted in Tierra del Fuego (Ballaré et al. 2001), and there are many exclusion studies carried out elsewhere showing that the ambient ultraviolet radiation, even in regions of the world where the ozone depletion has not been severe, is an important environmental factor for many terrestrial (as for aquatic) organisms. In most cases there is little or no direct inhibition of photosynthesis and biomass production by plants. The important effects instead arise in the interaction between organisms. Thus morphological changes in plants can lead to changes in the interspecies competion for light (Barnes et al. 1988). Changes in UV-B often results in changes in herbivory. In most cases increased UV-B leads to decreased herbivory due to changes in chemical composition of the plants (Gwynn-Jones 1999, Ballaré 2001, and reports cited therein). However, other cases are also known: Herbivory may be increased by increased UV-B (Lavola et al. 1998, Buck & Callaghan 1999), and it seems that some insects can perceive UV-B and react directly on changes (Buck & Callaghan 1999). Also interactions between microorganisms and other organisms can be changed by changes in UV-B, although the conclusion of Paul (2000) is that it is not proble that ozone depletion will cause any substantial change in the incidence or severity of crop diseases.

8. EFFECTS ON HUMAN EYES

Most medical effects of ultraviolet radiation take place via the skin, and such effects are dealt with in Chapter 13. The only medical effects of ultraviolet radiation that shall concern us is the present chapter will therefore be those on the eyes.

According to Bottner & Walter (1962) different wavebands of ultraviolet radiation incident on the cornea are absorbed in the various tissues of the eye as shown in Table 1.

Table 1. Percentage of radiation incident on the cornea over the pupil that is absorbed in various part of the eye

Wavelength, nm	Cornea	Aqueous layer	Lens	Vitreous body
280	100	0	0	0
300	92	6	2	0
320	45	16	36	1
340	37	48	48	1
360	34	52	52	2

The values in Table 1 should be regarded only as approximate values. Especially the lens transmission varies between individuals, and it is normal that it decreases with age (even if no cataract develops), and this decrease affects also the blue waveband, and thus decreases the eye's sensitivity to blue light. For further information on transmission of ocular media, see Polo et al. (1997).

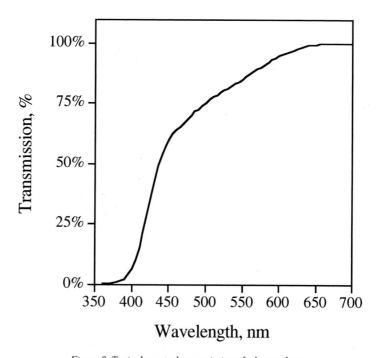

Figure 9. Typical spectral transmission of a human lens.

Because it does not penetrate any further, UV-C radiation can directly affect only the cornea and conjunctiva, and the most common type of UV-C radiation (254 nm-radiation from low pressure mercury lamps) only the outermost part of the cornea. Such radiation is, in a way, the least dangerous kind of ultraviolet radiation as far as the eyes are concerned, since the person who has been exposed will very quickly be aware of it, and will in the future avoid exposure. Within two to three hours after the radiation exposure very unpleasant symptoms will develop. It feels as if the eyelids are covered with sandpaper on the inside. If the exposure has not been very severe, the pain will disappear in a couple of days. The scientific name for this is photokeratitis, and it can also be caused by UV-B radiation in daylight. Well-known forms of photokeratitis are "snow-blindness" and "welder's flash". According to Podskochy et al. (2000) photokeratitis is caused by UV-induced apoptosis of cornea cells, resulting in a speeding up of the cell shedding, which normally takes place at a lower rate. This exposes subsurface nerve endings, which causes the gritty feeling and the pain. The action spectrum for ultraviolet damage to the cornea was determined by Pitts et al. (1977).

Cataract is a common and serious group of diseases, which lead to opacity of the lens, and ultimately to blindness. Cataracts are usually divided into cortical, nuclear and posteriour subcapsular cataracts. Of these nuclear cataracts do not seem to be caused by UV exposure, while the two other forms are (Taylor 1994, see also Zigman 1995). In the United States the probability of cataract surgery increases by 3% for each degree decrease of latitude (Javitt & Taylor 1994). Such surgery involves removal of the lens, and often replacement of it with an artificial (acrylate) lens. Nowadays a lens material is chosen that absorbs ultraviolet radiation efficiently, so radiation exposure of the inner parts of the eye will not be increased after the operation.

Action spectra for cataract induction in lenses of rat and swine have been determined by Merriam et al. (2000) and Oriowo et al. (2001). The highest sensitivity occurs around 295-300 nm, i.e. in the middle of the UV-B band.

UV-related cancer forms, such as melanoma (Michalova et al. 2001) and squamous cell carcinoma affect not only skin, but also the eye. Newton et al. (1996) estimate that the incidence of squamous cell carcinoma of the eye doubles for a 50 degree increase in latitude (from the U.K. to Uganda), but genetic factors have not been taken into account in this estimate.

Pterygium is an outgrowth of the conjunctiva over the cornea. It is strongly related to sun exposure (Threlfall & English 1999, and literature cited by Longstreth et al. 1998), but the spectral dependence is not known.

Regarding the role of ultraviolet (and visible sunlight) for eye disease the reader is also referred to WHO (1994), Taylor (1994), Young (1994), Sliney (1997), Scott (1998) and Longstreth et al. (1998).

REFERENCES AND FURTHER READING

A.-H.-Mackerness, S., Jordan, B.R. & Thomas, B. (1999). Reactive oxygen species in the regulation of photosynthetic genes by ultraviolet-B radiation (UV-B: 280-320 nm) in green and etiolated buds of pea (*Pisum sativum* L.). *J. Photochem. Photobiol. B: Biol. 148*, 180-188.

A.-H. Mackerness, S., John, C.F., Jordan, B. & Thomas, B. (2001. Early singaling components in ultraviolet-B responses: distinct roles for different reactive oxygen species and nitric oxide. *FEBS Lett., 489*, 237-242.

Barnes, P.W., Jordan, P.W., Gold, W.G., Flint, S.D., Caldwell, M.M. (1988) Competition, morphology, and canopy structure in wheat (*Triticum aestivum* L.) and wild oats (*Avena fatua* L.) exposed to enhanced ultraviolet-B radiation. *Functional Ecol., 2*, 319-330.

Benedick, R.E. (1991). Ozone diplomacy: *New directions in safeguarding the planet* (enlarged edition 1998), pp. XIX+449. Cambridge (MA): Harvard University Press. ISBN 0-674-65003-4.

Berrocal,-Tito, G., Saetz-Baron, L., Eichenberg, K., Horwitz, B.A. & Herrera-Estrella, A. (1999). Rapid blue light regulation of a *Trichoderma harzianum* photolyase gene. *J. Biol. Chem, 274,* 14288-14294.

Bischof, K., Hanelt, D. & Wiencke, C. (2002). UV radiation and arctic marine macroalgae. *In* Hessen, D. (ed.) *UV radiation and arctic ecosystems,* Chapt. 11, pp. 225-243.

Björn, L.O. (1969). Photoinactivation of catalases from mammal liver, plant leaves and bacteria. Comparison of inactivation cross sections and quantum yields at 406 nm. *Photochem. Photobiol., 10,* 125-129.

Björn, L.O. (1996). Effects of ozone depletion and increased UV-B on terrestrial ecosystems. *Int. J. Environ. Stud., 51/3(A),* 217-243.

Björn, L.O., Callaghan, T.V., Johnsen, I., Lee, J.A., Manetas, Y., Paul, N.D., Sonesson, M., Wellburn, A., Coop, D., Heide-Jørgensen, H.S., Gehrke, C., Gwynn-Jones, D., Johanson, U., Kyparissos, A., Lenzou, E., Nikolopoulos, D., Petropoulou, Y. & Stephanou, M. (1997). The effects of UV-B radiation on European heathland species. *Plant Ecology, 128,* 252-264.

Buck, N., Callaghan, T.V. (1999) The direct and indirect effects of enhanced UV-B on the moth caterpillar *Epirrita autumnata. Ecol. Bull. 47,* 68-76.

Cen, Y.-P. & Björn, L.O. (1994). Action spectra for enhancement of ultraweak luminescence by ultraviolet radiation (270-340 nm) in leaves of *Brassica napus. J. Photochem. Photobiol. B: Biology, 22,* 125-129.

Danon, A. & Gallois, P. (1998). UV-C radiation induces apoptotic-like changes in *Arabidopsis thaliana . FEBS Lett.* 437,131-136.

Day, T.A. (2001). Multiple trophic levels in UV-B assessments — completing the ecosystem. *New Phytol., 152,* 183-185.

Day, T.A., Ruhland, C.T. & Xiong, F.S. (2001). Influence of solar ultraviolet-B radiation on Antarctic terrestrial plants: results from a 4-year study. *J. Photochem. Photobiol. B: Biology, 62,* 78-87.

Deisenhofer, J. (2000). DNA photolyases and cryptochromes. *Mutat. Res., 460,* 143-149.

Gehrke, C., Johanson, U., Callaghan, T.V. , Chadwick, D. & Robinson, C.H. (1995). The impact of enhanced ultraviolet-B radiation on litter quality and decomposition processes in Vaccinium leaves from the Subarctic. *Oikos, 72,* 213-222.

Gibbs, P.E.M., Kilbey, B.J., Banerjee, S.K. & Lawrence, C.W. (1993). The frequency of and accuracy of replication past a thymine-thymine cyclobutane dimer are very different in *Saccharomyces cerevesiae* and *Escherichia coli. J. Bacteriol., 175,* 2607-2612.

Girotti, A. (2001). Photosensitized oxidation of membrane lipids: reaction pathways, cytotoxic effects, and cytoprotective mechanisms. *J. Photochem. Photobiol. B: Biology, 63,* 103-113.

Godar, D.E. (1999a). Light and death: Photons and apoptosis. *J. Invest. Dermatol. Symp. Proc., 4,* 17-23.

Godar, D.E. (1999b). UVA1 radiation triggers two different final apoptotic pathways. *J. Invest. Dermatol. 112,* 3-12.

Gwynn-Jones, D. (1999). Enhanced UV-B radiation and herbivory. Ecol. Bull., 47, 77-83

Hada, M., Iida, Y. & Takeuchi, Y. (2000). Action spectra of DNA photolyases for photorepair of cyclobutane pyrimidine dimers in sorghum and cucumber. *Plant Cell Physiol., 41,* 644-648.

Hausser, K.W. & v. Oehmcke, H. (1933). Lichtbräunung an Fruchtschalen. *Strahlentherapie, 48,* 223-229.

Helbling, W., Ballaré, C.L. & Villafañe, V.E. (2001a). *Impact of ultraviolet rdiation on aquatic and terrestrial ecosystems,* pp. ix+122. *J. Photochem. Photobiol. B: Biology, 62 (1-2),* 1-122.

Helbling, E.W., Buma, A.G.J., de Boer, M.K. & Villafañe, V.E. (2001). In situ impact of solar ultraviolet radiation on photosynthesis and DNA in temperate marine phytoplankton. Mar. Ecol. Prog. Ser., 211, 43-49.

Herndl, G.J., Brugger, A., Hager, S., Kaiser, E., Obernosterer, I., Reitner, B. & Slezak, D. (1997). Role of ultraviolet-B radiation on bacterioplankton and the availability of disolved organic matter. *Plant Ecology, 128,* 42-51.

Hessen, D.O. (ed) (2002a). *UV radiation and arctic ecosystems,* pp. xx+321. Ecological studies, vol. 153. Berlin: Springer.

Hessen, D.O. (2002b). UV radiation and arctic freshwater zooplankton. In Hessen, D.O. (ed). *UV radiation and arctic ecosystems,* Ecological studies, vol. 153, Chapt. 8, pp. 158-184.. Berlin: Springer.

Ikenaga, M., Kondo, S. & Fujii, T. (1974). Action spectrum for photoreactivation in maize. *Photochem. Photobiol., 19,* 109-113.

Javitt, J.C. & Taylor, H.R. (1994). Cataract and latitude. *Documenta Ophthalmol., 88,* 307-325.

Jiang, , N. & Taylor, J.-S. (1993). In vivo evidence that UV-induced C=>T mutations at dipyrimidine sites could result from the replicative bypass of *cis-syn* cyclobutane dimers or their deamination products. *Biochemistgry, 32*, 472-481.

Karentz, D., Cleaver, J.E. & Mitchell, D.L. (1991). Cell-survival characteristics and molecular responses of antarctic phytoplankton to ultraviolet radiation. *J. Phycol., 27*, 326-341.

Labitzke, K. & van Loon, H. (1997). Total ozone and the 11 yr-sunspot cycle. *J. Atmospheric Solar-Terrestrial Phys., 59*, 9-19.

Lamb, C. & Dixon, R.A. (1997). The oxidative burst in plant disease resistance. *Annu. Rev. Plant Physiol. Plant Mol. Biol., 48*, 251-275.

Laurion, I & Vincent, W.F. (1998). Cell size versus taxonomic composition as determinatnts of UV-sensitivity in natural phytoplankton communities. *Limnol. Oceanogr., 43*, 1774-1779.

Lavola, A., Julkunen-Tiitto, R., Roinenen, H., Aphalo, P. (1998) Host-plant preference of an insect herbivore mediated by UV-B and CO_2 in relation to plant secondary metabolites. *Biochem. Syst. Ecol., 26*, 1-12.

Longstreth, J., de Gruijl, F.R., Kripke, M.L., Abseck, S. Arnold, F., Slaper, H.I., Velders, G., Takizawa, Y. & van der Leun, J.C. (1998). Health risks. *J. Photochem. Photobiol. B: Biology, 46*, 20-39.

Merriam, J.C., Lofgren, S., Michael, R., Soderberg, P., Dillon, J., Zheng, L. & Ayala, M. (2000). An action spectrum for UV-B radiation and the rat lens. *Investig. Ophthalmol. Visual Sci., 41*, 2642-2647.

Michalova, K., Clemett, R., Dempster, A., Evans, J. & Allardyce, R.A. (2001). Iris melanomas: are they more frequent in New Zealand? *Brit. J. Ophthalmol., 85*, 4-5.

Moody, S.A., Paul, N.D., Björn, L.O., Callaghan, T.V., Lee, J.A., Manetas, Y., Rozema, J., Gwynn-Jones, D., Johanson, U., Kyparissis, A. & Oudejans, A.M.C. (2001). The direct effects of UV-B radiation on *Betula pubescens* litter decomposing at flur European sites. *Plant Ecology, 154*, 29-36.

Mostajir, B., Sime-Ngando, T., Demers, S., Belzile, C., Roy, S., Gosselin, M., Chanut, J.P., de Mora, S., Fauchot, J., Vidussi, F. & Levasseur, M. (1999a). Ecological implications of changes in cell size and photosynthetic capacity of marine Prymnesiphyceae induced by ultraviolet-B radiation. *Marine Ecol. Progr. Ser., 187*, 89-100.

Mostajir, B., Demers, S., de Mora, S., Belzile, C., Chanut, J.P., Gosselin, M., Roy, S., Villegas, P.Z., Fauchot, J., Bouchard , J., Bird, D., Monfort , P. & Levasseur, M. (1999b). Experimental test of the effect of ultraviolet-B radiation in a planktonic community. *Limnol. Oceanogr., 44*, 586-596.

Murphy, T.M. (1990). Effect of broadband and visible radiation on hydrogen peroxide formation by cultured rose cells. *Physiol. Plant. 80*, 63-68.

Murphy, T.M. & Huerta, A.J. (1990). Hydrogen peroxide formation in cultured rose cells in response to UV-C radiation. *Physiol. Plant., 78*, 247-253.

Nilsson, A. (1996). *Ultraviolet reflections: Life under a thinning ozone layer*, pp. viii+152. Chichester: Wiley. ISBN 0-471-95843-3.

Nishigaki, R., Mitani, H., and Shima, A. 1998. Evasion of UVC-induced apoptosis by photorepair of cyclobutane pyrimidine dimers. *Exp. Cell Res.* 244, 43-53.

Nolin, J. (1995). *Ozonskiktet och vetenskapen*, pp. 302. Stockholm: Almqvist & Wiksell. ISBN 91-22-01687-2.

Oriowo, O.M., Cullen, A.P., Chou, B.R. & Sivak, J.G. (2001). Action spectrum and recovery for *in vitro* UV-induced cataract using whole lenses. *Invest. Ophthalmol. Visual Sci., 42*, 2596-2602.

Paul, D.D. (2000). Stratospheric ozone depletion, UV-B radiation and crop disease. *Envir. Pollut., 108*, 343-355.

Pang, Q.S. & Hays, J.B. (1991). UV-B inducible and temperature-sensitive photoreactivation of cyclobutane pyrimidine dimers in *Arabidopsis thaliana*. *Plant Physiol., 95*, 536-543.

Paul, N.D., Callaghan, T.V., Moody, S., Gwynn-Jones, D., Johanson, U. & Gehrke, C. (1999). UV-B impacts on decomposition and biogeochemical cycling. *In* Rozema, J. (ed.) *Stratospheric ozone depletion: the effects of enhanced UV-B radiation on terrestrial ecosystems*, pp. 117-133. Leiden: Backhaus.

Pitts, D.G., Cullen, A.P. & Hacker, P.D. (1977). Ocular effects of ultraviolet radiation from 295 to 365 nm. *Investig. Ophthalmol., 16*, 932-939.

Podskochy, A., Gan ,L., Fagerholm ,P. (2000). Apoptosis in UV-exposed rabbit corneas. *Cornea, 19*, 99-103.

Polo, V., Pinilla, I., Abecia, E., Larrosa, J.M., Pablo, L.E. & Honrubia, F.M. (1997). Assessment of the ocular media absorption index. *Int. J. Ophthalmol., 20*, 1-3.

Ravanat, J.-L., Douki, T. & Cadet, J. (2001). Direct and indirect effects of UV radiation on DNA and its components. *J. Photochem. Photobiol. B: Biology*, *63*, 88-102.

Reid, G.C. (1999). Solar variability and its implications fo the human environment. *J. Atmospheric Solar-Terrestrial Phys.*, *61*, 3-14.

Ren, H.W. & Wilson, G. (1994). The effect of ultraviolet-B irradiation on the cell shedding rate of the corneal epithelium. *Acta Ophthalmol.*, *72*, 447-452.

Rozema J (ed) (1999) *Stratospheric ozone depletion: the effects of enhanced UV-B radiation on terrestrial ecosystems.* Backhuys, Leiden.

Rozema, J., van de Staaij, J., Caldwell, M.M. & Björn, L.O. (1997a). UV-B as an environmental factor in plant life: stress and regulation. *Trends Ecol. Evolution*, *12*, 22-28.

Rozema, J., Gieskes, W.W.C., van de Geijn, S.C., Nolan, C. & de Boois, H. (eds) (1997b). *UV-B and biosphere*, pp. 319. Dordrecht: Kluwer Academic Publ. ISBN 0-7923-4422-7.

Rozema, J., Manetas, Y. & Björn, L.O. (eds) (2001). *Responses of plants to UV-B radiation*, vii+278. Dordrecht: Kluwer Academic Publishers.

Rousseaux, J.C., Ballaré, C.L., Giordano, C.V., Scopel, A.L., Zima, A.M., Szwarccberg-Bracitta, M., Searles, P.S., Caldwell, M.M. & Diaz, S.B. (1999). Ozone depletion and UVB radiation: impact on plant DNA damage in southern South America. *Proc. Natl Acad. Sci. USA*, *96*, 15310-15315.

Saito, N. & Werbin, H. (1969). Action spectrum for a DNA-photoreactivating enzyme isolated from higher plants. *Radiation Botany*, *9*, 421-424.

Scheuerlein, R., Treml, S., Thar, B., Tirlapur, U.K. & Häder, D.-P. (1995). Evidence for UV-B-induced DNA degradation in *Euglena gracilis* mediated by activation of metal-dependent nucleases. - *J. Photochem. Photobiol. B: Biol.* 31, 113-123.

Scott, B.R. (1998). Ultraviolet radiation effects upon the eye: Problems of dosimetry. *Radiation Prot. Dosim.*, *76*, 277-277

Sinha, R.P., Klisch, M., Gröninger, A. & Häder, D.-P. (1998). Ultraviolet-absorbing Ultraviolet-absorbing/screening substances in cyanobacteria, phytoplankton and macroalgae. J. Photochem. Photobiol. B: Biol., 47, 83-94.

Sinha, R.P., Klisch, M., Gröniger, A. & Häder, D.-P. (2001). Responses of aquatic algae and cyanobacteria to solar UV-B. *Plant Ecology*, *154*, 221-236.

Sliney, D.H. (1997). Ultraviolet radiation effects upon the eye: Problems of dosimetry. *Radiation Prot. Dosim.*, *72*, 197-206.

Smith, C.C., Prézelin, B.B., Baker, K.S., Bidigare, R.R., Boucher, N.P., Coley, T.L., Karentz, D., MacIntyre, S., Matlick, H.A., Menzies, D., Ondrusek, M., Wan, Z., & Waters, K.J. (1992). Ozone depletion: ultraviolet radiation and phytoplankton biology in Antarctic waters. 952-959.

Takeuchi, Y., Murakami, M., Nakajima, N., Kondo, N. & Nikaido, O. (1998). The photorepair and photoisomerization of DNA lesions in etiolated cucumber cotyledons after irradiation by UV-B depends on wavelength. *Plant Cell Physiol.*, *39*, 745-750.

Taylor, H.R. (1994). Ocular effects of UV-B exposure. *Documenta Ophthalm.*, *88*, 285-293.

Threlfall, T.J. & English, D.R. (1999). Sun exposure and pterygium of the eye: A dose-response curve. *Am. J. Ophthalmol.*, *128*, 280-287.

Todo, T. (1999). Functional diversity of the DNA photolyase/blue light receptor family. *Mutat. Res.*, *434*, 89-97.

van der Leun , J., Tang, X. & Tevini, M. (eds) (1995). Environmental effects of ozone depletion: 1994 assessment. *Ambio*, *24*, 137-197.

van der Leun , J., Tang, X. & Tevini, M. (eds) (1998). Environmental effects of ozone depletion: 1998 assessment. *J. Photochem. Photobiol.B: Biology* 46, 1-108.

Van Donk, E., Faafeng, B.A., De Lange, H.J. & Hessen, D.O. (2001). Differential sensitivity to natural ultraviolet radiation among phytoplankton species in Arctic lakes (Spitsbergen, Norway). Plant Ecology, 154, 213-223.

Wu, F. & Deng, N.S. (2000). Photochemistry of hydrolytic iron (III) species and photoinduced degradation of organic compounds. A minireview. *Chemosphere*, *41*, 1137-1147.

WHO (1994). Environmental health criteria 160: Ultraviolet radiation. Geneva: World Health Organisation. ISBN 92-4-157160-8.

Yoon, J.-H., Lee, C.-S., O'Connor, T.R., Yasui, A. & Pfeifer, G.P. (2000). The DNA damage spectrum produced by simulated sunlight. *J. Mol. Biol.*, *299*, 681-693.

Young, R.W. (1994). The family of sunlight-related eye diseases. Optometry Vision Sci., 71, 125-144.

Zhao, X. & Taylor, J-S. (1996). Mutation spectra of TA*, the major photoproduct of thymidylyl-(3´5´)-deoxyadenosine, in *Escherichia coli* under SOS conditions. *Nucleic Acids Res.*, *24*, 1561-1565.

Zhao, X., Liu, J., Hsu, D.S., Zhao, S., Taylor, J.-S. & Sancar, A. (1997). Reaction mechanism of (6-4) photolyase. *J. Biol. Chem., 272,* 32580-32590.

Zigman , S. (1995). Environmental near-UV radiation and cataracts. Optometry Vision Sci., 72, 899-901.

LARS OLOF BJÖRN

12. VITAMIN D

PHOTOBIOLOGICAL AND ECOLOGICAL ASPECTS

1. INTRODUCTION

As a young boy I was forced to swallow a spoonful of cod liver oil every day. I was told that it contained vitamin D and that I had to eat it to get good bones in my body. I did not wonder why it was in the cod and how it got there. In fact I did not started to ask such questions until a few years ago. For some of the questions I have so far found no good answers. I learnt some surprising things, for instance that vitamin D is not really a vitamin in the strict sense, and that the cod can hardly make any of it at all.

The early research history relating to vitamin D has been recounted many times, for instance by DeLuca (1997), and only a short summary will be given here. Rickets was first described in England by Whistler (1645) and Glisson (1650), and became known as the English disease in many countries. Mellanby (1918) demonstrated that rickets (Fig. 1) could be prevented in dogs by supplanting their diet with cod liver oil, and Hess & Unger (1921) showed that rickets could be cured by sunlight. Hess & Weinstock (1924) and Steenbock & Black (1924) showed that the exposure of lettuce and several other foodstuffs to ultraviolet-C radiation would render them antirachitic.

The present review will not treat any medical aspects of vitamin D in any detail, but several recent treatises of this topic are available, e.g., Feldman & Glorieux (1997), Holick (1999).

2. CHEMISTRY AND PHOTOCHEMISTRY OF PROVITAMINS AND VITAMINS D

There are (at least) two kinds of vitamin D (also called calciferol), i.e. vitamin D_2 (ergocalciferol) and vitamin D_3 (cholecalciferol), with slightly different structures (Fig. 1). The reason that there is no vitamin D_1 is that the product which was first given this name turned out not to be a single compound, but a mixture containing vitamin D. There seem to be in some non-mammal vertebrates also other compounds acting in a similar way as vitamins D_2 and D_3 (Holick 1989), but they have not been chemically defined. In most organisms the synthesis of vitamin D requires ultraviolet-B radiation. Exceptions to this rule will be described in the following. Vitamins D are formed from the provitamins (provitamin D_2, also called ergosterol, and provitamin D_3, is also called 7-dehydrocholesterol. Ultraviolet-B

265

L.O. Björn (ed.), Photobiology, 265–280.
© 2002 *Kluwer Academic Publishers. Printed in the Netherlands.*

radiation can photoisomerize the provitamins to the corresponding previtamins, either *in vivo* without the mediation of any enzyme, or in solution. The previtamins are slowly converted by an non-enzymatic and non-photochemical reaction to the vitamins.

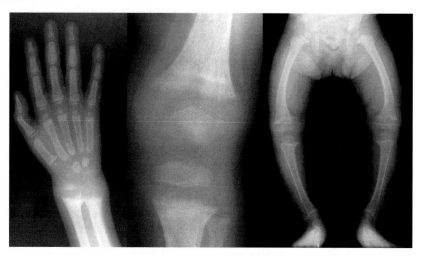

Figure 1. X-ray plates showing signs of vitamin D deficiency (rickets): incomplete bone-formation in wrist and knee, and malformed legs in a child. Courtesy Dr. Michael L. Richardson, University of Washington Department of Radiology.

The curing of rickets by sunlight can be explained by the fact that provitamin D_3 is synthesized in human skin cells. Exposure to sunlight converts it to previtamin D_3, which is in turn converted to vitamin D_3. Since a vitamin is defined as a substance necessary for health which cannot be synthesized by the body and must be ingested with the food, vitamin D is, strictly speaking, not a vitamin. Since, however, exposure to sunlight is often insufficient for maintaining health, and deficiency can be prevented by vitamin D in the food, the vitamin status is defendable.

On the other hand, it should be noted that the commonly recommended daily vitamin D intake (200 international units per day) is insufficient for prevention of deficiency if exposure to ultraviolet-B radiation does not supplement the supply (Glerup et al. 2000). Attempts to avoid the need for ultraviolet exposure by high daily intake cannot be recommended, as this can lead to vitamin D poisoning. Exposure even to high daily fluence of UV-B radiation can never lead to vitamin D overdosage, as will soon be explained.

As mentioned, the vitamin D precursor previtamin D is formed from provitamin D by a photochemical reaction (Fig. 2) driven by ultraviolet radiation (UV-B in the natural condition, but also UV-C can be used artificially). But this is far from the only ultraviolet driven reaction in the vitamin D context. The basic photochemistry of the vitamin D system was summarized already by Havinga (1973). Previtamin D is also sensitive to ultraviolet radiation and can undergo three different photoreversible photochemical reactions. It can either be reconverted to provitamin D, or converted to lumisterol or to tachysterol or and further irreversibly to products known under the common name of toxisterols (Boomsma et al. 1975). Also vitamin D is sensitive to ultraviolet radiation and can be photoconverted to three compounds known as 5,6-trans- vitamin D₃, suprasterol 1 and suprasterol 2 (Webb et al. 1989).

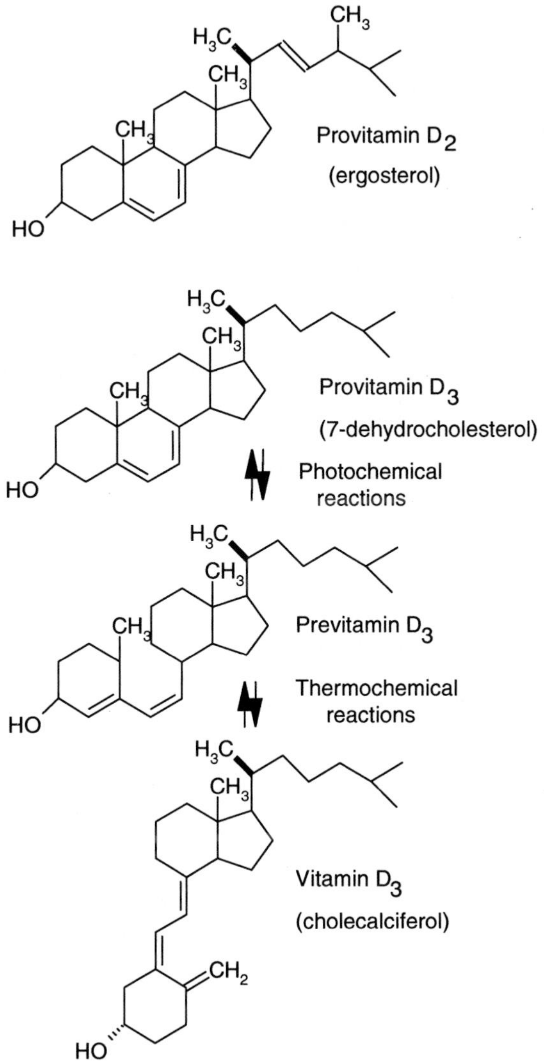

Figure 2. The structure of the two types of provitamin D, the reversible photoconversion of provitamin D₃ to previtamin D₃, and the reversible thermochemical conversion of previtamin D₃ to vitamin D₃.

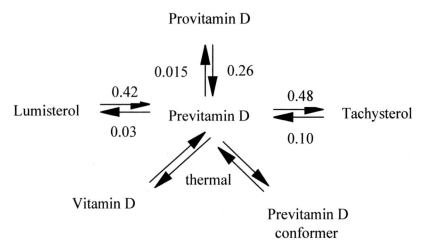

Figure 3. The reversible conversions of previtamin D with quantum yields of the photochemical reactions. Previtamin D can be photochemically converted also to various compounds termed toxisterols (not shown). From Havinga (1973), modified.

Havinga states that the quantum yields are independent of wavelength, or at least have the same values at 254 and 313 nm. There is, however, as will be detailed below, an important exception to this rule. Although some literature sources give values slightly different from those of Havinga shown in Fig. 3, there is nothing pointing to differences in quantum yield between the D_2 and D_3 series.

Provitamin D_3 is present in mammalian skin not only in free form, but also esterified with fatty acids, and the esterified provitamin is tranformed to esterified vitamin D3 upon exposure to ultraviolet radiation (Takada 1983). In fact, most of the provitamin and vitamin D in rat skin is in the esterified form.

A number of authors (reviewed by Dmitrenko et al. 2001) have found a curious behaviour of the quantum yield for photochemical ring closure of previtamin D (ring closure results in either conversion back to provitamin D, or to formation of lumisterol, which has the same structure as provitamin D except for the direction of a methyl group, which is up for previtamin D as shown in Fig. 1, down for lumisterol). This quantum yield increases slowly with wavelength from 295 to 302 nm, but then doubles from 0.08 at 302 nm to 0.16 at 305 nm, and then increases steadily to 0.29 at 325 nm. The quantum yield of cis-trans isomerisation to tachysterol decreases correspondingly over the same wavelength range. Various explanations for this behaviour have been advanced (see Dmitrenko et al. 2001 for further literature).

There are more complications to this photochemical system, which at first glance looks rather simple. It was found that the thermochemical step forming vitamin D following the photochemical conversion of provitamin D takes place much faster in cells than in solution (Tian et al. 1993, Holick et al. 1995). The reason for this is the existence of the conformer of previtamin D shown at lower right in Fig. 3. In solution this is the preferred conformer, and it cannot be converted directly to vitamin D. In a membrane, both a natural one and in artificial liposome membranes

(Tian & Holick 1999) the previtamin is held in the active, vitamin-producing conformer. The same effect can be achieved by complexing the previtamin with β-cyclodextrin (Tian & Holick 1995).

Provitamins, previtamins and vitamins D occur not only in free form, but in plants also also as glycosides and in mammal skin also as fatty acid esters (Takada et al. 1983). In rat skin at least 80% of the provitamin D_3 was found to be esterified, and upon exposure of the skin to ultraviolet radiation the provitamin D_3 ester is converted to vitamin D_3 ester.

The action spectrum for conversion of provitamin D_3 to previtamin D_3 in human skin has been determined by MacLaughlin et al. (1982). It has a single peak at 295 nm. This roughly corresponds to the long-wavelength absorption band of provitamin D_3 dissolved in n-hexane (Fig. 4). The absorption spectrum for provitamin D_3 is three-peaked, but the two short-wave absorption bands are lacking in the action spectrum. Two circumstances could contribute to this lack: (1) The stratum corneum of the skin could filter off the shorter wavelength components, and (2) since at the shorter wavelengths both provitamin and previtamin absorb, but at the longer wavelengths (around 295 nm) only provitamin, the conversion of provitamin to previtamin is favoured at long wavelengths, while at shorter wavelengths the back- and side-reactions of previtamin are important competitors. Probably, under the conditions the action spectrum was determined, the first reason is the more important one.

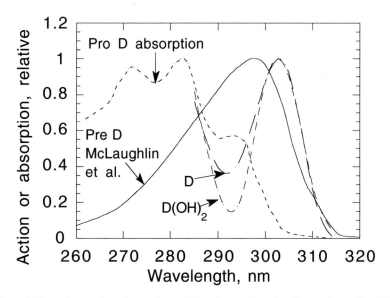

Figure 4. Absorption spectrum for provitamin D3, action spectrum for photosynthesis of previtamin D_3 according to McLaughlin et al. (1982), and action spectra for photosynthesis of vitamin D_3 and $1\alpha,25$-dihydroxyvitamin D_3 accoring to Lehmann et al. (2001).

An investigation by Nemanic et al (1985) does not contribute much to the knowledge of action spectra, but a recent one by Lehmann et al. (2001) is important. These investigators measured the action spectra for formation of vitamin D_3 as well as $1\alpha,25(OH)_2D_3$ ($1\alpha,25$-dihydroxyvitamin D_3) from provitamin D_3 in "artificial skin" containing cultured human keratinocytes. Remarkably, this action spectrum is displaced about 7 nm towards longer wavelength (peaking at about 302 nm) compared to the spectrum determined by MacLaughlin et al. (1982). Although no wavelength below 285 nm was tested, the spectrum indicates a rise from 290 nm towards shorter wavelengths. The minimum at ca 293 nm is deeper than what would be expected from the provitamin D_3 absorption spectrum.

3. TRANSPORT AND TRANSFORMATIONS OF VITAMIN D IN THE HUMAN BODY

In the human body the endogenous provitamin D_3 is present in the cell membranes of cells in the skin. After conversion to vitamin D_3 the molecule no longer fits well into the membrane; one of the hydroxyl groups is sticking out on the outside. This hydroxyl group attaches itself to a carrier protein present in the intercellular liquid, which carries the vitamin into the bloodstream. In the liver it is hydroxylated at carbon atom 25, and after transport to the kidneys also at carbon atom 1. The resulting $1\alpha,25(OH)_2D_3$ is the main active form, an important hormone which travels to different target cells in the body. There it interacts both with a nuclear receptor, regulating gene transcription in a tissue-specific manner, and with a receptor in the cell membrane, triggering other and generally more rapid responses. This double way of action is reminiscent of how auxin and phytochrome act in plant cells.

To some extent also the skin cells are able to transform the vitamin D formed in them to $1\alpha,25(OH)_2D_3$ (Nemanic et al. 1985, Lehmann et al. 2001).

4. PHYSIOLOGICAL ROLES OF 1,25-DIHYDROXYVITAMIN D IN VERTEBRATES

$1\alpha,25(OH)_2D_3$ (1,25-dihydroxy-vitamin D) regulates calcium transport and metabolism at several levels. In our bodies it probably stimulates calcium uptake into enterocytes by keeping calcium channels open (this phase of the calcium uptake is passive). The movement of calcium ions through these cells may also be stimulated by $1\alpha,25(OH)_2D_3$ via vitamin $1\alpha,25(OH)_2D_3$ dependent calbindin. We know for sure that the energy-requiring extrusion of calcium ions on the inside of the cells is stimulated by vitamin $1\alpha,25(OH)_2D_3$, probably by increasing expression of the calcium pump (Zelinski et al. 1991). $1\alpha,25(OH)_2D_3$ also regulates calcium excretion in the kidneys. It also regulates calcium fluxes to and from the skeleton, and bone formation. But apart from its obviously calcium-related effects, it has many other functions (in some cases also related to calcium, but in a less obvious way). Of special interest in understanding the evolution (see below) of the vitamin D system may be the effects on the immune system. It can stimulate the differentiation of bone marrow cells into macrophages (Abe et al. 1981), regulate T-cell proliferation (Nunn et al. 1986). Recently, however, the opinion has been voiced (Mathieu et al. 2001) that, although immune cells carry $1\alpha,25(OH)_2D_3$ receptors, and although calcium is involved in the functioning of the immune system, vitamin $1\alpha,25(OH)_2D_3$ is redundant in the context. At the same time others maintain a role for vitamin $1\alpha,25(OH)_2D_3$ in immune response (Panda et al. 2001, Griffin et al. 2001).

Vitamin D may protect from some forms of cancer (discussions by Garland et al. 1999, van Leeuwen et al. 1999, and Uskovic' et al. 1999). On the other hand, based on its photobiology, a role for vitamin D has been implicated in the development of melanoma (Braun & Tucker 1997).

5. EVOLUTIONARY ASPECTS

Why has Nature chosen, for the hormonal regulation of calcium metabolism and other bodily functions, as substance requiring the uncertain exposure to ultraviolet radiation for its synthesis? Because the answer to this question is not obvious, we shall take a look into our past to search for it.

One explanation, proposed by Chevalier et al. (1997) is that the formation of 1,25-dihydroxyvitamin D_3 from 7-dehydrocholesterol was originally a catabolic pathway, which has later been taken over for regulatory purposes. Arguments for this are (1) that vitamin D and related substances are rather toxic, and (2) that P450-type enzymes are involved both in hydroxylations that lead to detoxification and solubilization of known toxins and in several hydroxylation steps of vitamin D and its analogs (reviews of the vitamin D-related hydroxylations by Jones (1999) and Okuda & Ohyama (1999).

One way of probing into the past is to compare aminoacid sequences in proteins of living organisms. We that the vitamin D nuclear receptor (VDR) belongs to a class of nuclear receptors of very ancient origin. The nuclear receptor class can be divided into several subclasses, and the divergence into these subclasses occured at least 500 million years ago (Laudet et al. 1992). The closest known relative to the VDR is the ecdysone receptor in insects. One way of tracing the origin of the vitamin D regulation system is to track the evolution of the VDR more in detail. Until this has been done we can look for other cues.

In terrestrial vertebrates, i.e. birds, reptiles, and amphibians, the role of vitamin D seems to be in principle rather like that in mammals, although birds are not able to use vitamin D_2 efficiently and there may exist in lizards and frogs other provitamins and vitamins D than D_2 and D_3 (Holick 1989). When we dig further back into our evolutionary past the evidence starts to become, if the expression is excused, more fishy. Michael F. Holick has repeatedly voiced the view that in the marine environment the concentration of calcium was always so high that calcium could passively flow into the compartments where it was needed, and it was only when vertebrates embarked on their journey onto land that the situation became different, and calcium-pumping and regulation became a necessity. Holick (1999) opens a recent book on vitamin D with the sentence "Approximately 400 years ago, as vertebrates ventured from the ocean unto land, they were confronted with a significant crisis. As they had evolved in the calcium-rich ocean environment .... However, on land, the environment was deficient in calcium; as a result, early marine vertebrates that ventured onto land needed to develop a mechanism to utilize and process the scarce amounts of calcium in their environment...".

This view is supported by several investigations which show that various fishes are doing well without vitamin D, and this is true also for at least one freshwater fish (Ashok et al. 1998, 1999). There are, however, an even larger number of investigations pointing to a function for vitamin D in other fish species (Barnett et al. 1979 and Brown & Robinson 1992, Larsson 1999 and sources cited therein). It is also not clear why regulation would be unnecessary if calcium uptake does not require energy. There must be some mechanism then to avoid too high calcium concentration in the cytosol.

The first vertebrates were the jawless Heterostraci and Osteostrachi, which had their bodies covered by bony plates. They were followed in evolution by the first true fishes, the sharklike Placodermi. These, as belonging to the Elasmobranchiomorphi (cartilagous fishes) had no bones inside their bodies, but they were also covered with a bony armour. Could it possibly be that early in evolution the deposition of calcium phosphate and calcium carbonate served as a protection against ultraviolet radiation, and that therefore it was reasonable that its deposition was regulated by radiation? The hypothesis may seem far-fetched, but is at least in principle testable. One could try to find out whether the thickness of the armour varied with latitude (and thus with UV-B exposure), of course taking continental drift and polar migration into account.

Even the earliest vertebrates mentioned lived less than 550 million years ago, at a time when the protecting ozone shield is thought to have afforded almost the same protection as today (it evolved about 2200 million years ago). But could the regulation of calcium metabolism by vitamin D be of even more ancient origin than the vertebrates? There are a few investigations pointing in this direction.

Coccolithophorids are microscopic organisms regarded as protozoans by zoologists and as a kind of golden algae by botanists. They are covered by plates of calcium carbonate, and it has been reported that they loose their ability to form plates when grown in indoor cultures (Braarud 1954, Manton & Peterfi 1969, and others). If the cultures are exposed to "fluorescent light" the ability to deposit calcium carbonate was restored (Dorigan & Wilbur 1973), but the spectral composition of the fluorescent light was not stated. Vitamin D_2 is reported to have, in combination with bacteria and other food, a beneficial effect on the growth of a copepod (Guerin et al. 2001). In a kind of coral (incidentally a relative of the kind of red coral used for gems) ultraviolet radiation favoured the development of normal spicules, structures containing collagen and calcium carbonate. The animal was also shown to produce 1,25-dihydroxy vitamin D in a UV dependent manner (Kingsley et al. 2001).

The most compelling evidence, however, for the ancient origin of vitamin D as a calcium regulator comes from experiments with snails (Kriajev & Edelstein 1994, 1995, Kriajev et al. 1994). In these animals certain vitamin D-like compouns elevates inracellular exchangable calcium and suppresses alkaline phosphatase activity. The authors conclude in their latest paper that the snails adapt to light conditions via the vitamin D endocrine system. The evolutionary lines leading to molluscs and to vertebrates can be estimated to have diverged about 720 million years ago (with great uncertainty, from information in Van de Peer et al. 2000 combined with the divergance time of 833 MbP for vertebrates and arthropodes in Nei et al. 2001). If more compelling evidence of vitamin D regulation in copepods, corals, and coccolithophorids turns up, a much higher age for the regulation system would have credence. If we assume that also the vitamin D elicited induction of calmodulin in plant roots (see section 7 below) has an evolutionary origin common with the regulation of calcium metabolism in animals, then this origin lies more than a one and a half billion years back in time (Nei et al. 2001).

There remains the unlikely possibility of convergent evolution, that distantly related organism have independently chosen vitamin D as their calcium regulator. If this is the case, what is so special with vitamin D; why is it the best choice?

Calcium carbonate itself is a poor absorber for ultraviolet radiation and may appear to be a bad choice for production of a radiation shield. Even a cm-thick layer absorbs only half of the incident radiation at the DNA absorption maximum, 260 nm) as calculated from data for clear calcite crystals (Washburn et al. 1929). However, to this should be added the scattering effect and, above all, the absorption by proteins and other substances always associated with calcium carbonate shells and other calcified structures.

If the reason for the choice of the ultraviolet sensitive vitamin D system is not regulation of ultraviolet shielding, what could it be? In Chapter 13 is described how our immune defence is modulated by ultraviolet radiation, but the evolutionary pressure that has selected for this modulation is obscure. It is likely to be relevant, since it occurs through two different mechanisms, via UV absorption in urocanic acid, and via absorption in DNA. Could there be that the original function of the vitamin D system was to modulate the immune defence, a function that to some extent seems still to exist?

6. DISTRIBUTION OF PROVITAMINS AND VITAMINS D IN THE PLANT KINGDOM

Among microalgae several (but not all) species of the green algae *Chlorella* (Patterson 1971) and *Chlamydomonas reinhardtii* (Patterson 1974) contain ergosterol. This provitamin has also been found in the diatom *Skeletonema menzelii* and the coccolithophorid *Emaliana huxlei* (Holick 1989), and the chrysophycean *Ochromonas danica* (Gershengorn et al. 1968). In addition there are numerous investigations on phytoplankton of mixed composition. Of special interest is perhaps a case in which a correlation with the probable ultraviolet exposure has been established, using season as radiation exposure proxy (Takeuchi et al. 1991; see also Tables 1a and 1b of Björn & Wang 2001). Among macroalgae not only ergosterol, but also provitamin D_2 and vitamins D_2 and D_3 have been found in the brown alga *Fucus vesiculosus* grown under natural conditions with higher content of the vitamins at a lower (southern Sweden) than at a higher (northern Norway) latitude (Björn & Wang unpublished), and provitamin D_3 is reported present in the gametophyte of the red alga *Chondrus crispus*, while the sporophyte of the same species contained the isomer 22-dehydrocholesterol.

Higher plants generally contain provitamins and vitamins D_2 and D_3 in their leaves (Napoli et al. 1977, Rambeck et al. 1981, Horst et al. 1984, Prema & Raghuramulu 1994, 1996) and in general vitamins are present only after exposure to ultraviolet radiation (Hess & Weinstock 1924, Wasserman et al. 1976, Zucker et al. 1980, Skliar et al. 2000, Björn & Wang 2001), but there are exceptions (see section 10 below). Some plants even form the hydroxylated forms of vitamin D (Napoli et al. 1977, Skliar et al. 2000).

7. PHYSIOLOGICAL EFFECTS OF PROVITAMINS AND VITAMINS D ON PLANTS AND ALGAE

Fries (1984) showed that growth of the green macroalga *Enteromorpha compressa*, the red alga *Nemalion helminthoides*, and the brown alga *Fucus spiralis* is stimulated by vitamins and provitamins D. Vitamin D applied to herbaceous and woody plants stimulates initiation of adventitious roots (Buchala & Schmid 1979, Jarvis & Booth 1981, Moncousin & Gaspar 1983). Vitamin D3 at a nanomolar concentration inhibits root elongation in *Phaseolus vulgaris* and promotes germination of light sensitive lettuce seed in darkness (Buchala & Pythoud 1988). Vitamin D_3 induces the synthesis of the calcium-binding signalling protein calmodulin in bean roots (Vega & Boland 1986), and a vitamin D_3-binding protein has been detected in the root cells (Vega & Boland 1988. 1989).

8. WHAT IS THE ROLE OF PROVITAMIN D IN PLANTS?

In most plant parts and in algae ultraviolet radiation seems to be an obligatory requierement for vitamin D formation. It has been proposed (see Chapter 8) that in those cases provitamin D could function as a UV-B photoreceptor.

The exceptional plant *Solanum glaucophyllum* forms so large amounts of the active vertebrate hormone form, $1\alpha,25(OH)_2D_3$, that grazing animals are poisoned (see Curino et al. 1998 for literature). In this case one can assume a protective function.

9. BIOGEOGRAPHICAL ASPECTS

Human complexion tends to be darker the higher the ultraviolet radiation in the environment. This is an inherited ("racial") trait, but we also are able to acclimatize phenotypically to some extent (i.e. the skin of an individual forms pigment in response to ultraviolet radiation, see chapter 13). We also know that ultraviolet radiation can cause skin cancer and other problems, and that these effects are particularly frequent for people poorly adapted for the high environmental radiation they are exposed to, such as people of European origin living in South Africa and Australia. Thus, clearly, the pigment works as protection against high radiation. Although vitamin D is toxic at too high a concentration, it has been shown by Holick et al. (1981) that skin pigment is not necessary to prevent its overaccumulation; the photochemical system is self-regulating. The reason for this is the low rate of conversion of previtamin to vitamin, in combination with the photochemical side- and back reactions of previtamin D. Thus poisoning can occur only by excessive intake (e.g., Koutkia et al. 2001).

There is, however, another connection between complexion and vitamin D. All humans are thought to originate from Africa, and presumably we are all descendents of black people, although at a prehuman furred stage they may have had lighter skin, as chimpanzees (Jablonski & Chaplin 2000). But as our forefathers emigrated to higher and higher latitudes, they became paler (Fig. 5), and the selection pressure for this is clear: it was avoidance of vitamin D deficiency (Clemens et al. 1982). African people who have emigrated north in historic time are known to suffer just from such deficiency (Shewakramani et al. 2001). Inuits may have more pigment than what one would expect from their northern habitat, but their traditional food is from the sea, and mostly rich in vitamin D (because the sea currents bring vitamin D from lower and more sunny latitudes), so they have not been exposed to the same selection pressure as people with more terrestrial habits.

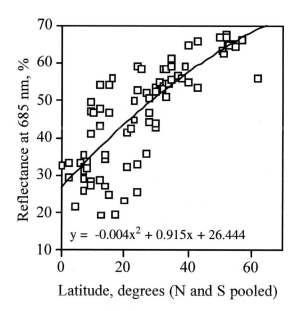

$$y = -0.004x^2 + 0.915x + 26.444$$

Latitude, degrees (N and S pooled)

Figure 5. The relation between skin colour (measured as reflectance at 685 nm) and latitude for 85 samples of "indigenous populations" from different parts of the world. Thus darker skin colour is lower in the diagram, and on the abscissa 0 stands for the equator. In the regression equation at the bottom of the graph y stands for reflectance in percent, and x for latitude (either north or south) in degrees. The graph reflects a clear trend of darker skin colour towards the equator. The great variation around the regression curve has several causes. Even "indigenous populations" have migrated and settled in their present regions within a time-span which is often too short to allow complete adaptation to the environment. The way of life also modulates the need for sunlight. Thus, for example, the square at the highest latitude (a little above 60 degrees) lies far below the regression curve. It represents inuits in southern Greenland. It is thought that they, due to their vitamin D-rich food from the sea, have a lesser need for vitamin D from photochemical conversion in the skin than most other populations. Data adapted from those compiled by Jablonski & Chaplin (2000).

Jablonski & Chaplin (2000) found that in that in all human populations where data were available, the complexion of women is lighter than that of men. It may be difficult to separate out the acclimation component due to different ways of life, but the authors believe that it could be an adaptation to the greater need for calcium and vitamin D during pregnancy and lactation.

The question remains: how do non-human terrestrial vertebrates manage at high latitudes? As they are often covered with hair or plumage, or are "cold-blooded" (poikilothermic), they would have difficulties in producing their own vitamin D by having either inefficient photochemical conversion of pro- to previtamin, or inefficient thermochemical conversion of previtamin to vitamin. In fact, amphibians, and reptiles in particular, decline in frequency with increasing latitude. The arctic dinosaurs may, in fact, have been homeothermic (thermoregulating).

According to an old and abandoned theory birds produce provitamin D in their uropygial gland, distribute it over their plumage when preening, thus expose it to sunlight and convert it to vitamin D and ingest it at the next preening. Later investigations with more modern methods of analysis have failed to establish with certainty provitamin D in the uropygial secretion. Disregarding very old and unreliable findingts, there is a single paper (Uva et al. 1978) stating the presence of provitamin D_3 in the uropygial gland of domestic fowl (*Gallus*). The analytical methods used are good for their time. Nevertheless the investigation should be repeated using HPLC, NMR and absorption spectroscopy, since the identification of provitamin D_3 among all the steroids present in uropygial secretion is no easy matter, and if its presence can be established the analysis should be extended to other kinds of bird. It should also be mentioned that Holick (1989), referring to unpublished observations by himself and M.A. St. Lezin) found no provitamin D_3 in chicken feathers. On the other hand it is well established that fowl can use UV absorbed by head and legs to improve their vitamin D and calcium status and egg production. It must be assumed that birds like arctic owls and ptarmigans are totally dependent on vitamin D in the food for covering their requirements. Birds cannot efficiently use vitamin D_2, only D_3

For arctic mammals, as reindeer, the situation appears grim. Reindeer need much calcium, not only for the skeleton, but also for the yearly production of antlers. They are covered with fur and do not have an uropygial gland, so the only vitamin D source is the food. The critical time is the dark winter, and the most important winter food is reindeer lichen. We (Wang et al. 2001) have investigated one species of reindeer lichen from different latitudes (Fig. 4) and found that it contains both vitamins D_2 and D_3, and in a strongly latitude dependent manner, with the lowest values in northern Scandiavia even by the end of the summer. Still, 10 g of the lichen from northern Scandinavia would provide a man with the necessary daily amount. Wild reindeer survive even on Spitsbergen island at 78°N, where the vitamin D content in their food must be lower than this. Could there be another source? It would perhaps be worth looking at what the rumen bacteria can produce.

10. NON-PHOTOCHEMICAL PRODUCTION OF VITAMIN D

But in the rumen there is no sunlight. Can vitamin D be produced non-photochemically? The answer is yes. Curino et al. (1998, 2001) have shown that cells of *Solanum glaucophyllum* grown in culture in darkness forms $1\alpha,25(OH)_2D_3$, albeit at lower concentrations than the plant under sun-exposed field conditions. The mechanism is not known, but one has been proposed by Norman & Norman (1993).

The mechanism was proposed to explain how animals like subterranean mole rats, living in darkness from underground plant parts can obtain their requirement. It is, however, (1) doubtful whether these animals need vitamin D (Pitcher et al. 1994a,b, Buffenstein et al. 1991, 1994, 1995, Buffenstein 1996) and (2) that their diet is really completely vitamin D-free, as we have found small amounts of vitamins D_2 and D_3 in carrot roots not exposed to ultraviolet radiation (Wang & Björn in preparation). The same holds for nocturnal (Oppermann & Ross 1990, Kwiecinski et al. 2001). Larsson (1999) has erroneously claimed that Oppermann & Ross (1990) found that the nocturnal fruit-eating bat *Rousettus aeghypticus* can form 7-dehydrocholesterol from mevalonate.

REFERENCES

Abe, E., Miyaura, C., Sakagami, H., Takeda, M., Konno, K., Yamazaki, ?, Yoshiki, S. & Suda, T. (1981). Differentiation of mouse myeloid leukemia cells induced by 1α,25-dihydroxyvitamin D_3. *Proc. Acad. Sci. USA, 78,* 4990-4994..

Ashok A., Rao, D.S. & Raghuramulu, N. (1998). Vitamin D is not an essential nutrient for rora (*Labeo rohita*) as a representative of freshwater fish. *J. Nutrit. Sci. Vitaminol., 44,* 195-205.

Ashok, A., Rao, D.S., Chennaiah, S. & Raghuramulu, N. (1999). Vitamin D_2 is not biologically active for rora (*Labeo rohita*) as vitamin D_3. *J. Nutrit. Sci. Vitaminol., 45,* 21-30.

Barnett, B.J., Cho, C.Y. & Slinger, S.J. (1979). The essentiality of cholecalciferol in the diets of rainbow trout (*Salmo gairdneri*). Comp. Biochem. *Physiol., 63A,* 291-297.

Björn, L.O. & Wang, T. (2001). Is provitamin D a UV-B receptor in plants? *Plant Ecology, 154,* 3-8.

Boomsma, F., Jacobs, H.J.C., Havinga, E. & van der Gen, A. (1975). Studies of vitamin D and related compounds, part 24. New irradiation products of pre-vitamin D_3. *Tetrahedron Lett. 7,* 427-430.

Brown, P.B. & Robinson, E.H. (1992). Vitamin D studies with channel catfish (*Ictalurus punctuatus*) reared in calcium-free water. *Comp. Biochem. Physiol., 103A,* 213-219.

Brown, M.M. & Tucker, M.A. (1997). A role for photoproducts of vitamin D in the etiology of cutaneous melanoma. *Medical Hypotheses, 48,* 351-354.

Buchala, A.J. & Pythoud, F. (1988). Vitamin D and related compounds as plant growth substances. *Physiol. Plantarum, 74,* 391-396.

Buchala, A.J. & Schmid, A. (1979). Vitamin D and its analogues as a new class of plant growth substances affecting rhizogenesis. *Nature, 280,* 230-231.

Chevalier, G., Baudet, C., AvenelAudran, M., Furman, I. & Wion, D. (1997). Was the formtion of 1,25.dihydroxyvitamin D initially a catabolic pathway? *Medical Hypotheses, 48,* 325-329.

Clemens, T.L., Henderson, S.L., Adams, J.S. & Holick, M.F. (1982). Increased skin pigment reduces the capacity of skin to produce vitamin D in response to ultraviolet irradiation. *Lancet, 9,* 74-76.

Curino, A., Skliar, M. & Boland, R. (1998). Identification of 7-dehydrocholesterol, vitamin D_3, 25(OH)-vitamin D_3 and 1,25(OH)$_2$-vitamin D_3 in *Solanum glaucophyllum* cultures grown in absence of light. *Biochim. Biophys. Acta, 1425,* 485-492.

Curino, A., Milanesi, L., Benassati, S., Skliar, M. & Boland, R. (2001). Effect of culture concitions on the synthesis of vitamin D_3 metabolites in *Solanum glaucophyllum* grown in vitro. *Phytochemistry, 58,* 81-89.

DeLuca (1997). Historical overview. *In* Feldman, D., Glorieux, F.H. & Pike J.W. (eds): *Vitamin D,* pp. 3-12. New York: Academic Press. ISBN 0-12-252685-6.

Fries, L. (1984). D-vitamins and their precursors as growth regulators in axenically cultivated marine macroalgae. *J. Phycol., 20,* 62-66.

Feldman, D., Glorieux, F.H. & Pike J.W. (eds) (1997). *Vitamin D,* pp. ix+450. New York: Academic Press. ISBN 0-12-252685-6.

Garland, C.F., Garland, F.C. & Gorham, E.D. (1999). Epidemiology of cancer risk and vitamin D. *In* Holick, M.F. (ed.) *Vitamin D: Physiology, molecular biology, and clinical applications,* pp. 375-391. Totowa, N.J.: Humana Press. ISBN 0-89603-467-4.

Gershengorn. M.C., Smith, A.R.H., Goulston, G., Goad, L.J., Goodwon, T.W. & Haines, T.H. (1968). The sterols of *Ochromonas danica* and *Ochromonas malhamensis. Biochemistry, 7,* 1698-1706.

278

Glerup, H., Mikkelsen, K., Poulsen, L., Hass, E., Overbeck, S., Thomsen, J., Charles, P. & Eriksen, E.F. (2000). Commonly recommeded daily intake of vitamin D is not sufficient if sunlight exposure is limited. *J. Internal Med.247*, 260-268.

Glisson, F. (1650). *De Rachitide sive morbo puerili, qui vulgo The Rickets diciteur*. Pp. 414.

Griffin, M.D., Lutz, W., Phan, V.A., BAchman, L.A., McKean, D.J. & Kumar, R. (2001). Dendritic cell modulation by 1α,25 dihydroxyvitamin D₃ and its analogs: A vitamin D receptor-dependent pathway that promotes a persistent state of immaturity in vitro and in vivo. *Proc. Natl. Acad. Sci. USA, 98*, 6800-6805.

Guerin, J.P., Kirchner, M. & Cubizolles, F. (2001). Effects of *Oxyrrhis marina* (Dinoflagellata), bacteria and vitamin D₂ on population dynamics of *Tisbe holothuris* (Copepoda). *J. Exp. Marine Biol. Ecol., 261*, 1-16.

Havinga, E. (1973). Vitamin D, example and challenge. *Experientia, 29*, 1181-1193.

Hess, A.F. & Unger, L.G. (1921). Cure of infantile rickets by sunlight. *J. Am. Med. Assoc.* 77, 39.

Hess, A.F. & Weinstock, M. (1924). Antirachitic properties imparted to inert fluids and green vegetables by ultraviolet irradiation. *J. Biol. Chem. 62*, 301-313.

Holick, M.F. (ed.) (1999). *Vitamin D: Physiology, molecular biology, and clinical applications*, pp. xii+458. Totowa, N.J.: Humana Press. ISBN 0-89603-467-4.

Holick, M.F. (1989).Phylogenetic and evolutionary aspects of vitamin D from phytoplankton to humans. *In* Pang, P.K.T. & Schreibman, M.P. *Vertebrate endocrinology: Fundamentals and biomedical implications*, vol. 3, pp. 7-43. Orlando: Academic Press.

Holick, M.F., MacLaughlin, J.A. & Doppelt, S.H. (1981). Regulation of cutaneous previtamin D3 photosynthesis in man: skin pigment is not an essential regulator. *Science, 211*, 590-592.

Holick, M.F., Tian, X.Q. & Allen, M. (1995). Evolutionary importance for the membrane enhancement of the production of vitamin D₃ in the skin of poikilothermic animals. *Proc. Natl Acad. Sci. USA, 98*, 3124-3126.

Holick, M.F. (ed.) *Vitamin D: Physiology, molecular biology, and clinical applications, pp. xii+458*. Totowa, N.J.: Humana Press. ISBN 0-89603-467-4.

Jablonski, N.G. & Chaplin, G. (2000). The evolution of human skin coloration. *J. Human Evol., 39*, 57-106.

Jarvis, B.C. & Booth, A. (1981). Influence of indole-butyric acid, boron, *myo* -inositol, vitamin D₂ and seedling age on adventitious root developmant in cuttings of *Phaseolus aureus*. *Physiol. Plantarum, 53*, 213-218.

Jones, G. (1999). Metabolism and catabolism of vitamin D, its metabolites and clinically relevant analogs. *In* Holick, M.F. (ed.) *Vitamin D: Physiology, molecular biology, and clinical applications, pp. 57-84*. Totowa, N.J.: Humana Press. ISBN 0-89603-467-4.

Kingsley, R.J., Corcoran, M.L., Krider, K.L. & Kriechbaum, K.L. (2001). Thyroxine and vitamin D in the gorgonian *Leptogorgia virgulata*. *Comp. Biochem. Physiol. A, 129*, 897-907.

Kwiecinski, G.G., Lu, Z.R., Chen, T.C. & Holick, M.F. (2001). Observations on serum 25-hydroxyvitamin D and calcium concentrations form wild-caught and captive neotropical bats, *Artibeus jamaicensis*. *Gen. Comp. Endocrinol., 122*, 225-231.

Koutkia, P., Chen, T.C. & Holick, M.F. (2001). Vitamin D intoxication associated with an over-the-counter supplement. *New England J. Medicine, 345*, 66-67.

Kriajev, L. & Edelstein, S. (1994). Vitamin D metabolites and extracellular calcium currents in hemocytes of land snails. *Biochem. Biophys. Res. Commun., 204*, 1096-1101.

Kriajev, L. & Edelstein, S. (1995). Effect of light and nutrient restriction on the metabolism of calcium and vitamin D in land snails. *J. Exp. Zool. 272*, 153-158.

Kriajev, L., Otremski, I. & Edelstein, S. (1994). Calcium shells from snails: Response to vitamin D metabolites, *Calcified Tissue Internat., 55*, 204-207.

Larsson, D. (1999). *Vitamin D in teleost fish: Non-genomic regulation of intestinal calcium transport*. Diss. Göteborg Univ., Dept of Zoophysiology. ISBN 91-628-3681-1.

Laudet, V. (1997). Evolution of the nuclear receptor superfamily: early diversification from an ancestral orphan receptor. *J. Molec. Biol., 19*, 207-226.

Lehmann, B., Genehr, T., Pietzsch, J. & Meurer, M. (2001). UVB-induced conversion of 7-dehydrocholesterol to 1α,25-dihydroxyvitamin D₃ in an in vitro human skin equivalent model. *J. Investig. Dermatol., 117*, 1179-1185.

Mathieu, C., Van Etten, E., Gysemans, C., Decallone, B., Kato, S., Laureys, J., Devovere, J., Valcx, D., Verstuyf, A. & Bouillon, R. (2001). *In vitro* and *in vivo* analysis of the immune system of vitamin D receptor knockout mice. *J. Bone Mineral Res., 16*, 2057-2065.

MacLaughlin, J.A., Anderson, R.R. & Holick, M.F. (1982). Spectral character of sunlight modulates photosynthesis of previtamin D_3 and its photoisomers in human skin. *Science, 216,* 1001-1003.

Mellanby, E. (1918). The part played by an "accessory factor"in the production of experimental rickets. *J. Physiol. (Lond.), 52,* 11-14.

Moncousin, C. & Gaspar, T. (1983). Peroxidase as a marker for rooting improvement of *Cynara scolymus* L. cultured in vitro. Biochem. *Physiol. Pflanzen, 178,* 263-271.

Nemanic, M.K., Whitney, J. & Elias, P.M. (1985). In vitro synthesis of vitamin D-3 by cultured keratinocytes and fibroblasts: action spectrum and effect of AY-9944. *Biochim. Biophys. Acta, 841,* 267-277.

Nunn, J.D., Katz, D.R., Barker, S., Fraher, L.J., Hewison, M., Hendy, G.N. & O'Riordan, J.L.H. (1986). Regulation of human tonsillar T-cell proliferation by the active metabolite of vitamin D_3. *Immunology, 59,* 479-484.

Okuda, K.-I. & Ohyama, Y. (1999). The enzymes responsible for metabolizing vitamin D. *In* Holick, M.F. (ed.) *Vitamin D: Physiology, molecular biology, and clinical applications, pp. 85-107.* Totowa, N.J.: Humana Press. ISBN 0-89603-467-4.

Opperman, L.A. & Ross, F.P. (1990). The adult fruit bat (Rousettus aegypticus) expresses only calbindin-D9K (vitamin D-dependent calcium-binding protein) in its kidney. Comp. Biochem. Physiol. B: Biochem. *Molec. Biol., 97,* 295-299.

Panda, D.K., Miao, D., Tremblay, M.L., Sirois, J., Faroohki, R., Hendy, G.N. 6 Goltzman, D. (2001). Targeted ablation of the 25-hydroxyvitamin D 1α-hydroxylase enzyme: evidence for skeletal, reproductive, and immune dysfunction. *Proc. Natl Acad. Sci. USA, 98,* 7498-7503.

Patterson, G.W. (1971). The distribution of sterols in algae. *Lipids, 6,* 120-127.

Patterson, G.W. (1974). Sterols of some green algae. *Comp. Biochem. Physiol. B, 47,* 453-457.

Rambeck, W.A., Kreutzberg, O., Bruns-Droste, C. & Zucker, H. (1981). Vitamin D-like activity of *Trisetum flavescens. Zschr. Pflanzenphysiol., 104,* 9-16.

Shewakramani, S., Rakita, D., Tangpricha, V. & Holick, M.F. (2001). Vitamin D insufficiency is common and under-diagnosed among African American patients. *J. Bone Mineral Res., 16,* S512.

Steenbock, H. & Black, A. (1924). The induction of growth-promoting and calcifying properties in a ration by exposure to ultra-violet light. *J. Biol.Chem., 64,* 263-298.

Takada, K. (1983). Formation of fatty acid esterified vitamin D_3 in rat skin by exposure to ultraviolet radiation. *J. Lipid Res., 24,* 441-448.

Takeuchi, A., Okano, T., Tanda, M. & Kobayashi, T. (1991). Possible origin of extremely high contents of vitamin D_3 in some kinds of fish liver. *Comp. Biochem. Physiol. 100A,* 483-487.

Tasende, M.G. (2000). Fatty acid and sterol composition of gametophytes and sporophytes of *Chondrus crispus (*Gigartinaceae, Rhodophyta). *Scientia Marina, 64,* 421-426.

Tian, W.Q. & Holick, M.F. (1995). Catalyzed thermal isomerization between previtamin D_3 and vitamin D_3 via β-cyclodextrin complexation. *J. Biol. Chem., 270,* 8706-8711.

Tian, W.Q. & Holick, M.F. (1999). A liposomal model that mimics the cutaneous production of vitamin D_3. *J. Biol. Chem., 274,* 4174-4179.

Tian, X.Q., Chen, T.C., Matsuoka, L.Y., Wortsman, J. & Holick, M.F. (1993). Kinetic and thermodynamic studies of the conversion of previtamin D_3 to vitamin D_3 in human skin. *J. Biol. Chem., 268,* 14888-14892.

Uskovic', M.R., Johnson, C.S., Trump, D.L. & Getzenberg, R.H. (1999). Anticancer activity of vitamin D analogs.*In* Holick, M.F. (ed.) *Vitamin D: Physiology, molecular biology, and clinical applications,* pp. 431-445. Totowa, N.J.: Humana Press. ISBN 0-89603-467-4.

Uva, B.M., Ghiani, P., Deplano, S., Madich, A., Vaccari, M. & Vallarino, M. (1978). Occurrence of 7-dehydrocholesterol in the uropygial gland of domestic fowls. *Acta Histochem. 62,* 237-243.

van Leeuwen, P.T.M., Vink-van Wijngaarden, T. & Pols, H.A. (1999). Vitamin D and breast cancer. *In* Holick, M.F. (ed.) *Vitamin D: Physiology, molecular biology, and clinical applications,* pp. 411-429. Totowa, N.J.: Humana Press. ISBN 0-89603-467-4.

Vega, M.A. & Boland, R.L. (1986).Vitamin D-3 induces the novo synthesis of calmodulin in Phaseolus vulgaris root segments in vitro. *Biochim. Biophys. Acta, 881,* 364-374.

Vega, M.A. & Boland, R.L. (1988). Presence of sterol-binding sites in the cytosol of French-bean (*Phaseolus vulgaris*) roots. *Biochem. J., 250,* 565-569.

Vega, M.A. & Boland, R.L. (1989). Partial characterization of the sterol binding macromolecule of *Phaseolus vulgaris* roots. *Biochim. Biophys. Acta, 1012,* 10-15.

Wang, T., Bengtsson, G., Kärnefelt, I. & Björn, L.O. (2001). Provitamins and vitamins D_2 and D_3 in *Cladina* spp. over a latitudinal gradient: possible correlation with UV levels. *J. Photochem. Photobiol. B: Biology, 62,* 118-122.

Washburn, E.W. et al. (eds) *International critical tables of numerical data, physics chemistry and technology,* vol. V, p. 270. New York:McGraw-Hill.

Webb, A.R., de Costa, B. & Holick, M.F. (1989). Sunlight regulates the cutaneous production of vitamin D_3 by causing its photodegradation. *J. Clin. Endocrin. Metab., 68,* 882-887.

Whistler, D. Morbo puerili Anglorum, quem patrio idiomate indigenae vocant The Rickets. *Lugduni Batavorum* 1-13.

Zelinski, J.M., Sykes, D.E. & Weiser, M.M. (1991). The effect of vitamin D on rat intestinal plasma membrane Ca-pump mRNA. *Biochem. Biophys. Res. Commun., 179,* 749-755.

MARY NORVAL

13. THE PHOTOBIOLOGY OF HUMAN SKIN

1. INTRODUCTION

The skin is the largest organ of the body and the one which is most exposed to external insults, such as chemicals, infecting microorganisms and mechanical trauma. These insults include UV radiation from the sun. The structure of skin is complex: it is composed of three basic layers, the epidermis, dermis and subcutis, each comprising a variety of cell types. It has been recognised (Streilein 1978) that the skin contains its own immune system, first called skin associated lymphoid tissues, which generally acts very effectively to deal with any local disturbances. However UV radiation poses two unique and potentially dangerous consequences for the skin. It induces genotoxic changes, these mutations leading on some occasions to the development of skin cancers, and it can also suppress cell-mediated immune responses to a variety of antigens. The reason for the latter change may be to prevent excessive inflammation in sun-exposed skin but, if it occurs at the same time as, say, an infection or oncogenesis, then there may be disadvantages for the host. Solar UV radiation does not always cause harmful effects in the skin, and one case where it is beneficial is in promoting the synthesis of vitamin D, essential for calcium metabolism and a healthy skeleton. This aspect is covered in Chapter 12.

The following section outlines the structure of the skin and describes the skin immune system. Consideration is then given to UV radiation in the context of cutaneous pigmentation, sunburn and photoageing. Sections on photocarcinogenesis and UV-induced immunomodulation follow, and the final part of the Chapter outlines some photosensitivity disorders which can occur in human subjects.

2. THE STRUCTURE OF SKIN AND THE SKIN IMMUNE SYSTEM

2.1. Skin structure

The outermost layer of the skin is the epidermis which is separated from the dermis by a basement membrane, and the layer underlying the dermis is the subcutis. UV-B (280-315 nm) penetrates into the epidermis and UV-A (315-400 nm) deeper into the dermis. These layers are transversed vertically by the skin appendages, such as the sweat glands, hair follicles and sebaceous glands. The appendages are rarely affected by UV radiation and are not considered in this Chapter.

The epidermis is composed mainly of keratinocytes which are formed in the basal layer and migrate upwards to terminally differentiate at the skin surface.

L.O. Björn (ed.), Photobiology, 281–298.

It takes about four weeks to complete this process. The appearance of the keratinocytes at each stage divides the epidermis into four layers: the basal layer where the cells divide intermittently giving rise to one daughter cell remaining in the basal layer and one which begins to differentiate and move upwards, the prickle cell layer, so-called as the keratinocytes have distinct interconnecting junctions, the granular layer where the keratinocytes begin to flatten and contain keratohyalin granules and degenerating organelles, and the stratum corneum where the keratinocytes die and are sloughed off. The thickness of each layer depends on the location in the body. Other important cell types contained within the epidermis are the melanocytes and the Langerhans cells. The former give the skin its colour and are found mainly at the dermato-epidermal junction. The latter are dendritic cells whose processes form a network throughout the epidermis. There are also some scattered lymphocytes. Figure 1 illustrates the cellular structure of the epidermis.

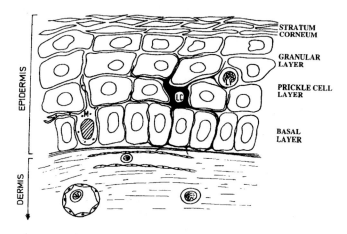

Figure 1. Diagram of the epidermis indicating the keratinocyte layers and a representative Langerhans cell (LC), melanocyte (M) and lymphocyte (LY). Part of the dermis is also shown illustrating collagen fibres and a small blood vessel.The dermis is very different from the epidermis consisting mainly of collagen fibres with fibroblasts and elastic tissue throughout. There are blood vessels, nerve fibres and some smooth muscle, together with small numbers of dendritic cells and tissue macrophages. Mast cells are also present which contain histamine and other inflammatory mediators.

2.2. The skin immune system

The skin immune system consists of contributions from many cell types, some resident in the epidermis or dermis, and others highly mobile, frequently connecting the cutaneous environment with the blood or lymph (reviewed in Bos, 1997). The Langerhans cells are the initiators of the process, acting as the major antigen presenting cells of the epidermis. On contact with an antigen, they internalise it, often by phagocytosis, and process it into smaller peptides for presentation on the cell surface. There are associated changes around the Langerhans cells due to a variety of cytokines expressed by the local keratinocytes including tumour necrosis factor (TNF)-a and interleukin (IL)-1b. In response to these and other mediators, the

Langerhans cells migrate from the epidermis. They move down the afferent lymph and enter the draining lymph node. This process takes around 18 hours following cutaneous application of a contact sensitiser. During the migration, the Langerhans cells mature into dendritic cells, as indicated by the changed expression of various adhesion and co-stimulatory molecues on their surfaces. In the paracortical area of the lymph node, they then present the processed antigenic peptides to specific CD4-positive T cells. These T cells are stimulated to proliferate and to express a particular cytokine profile. The T cells can be divided into T helper 1 (Th1) and Th2 subsets on the basis of the cytokines they produce: the Th1 cells secrete cytokines such as IL-2 and interferon (IFN)-g, while the Th2 cells secrete cytokines such as IL-4 and IL-10. In some instances the activated T cells home preferentially to the skin as they express particular surface markers. They leave the lymph node in the efferent lymph, enter the bloodstream through the thoracic duct and migrate via the blood to the site of antigen application. They extravasate through the high endothelial venules, and enter the dermis or epidermis to act as effector cells locally. The cytokine profile of the T cells largely determines the type of immune response generated and its efficiency.

2.3. Contact and delayed type hypersensitivity

Contact hypersensitivity (CHS) is frequently used to assess immune responses in the skin under experimental conditions. In brief, it is divided into two phases: the first is induction/sensitisation (afferent) and the second is elicitation (efferent). During sensitisation small, structurally simple haptens are placed on the skin. On their own haptens are incapable of generating immune responses but they react with proteins locally in the skin to create immunologically relevant hapten-derivitised proteins. These are taken up by Langerhans cells in the epidermis, as described above, carried to the draining lymph nodes and presented there to specific T cells. In the elicitation phase, re-challenge with the same contact sensitiser occurs. Antigen-specific T cells are recruited or activated *in situ*, leading to the release of particular cytokines, infiltration of macrophages and neutrophils, and inflammation. This can be quantified at 24-72 hours post-challenge either by a colour change, as is frequently used in human studies, or by swelling, such as of an ear, in the case of mice. The majority of contact sensitisers promote a Th1-like cytokine profile. A summary of the events in CHS is given below:

Induction/sensitisation: ----------------------->   Elicitation:

Induction/sensitisation	Elicitation
first skin contact with antigen	second skin contact with same antigen
interaction of antigen with Langerhans cells	interaction with local antigen presenting cells
local changes in cytokines in the epidermis	cytokines induced locally
local changes in adhesion molecules	stimulation of antigen-specific Th1 cells
migration of Langerhans cells from the epidermis	local inflammatory response
arrival of Langerhans cells in the draining lymph nodes as dendritic cells	
interaction of dendritic cells with antigen-specific T cells	
proliferation of T cells	
release of Th1 cytokines	

Delayed type hypersensitivity (DTH) has the same end-points but, in contrast to CHS, the antigens are complex, such as microbial or tumour antigens. In natural circumstances and in some animal models, these antigens are often derived from infecting microorganisms or developing tumours. Under experimental conditions, they are frequently injected subcutaneously or intradermally. They are therefore taken up, processed and presented differently from the simple haptens applied epicutaneously.

Exposure of human skin to solar UV radiation results in a number of effects which are outlined in Sections 3-7 below. In some instances the radiation of the wavelength which promotes an individual effect with maximal efficiency has been derived from the action spectrum, and these results are shown in summary form in Table 1 with more detail in the text.

Table 1. Selected skin responses to UV radiation
(See text for more details. The wavelength is that of maximum effectiveness)

	Wavelength (nm)	
Induction of non-melanoma skin cancer (mouse)	293	de Gruijl *et al.* (1993)
Immediate pigment darkening (human)	350	Irwin *et al.* (1993)
Melanogenesis (human)	290	Parrish *et al.* (1982)
Erythema (human)	300	McKinley & Diffey (1987)
Decrease in epidermal Langerhans cells (mouse)	270-290	Noonan *et al.* (1984)
Systemic suppression of contact hypersensitivity (mouse)	260-280	De Fabo & Noonan (1983)
Induction of cyclo-butane pyrimidine dimers (mouse)	290	Cooke & Johnson (1978)
Isomerisation from *trans* to *cis*-urocanic acid (mouse)	300-315	Gibbs *et al.* (1993)

3. PIGMENTATION AND SUNBURN

3.1. Pigmentation and photorypes

Skin colour is determined by cutaneous pigments particularly melanin, by blood circulating through the skin and by the thickness of the stratum corneum. Melanin is synthesised by the melanocytes and is released as granules (melanosomes) by exocytosis, which are then taken up by the adjacent keratinocytes. Melanin is present in at least two forms: eumelanin, a brown polymer predominating in darkly pigmented skin, and pheomelanin, a reddish-yellow pigment found in lighter skin types. The production of melanin is influenced by genetic factors, hormones and exposure to UV radiation. Racial variation in skin colour - black, brown, yellow and while - is not determined by the absolute number of melanocytes but by their activity in producing melanosomes. As might be expected, the melanosomes are larger and more numerous in black-skinned people. On exposure of the skin to UV, the melanocytes are stimulated to produce melanin which gives the skin its tan. Individuals vary enormously in this response. Those with fair non-pigmented skin, who cannot or are hardly able to synthesis melanin in response to sun exposure, suffer more cutaneous damage than those who tan, mainly because more UV radiation reaches the dermis. Similarly the ability to tan correlates with less risk of burning after solar UV. There are six phototypes, originally recognised by Fitzpatrick, as shown below:

	Ethnicity	UV-sensitivity	Sunburn/tan
I	white Caucasian	extremely sensitive	always burns, never tans
II	white Caucasian	very sensitive	burns readily, tans slowly and with difficulty
III	white Caucasian	moderately sensitive	can burn after high exposure, tans slowly
IV	white Caucasian, often S. Mediterranean	relatively tolerant	burns rarely, tans easily
V	brown, Asian/middle Eastern	variable	can burn easily, difficult to assess as pigment already is present
VI	black, Afro-Caribbean	relatively insensitive	rarely burns

Skin pigmentation induced by UV exposure occurs in two phases called immediate pigment darkening and delayed tanning. The former is thought to result from oxidation and the redistribution of melanin in the skin. It begins during irradiation and is maximal immediately afterwards. The action spectrum for immediate pigment darkening shows maximal effectiveness in the UV-A waveband (peak at 350 nm) (Irwin *et al.* 1993) which is similar to the absorption spectrum of epidermal melanin. Delayed tanning (melanogenesis) does not become apparent until about three days after the UV exposure. In contrast to immediate pigment tanning, the UV-B wavelengths are more effective than UV-A at inducing melanogenesis with a peak at 290 nm (Parrish *et al.* 1982), mimicking the peak for erythema (see Section below). There is an increase in the number of melanocytes and in their melanocytic activity with enhanced transfer of melanosomes to keratinocytes. Similar, although lesser, changes occur in unexposed areas of the body, suggesting that secreted factors such as cytokines could affect melanogenesis to some extent.

3.2. Sunburn and minimal erythema dose

Sunburn is recognised by erythema and blistering and is caused most effectively by radiation of wavelengths around 300 nm in the UV-B waveband. It is a delayed response, being maximal 8-24 hours after exposure, which gradually resolves with subsequent skin dryness and peeling. Solar lentigines or freckles can be formed after only one or two episodes of acute solar burning and are often seen on the shoulders, especially of men. Factors which are involved in the vasodilatation which characterises sunburn include direct effects of UV on the vascular endothelium especially endothelial cell enlargement, the loss of epidermal Langerhans cells, the release of epidermal inflammatory mediators such as TNF-a, and the secretion of vasoactive substances from mast cells, for example histamine and prostaglandins (Gilchrest *et al.* 1981).

In sun-burnt skin, so-called sunburn cells are seen in the epidermis. They are thought to represent apoptotic keratinocytes and are characterised by a glassy eosinophilic cytoplasm and a pyknotic nucleus (Young 1986) They are induced in a UV-dose dependent manner, most efficiently by radiation towards the lower end of the UV-B waveband, and are found maximally at 24 hours post-irradiation. They are removed either by desquamation or by keratinocyte-mediated phagocytosis.

It is sometimes necessary to ascertain the minimal erythema dose (MED) of an individual. This is defined as the smallest dose of radiation that results in just detectable erythema, usually assessed at 24 hours after exposure. It is determined by irradiating the normal, untanned skin, often on the back or the inner upper arm, with a graded series of UV doses. The MED for skin type I is about 150-300 effective J m^{-2}, while for skin type IV, it is 450-600. A monochromatic source is frequently used so that the erythemal response can be investigated at several wavelengths, for example 300, 320, 350 and 400 nm. Several methods can be used to determine erythema, such as visually, by reflectance spectrophotometry or by scanning laser Doppler velocimetry.

4. PHOTOAGEING

Photoageing is associated with chronically-exposed skin and can be distinguished from the more subtle changes which occur during intrinsic ageing (Taylor *et al.* 1990). Animal models which have been used to investigate photoageing and its possible repair include the micro-pig and the hairless mouse. The UV-B waveband induces most of the changes observed. Photoageing is found on the body sites most frequently exposed to sunlight such as the face, and the back of the hands and neck. The last site was first recognised over a hundred years ago and the condition called "farmer's neck" as the heavily wrinkled nape of the neck was seen in farmers and sailors who work outdoors predominantly.

The characteristic features of photoageing include coarse and fine wrinkles, age spots (actinic lentigines containing increased numbers of dermal melanocytes), mottled hyperpigmentation and freckles, elastosis, leathery and thickened skin with surface roughness and actinoic keratoses which are small scaly lesions, often multiple and persistent. In contrast unexposed skin or intrinsically aged skin is pale, smooth and relatively unwrinkled.

The first stages of photoageing are inflammatory where mast cells, monocytes and neutrophils invade the dermis (Lavker *et al.* 1988). An increase in dermal elastin is found where the fibres are not laid down in an orderly fashion but are amorphous and severely truncated (Sams *et al.* 1961). Glycosaminoglycans, such as hyaluronate, are deposited in the dermis. The dermal vasculature is reduced and more liable to damage from trauma. The epidermis is initially increased in thickness but then becomes atrophic. The melanin content varies from none to increased in different areas. Loss of collagen within the papillary dermis probably leads to the wrinkled appearance. On each exposure to UV, collagen catabolism is induced, followed by repair, but the repair process is not absolute and, over time, a net loss in collagen occurs, with the skin becoming more fragile as a result (Fisher *et al.* 1997).

One of the main concerns regarding photoageing is that it is related to the risk of developing basal cell carcinoma (BCC) and squamous cell carcinoma (SCC) (see next section). Solar elastosis of the back of the neck is one of the most accurate predictors for the risk of developing either of these cutaneous tumours.

5. PHOTOCARCINOGENESIS

The three skin tumours associated with sun exposure are malignant melanoma (MM), arising from the melanocytes in the epidermis, and SCC and BCC, together called the non-melanoma skin cancers (NMSCs), both of which arise from keratinocytes, also in the epidermis, the latter probably from a pluripotential epidermal stem cell.

5.1. Non-melanoma skin cancer

SCCs are found as persistent red crusted lesions on sun-exposed areas of the body, most frequently on the face and scalp (Figure 2). They metastasise more readily than BCCs and are sometimes fatal. The incidence of SCCs is about 25% that of BCCs in immunocompetent subjects, but SCCs are 15 times more common than BCCs in immunosuppressed subjects, such as those receiving organ transplants. SCCs are found frequently in such patients, with the incidence rising in direct proportion to the time since transplantation and there is an association with papillomavirus infection. For example, in a study of renal allograft recipients in Edinburgh, 20% had papillomas and 2% had SCCs by 5 years post-transplantation, while 77% had papillomas and 13% had SCCs by 22 years post-transplantation (Barr *et al.* 1989).

BCCs present as raised translucent nodules (Figure 3) which develop slowly over a period of months or years. The central area of the face, often around the eyes, is the most likely site to be affected. BCCs are sometimes called rodent ulcers as they have the property of relentless local spread and destruction of large areas of skin, cartilage and even bone, if left untreated. They can be cured by surgery or radiotherapy, and rarely recur.

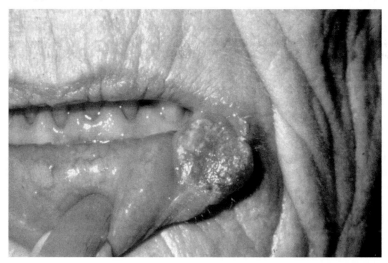

Figure 2. Squamous cell carcinoma at the corner of the mouth shown as a crusted lesion with ulcerated centre.

Both BCCs and SCCs are very common. For example NMSC accounted for 20% of registered maligancies in Scotland in 1997, which is probably a substantial under-estimation, but for less than 0.5% of cancer deaths. The highest recorded incidences of NMSC are in Australia. One study in W. Australia in 1987-92 in people aged 40-64 recorded incidence rates of 7.1 per 100 per annum in men and 3.4 per 100 in women for BCC, and 0.8 and 0.5 per 100 per annum respectively for SCC (English *et al.* 1997). Although the numbers of cases of NMSC in a population are difficult to assess accurately, various reports indicate an increasing incidence of both BCC and SCC in whites in temperate countries and in places nearer to the equator. In Norway, for example, SCCs have more than trebled in men and more than quadrupled in women between 1966/70 and 1991/95 (Iversen & Tretli 1999). The rate of increase in BCCs is not so large in general.

NMSC has a much higher incidence in white than in non-white individuals, and those with skin type I are at particular risk. In people with evidence of long-term cutaneous sun-damage, such as elastosis of the neck and a large number of solar keratoses, there is also a raised risk of both SCC and BCC. The involvement of DNA damage in the induction of NMSC is shown most clearly in patients with the rare genetic disorder, xeroderma pigmentosum (XP), in which there is a defect in the ability to repair DNA following UV exposure and a clinical hypersensitivity to UV radiation. These individuals are at greatly increased risk (5000 times) of developing NMSC compared with normal people, and almost all of the tumours are on constantly exposed sites (Kraemer *et al.* 1987). A recent preliminary study indicates that application of a cream containing a DNA repair enzyme to sun-exposed areas of the body lessens the chances of the XP patients developing the tumours (Yarosh *et al.* 2001).

The risk of NMSC in white populations of similar ethnicity increases with decreasing latitude. Positive correlations have been demonstrated between the incidence/mortality of NMSC and solar UV irradiation in the same location. In most studies outdoor workers have a higher incidence of NMSC than indoors workers. In addition the body sites where the tumours most frequently occur, the face and neck, indicate the importance of sun exposure in their aetiology.

However the pattern of exposure seems to be somewhat different: for SCC the risk increases with increasing cumulative lifetime dose of UV radiation (Kricker *et al.* 1994), but for BCC the relationship is more complex and, although the cumulative dose of UV matters, exposure early in life or intermittent intense exposure such as may be experienced by sunbathing or outdoor recreational activites, may be equally important (Kricker *et al.* 1995). A recent study in Queensland has revealed that sunscreens applied daily over a period of 4.5 years could prevent the development of SCC and that the dose of solar UV experienced as recently as in the past 5 years can affect the risk of developing SCC (Green *et al.* 1999). The incidence of BCC was not affected by the sunscreen use.

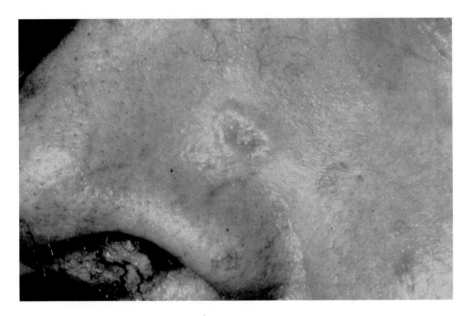

Figure 3. Basal cell carcinoma on the side of the nose showing a raised rolled edge and translucent appearance.

5.2. Malignant melanoma

Although MM is much less common than BCC or SCC, it causes 80% of the deaths associated with skin cancer. The most frequent type is called superficial spreading, seen as small brown or black lesions characterised by irregular lateral edges (Figure 4), and occurring predominantly on the legs of women and the trunk of men. Survival after surgical removal of a primary melanoma is directly related to the thickness of the tumour which signals how far the melanocytes have invaded into the underlying dermis. In many areas of the world, the incidence of cutaneous melanoma increases as the latitude decreases, with the highest recorded incidence being in Queensland at 51/100,000 per year in men and 41/100,000 per year in women. However in Europe the rates are higher in the north than in the south. This may reflect the propensity of northern Europeans to take sunshine holidays in more southern countries. In addition skin phototype may be an important factor as the risk of melanoma is highest in people who burn and do not tan in response to solar UV. The incidence of MM and mortality rates have risen in recent decades in whites, although not in blacks. For example, in Scotland from 1960-64 to 1985-90, the incidence in both males and females quadrupled, but it seems to have stabilised since than, perhaps due to public health campaigns (MacKie *et al.* 1997).

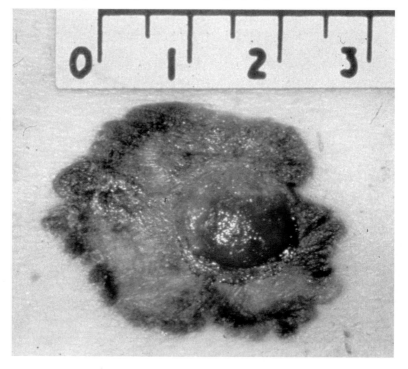

Figure 4. Superficial spreading malignant melanoma with irregular lateral edge and varying pigments within.

The strongest risk factor for MM discovered to date is large numbers of atypical naevi (moles) (Swerdlow *et al.* 1986}. It is possible that at least some MM arise from pre-existing benign melanocytic naevi, presumably by further genetic changes. There is an association between early childhood solar exposure and the development of naevi, particularly episodes of sunburn or intense sun exposure, and the number of naevi is also determined, in part, by genetic factors. Studies which attempt to correlate cumulative UV exposure in white populations with the rise in the number of cases of MM have not yielded consistent results. The consensus view at present is that intermittent recreational exposure to the sun may be critical (Elwood & Jopson 1997; Rosso *et al.* 1998). Certainly there are life-style trends in recent decades such as the fashion to be tanned, many people holidaying in the sun especially with the advent of cheap charter flights, and minimal clothing being socially acceptable, all of which could lead to intense solar radiation exposure of untanned skin. If this hypothesis is true, then tanning and skin thickening should protect, at least to some extent, against the mutagenic effects of solar UV radiation on melanocytes. The age at which acute solar exposure is experienced may be a factor and, in general, as for the development of naevi, there is a greater risk in childhood compared with adult UV irradiation.

5.3. Animal studies of skin cancer

Many animal studies have invesigated the induction of SCC, BCC and MM, and their association with UV exposure. Mice, particularly hairless mice, have been used most frequently and protocols where irradiation is given daily for weeks or months. SCCs are most readily formed in rodents and BCCs very rarely. Quantitative experiments in hairless mice have revealed the action spectrum and dose-dependence for the induction of SCC (de Gruijl et al. 1993). Radiation of wavelengths around 290 nm is the most effective, but there is also a second, lesser peak in the UV-A waveband, about 390 nm. Mutations in the tumour suppressor gene, p53, are implicated as early events in the aetiology of SCC (Berg et al. 1996). For BCC, UV-induced mutations in the sonic hedgehog signalling pathway, particularly of the patched gene, are thought to be important in the initiation of the tumour (Daya-Grosjean & Sarasin 2000). This pathway plays a role in embryonic development and is involved in oncogeneic transformation. Melanin-like tumours can be induced in hybrid fish (Setlow et al. 1989) and in a marsupial (a South American opossum) (Ley 1997) following UV radiation, but not in rodents, unless given at the same time as a chemical carcinogen. However some transgenic mouse melanoma models have become available recently and these, together with models in which full thickness human skin is grafted onto immunocompromised mice, should enable rapid progress to be made regarding the involvement of UV radiation in MM.

6. IMMUNOSUPPRESSION

6.1. UV-induced immunosuppression

In addition to the mutagenic properties of UV, irradiation also leads to suppression of cell-mediated immune responses. The immunomodulation was first recognised two decades ago in a study of the induction of skin tumours in mice by chronic UV-B exposure when a decrease in immunosurveillance against the tumour cells was observed (Fisher & Kripke 1977). The suppression is considered local when the antigen is applied to the site of the UV exposure or it is considered systemic if the antigen is applied to a distant unexposed site. A variety of antigens have been tested, including microorganisms and tumour cells.

It is envisaged that a cascade of reactions occurs when the skin is exposed to UV which leads to the immunosuppression (reviewed for human skin in Duthie et al. 1999). Exactly what happens may depend to some extent on the dose and wavelength of UV and the antigen in question. Due to the poor penetrating power of UV, the initiating event is thought to be absorption by chromophores at the body surface which change their structure as a result, thus leading to the production of various immune mediators locally. Changes in cell phenotype and function follow with subsequent abnormal antigen presentation in the local lymph node or skin, and an aberrant T cell response. The T cell population so induced is capable of suppressing antigen-specific immune responses if transferred to naive animals in various model systems. There is some evidence for promotion of Th2 cytokine production and down-regulation of Th1 cytokine production. The Th1 cytokines are known to be critical in the control of many microbial infections and of tumour growth. A summary of these steps leading to UV-induced immunosuppression is shown below:

UV radiation --> photoreceptors:--->*increases in local mediators*:-->suppression of hypersensitivity,
                DNA, urocanic acid, *prostaglandins, histamine,* induction of T suppressor cells,
                membrane damage   *IL-6, IL-10, TNF-a*   promotion of Th2 cytokine and
                                     changes in antigen presenting abrogation of Th1 cytokines
                                     populations and activity

In the initiating step of UV-induced immunosuppression, DNA and *trans*-urocanic acid (UCA), both major absorbers of photons in the epidermis, act as important chromophores. In addition membrane damage and induction of cytokplasmic transcription factors may play cirítical roles. Brief consideration is now given to each of these.

On UV irradiation of the skin, various types of DNA damage occur of which the most common is the formation of cyclo-butane pyrimidine dimers (CPDs). CPDs have been located in keratinocytes and Langerhans cells after UV exposure, and also in dendritic cells in lymph nodes draining irradiated sites. The CPDs can be repaired artificially using specific enzymes and thus experiments can be performed in mice to demonstrate that repair leads to the restoration of the immune responses normally suppressed by the UV radiation (Vink *et al.* 1998). It is not known precisely how the DNA damage affects immunity but one mechanism might be through activation of particular cytokine genes. It has been shown, for example, the CPD formation leads to the release of both IL-6 and IL-10 from keratinocyte cultures, both important immunoregulatory molecules.

UCA is formed as the *trans*-isomer from histidine as the enzyme histidase is activated in the stratum corneum. It accumulates in that site as no urocanase is found there to catabolise it. On UV exposure of the skin, there is conversion from *trans* to *cis*-UCA, a reaction which is dose-dependent until the photostationary state is reached with approximately equal quantities of the two isomers. UCA was first suggested in 1983 (De Fabo & Noonan) to be an initiator of UV-induced immunosuppression as its absorption spectrum matched the action spectrum for the suppression of CHS. Since that time, a variety of approaches have confirmed this role (reviewed in Mohammad *et al.* 1999). For example, if *cis*-UCA is applied to the skin of mice before infection with a microorganism, the subsequent DTH response to that microorganism is considerably reduced. Furthermore, treating mice with a monoclonal antibody with specificity for *cis*-UCA at the time of UV exposure followed by infection with a microorganism, then challenge with the microorganism led to the restoration of the DTH response, normally suppressed by the UV radiation. The mechanism of action of *cis*-UCA is not clear at present, and, indeed, where it acts. There is some evidence that it may induce mast cells in the dermis to degranulate thus releasing mediators such as histamine, or it may alter the function of antigen presenting cells in the epidermis, or even interact with particular neuropeptides of the sensory nerve system in the skin.

UV-B radiation has been shown to cause the clustering and internalisation of several cell surface receptors, for example TNF, which may be of functional significance in the skin. In addition it affects signal transduction events, acting perhaps through phosphatases located within or near to the cell membrane. These membrane changes may represent another initial target of UV in the immunosuppression cascade (reviewed in Schwarz 1998).

As a result of the above events, there are changes in the cytokine profiles of several cell types, mainly in the epidermis, which have both pro-inflammatory and anti-inflammatory activites (reviewed in Takashima & Bergstresser 1996). For example, keratinocytes begin to produce IL-1a and IL-6. There are also changes in other mediators, such as histamine, neuropeptides and prostanoids, which are equally important in determining the type of immune response generated. As well as these mediators expressed locally within irradiated skin, there are modulations in the function and phenotype of the antigen presenting cells. One of the most striking effects is the decrease in the number of Langerhans cells in the epidermis with loss of the interdigitating network (Toews *et al.* 1980). Many migrate to the draining lymph node while others may undergo apoptosis. If antigen is applied during this time, the dendritic cells which arrive in the lymph node bearing antigen cluster abnormally with the T cells due to modulation in the expression of adhesion and co-stimulatory molecules on their surface, and perhaps additional changes in antigen internalisation and processing. This is thought to lead to the preferential stimulation of Th2 cytokine release, with concomitant down-regulation in Th1 cytokine production (Ullrich 1996). For many infectious diseases and tumours, the Th1 cytokines are thought to be protective and to confer resistance, so a shift to a Th2 response may be adverse. There is also generation of antigen-specific T cells which can act as suppressor cells: this has been shown in mouse models where they have been purified and transferred to naive mice, thereby down-regulating the immune response to that antigen when tested. Finally, within 2-3 days of UV exposure, a new population of antigen presenting cells enters the dermis. These are macrophage in nature, rather than dendritic. They express high levels of particular cytokines such as IL-10 and different co-stimulatory molecules from Langerhans cells, thereby promoting immunosuppression in the skin.

6.2. UV-induced immunosuppression and tumours

Skin cancers, induced by chronic UV exposure of mice, are highly antigenic and are likely to be just as antigenic in human subjects. It is hypothesised that a neoantigen is formed in the skin by UV at a time when the antigen presenting cells are altered, resulting in the activation of T cells which suppress the normal immune responses to the tumour antigens (Kripke 1981). Therefore the interference by UV of the normal host defence mechanisms may be critical. This can be seen most clearly in immunosuppressed individuals who are at significantly increased risk of developing cutaneous malignancies, particularly SCC. It has also been demonstrated experimentally in mice in several ways. For example, pre-irradiation of one site led to enhanced primary tumour growth at a second irradiated site, with the promotion phase of carcinogenesis being most affected (de Gruijl & van der Leun 1982). In addition mice which received T suppressor cells prepared from UV-irradiated animals during the course of chronic UV exposure developed skin tumours earlier than mice receiving T cells from control unirradiated mice. Skin phototype may be an important variable as it has been revealed recently that people with skin types I/II demonstrate an increased susceptibility to UV-induced immunosuppression compared to people with skin types III/IV (Kelly *et al.* 2000). This may help to explain why the former group are at higher risk of developing skin cancer than the latter group.

6.3. UV-induced immunosuppression and microbial infections

Investigations into the impact of UV radiation on infectious diseases are relatively few. However it is known that two such diseases are affected by natural exposure to sunlight. The first is the vesicular lesions (cold sores) in the orofacial region caused by herpes simplex virus (HSV). Here UV is recognised as being one of the commonest triggering factors for recrudescence from the latent state. UV-induced reactivation of the virus has also been shown experimentally in volunteers (for example Spruance 1985). It is speculated that one mechanism could involve the delay in, or the down-regulation of, the local HSV-specific cell mediated immunity caused by the UV which would allow the virus in the cutaneous site time to replicate and to cause the cold sore before recovery of the immune system occurred. The second example is the interaction between particular papillomavirus types and the development of SCC in immunosuppressed individuals. The tumours develop almost entirely on areas of the body most exposed to sunlight (face and backs of the hands) (Hartevelt *et al.* 1990) and the prevalence is highest in sunny climates. The virus itself may reduce the effectiveness of the local immune response, in addition to causing abnormal proliferation of keratinocytes. On top of this, the UV irradiation could cause mutagenic changes in epidermal cells as well as an alteration in cutaneous immunity. The last factor may be a critical one, even in subjects who are already severely immunosuppressed.

About 15 models of infection in rodents have been examined in terms of UV and immunity with the organisms ranging from viruses through bacteria and yeasts to worms (reviewed in Halliday & Norval 1997). In practically all cases, suppression of immunity resulted, together with a decreased ability to clear the infectious agent and sometimes increased severity of symptoms. Calculations have been made to relate the results obtained in the animal models with the human situation and it was concluded that people could receive sufficent solar UV in about 100 minutes or less at mid-latitudes around noon to suppress their immune responses by 50% (Garssen *et al.* 1996). Therefore sunlight irradiance is biologically relevant to the effectiveness of the immune response against microbial agents. Questions arise from this conclusion concerning many infectious diseases, particularly the persistent infections where the organisms are not cleared from the body following the primary infection. In addition vaccination policies need to be examined to determine whether it might be inadvisable to vaccinate an individual who showed evidence of recent exceptional sun exposure such as on the return from a sunshine holiday. The design and testing of sunscreens should also be considered in this context, i.e. to evaluate whether they should, or do, protect against immunomodulation in addition to protecting against erythema.

7. PHOTODERMATOSES

The photodermatoses represent a diverse group of conditions associated with abnormal skin responses to UV and/or visible radiation. These diseases have differing aetiologies and symptoms and can be divided into four main categories, each of which will be described briefly below.

7.1. Genodermatoses: xeroderma pigmentosum

The first is the genodermatoses which are usually inherited. One example of this group, already mentioned above, is XP, a rare autosomal recessive disorder where the repair of DNA after UV exposure is defective, most frequently because of a mutation in the nucleotide excision repair process (reviewed in Copeland *et al.* 1997). The subjects suffer from extreme photosensitivity, burning after even minimal sun exposure. They are at greatly increased risk of developing all forms of skin cancer, but especially SCC and BCC. In former times they died before the age of 30 from a metastatic malignancy, usually SCC or MM. More recently, better sun protection is offered and their life expectancy has increased.

7.2. Idiopathic photodermatoses: polymorphic light eruption

The second group is the idiopathic photodermatoses, comprising a variety of diseases which probably all have an immunological basis. One of the commonest is polymorphic light eruption (PLE), estimated to affect 15% of the population in Britain, for example. PLE is thought to have a higher incidence in temperate climates than nearer the equator, and is found in females 3-9 times more frequently than in males. Recently it has been demonstrated from family histories that PLE has a major genetic component with contributing environmental components, mainly UV exposure. PLE is manifest as an intermittent pruritic skin eruption which occurs several hours after sun exposure and takes the form of multiple small red papules which resolve without scarring after a few days to weeks (Figure 5). An influx of first CD4-positive lymphocytes, then CD8- positive lymphocytes, are found in the exposed skin, together with increased numbers of Langerhans cells in the epidermis and dermis.

Many years ago it was suggested that a photoallergen could be induced in the skin of PLE patients which then stimulates a cell-mediated immune response, thus precipitating the development of the lesions. However such a "self" photoallergen has not been identified as yet.

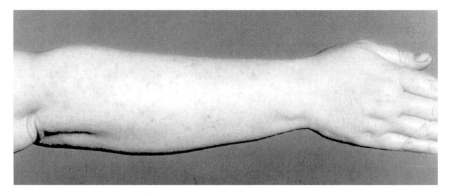

Figure 5. Polymorphic light eruption showing multiple small papules.

7.3. Cutaneous porphyrias

The third group comprises the cutaneous porphyrias in which there are inherited enzymic defects in haem synthesis leading to the accumulation of photoreactive porphyrins which then photosensitise the skin. However the porphyrias are precipitated by visible light (around 410 nm) rather than UV and are not considered further here.

7.4. Photoallergic contact dermatitis

The fourth group comprises photosensitisation by systemic and tropical drugs and by chemicals. Such compounds are increasing in number at the moment with the development of many novel pharmaceutical agents. When these substances get into the skin, they can act as chromophores to absorb UV, thus becoming either phototoxic or triggering a range of biochemical or immunological responses in a proportion of individuals. Phototoxicity is considered in Chapter 10. At low concentrations, some drugs and chemicals cause topical photoallergic contact dermatitis where an eczematous eruption occurs on UV-exposed skin sites, activated mainly by UV-A. The drug is converted to a photoproduct which then binds to proteins or to cells in the skin, forming a novel antigen which then triggers a DTH response. More than one, and frequently many exposures to the photoallergen in the presence of UV are required before this response is induced. As one example, 4-para-aminobenzoic acid has been implicated in photoallergic reactions and its inclusion in sunscreen preparation has declined as a result. Photoallergic contact dermatitis can be diagnosed by duplicate photopatch testing of subjects, in which the substance is assessed with and without exposure to UV-A. Recently the structural features of some classes of chemicals which make them active as photoallergens have been identified as part of the European Phototoxicology Project (Barratt *et al.* 2000).

REFERENCES

Barr, B.B., Benton, E.C., McLaren K., Bunney, H., Smith, I.W., Blessing, K. & Hunter, J.A.A. (1989) Human papillomavirus infection and skin cancer in renal allograft recipients. *Lancet, i,* 124-128.

Barratt, M.D., Castell, J.V., Miranda, M.A. & Langowski, J.J. (2000) Development of an expert system rulebase for the prospective identification of photoallergens. *J. Photochem. Photobiol. B: Biol., 58,* 54-61.

Berg, R.J., van Kranen, H.J., Rebel, H.G., de Vries, A., van Voten, W.A., van Kreijl, C.F., van der Leun, J.C. & de Gruijl, F.R. (1996) Early p53 alterations in mouse skin carcinogenesis by UVB radiation: immunohistochemical detection of mutant p53 protein in clusters of preneoplastic epidermal cells. *Proc. Natl Acad. Sci. USA, 93,* 274-278.

Bos, J.D. (1997) *Skin Immune System (SIS),* second edition. CRC Press, New York.

Cooke, A. & Johnson, B.E. (1978) Dose response, wavelength dependence and rate of excision of ultraviolet radiation-induced pyrimidine dimers in mouse skin DNA. *Biochim. Biophys. Acta, 517,* 24-30.

Copeland, N.E., Hanke, C.W. & Michalak, J.A. (1997) The molecular basis of xeroderma pigmentosum. *Dermatol. Surg., 23,* 447-455.

Daya-Grosjean, L. & Sarasin, A. (2000) UV-specific mutations of the human *patched* gene in basal carcinomas from normal individuals and xeroderma pigmentosus patients. *Mut. Res., 450,* 193-199.

De Fabo, E.C. & Noonan, F.P. (1983) Mechanism of immune suppression by ultraviolet irradiation in vivo. I. Evidence for the eistence of a unique photoreceptor in skin and its role in photoimmunology. *J. Exp. Med., 157,* 84-98.

de Gruijl, F.R., Sterenborg, H.J., Forbes, P.D., Davies, R.E., Cole, C., Kelfkens, G., van Weelden, H., Slaper, H & van der Leun, J.C. (1993) Wavelength dependence of skin cancer induction by ultraviolet irradiation of albino hairless mice. *Cancer Res., 53,* 53-60.

de Gruijl, F.R. & van der Leun, J.C. (1982) Systemic influence of pre-irradiation of a limited area on UV-tumorigenesis. *Photochem. Photobiol., 35,* 379-383.

Duthie, M.S., Kimber, I. & Norval M. (1999) The effects of ultraviolet radiation on the human immune system. *Br. J. Dermatol., 140,* 995-1009.

Elwood, J.M. & Jopson, J. (1997) Melanoma and sun exposure: an overview of published studies. *Int. J. Cancer, 73,* 198-203.

English, D.R., Kricker, A., Heenan, P.J., Randell, P.L., Winter, M.G. & Armstrong, B.K. (1997) Incidence of non-melanocytic skin cancer in Geraldton, Western Australia. *Int. J. Cancer, 73,* 629-633.

Fisher, M.S. & Kripke, M.L. (1977) Systemic alteration induced in mice by ultraviolet light irradiation and its relationship to ultraviolet carcinogenesis. *Proc. Natl Acad. Sci USA, 74,* 1688-1692.

Fisher, G.J., Wang, Z.Q., Datta, S.C., Varani, J., Kang, S. & Vourhees, J.J. (1997) Pathophysiology of premature skin ageing induced by ultraviolet light. *N. Engl. J. Med., 337,* 1419-1428.

Garssen, J., Goettsch, W., de Gruijl, F., Slob, W & Van Loveren, H. (1996) Risk assessment of UVB effects on resistance to infectious diseases. *Photochem. Photobiol., 64,* 269-274.

Gibbs, N.K., Norval, M., Traynor, N.J., Wolf, M., Johnson, B.E. & Crosby, J. (1993) Action spectra for the *trans* to *cis* photoisomerisation of urocanic acid *in vitro* and in mouse skin. *Photochem. Photobiol., 57,* 584-590.

Gilchrest, B.A., Soter, N.A., Stoff, J.S. & Mihm, M.C. 1981. The human sunburn reaction: Histologic and biochemical studies. *J. Am. Acad. Dermatol., 5,* 411-422.

Green, A., Williams, G., Neale R., *et al.* (1999) Daily sunscreen application and betacarotene supplementation in prevention of basal-cell and squamous-cell carcinoma of the skin: a randomised controlled trial. *Lancet, 354,* 723-729.

Halliday, K.E. & Norval, M. (1997) The effects of ultraviolet radiation on infectious diseases. *Rev. Med. Microbiol., 8,* 179-188.

Hartevelt, M.M., Bouwes Bavinck, J.N., Kootte A.M., Vermeer, B-J. & Vandenbroucke, J.P. (1990) Incidence of skin cancer after renal transplantation in the Netherlands. *Transplantation, 49,* 506-509.

Irwin, C., Barnes, A., Veres, D. & Kaidbey, K. (1993) An ultraviolet action spectrum for immediate pigment darkening. *Photochem. Photobiol., 57,* 504-507.

Iversen, T. & Tretli, S. (1999) Trends for invasive squamous cell neoplasia of the skin in Norway. *Br. J. Cancer, 81,* 528-531.

298

Kelly, D.A., Young, A.R., McGregor, J.M., Seed, P.T., Potten, C.S. & Walker, S.L. (2000) Sensitivity to sunburn is associated with susceptibility to ultraviolet radiation-induced suppression of cutnaeous cell-mediated immunity. *J. Exp. Med., 191,* 561-566.

Kraemer, K.H., Lee, M.M. & Scotto, J. (1987) Xeroderma pigmentosum. Cutaneous, ocular, and neurologic abnormalities in 830 published cases. *Arch. Dermatol., 123,* 241-250.

Kricker, A., Armstrong B.K. & English, D.R. (1994) Sun exposure and non-melanocytic skin cancer. *Cancer Causes Control, 5,* 367-392.

Kricker, A., Armstrong, B.K., English, D.R. & Heenan, P.J. (1995) Does intermittent sun exposure cause basal cell carcinoma? A case-control study in Western Australia. *Int. J. Cancer, 60,* 489-494.

Kripke, M.L. (1981) Immunologic mechanisms in UV radiation carcinogenesis. *Adv. Cancer Res., 34,* 69-106.

Lavker, R.M. & Kligman, A.M. (1988) Chronic heliodermatitis: a morphologic evaluation of chronic actinic dermal damage with emphasis on the role of mast cells. J. Invest. Dermatol. 90:325-330.

Ley, R.D. (1997) Ultraviolet radiation A-induced precursors of cutaneous melanoma in *Monodelphus domestica. Cancer Res., 57,* 3682-3684.

MacKie, R.M., Hole, D., Hunter, J.A.A., *et al.* (1997) Cutaneous malignant melanoma in Scotland: incidence, survival and mortality, 1979-94. *Br. Med. J., 315,* 1117-1121.

McKinlay, A.F. & Diffey, B.L. (1987) A reference action spectrum for ultraviolet induced erythema in humans skin. *In* Passhier, W.F. & Bosnjakovic, B.F. (eds) *Human exposure to ultraviolet radiation: risks and regulation,* pp 45-52. Amsterdam: Elsevier,.

Mohammad, T., Morrison, H & HogenEsch, H. (1999) Urocanic acid photochemistry and Photobiology. *Photochem. Photobiol., 69,* 115-135.

Noonan, F.P., Bucana, C., Sauder, D.N. & De Fabo, E.C. (1984) Mechanism of systemic immune suppression by UV radiation *in vivo.* II. The UV effects on number and morphology of epidermal Langerhans cells and the UV-induced suppression of contact hypersensitivity have different wavelength dependencies. *J. Immunol., 132,* 2408-2416.

Parrish, J.A., Jaenicke, K.F. & Anderson, R.R. (1982) Erythemas and melanogenesis action spectra of normal human skin. *Photochem. Photobiol., 36,* 187-191.

Rosso, S., Zanetti, R., Pippioni, M. & Sancho-Garnier, H. (1998) Parallel risk assessment of melanoma and basal cell carcinoma: skin characteristics and sun exposure. *Melanoma Res., 8,* 573-583.

Sams, W.M. & Smith, J.G. (1961) The histochemistry of chronically sun damaged skin. *J. Invest. Dermatol., 37,* 447-452.

Schwarz, T. (1998) UV light affects cell membranes and cytoplasmic targets. *J. Photochem. Photobiol. B: Biol., 44,* 91-96.

Setlow, R.B., Woodhead, A.D. & Grist, E. (1989) Animal model for ultraviolet radiation-induced malignant melanoma. *Proc. Natl. Acad. Sci., USA 86,* 8922-8926.

Spruance, S.L. (1985) Pathogenesis of herpes simplex labialis: experimental induction of lesions with UV light. *J. Clin. Microbiol., 22,* 366-368.

Streilein, J.W. 1978. Lymphocyte traffic, T cell malignancies and the skin. *J. Invest. Dermatol., 71,* 167-171.

Swerdlow, A.J., English, J., MacKie, R.M., O'Doherty, C.J., Hunter, J.A.A., Clark, J. & Hole, D. (1986) Benign melanocytic naevi as a risk factor for malignant melanoma. *Br. Med. J., 292,* 1555-1559.

Takashima, A & Bergstresser, P.R. (1996) Impact of UVB radiaiton on the epidermal cytokine network. *Photochem. Photobiol., 63,* 397-400.

Taylor, C.R., Stern, R., Leyden, J.J. & Gilchrest, B.A. (1990) Photoageing, photodamage and photoprotection. *J. Am. Acad. Dermatol., 22,* 1-15.

Toews, G.B., Bergstresser, P.R. & Streilein, J.W. (1980) Epidermal Langerhans density determines whether contact hypersensitivity or unresponsiveness follows skin painting with DNFB. *J. Immunol., 124,* 445-453.

Ullrich, S.E. (1996) Does exposure to UV radiation induce a shift to a Th-2-like immune reaction? *Photochem. Photobiol., 66,* 254-258.

Vink A.A., Schreedhar, V., Roza, L., Krutmann, J & Kripke, M.L. (1998) Cellular target of UVB-induced DNA damage resulting in local suppression of contact hypersensitivity. *J. Photochem. Photobiol. B: Biol., 44,* 107-111.

Yarosh, D., Klein, J., O'Connor, A. *et al.* (2001) Effect of topically applied T4 endonuclease V in liposomes on skin cancer in xeroderma pigmentosum. *Lancet, 357,* 926-929.

Young, A.R. (1986) The sunburn cell. *Photodermatology, 4,* 127-134.

JIM L. WELLER and RICHARD E. KENDRICK

14. PHOTOMORPHOGENESIS AND PHOTOPERIODISM IN PLANTS

1. INTRODUCTION

It has long been observed that light affects the way plants grow. Effects of light can be observed on processes and phenomena throughout the plant life cycle, including seed germination, apical hook opening, stem elongation, leaf expansion, the synthesis of photosynthetic and protective pigments, lateral branching, bud dormancy, and flowering. The vast majority of these effects are unrelated to the use of light for photosynthesis, and are mediated through a specialized system of photoreceptors that informs the plant about its surroundings and directs it to develop appropriately.

Photomorphogenesis is a general term encompassing all responses to light that affect plant form. Two specific classes of photomorphogenic response are sometimes distinguished. *Phototropic* responses involve the reorientation of plant organs with respect to an asymmetry in the incident light, as in the case of shoot tips bending to grow towards the light. *Photoperiodic* responses are those in which various aspects of development are modified in response to changes in the daily light/dark cycle, and involve a circadian timing mechanism. Developmental features commonly subject to photoperiodic control include flowering, bud dormancy and leaf senescence.

This chapter will give an overview of our current knowledge about the way in which these different responses are achieved. We will discuss the discovery and nature of the photoreceptors involved in these phenomena, their physiological roles as determined in the laboratory, and their possible significance in the natural environment. We will also summarize what is known about transduction of the signals arising from photoreceptor activation. Although lower plants also show clear photomorphogenic responses, they have in general been less intensively studied, and we will restrict this discussion to higher plants.

2. PHOTOMORPHOGENIC PHOTORECEPTORS

As with other photobiological responses, an initial step in the investigation of photomorphogenic responses has been the determination of action spectra. Early measurements identified the blue (BL), red (R), and far-red (FR) regions of the

299

L.O. Björn (ed.), Photobiology, 299–334.
© 2002 *Kluwer Academic Publishers. Printed in the Netherlands.*

spectrum as being particularly important for the control of plant growth (e.g. Went 1941, Parker et al. 1949), and formed the point of departure in the search for specific photoreceptor pigments for light in these wavebands (Fig. 1). Relatively rapid progress was made in biochemical characterization of the photoreceptor responsible for R and FR responses (Sage 1992). In contrast, progress towards identification of a specific BL photoreceptor was limited due to the large number of different BL-absorbing compounds in the plant with the potential to serve as a photoreceptor chromophore, and to the lack of a distinctive photophysiological assay.

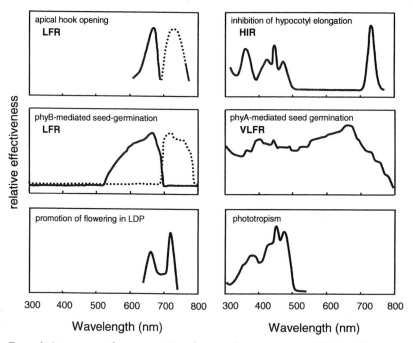

Figure 1. Action spectra for representative photomorphogenic responses. LFR- low-fluence response, HIR, high-irradiance response, VLFR, very-low-fluence response. Broken lines represent reversal of response. Redrawn from Withrow et al. (1957), Hartmann (1967), Baskin and Iino (1987), Carr-Smith et al. (1989) and Shinomura et al. (1996).

However, the advent of a molecular genetic approach has brought rapid developments in our understanding of the nature, diversity and functions of the photoreceptor pigments involved in informational light sensing. Three classes of photoreceptors have now been characterized in higher plants; the *phytochrome* family of R- and FR-absorbing photoreceptors, and two different photoreceptor families mediating responses in the BL and UV-A regions of the spectrum; the *cryptochrome* and *phototropin* families. We will discuss each of these in turn.

2.1 Phytochrome

2.1.1 Isolation

The main impetus in the early search for photoreceptors came from observations that the inductive effects of R on several aspects of plant development could be reversed by irradiation with FR (Fig. 1). The fact that this R/FR reversibility occurred for diverse responses suggested that it might be a property of a single photoreceptor (Withrow et al. 1957) . This was proven by the purification of a protein that exhibited R/FR reversible absorption changes. The protein was named *phytochrome*, a name derived from the Greek words for *plant* and *colour*. The two forms of phytochrome are characterized by absorption peaks at around 660 nm and 730 nm, and are referred to as Pr and Pfr, respectively (Fig. 2). Both Pr and Pfr also have a minor absorption peak in the BL region of the spectrum. These two forms of phytochrome can be repeatedly interconverted by light pulses, and continuous light establishes a dynamic equilibrium between them that depends on the composition of the light.

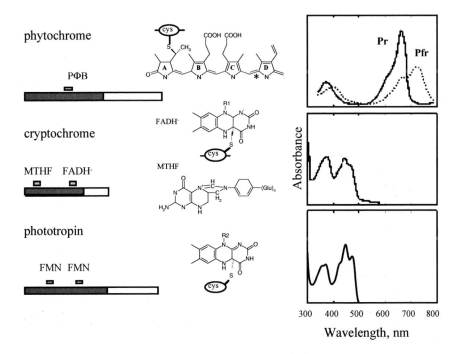

Figure 2. Diagram showing protein domain structures for generic phytochrome, cryptochrome and phototropin photoreceptor proteins. The structures and approximate attachment sites of chromophores are shown for each protein. Absorption spectra redrawn from Butler et al. (1964), Lin et al. (1995) and Christie et al (1999). The asterix indicates the double bond involved in photoisomerization of the phytochrome chromophore.

2.1.2 Genes and gene surveys

The first phytochrome-encoding gene was identified from oat seedlings in 1984 by expression screening using antibodies raised against purified phytochrome Hershey et al., 1984). Phytochrome genes were subsequently identified in other flowering plants, as well as gymnosperms, ferns, mosses and algae (Schneider-Poetsch et al. 1998). More recently, phytochrome-related sequences have also been found in cyanobacteria and even in non-photosynthetic bacteria, suggesting a very ancient origin (Kehoe and Grossman 1996, Davis et al., 1999). Many higher plants are now known to contain several different forms of phytochrome, which are encoded by a small gene family. The two best-studied species, *Arabidopsis* and tomato, each have five expressed phytochrome genes, including two closely related phyB-type phytochromes (Clack et al., 1994, Hauser et al. 1995). *In vitro* studies indicate that these phytochromes have rather similar absorption spectra, and suggest that they may share a common chromophore. Phylogenetic studies suggest that an original phytochrome progenitor gene may have duplicated at around the time of origin of seed plants. Further duplications appear to have occurred soon after the origin of flowering plants, giving rise to four subfamilies corresponding to phyA, phyB/D, phyC and phyE in *Arabidopsis*. More recent duplications within the phytochrome A (phyA) and phytochrome B (phyB) sub-families have occurred independently within various taxa (Mathews and Sharrock 1997).

2.1.3 Gene/protein structure

The generic phytochrome apoprotein has a molecular mass of around 125 kDa and consists of two domains (Fig. 2); an N-terminal domain of 75 kDa that binds the chromophore, and a C-terminal domain of 55 kDa that consists of two regions with homology to histidine kinases (Quail 1997). The first of these C-terminal regions contains two PAS domains, which are implicated in protein-protein interactions. This region also contains a small domain that is essential for the regulatory activity of the molecule. Gene and protein structure are in general highly conserved across the phytochrome family, with the most notable difference being small poorly conserved N- and C-terminal extensions in phyB-type phytochromes. Phytochromes are synthesized in the cytosol in the Pr configuration, and form dimers *in vivo*. Deletion and point mutation studies have given some indication of areas of importance for determination of the absorption spectrum, photochromicity, dimerization and signal transduction (Quail 1997).

2.1.4 Expression

Two features considered characteristic of phytochrome in early studies were its presence at much higher levels in dark-grown than in light-grown seedlings, and its rapid disappearance after exposure to light. It was clear that a major proportion of the phytochrome present in dark-grown seedlings was strongly regulated by light. However, physiological and spectrophotometric experiments also defined a small light-stable pool of phytochrome (Furuya 1989).

It is now known that the light-lability of phytochrome mainly reflects the characteristics of phyA. The *PHYA* gene is expressed at a high level in darkness, and phyA accumulates as Pr in the cytosol. Conversion to Pfr after light exposure initiates both a rapid ubiquitin-mediated degradation of the protein, and a rapid down-regulation of *PHYA* transcription (Quail 1994, Clough and Vierstra 1997). In addition, a proportion of the Pfr pool appears to be able to revert spontaneously to Pr in darkness (Fig. 3). In contrast to phyA, the message and protein levels of other phytochromes are much lower in darkness. Although they do not exhibit the strong and rapid down-regulation in light shown by phyA, they are nevertheless regulated, by light and by the circadian clock (Hauser et al. 1998, Toth et al. 2001)

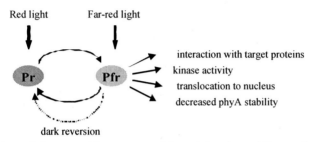

Figure 3. Summary of the activities and fates of phytochrome following photoconversion.

Immunochemical detection of phyA apoprotein in dark-grown seedlings has shown that it is predominantly localized in the apical hook region, the root tip, and the epidermis of young leaves (Hisada et al. 2000). The distribution of other phytochromes has been investigated using promoter-GUS fusions, and quantitation of native transcripts in various tissues (Goosey et al. 1997, Hauser et al. 1997). The early fate of phytochrome within the cell after photoconversion has also been investigated, using promoter-GFP fusions and immunochemical techniques (Kircher et al. 1999, Hisada et al. 2000). Both phyA and phyB are synthesized in the cytosol in darkness, and differentially translocated to the nucleus under specific irradiation regimes. Under R, phyA becomes rapidly localized in the cytosol before appearing in the nucleus within 10 minutes of exposure. FR is also effective for nuclear import of phyA. PhyB translocation is induced by R, but more slowly, and in a FR–reversible manner.

2.1.5 Chromophore

Early suggestions that the phytochrome chromophore might be a tetrapyrrole were confirmed after its release from purified phytochrome and chemical analysis. The chromophore, phytochromobilin (PΦB) is an open chain bilitriene (Fig. 3), similar to the chromophore for the photosynthetic pigment C-phycocyanin in cyanobacteria. Feeding studies combined with analysis of chromophore-deficient mutants have shown that phytochromobilin is formed in plastids by a pathway which branches from the pathway for chlorophyll synthesis with the chelation of Fe^{2+} rather than Mg^{2+} to protoporphyrin IX (Terry et al. 1993). Heme is then oxygenized to biliverdin resulting in the opening of the tetrapyrrole ring. This is followed by reduction of the A-ring to form PΦB. Free PΦB assembles autocatalytically to the phytochrome apoprotein, and attaches at its C3 position to a cysteine residue in the middle of the N-terminal domain via a thioether linkage (Terry et al. 1993). Photoreversibility of the phytochrome holoprotein derives from isomerization of bound PΦB about the double bond between rings C and D. It is worth noting that the absorption peaks of the Pr and Pfr are red-shifted 35 and 100 nm respectively relative to that of free phytochromobilin conformers. This illustrates the importance of the protein environment for the light-absorbing properties of the chromophore and hence for the spectral characteristics of the photoreceptor. PΦB–deficient mutants have been isolated and shown to carry mutations in structural genes for enzymes in PΦB biosynthesis (Muramoto et al. 1999, Kohchi et al. 2001). As a consequence of PΦB deficiency, they have reduced levels of spectrally active phytochrome, and exhibit strong defects in responses to light attributable to reduced activity of multiple members of the phytochrome family (Parks and Quail 1991, Weller et al. 1997c).

2.2. Cryptochrome

The contribution of a BL-specific photoreceptor system to de-etiolation was first inferred from observations that BL could promote de-etiolation, even in plants grown under continuous red light (i.e. under conditions which are saturating for phytochrome activity). Detailed kinetic studies of growth inhibition also suggested that a photoreceptor other than phytochrome was responsible for a rapid component. In addition, mutants strongly deficient in phytochrome chromophore synthesis (and hence in the activities of all phytochromes) were shown to retain substantial responsiveness to BL (Briggs and Huala 1999). This unidentified pigment was often referred to as *cryptochrome*, from the Greek words for *hidden* and *colour*, reflecting its elusive nature. The primary BL photoreceptor in de-etiolation was finally identified in 1994, after cloning of the defective gene in an *Arabidopsis* mutant showing impaired de-etiolation responses to BL (Ahmad and Cashmore 1994). This photoreceptor is now known as cryptochrome 1 (cry1). A second member of the cryptochrome family, cryptochrome 2 (cry2), was identified by its homology to cry1 (Lin et al. 1998).

The cryptochrome apoproteins are around 75 kDa in molecular mass and have two distinct parts (Fig. 2). The N-terminal half shows similarity to enzymes called photolyases, which are activated by BL and UV light to repair certain kinds of damage to DNA. When expressed in *E. coli*, this part of the molecule binds the same two chromophores as photolyases; a flavin, (flavin adenine dinucleotide) and a pterin (methenyltetrahydrofolate) (Lin et al. 1995, Malhotra et al. 1995) although it has yet to be confirmed that the latter chromophore is utilized *in planta*. It has been speculated that the pterin chromophore may serve as a kind of antenna, and may predominantly determine the UV-A/BL-absorbing properties of the molecule, while the flavin chromophore may be essential to the initial signalling reaction and may extend the absorption spectrum into the green region (Cashmore et al. 1999). The C-terminal halves of cry1 and cry2 show only a very low degree of similarity to each other, and to other known proteins. One interesting feature of cry2 is its instability under high-irradiance BL (Lin et al. 1998). This is reminiscent of the rapid light-induced degradation of phyA. However, unlike phyA, light does not appear to affect *CRY2* transcription. Both cry1 and cry2 have been shown to localize to the nucleus (Ahmad 1999, Kleiner et al. 1999), although studies of a number of BL-induced phenomena involving changes in ion fluxes across cell membranes (see below) suggest that cryptochromes could also be involved in light-driven redox reactions outside the nucleus.

2.3. Phototropin

Despite the substantial problems encountered in the biochemical search for a BL photoreceptor, this approach did prove successful in the identification of the photoreceptor for BL-induced phototropism. Work in the lab of Winslow Briggs identified a 120 kDa membrane protein that underwent autophosphorylation after irradiation with BL (Short and Briggs 1994). The action spectrum and various kinetic aspects of this reaction showed a close correlation to those for phototropism, suggesting that the protein itself might function as a photoreceptor. The physiological significance of the protein was confirmed by its absence in aphototropic *nph1* mutants of *Arabidopsis* (Liscum and Briggs 1995), and cloning of the *NPH1* gene in 1997 subsequently revealed a protein with clear characteristics of BL receptor (Huala et al. 1997). This protein is now known as phototropin 1 (phot1).

The phot1 molecule consists of two distinct halves; a C-terminal domain with clear homology to classical serine/threonine kinases, and an N-terminal half containing two domains that each bind a flavin mononucleotide (FMN) chromophore (Fig. 2). These domains have been termed LOV domains for their presence in a range of proteins involved in the sensing of *l*ight, *o*xygen or *v*oltage (Christie et al. 1999). Both FMN chromophores undergo a photocycle in which BL absorbance is lost after light exposure and recovered in darkness. *NPH1* gene expression shows circadian regulation in *Arabidopsis*, and expression of the closest *NPH1* homologue in rice is strongly down-regulated in response to light (Kanegae et al. 2000). Several *NPH1*-related genes have now been identified in *Arabidopsis*, including a close homologue *NPL1* and four other more distantly related sequences. Analysis of null mutants for NPL1 show that it also encodes an active phototropin photoreceptor (phot2) that functions together with phot1 in the BL-regulation of phototropism, chloroplast movement and stomatal opening (Kagawa et al. 2001, Kinoshita et al. 2001, Sakai et al. 2001)

2.4. Other photoreceptors

Sequence homologies also hint at the possible existence of additional BL receptors in the plant. For example, in addition to phototropin homologues mentioned above, several other LOV-domain proteins have been identified in the *Arabidopsis* genome. It is not yet known whether these attach a chromophore or function as photoreceptors, but mutant phenotypes for three of them (LKP2, FKF1 and ZTL) indicate a possible role in light input to the circadian clock (Nelson et al. 2000, Somers et al. 2000, Schultz et al. 2001).

In addition to the effects of BL and UV-A mediated by the phytochrome and cryptochrome families, shorter wavelength UV-B also affects plant growth. At high irradiances this is due to the effects of DNA damage, but at low irradiances photomorphogenic effects are also observed (Kim et al. 1998). In some cases phyA and phyB contribute to these low-irradiance effects. In other cases, they are independent of phytochrome and of cry1, suggesting the existence of a distinct photoreceptor for UV-B.

3. PHYSIOLOGICAL ROLES OF PHOTORECEPTORS

Now that we have some idea of the number and nature of plant photoreceptors, it is easy to see how early attempts to interpret physiological observations of photomorphogenesis were hampered by the diversity and functional overlap of the photoreceptors involved. The identification of photoreceptor-specific mutants has been essential for the characterization of the functions and interactions of the photoreceptors, and mutants continue to be an important tool in dissection of signalling pathways. At the time of writing, null mutants have been identified for seven of the eight known photoreceptors in *Arabidopsis*, with phyC the exception. Photoreceptor-specific mutants have also been identified in other higher plant species, notably tomato and pea. These have been useful in testing generalizations about photoreceptor function, and in studying processes not easily studied in *Arabidopsis*. Sense and antisense transgenic lines expressing altered levels of specific photoreceptors have also been of use in exploring photoreceptor functions where no mutants have been available, and in species not convenient for mutant analysis.

3.1. Germination

In species that exhibit seed dormancy, germination can often be induced by a light treatment given to imbibed seed. In general, small-seeded species are more responsive than large-seeded ones. In studies of the effects of light on lettuce seed germination in the 1930s, Flint and MacAlister found that R was particularly effective at inducing germination, while FR and BL were inhibitory (Sage 1992). It was shown subsequently that the effect of R could be reversed by FR. In some highly sensitive seeds, including *Arabidopsis*, a distinct non-FR-reversible phase can be identified. Three hours after imbibition, germination of *Arabidopsis* seeds can be induced by R in a fully FR-reversible manner (Fig. 1), and this response is absent in the *phyB* mutant (Shinomura et al. 1996). After longer periods of imbibition, the sensitivity to light increases, and at 48 h germination can be induced by very small amounts of light, including FR, and is therefore no longer FR-reversible (Fig. 1). This second phase is absent in the *phyA* mutant (Shinomura et al. 1996).

These two responses illustrate some more general features of phyA and phyB function. PhyB controls responses which can be induced by low-fluence R in the order of 1-1000 μmol m^{-2} and which are reversible by FR (Fig. 1). These are called low-fluence responses (LFR) and are a function of the amount of phyB in the Pfr form. PhyA-mediated responses are much more sensitive to light, and have a range in threshold fluence approximately four orders of magnitude lower than LFR (0.1-100 nmol m^{-2}). These very-low-fluence responses (VLFR) require only a very small proportion of phyA (<0.1%) to be converted to the Pfr form. They can therefore be induced by light of any wavelength from 300 to 750 nm, and are not reversible by FR (Fig. 1). The molecular basis for the difference between these two forms of response is not yet understood.

3.2. Seedling Establishment

The development of the germinating seedling and its establishment as a fully autotrophic plant require the co-ordination of several different light-regulated processes. These include inhibition of stem or hypocotyl elongation, apical hook opening, opening and expansion of cotyledons and leaves, and the induction of accumulation of photosynthetic and protective pigments. The regulation of these processes by light, which is often termed "de-etiolation", can be dramatically demonstrated by the exposure of dark-grown or "etiolated" seedlings to brief R pulses. Experiments of this kind show roles for phyA and phyB that are generally consistent with the LFR and VLFR response modes identified in the control of germination. These responses also manifest in co-ordinated changes in the expression of many different genes, both nuclear and plastidic. Prominent examples are the induction of genes involved in light-harvesting (e.g. chlorophyll *a/b*-binding (*CAB*)) carbon fixation (ribulose-1,5-bisphosphate carboxylase/oxygenase small subunit (*RBCS*)) and anthocyanin biosynthesis (e.g. chalcone synthase (*CHS*)). Other genes such as lipoxygenase and *PHYA* are repressed by light (Kuno and Furuya 2000).

Light responses during seedling establishment are also studied by growing plants under different irradiances of continuous monochromatic light (Fig. 4). The responses induced under these conditions are often stronger than those observed in response to a single light pulse, and are termed high irradiance responses (HIR). HIRs to continuous BL, R and FR have all been reported. The FR-HIR is controlled entirely by phyA, whereas the R-HIR is controlled by the phyB-type phytochromes (phyB and phyD in *Arabidopsis*, phyB1 and phyB2 in tomato.) PhyA can also act under continuous R, but its contribution can often only be seen at lower irradiances (Fig. 5), where the phyB-type phytochromes are not active (Kerckhoffs et al. 1997, Mazzella et al. 1997).

The HIRs can be considered as a series of responses to pulses of light, and the fundamental photoreactions have been explored by replacing the continuous irradiation with intermittent pulses. R-HIR can be effectively induced by a R pulse every 4 h in a FR-reversible manner, suggesting that the R-HIR is effectively a continuous activation of LFR. In contrast, the FR-HIR can be replaced only by FR pulses given every 3 min. Under this regime, the effect of the FR pulses is reversible by R (Shinomura et al. 2000). The FR-HIR is therefore distinct from the phyA-mediated VLFR, and operates by a mechanism fundamentally different from phyB-mediated LFR. Once again, the molecular bases for these differences are not yet understood.

De-etiolation can also be induced by BL. Under high-irradiance continuous BL, cry1 is the predominant photoreceptor for de-etiolation responses in both *Arabidopsis* and tomato, with a threshold for activity of around 5 μmol.m^{-2}.s^{-1} At lower irradiances, phyA becomes the predominant photoreceptor for BL (Fig. 5). PhyB-type phytochromes also make a minor contribution under high irradiance BL

(Poppe et al. 1999, Weller et al. 2001b). In *Arabidopsis*, the second cryptochrome, cry2 is also reported to have a minor role in the control of seedling light responses under lower-irradiance BL (Lin et al. 1998).

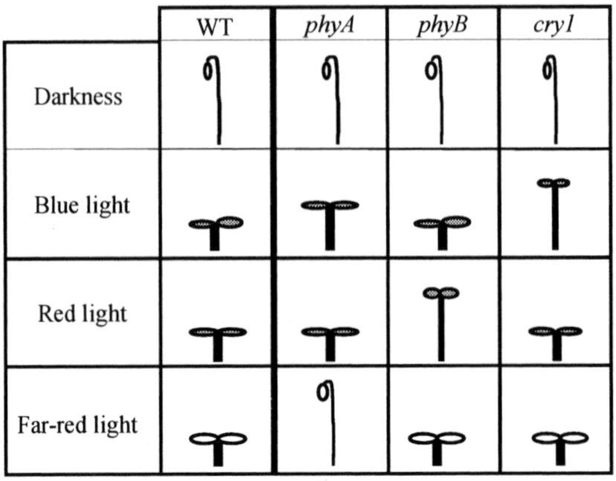

	WT	*phyA*	*phyB*	*cry1*
Darkness				
Blue light				
Red light				
Far-red light				

Figure 4. Diagram illustrating the phenotypes of phyA, phyB and cry1 photoreceptor-deficient mutants under monochromatic blue, red or far-red light. Grey shading of cotyledons indicates presence of chlorophyll.

It is clear that photoreceptors overlap in their sensitivity to both irradiance and wavelength. In some cases their effects are merely additive. However, there are an increasing number of cases where synergistic or antagonistic interactions between specific photoreceptors have been identified (Casal, 2000).

3.3. Phototropism

Detailed action spectra indicated the photoreceptor responsible for seedling phototropism has an absorption peak in the UV-A and a 3-component peak in the BL region of the spectrum (Fig. 1). Fluence-response curves for induction of phototropism by BL pulses resolved two components (Liscum and Stowe-Evans 2000). The "first-positive" component can be induced by fluences of 0.1-500 μmol m^{-2} and shows reciprocity within a certain fluence range. The "second-positive" component has a similar fluence threshold, but is also time dependent, with a minimum time requirement of around 10 min. Phot1-deficient *nph1* mutants lack the first positive response, and are completely aphototropic under low irradiances of continuous BL, suggesting that they are also deficient in the second-positive response. However, under continuous BL of higher irradiance (>10 μmol m^{-2}s^{-1}) the *nph1* mutant shows a normal phototropic response (Sakai et al. 2000), implying the action of another photoreceptor, recently shown to be phot2 (NPL1) (Sakai et al. 2001) The function of phototropin in the control of hypocotyl elongation is apparently restricted to the perception of unilateral BL, because *nph1* plants exhibit normal de-etiolation responses (Liscum and Briggs 1995).

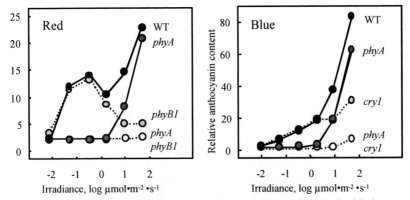

Figure 5. Irradiance dependence for anthocyanin accumulation in hypocotyls of dark-grown wild-type (WT) and photoreceptor-deficient mutant seedlings of tomato exposed for 24 h to monochromatic red or blue light. Redrawn from Kerckhoffs et al. (1997) and Weller et al. (2001b).

Although phyA, phyB, cry1 and cry2 have all been proposed to contribute to the BL phototropic response, it is now clear that these photoreceptors are neither necessary nor sufficient for directional light sensing, at least in *Arabidopsis* hypocotyls. Nevertheless they can modulate expression of the phototropic response, by increasing its amplitude or speeding up its development. In addition, absorption of R by phytochrome can enhance the subsequent phototropin-mediated response to unilateral BL (Parks et al. 1996).

3.4. Shade-avoidance

Light responses in established, fully autotrophic seedlings are often referred to as shade-avoidance responses. In response to shading, stem elongation increases, development of lateral organs such as leaves and branches is suppressed, and flowering is accelerated (Smith and Whitelam 1997). Vegetational shading involves changes in both irradiance and spectral quality of the light reaching the plant (Ballare 1999). The main difference in spectral quality is an effective enrichment for FR, which is due to the fact that leaves transmit FR but absorb BL and R. The term shade-avoidance as applied to laboratory experiments refers specifically to responses induced by manipulation of the FR content against a constant background of photosynthetically active radiation (Fig. 6).

Since shade-avoidance responses occur in fully green, de-etiolated plants, it was often assumed that phyA could not be important, since levels of phyA are very low in light-grown relative to dark-grown plants, and seedlings of phyA-deficient mutants show no substantial difference from WT seedlings when grown in white light. In contrast, *phyB* mutants have the appearance of strongly shade-avoiding plants, indicating an important role for phyB inactivation in the shade-avoidance response. In *Arabidopsis*, phyD and phyE also make an important contribution to perception of R:FR photon ratio (Devlin et al. 1998, 1999). These observations suggest that phyB, phyD and phyE all have a similar mode of action, which can be understood in terms of LFR and phytochrome acting as a simple developmental switch. Light of high R:FR converts a large proportion of phytochrome into its Pfr form, which actively initiates photomorphogenic responses including leaf expansion and the inhibition of stem elongation. Increasing FR supplementation drives the photoequilibrium back towards the Pr form, reducing the level of the active Pfr form, and thus reducing the extent of the photomorphogenic response (Smith 2000).

white   white
              +
           far-red

Figure 6. Shade-avoidance response of wild-type tomato seedlings simulated by addition of high irradiance far-red light to the white light source, lowering the ratio of red to far-red light from 6.3 to 0.1.

More detailed investigations also show that phyA, although not able to perceive changes in R:FR in the manner of the phyB-type phytochromes, does influence the shade-avoidance response by promoting de-etiolation under light of low R:FR ratio (McCormac et al 1992, Weller et al. 1997a). This response opposes the reduction in activity of phyB under the same conditions, and it is clear that the balance of phyA to phyB is therefore important in determining the degree of responsiveness to changes in R:FR. Constitutive overexpression of phyA can enhance the photomorphogenic effect of FR-rich light to the extent that the normal shade-avoidance response acting through phyB is suppressed (Robson et al. 1996).

4. PHOTORECEPTOR SIGNAL TRANSDUCTION

Most of the classical photomorphogenic responses listed above are whole-plant responses, and occur on a scale of hours to days after first exposure to the light stimulus. Recent molecular and genetic dissections have begun to reveal some of the shorter-term cellular and molecular events underlying these responses. It is now clear that most responses require changes in gene expression, and the signalling pathways from photoreceptor to the transcriptional complexes that regulate gene expression are now the subject of intense interest. Other responses to light occur much more rapidly and do not seem to involve changes in gene expression, thus suggesting that some light signalling pathways may be located entirely in the cytoplasm.

Attempts to understand the transduction of signals from photoreceptor to responses have taken four main approaches. These have involved studies of the nature and primary reactions of photoreceptor, two-hybrid screens for interacting factors, mutant screens for downstream genes, and pharmacological manipulations of putative signalling intermediates (Christie and Briggs 2001, Quail 2000, Hudson 2000, Jenkins et al. 2001). These approaches clearly overlap and are converging, but it is perhaps still useful to distinguish them here.

4.1. Primary reactions of photoreceptors

4.1.1. Phytochrome

Early speculations about the initial reactions of phytochrome considered the possibility that it might possess enzymatic activity, and more specifically, that it might function as a kinase (Sage 1992). Early studies reported a ser/thr kinase activity in purified phyA preparations, but these were considered inconclusive due to the possibility of a copurifying activity. Sequence comparisons show homology between a domain in the C-terminus of phytochrome and histidine kinases, but mutation of conserved histidines in this domain was found to have no effect on phytochrome function (Quail, 1997). The question has now been largely resolved by the demonstration that recombinant eukaryotic phytochrome shows a light-dependent autophosphorylation characteristic of ser/thr kinase activity (Yeh and Lagarias 1998). Several targets for phosphorylation by phytochrome have been identified, and including cry1 (Ahmad et al 1998), members of the Aux/IAA family of transcription factors (Colon-Carmona et al. 2000), and PKS1, a phyB-interacting protein (Fankhauser et al. 1999).

4.1.2. Cryptochrome

To date, there is very little known about the primary reactions of cryptochrome, although as a flavoprotein, it is presumed that some form of electron transfer is involved. The C-terminal domains of cry1 and cry2 each confer light independent photomorphogenesis when constitutively expressed in transgenic plants (Yang et al. 2000). This suggests that activity of this domain in the full-length molecule is constrained in darkness by the N-terminal domain, and that this constraint is somehow released following light absorption. Cryptochrome activation therefore appears to involve communication between two halves of the molecule, and the interaction of reaction partners with the C-terminal domain.

4.1.3. Phototropin

The primary structure of phototropin clearly identifies it as a classical ser/thr kinase. Although phototropin is autophosphorylated in response to BL (Huala et al. 1997), no targets for transphosphorylation have yet been identified.

4.2. Mutants and interacting factors

A large number of genetic screens have been performed in *Arabidopsis* to identify genes that function downstream of photoreceptors in the mediation of light responses. Physiological and genetic studies with these mutants have attempted to assign them to the signalling pathways of one or more photoreceptors (Fig. 7). Several of these putative signalling components have now been cloned, and their functions are currently being explored.

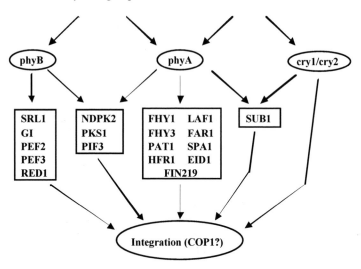

Figure 7. A number of proteins important in photoreceptor signal transduction have been identified from genetic and protein-protein interaction screens. These proteins can be classified according to their known interactions and the specificity of their null mutant phenotypes.

4.2.1. Phytochrome

Phytochrome was long thought to be a cytosolic protein, and it was therefore presumed that the initial steps in phytochrome signalling took place in the cytoplasm. The more recent demonstrations of rapid light-induced nuclear import of phytochrome have therefore been somewhat surprising, and suggest that phytochrome might have a more direct role in the regulation of gene expression than previously considered. This has been confirmed by identification of PIF3, a phytochrome-interacting basic helix-loop-helix (bHLH) protein which is nuclear localized and binds to specific light-responsive elements in the promoters of light-regulated. genes. PhyB has been shown to bind to PIF3 *in vivo* and to control its ability to activate transcription in a light-dependent manner (Martinez-Garcia et al. 2000). PhyA can also bind to PIF3 but with much lower affinity (Zhu et al. 2000). Loss of a related bHLH protein, HFR1, specifically impairs de-etiolation under FR, indicating that it acts in a signalling pathway specific for phyA (Fairchild et al. 2000). Although transcriptional activation activity has not yet been demonstrated for HFR1, it can form a heterodimer with PIF3. This raises the possibility that the specificities of phyA and phyB may be partially determined by their differential affinities for different transcription factors. Loss of EID 1, a nuclear F-box protein, specifically enhances phyA responses. EID1 participates as a subunit of ubiquitin E3 ligase and may target activated components of the phyA signalling pathway for ubiquitin-dependent degradation (Dieterle et al. 2001). FAR1 and SPA1 are two other nuclear proteins for which mutant phenotypes indicate a specific role in phyA signalling (Hudson et al. 1999, Hoecker et al. 1999). However, cytoplasmic proteins with specific role in phyA (PAT1, FIN219) and phyB (PKS1) signalling have also been identified (Hudson 2000). This suggests that a separate pathway for phytochrome signalling may exist in the cytoplasm, or that certain cytoplasmic components are necessary to facilitate the movement of phytochromes to the nucleus after light exposure. A number of other phyA-specific (*fhy1, fhy3, fin2*) and phyB-specific (*red1, srl1, pef2, pef3*) mutants have been isolated (Hudson 2000), but cloning of the corresponding genes has not yet been reported (Fig. 7).

4.2.2 Cryptochrome

So far no genes have been identified which are specifically involved in BL signalling. Some screens have identified mutants with impaired BL responses (e.g. *hy5*, Koornneef et al. 1980; *sub1*, Guo et al 2001) but these are also impaired in response to R and/or FR and may define signalling components common to phytochromes and cryptochromes (Fig. 7). It is possible that the absence of specific cryptochrome signalling mutants may reflect a very close relationship between cryptochrome and the phytochrome signalling pathway. This explanation is supported by the recent demonstration by Wang et al. (2001) that the C-terminal domains of both cry1 and cry2 interact directly with the key regulatory protein COP1 (see below).

4.2.3 Phototropin

Mutants impaired in phototropism have been identified by several groups, and define four distinct loci (*NPH2, NPH3, NPH4, RPT2*) in addition to *NPH1*(Liscum and Briggs 1996, Sakai et al. 2000). Three of these genes have now been cloned. *NPH3* encodes a protein that appears to interact physically with the phototropin photoreceptor. The NPH3 protein contains no motifs suggestive of a specific biochemical activity, but the presence of several protein-protein interaction domains has prompted the suggestion that it may function as a molecular scaffold, bringing together other components of phototropin signalling (Motchoulski and Liscum 1999). The RPT2 protein is also essential for normal phototropism of root and shoot, and shows sequence similarity to NPH3, suggesting that it may function in a similar manner (Sakai et al. 2000). Mutants in the *NPH4* gene are aphototropic in low fluence rate BL, and unlike mutants at the other phototropism loci, also show impairment of gravitropic responses. *NPH4* encodes an auxin-regulated transcriptional activator, indicating that auxin-regulated changes in gene expression occur in response to phototropin activation and are necessary for normal tropic responses generally (Harper et al. 2000).

4.2.4 Common light-signalling components

Several different screens have identified a number of mutants expressing constitutive photomorphogenic responses even when grown in darkness. The light-independent phenotype of these mutants has implied the existence of proteins that normally function to repress the de-etiolated state in darkness. It also suggests that photomorphogenesis can be viewed as the default pathway for seedling development, and that the action of light is to overcome the active repression of this pathway.

One of the key proteins involved in this repression is COP1, a protein containing three domains recognized as important for protein-protein interaction (Torii and Deng 1997). COP1 accumulates in the nucleus in darkness, but is excluded from the nucleus in light. Both phytochrome and cryptochrome signalling can contribute to the nuclear exclusion of COP1 (Osterlund and Deng 1998). The importance of COP1 function in the nucleus has recently become clearer with reports that it interacts physically with HY5, a bZIP transcription factor which binds to LREs in promoters of light regulated genes and is necessary for normal de-etiolation responses. In darkness, COP1 is enriched in the nucleus, where it binds HY5 and targets it for degradation (Fig. 8). This is suggested to be an important mechanism for the regulation of HY5-dependent gene expression (Osterlund et al. 2000).

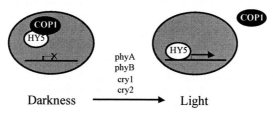

Figure 8. Proposed mechanism for the integration of light signals through the COP1 protein (Osterlund and Deng 2000). In darkness, COP1 accumulates in the nucleus where it interacts with the transcription factor HY5 and targets it for degradation. Photoreceptor activation by light results in the exclusion of COP1 from the nucleus, and the accumulation of HY5, which activates transcription of target genes.

Several proteins necessary for light regulation of COP1 activity have also been identified, and these may provide the first links between specific photoreceptors and COP1 activity. FIN219 is a cytoplasmic protein that is necessary for phyA activity, and may contribute to the negative regulation of COP1 activity by phyA. *FIN219* expression is strongly upregulated by auxin, providing further evidence for the involvement of auxin in phytochrome responses (Hsieh et al. 2000). SUB1 functions downstream of cry1 and phyA to negatively regulate HY5 abundance, possibly through positive effects on COP1 (Guo et al. 2001). The identification of SUB1 as a Ca^{2+} binding protein and its location in the nuclear envelope raise the intriguing possibility that it may be involved in the regulation of nuclear translocation by Ca^{2+}.

Other proteins with similar mutant phenotypes to COP1, such as COP9, COP8, FUS5 and FUS6, participate in a large multimeric complex referred to as the "COP9 complex". Subunits of the complex show a one-to-one correspondence with those making up the lid complex of the 26S proteasome, and interaction of the complex with ubiquitin E3 ligase has recently been demonstrated, suggesting a probable role in the regulation of protein degradation (Schweichheimer et al. 2001). In plants, the complex is necessary to maintain COP1 in the nucleus in darkness, an observation that may explain the COP1-like phenotype of mutants lacking individual complex components. Other proteins with similar mutant phenotypes, such as COP10 and DET1, are not part of the COP9 complex, but still seem to be involved in regulation of COP1 localization (Chamowitz and Deng 1997).

4.3. Pharmacological approaches

The involvement of putative signalling intermediates in light-signal transduction have been investigated in a number of different systems, including cell cultures and protoplasts, and by direct injection into hypocotyl cells of a phytochrome-deficient tomato mutant.

Microinjection studies identified two distinct pathways involved in phytochrome regulation of gene expression in tomato hypocotyl cells (Mustilli and Bowler 1997). Induction of *CHS* expression was found to require cGMP and his/tyr kinase activity, whereas induction of *CAB* expression was specifically blocked by Ca^{2+}/calmodulin (CaM) antagonists. G-protein inhibitors blocked both responses, and the full development of chloroplasts only occurred if both pathways were intact. Results using a soybean cell culture essentially confirmed these conclusions about signalling in plastid development. In contrast, the induction of *CHS* by BL/UV-A and UV-B light in *Arabidopsis* cell cultures is not influenced by cGMP or his/tyr kinase antagonists, but uses a different signalling pathway involving redox reactions at the plasma membrane, specific Ca^{2+} channels, and ser/thr kinase activity (Jenkins et al. 2001). The UV-B effects are distinguished from the BL/UV-A effects by the involvement of CaM.

Another BL response that has been subjected to pharmacological dissection is the rapid membrane depolarization in hypocotyl cells and protoplasts. This response depends on the action of specific anion channels that are sensitive to the inhibitor 5-nitro-(3-phenylpropylamino)-benzoic acid (NPPB). BL-induced membrane depolarization is dramatically reduced in *cry1* mutants (Parks et al. 1998), and is probably related to subsequent cry1-mediated growth inhibition in intact *Arabidopsis* hypocotyls. Cry1-dependent anthocyanin accumulation is also inhibited by NPPB. These results indicate that ion fluxes at the plasma membrane may also play an important role in the early steps of cry1 signal transduction.

The involvement of calcium in BL responses has also been investigated in transgenic plants expressing the luminescent reporter protein aequorin (Baum et al. 1999). Targeting of aequorin to different cellular compartments has shown that BL pulses specifically induce Ca^{2+} transients in cytoplasm. The loss of cry1 or cry2 had no effect on this response, whereas it was somewhat impaired in the phototropin-deficient *nph1* mutant, suggesting a potential role for Ca^{2+} in phototropin signalling. This is strengthened by observations that the effects of R and BL pretreatment on phototropism in tobacco and *Arabidopsis* correlate well with their effects on the Ca^{2+} transient.

It is clear that although some promising progress has been made in identifying potential components of different light-signalling pathways, these have yet to be linked with transduction components identified by genetic means, and we are still

316

far from understanding the complete sequence of events initiated by photoreceptor activation.

5. PHOTOPERIODISM

The importance of the duration of the daily photoperiod for plant development was first noted over 90 years ago. Using changes in daylength, plants can monitor the time of year and predict seasonal change in other environmental variables. A number of processes exhibit photoperiodic regulation, including the induction of flowering, the formation of storage organs such as bulbs and tubers, and the onset of bud dormancy. Of these, it is the induction of flowering that has been studied most intensively. Clear differences in photoperiodic responsiveness were first documented by Garner and Allard (1920), who classified plants according to whether flowering was preferentially induced under long days (long day plants or LDP) or short days (short-day plants or SDP) (Fig. 9). Other early observations indicated that flowering responses could be dramatically altered by low irradiances or short exposures to light. This suggested that the light served as a source of information rather than of energy, and showed photoperiodism to be a truly photomorphogenic phenomenon.

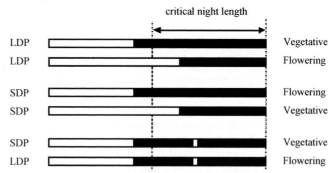

Figure 9. Summary of differences in flowering responses of short day plants (SDP) and long-day plants (LDP). Open and filled bars represent light and dark periods, respectively.

5.1. Light and the circadian clock

Endogenous rhythms have been observed in plants for more than two hundred years. However, the importance of an endogenous circadian rhythm for the timekeeping aspect of photoperiodism was first suggested in the 1930s by Erwin Bünning (Bünning 1964). In fact, photoperiodism can be thought of as the adaptation of circadian timekeeping to the measurement of daylength. As such it must involve interaction of light signalling ("input") and specific flower induction ("output") pathways with a circadian oscillator or "clock". One characteristic feature of circadian rhythms is that they show a free-running period that is close to, but not exactly 24h. However, under daily light/dark (L/D) cycles, the rhythmic outputs become synchronized or *entrained* to a period of 24 h exactly.

There are generally considered to be two basic models for the way in which light might interact with the clock (Thomas, 1998). In the *internal coincidence* model, photoperiodic induction results from the increasing overlap in phase of two distinct circadian output rhythms. Light interacts with the induction process solely by controlling the phase and/or period (i.e. entrainment) of the two rhythms. In the *external coincidence* model, developed from ideas first proposed by Bünning, the circadian clock generates an output rhythm in light sensitivity and photoperiodic induction results from the coincidence of an inductive light signal with the light-sensitive phase of this rhythm. In this model, light has two roles; entrainment of the clock, and direct interaction with downstream components necessary for the response.

A third effect of light in photoperiodism may be to influence output responses directly, without the involvement of a timing component. The challenge is to understand how the plant is able to integrate these signals and generate the appropriate response.

5.1.1. Physiological approaches

Detailed physiological investigations of the relationship between light and the circadian clock have been performed across a wide variety of different species, both SDP and LDP (see Thomas and Vince-Prue, 1997). These studies have generated a large amount of complex and often contradictory literature. However, there is reasonable agreement on some of the more general conclusions, and these are summarised below.

5.1.1.1. Short-day plants

In a fixed daily cycle, it is clear that changes in day length could in theory be detected either as changes in length of the light period or of the dark period. It has been established that for many SDP it is mainly the length of the night that is measured, suggesting that processes necessary for floral induction can only take place if the night is longer than a certain *critical night length*. Interruption of an inductive long night with a short light treatment prevents its effect and delays flowering (Fig. 9). In many species this night-break (NB) response is relatively sensitive and has thus been amenable to pulse experiments and a detailed photobiological analysis (Thomas and Vince-Prue, 1997).

NB responses in SDP show action spectra typical of phytochrome-mediated LFR, for which R is inhibitory to flowering and subsequent FR cancels this inhibition. In this response, phytochrome in its Pfr form is clearly acting to inhibit flowering. In addition to the light quality of the NB, its timing can also be important, and several different SDP species show circadian rhythmicity in the responsiveness to a NB (Fig. 10), consistent with the external coincidence model (Thomas and Vince-Prue, 1997). Other light treatments can reset the phase of this rhythm. In some cases the phase-setting effects of light were shown to occur independently of effects on flower induction, and R was also the most effective wavelength for inducing phase shifts. Phase shifting has generally been found to require longer exposures to light than the NB response. With even longer periods of light exposure

318

(>6 h), the phase of the rhythm is no longer shifted but suspended and only released approximately 9 h after transfer to darkness (Thomas and Vince-Prue, 1997).

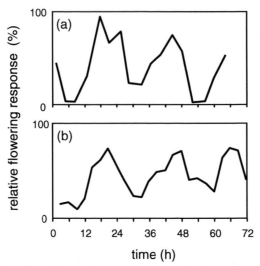

Figure 10. Circadian rhythms in flowering responses to light treatments in SDP and LDP. (a) Rhythmic response of the SDP Glycine max (soybean) to a 4 h night-break with white light given at various times during an extended night following an 8-h photoperiod (redrawn from Coulter and Hamner, 1964, (b) Rhythmic response of the LDP Hordeum vulgare (barley) to 6 h of far-red light added at various times during an extended photoperiod of continuous white light (redrawn from Deitzer et al. 1982).

Other studies have shown that in addition to the strong inhibitory effect of R shown in the NB response, FR is also effective for inhibition of flowering when given at the end of the day or early in the dark period. This clearly suggests the action of a second phytochrome, which promotes flowering in its Pfr form. This response does not affect the timing of NB sensitivity.

5.1.1.2. Long-day plants
    In general, less is known about light requirements in LDP photoperiodism. As for SDP, light reactions governing flowering in LDP occur in both light and dark periods, although one or the other may predominate in any one species. The concept of a critical night length is again relevant, but for LDP, nights must be shorter than the critical length for plants to flower. In some LDP, R NB are effective for promotion of flowering. Their effectiveness varies during the night, and the response can be partially reversed by FR. However, unlike in SDP, a clear rhythmicity in responsiveness is not observed, and in general, longer periods of light are required to elicit a response (Thomas and Vince-Prue, 1997).
    Light reactions during the photoperiod have also been demonstrated in LDP. For example, FR added to a photoperiod of R or white light (WL) can promote flowering, with rhythmic variation in effectiveness (Fig. 10). Although phase shifting experiments are much more difficult to perform in LDP and less conclusive,

light-induced changes in phase of the rhythm of FR responsiveness have been reported (e.g. Deitzer et al. 1982).

Photoperiodic responses in LDP have more often been investigated using extensions of a short, non-inductive photoperiod. Action spectra for the promotion of flowering by photoperiod extensions most commonly show peaks at around 710-720 nm, well above the absorbance peak of Pr and clearly below that of Pfr (Thomas and Vince-Prue 1997). This peak is similar to that seen for the FR-HIR in seedling de-etiolation, suggesting the involvement of phyA. However, in other species, action spectra with peaks in both R and FR have also been reported (Carr-Smith et al. 1989), indicating that a second phytochrome is probably involved (Fig. 1). In some species, notably crucifers, BL is also effective as a day-extension.

While it is not clear exactly how the light responses of LDP and SDP may be compared, it is clear that in each response type, two distinct types of reaction can be distinguished. One involves a promotive effect of FR (seen in end-of-day (EOD) responses of LDP and reversal of NB in SDP). The other involves a promotive effect of R (seen in day extensions and NB in LDP, and reversal of EOD-FR effects in SDP). These clearly suggest the opposing actions of two separate phytochromes (Fig. 11).

5.1.2. Genetic approaches

Many of the species used in classical studies of photoperiodism have, for one reason or another, not proven suitable for genetic analyses. Thus, although the physiology of photoperiodism has been most thoroughly characterized in SDP, the genetic dissection of photoperiodism has progressed most rapidly in LDP. This is of course due mainly to the prominence of the LDP *Arabidopsis*. Extensive mutational analysis of flowering in an SDP has yet to be performed, although the increasing prominence of rice as a model system for molecular genetics and the recent cloning of several quantitative trait loci for flowering qualify it as the SDP system of choice for the near future (Yano et al. 2001). A number of flowering mutants or genetic variants have also been characterized in other photoperiodic species, including garden pea, wheat, barley (LDP), soybean and tobacco (SDP).

Among mutants known to affect photoperiodic responses in *Arabidopsis* are mutants lacking various photoreceptors, in addition to those that affect light signalling to the circadian clock, maintenance of the circadian clock itself, or specific output pathways for photoperiodic flower induction (Figure 12). Molecular and physiological analyses of these mutants are providing invaluable information about molecular components important for photoperiodic responsiveness. However, points of comparison between these studies and the classical physiological studies are at the moment rather few, although this is likely to change dramatically over the next decade.

5.1.2.1. Photoreceptor mutants

In general, LDP flower early under low R:FR. By analogy with other shade-avoidance responses this represents the switching off of a phytochrome necessary for inhibition of flowering under high R:FR (Figure 11). The early-flowering phenotype of phyB, phyD and phyE mutants under high R:FR indicates that each of these phytochromes acts to inhibit flowering in its Pfr form (Devlin et al. 1998, 1999). This conclusion is supported by the phenotype of *phyB* mutants in other LDP species. PhyB is therefore unlikely to mediate the R NB effect in LDP, because this would require a promotive effect of Pfr. The promotion of flowering by EOD-FR treatments is also mediated by phyB, suggesting that after the light period, phyB is maintained in its inhibitory Pfr form during darkness. Inhibition may be directly related to length of time that phyB (and other phyB types) are present as Pfr, and not directly related to timing (Fig. 11). This is consistent with observations that phyB can affect flowering under both LD and SD. Interestingly, a phyB mutant in the SDP *Sorghum* also flowers early in both LD and SD (Childs et al. 1997), suggesting that phyB may have a similar photoperiod-independent inhibitory function in both SDP and LDP.

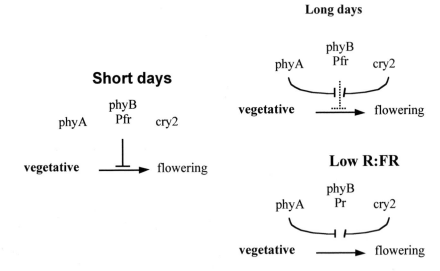

Figure 11. Photoreceptor interaction in the control of flowering in long-day plants. Under short days, phytochrome B in the Pfr form inhibits flowering. This inhibition is antagonised by phyA and cry2 under long days, resulting in early flowering. PhyB-mediated inhibition is reduced by light of low R:FR ratio given during the photoperiod or immediately before the dark period.

The fact that relatively long periods of light are necessary for promotion of flowering in LDP, and the similarity of action spectra for this promotion to that seen for the FR-HIR in seedling de-etiolation, have for some time suggested that phyA might have a significant flower-promoting role in LDP. This has been confirmed in *phyA* mutants of *Arabidopsis* and pea. In *Arabidopsis*, phyA contributes to the promotive effect both of WL NB and FR-rich day-extensions (Johnson et al 1994, Reed et al, 1994). In pea, phyA is also clearly needed for perception of LD, and promotes flowering in response to day extensions of either high or low R:FR (Weller et al. 1997a). This promotion of flowering is dependent on the presence of phyB (Figure 11). Pea *phyA* mutants also retain a strong response to low R:FR NB, but are completely deficient in response to high R:FR NB (Weller et al. 2001a)

These results show that in pea, phyA is also active under white light and phyA effects on flowering cannot be explained solely as a classical FR-HIR.

Null mutations in the rice phyA gene have recently been isolated, but have no apparent effect on flowering under LD or SD (Takano et al. 2001). This suggests that in SDP, unlike LDP, phyA does not contribute to the promotion of flowering under inductive conditions. The only clear evidence of a role for phyA in control of a photoperiodic response in SDP has come from studies of tuberization in potato. Potato plants expressing an antisense *PHYA* transgene and possessing reduced levels of phyA tuberize early under non-inductive LD, showing that phyA acts to inhibit tuberization (Yanovsky et al. 2000b). While it is difficult to relate this response to the photoperiodic flowering response, it is of interest that phyB in potato also acts to inhibit tuberization (Jackson et al. 1996). PhyA and phyB thus appear to act in the same direction to control tuberization, rather than antagonistically as in the control of LDP flowering.

In *Arabidopsis*, an important role in the photoperiodic control of flowering has also been demonstrated for the cryptochrome photoreceptor family. Cry2 in particular has a relatively minor effect on seedling photomorphogenic responses, but has a strong promotive effect on flowering under LD (Guo et al. 1998), that is dependent on the activation of phyB (Mockler et al. 1999) (Fig. 11). These findings suggest that the perception of daylength has become the dominant function of this photoreceptor in *Arabidopsis*. However, in general, relatively few species show strong control of flowering by BL, and the importance of cry2 in photoperiodism may not be widespread.

5.1.2.2. Photoreceptor mutants and the circadian rhythm

The most thorough application of mutants in the exploration of phenomena related to photoperiodism has been the investigation of photoreceptor effects on clock period (Somers et al. 1998a, Devlin and Kay 2000). In general, the period-shortening effects of continuous light are thought to represent a continuous entrainment response, and are thus expected to be relevant in plants grown under L/D cycles. Results from these studies show that photoreceptor roles in the regulation of circadian period closely parallel their roles in the control of de-etiolation. In *Arabidopsis*, phyB, phyD and phyE are responsible for period shortening under high-irradiance R, while phyA acts under low-irradiance R and BL. Cry1 and cry2 act redundantly to shorten the circadian period under BL across a wide irradiance range. PhyA has also been shown to mediate the phase-advancing effect of FR on leaf-movement rhythms in potato (Yanovsky et al. 2000b). These results suggest that all of these photoreceptors may have the ability to contribute to entrainment of the clock under certain circumstances. However, the significance of this for plants growing under high WL irradiances is unclear. Under such conditions, phyB and cry1 would be expected to be the predominant photoreceptors controlling clock period, although a quadruple mutant lacking phyA, phyB, cry1 and cry2 can entrain normally to WL/D cycles (Yanovsky et al 2000a). It is interesting that phyA and cry2, which both have profound effects on flowering, are reported to have only minor effects on clock period, suggesting that their primary function may be interaction with an output rhythm.

5.1.2.3. Circadian rhythm mutants

Several genes essential for correct maintenance of circadian rhythms under light/dark cycles have now been identified in *Arabidopsis* (Barak et al. 2000) Although they have been studied in slightly different ways, these genes have all been shown to affect rhythmic control of gene expression and leaf movement under some circumstances. In addition, they all have dramatic effects on the photoperiodic induction of flowering, emphasizing the importance of normal circadian regulation for photoperiodism,

Mutations in the *TOC1* gene shorten the free-running period of multiple circadian outputs in both light and darkness. Rhythms in *toc1* plants show normal light and temperature sensitivity, suggesting that TOC1 is involved with central oscillator function rather than light or temperature input (Somers et al. 1998b). The TOC1 protein shows homology to response regulators (Strayer et al. 2000). Null *toc1* mutants flower early under SD conditions, and approach photoperiod insensitivity. However, when the cycle length is shortened to more closely approximate the short period of the mutant, *toc1* plants show a normal delay of flowering (Strayer et al. 2000). CCA1 and LHY are closely related myb-like transcription factors that may also have a central role in clock function (Wang (Wang and Tobin 1998, Schaffer et al. 1998). The expression of the *CCA1* and *LHY* genes normally oscillates with a circadian rhythm and is maximal at dawn. Overexpression of either gene results in plants which are arrhythmic in both L and D for multiple output responses, are late flowering under LD, and have dramatic defects in the inhibition of hypocotyl elongation. Loss of CCA1 shortens the clock period and results in constitutively early flowering (Green and Tobin 1999).

Mutants for the *ELF3* gene were originally identified on the basis of an extremely early-flowering, photoperiod-insensitive phenotype. Null mutants of *ELF3* have a relatively normal circadian rhythm in continuous darkness, but show complete suppression of rhythmicity under continuous light (Hicks et al. 1996). Under L/D cycles, *elf3* plants have a relatively normal diurnal rhythm in SD, but show an increasingly aberrant rhythm in longer photoperiods. These observations suggest that *ELF3* may function in the pathway for input of light signals to the clock (Fig. 12), rather than in the central clock mechanism (McWatters et al. 2000). Similar conclusions have been reached for the *FKF1, ZTL,* and *GI* genes (Park et al. 1999, Nelson et al. 2000, Somers et al. 2000). Mutants for all three genes are late flowering in long days, and alter circadian gene expression in an apparently light-dependent manner. Null mutations in *GI* and *ZTL* both lengthen the clock period, whereas the *fkf* mutant shows a phase delay. Like *TOC1, GI* and *FKF1* show rhythmic expression with peaks late in the subjective day. Interestingly, *FKF1* and *ZTL* encode related proteins that both contain LOV domains similar to those found in the BL photoreceptor phototropin. Although the function of these domains has yet to be explored, their presence raises the intriguing possibility that the *FKF1* and *ZTL* proteins may represent a novel form of photoreceptor (Somers et al. 2000).

It is clear that results from these studies have yet to be integrated into a coherent model. However, the development of techniques for convenient monitoring of diverse rhythmic outputs has considerably improved the ease with which the circadian system can be investigated. The fact that we now know the molecular identity of some important components of the circadian system is likely to bring further rapid progress as the interactions among these are analyzed.

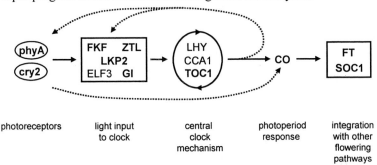

| photoreceptors | light input to clock | central clock mechanism | photoperiod response | integration with other flowering pathways |

Figure 12. A proposed model for the genetic control of flowering by photoperiod in Arabidopsis. *Light perceived by phyA and cry2 is thought to interact with a circadian timing mechanism to control the level and/or activity of the flower-promoting CONSTANS (CO) protein. CO is thought to be a specific regulator of the photoperiod response, and induces the expression of downstream genes FT and SOC1 which are regulated by other environmental factors in addition to photoperiod.*

The diurnal patterns in expression of *GI* have also been investigated (Park et al. 1999). Under 10-h SD cycles, *GI* expression levels peaked 8 h after dawn and declined rapidly at dusk. However, in 18-h LD cycles, the peak of GI expression occurred 2 h later and was much broader, only reaching its lowest level at 16 h after dawn. These results give the first indication that the expression of genes important for flowering may show a differential pattern of diurnal regulation in SD and LD.

5.2. Signalling in photoperiodism

The sensitivity to flower-inducing light signals varies tremendously among different species. In many cases the site at which light is perceived can be separated from the eventual site of flower formation, indicating the existence of some form of long-distance communication. This has been studied in physiological experiments involving grafting, leaf-removal, and differential exposure of different parts of the plant. Experiments of this kind have provided evidence for both a promoter (sometimes referred to as "florigen") and an inhibitor of flowering ("anti-florigen"). In general, the evidence for a promoter is clearer in SDP, whereas evidence for an inhibitor is clearer in LDP. However, in some species both influences have been shown to co-exist (Thomas and Vince-Prue, 1997). In several SDP, the rate of transport of the floral stimulus out of an induced leaf is similar to the rate of phloem movement, suggesting that it may be transported via the phloem. Despite the wealth of convincing evidence for the existence of specific, mobile regulators of flowering, we still have no idea what they are, or indeed if they represent a single substance, several interacting substances or merely reflect a change in the general flow of nutrients (Bernier, 1988). Speculations about the nature of the floral regulators have considered known plant hormones (gibberellins, cytokinins), various metabolites (sugars, polyamines), and specific RNA molecules.

Signalling in photoperiodism has also been subject to genetic dissection. In *Arabidopsis*, several flowering mutants have been identified which have photoperiod-specific phenotypes and no apparent effect on clock or light signaling (Figure 12). The best-characterized of these is *CO*, which encodes a protein with motifs suggestive of a role in regulation of transcription or protein-protein interaction. *CO* expression is low under SD conditions, and upregulation of expression occurs during the induction of flowering by LD (Guo et al. 1998). The reduced responsiveness to LD in the *cry2* mutant is associated with a failure to upregulate *CO* expression levels (Guo et al. 1998). Direct targets of CO include two genes (*FT* and *SOC1*) that are also required for normal vernalization responses (Samach et al. 2000), suggesting that CO may act close to the point at which signals from several flowering pathways are integrated, and therefore close to the end of the photoperiod-specific pathway (Figure 12). As seen for *GI*, the diurnal pattern of *CO* expression is also reported to differ under LD and SD cycles (Suarez-Lopez et al. 2001). These observations have prompted the speculation that the coincidence of light with a specific phase in the rhythm of *CO* and/or *GI* expression under one photoperiod but not the other might determine the differential flowering response, in keeping with the external coincidence model (Samach and Coupland 2000).

Genes with photoperiod-specific effects on flower induction have also been identified in a number of other species (Thomas and Vince-Prue, 1997). In pea, recessive mutant alleles at four different loci (*SN, DNE, PPD* and *HR*) each impair full expression of the photoperiod response (Weller et al. 1997b). Grafting experiments have shown that these genes control the level of a graft-transmissible factor necessary for normal inhibition of flowering under non-inductive SD. The late flowering phenotype of *phyA* mutants in LD results from failure to down-regulate this inhibitor (Weller et al. 1997a). It is not yet clear whether the pea genes are specific for photoperiodic output, or whether they might primarily affect clock function. Although described as a inhibitor of flowering, the factor defined by the pea photoperiod genes also acts to promote aspects of vegetative growth under SD, suggesting that it affects all aspects of photoperiodism rather than being specific for flowering (Weller et al. 1997b). Similar pleiotropic effects of a transmissible flower inhibitor are also reported from grafting experiments with tobacco (Thomas and Vince-Prue, 1997).

Despite recent rapid progress in identifying genes necessary for photoperiodic responses in *Arabidopsis*, it is still not possible to integrate this information with the large body of physiological evidence for mobile regulators of the photoperiodic response in other species. Evidence for a mobile floral stimulus in *Arabidopsis* has been demonstrated in leaf removal experiments (Corbesier et al. 1996), although the effects of genes in the photoperiod pathway in this system have yet to be reported. Further progress will depend on functional analyses of the *Arabidopsis* genes, either in this leaf-removal system, or by leaf-specific expression, or by identification of homologous mutants in other species more suitable for grafting.

6. PHOTOMORPHOGENESIS AND PHOTOPERIODISM IN THE
NATURAL ENVIRONMENT

In discussing the significance of photomorphogenesis in the natural environment, several things must be kept in mind. We need to consider the changes that may occur in the properties of light reaching the plant, the kind of information plants may extract from these changes, and the way in which this information might be converted into an appropriate developmental response.

Most plants have adopted a sedentary habit and are therefore committed to adapt to changes in their environment by developmental plasticity. Natural selection is therefore likely to have favoured modifications of development that maximize energy capture, or that improve the ability of the plant to resist detrimental effects of light. In addition, correlation between changes in light environment and other environmental variables such as cold or drought are also likely to have favoured cross-talk between light signalling and other signalling pathways and the exploitation of light information as a predictive signal. Conditions of continuous selection would also result in pressure to extract increasing amounts of information from light, through an ability to monitor more subtle changes. This in turn could conceivably have supported the evolution of multiple photoreceptors with diverse light sensing properties.

Speculations about the importance of photoreceptors in the natural environment have been largely based on studies of mutants and transgenic lines grown as single plants in controlled-environment conditions, combined with an intuitive appreciation for the developmental predicament of the plant. However, they have recently begun to receive solid support from experiments conducted under natural and/or competitive conditions (Ballare 1999).

6.1. Improving energy capture

In general, higher plants have two strategies to increase their capture of energy. They can either gain access to more energy by extending their growth into areas of stronger light, or they can more efficiently capture the energy already reaching them. Measurements of light quality in the natural environment have shown that changes in the amount of light due to cloud cover or the time of day are accompanied by relatively small changes in its spectral distribution (Smith 1994). In contrast, chlorophyll-containing tissues absorb efficiently in the R and B region of the spectrum, but transmit and reflect a substantial proportion of light in the FR waveband. Thus the presence of plants affects the local light environment by causing a measurable decrease in the ratio of R to FR light energy. This may occur by the filtering out of R and/or by increased lateral reflection of FR. However, it is particularly significant that increases in lateral reflection of FR from neighbouring plants occur prior to any reduction in the amount of photosynthetically active R and BL wavelengths (Ballaré et al. 1990). An increase in the amount of FR is thus an unambiguous signal to the plant that potential competitors are growing nearby. Where this signal is unidirectional, the appropriate response of the plant is obviously to redirect its growth away from the other plants. In a denser population, the gradient in FR will not be as great, and the response may also include an increase in overall growth rate. Although the horizontal and lateral components of the response to shading are often treated separately as phototropism and shade-avoidance, it is more appropriate to consider both responses as aspects of a strategy in which the plant is actively "foraging" for light (Ballare et al. 1997).

A negative phototropism in response to increases in lateral FR reflected from neighbouring plants has been demonstrated in cucumber. This response is completely lacking in a phyB-deficient mutant, indicating that in addition to its role in shade-avoidance, phyB is also important for the detection of non-shading neighbours (Ballaré et al. 1992). The existence of additional phyB-type phytochromes with differing roles at different stages of development suggests that plants are still evolving to fine-tune their capacity for shade-avoidance and neighbour detection, and emphasizes the importance of these responses for the plant.

As the canopy closes or population density increases, increases in the leaf area index also occur, and as a result the light energy in the R and BL wavebands decreases. Under such circumstances, a plant may also be exposed to a lateral gradient in BL, and show a positive phototropic response. The phyB-deficient mutant of cucumber retains the ability to respond to such a gradient (Ballare et al. 1991), implying the action of a BL photoreceptor, which is most probably phototropin. Under conditions of deep shade, where it is likely that the observed growth responses result from a reduction in activity of several photoreceptors including the phyB-type phytochromes, cry1 and phototropin (Ballare et al. 1999). Without phyA however, seedlings cannot sense the FR transmitted through the canopy and do not de-etiolate, indicating that phyA may be essential for maintaining a degree of de-etiolation in highly competitive situations (Yanovsky et al. 1995).

Under shade conditions many plants may also increase their efficiency of light capture, by modifying various features of their photosynthetic organs. Depending on the species, this may include modification of light harvesting complex composition, chloroplast organization, orientation or position, and leaf shape, size or thickness (Terashima and Hikosaka 1995, Vogelmann et al. 1996, Mullineaux and Karpinski 2002). Acclimation to shade is therefore a complex phenomenon, but does at least in some cases involve responses that can be considered photomorphogenic. The phototropin photoreceptors have recently been shown to mediate BL control of chloroplast orientation and stomatal aperture (Sakai et al. 2001, Kinoshita et al. 2001), and may thus play an important role in the regulation of photosynthesis. However, in general the contribution of photomorphogenic photoreceptors to photosynthetic acclimation is not yet well understood.

6.2. Light and the seed habit

Other adaptive responses to light seem to have arisen with the development of the seed habit. Control of seed germination important to allow seedling to develop in favourable light environment. Particularly for small-seeded species germinating under the soil surface, seedlings must be able to emerge into full light before the seed energy reserves are exhausted. The extremely low fluences of light needed to induce germination of some species can be understood as a signal to the seed that it is near the soil surface before making the irreversible commitment to germinate. In other species, the higher light requirement and R/FR reversibility of germination may reflect a strategy to preferentially promote germination under gaps in the canopy (Casal and Sanchez 1998).

Investment of energy in the seed has allowed a period of time in which seedlings can develop independently of the need to photosynthesize. In effect this provides a longer period of time over which the seedling can integrate information about its light environment and adjust its development appropriately. In combination with the seed habit, many species have developed a growth strategy of etiolation in which they are able to suppress normal light-regulated leaf development and elongate rapidly growing in darkness. This could conceivably have been favoured in evolution because it increases the efficiency with which the plant uses stored seed reserves, but also because its rapid emergence into the light after germination will maximize competitive advantage. However, along with this strategy comes the need for anticipation of imminent emergence into the light environment, and a rapid response immediately following emergence. PhyA does appear to serve this purpose under natural conditions, and it is conceivable that some of the distinct features of phyA could have arisen in response to this pressure.

6.3. Avoidance or survival of unfavourable conditions

In addition to useful light for photosynthesis, sunlight also contains potentially damaging UV wavelengths. Various phenylpropanoid pigments including sinapates and flavonoids absorb UV and reduce its damaging effects on the plant (Bieza and Lois, 2001). The production of these pigments in many cases is strongly induced by light as a result of increased expression of certain genes in phenylpropanoid metabolism (Shirley, 1996, Ryan et al. 2001). In many cases, this induction is strongest in young seedlings and immature leaves, which are more susceptible to UV damage.

It is also possible that light can serve as an indirect signal of other adverse aspects of the environment. For example, the fact that in some species light can act to inhibit germination can be understood in terms of the need for a damp environment and a correlation between reduced water availability near the soil surface and increased light levels. A similar association could explain the negative phototropism of some roots.

A more complicated kind of correlative selection may underlie many seasonal responses. Factors such as temperature and water-availability can clearly become limiting at certain times of year in some environments, and many plants are able to avoid the deleterious effects of these seasonal extremes through suppression of normal growth and the adoption of various survival strategies. These include seed and bud dormancy, formation of storage organs, and initiation of flowering, which can be timed so as to allow the reproductive cycle to be completed before unfavourable conditions return. Changes in limiting factors of temperature and water availability can act directly as triggers for these changes (as seen in cold-temperature requirements for germination or flowering). However although seasonal, these factors are also subject to irregular short-term variation, whereas change in daylength is a much more constant and reliable indicator of season from year to year. In general, the degree of photoperiod responsiveness is an important aspect of adaptation to growth at a given latitude (Thomas and Vince-Prue, 1997).

7. Concluding remarks

Considering that plants are fixed in one place and dependent on light as an energy source, it is not surprising that sophisticated mechanisms have evolved enabling them to modify their development in response to light and thus to better compete with their neighbours. Persistent selective pressures for extracting more subtle information from the light environment have favoured the evolution of several distinct photoreceptor systems. For the most part, these systems act synergistically, increasing the general sensitivity of the plant to light. However, some photoreceptors have developed discrete light sensing abilities, which may be linked to specific physiological and ecological roles.

The past 20 years have seen major advances in our understanding of the photoreceptors involved in photomorphogenesis, and the definitive identification of some of the molecular and cellular processes required for expression of light responses. We still do not understand the complexities of signal transduction, or the network of interactions between light, plant hormones and other factors. Nevertheless, the genetic and molecular tools are now available to enable a thorough analysis of these aspects of photomorphogenesis over the coming years. Greater understanding of photomorphogenesis will bring an increased ability to manipulate plant light responses for practical purposes, such as in the control of density-dependent shading, flowering and yield in horticultural and crop plants. It will also help us to better understand the origins and adaptive significance of natural variation in growth habit and flowering phenology.

8. REFERENCES

Ahmad, M. (1999). Seeing the world in red and blue: insight into plant vision and photoreceptors. *Curr. Opin. Plant Biol. 2*, 230-235.

Ahmad, M. & Cashmore, A.R. (1994). *HY4* gene of *Arabidopsis thaliana* encodes a protein with characteristics of a blue-light photoreceptor. *Nature 366*, 162-166.

Ahmad, M., Jarillo, J.A., Smirnova, O. & Cashmore, A.R. (1998). The CRY1 blue light photoreceptor of *Arabidopsis* interacts with phytochrome A in vitro. *Mol.Cell 1*, 939-948.

Ballare, C.L., Scopel, A. L. & Sanchez, R.A. (1990). Far-red irradiation reflected from adjacent leaves: an early signal of competition in plant canopies. *Science 247*, 329-332.

Ballare, C.L., Scopel, A.L., Radosevich, S.R. & Kendrick, R.E. (1992). Phytochrome-mediated phototropism in de-etiolated seedlings. Occurrence and ecological significance. *Plant Physiol. 100*, 170-177.

Ballare, C.L., Scopel, A.L. & Sanchez, R.A. (1997). Foraging for light – photosensory ecology and agricultural implications. *Plant Cell Environ. 20*, 820-825.

Ballare, C.L. (1999) Keeping up with the neighbours: phytochrome sensing and other signalling mechanisms. *Trends Plant Sci. 4*, 97-102.

Barak, S., Tobin, E.M., Andronis, C., Sugano, S. & Green, R.M. (2000). All in good time: the *Arabidopsis* circadian clock. *Trends Plant Sci. 5*, 517-522.

Baskin, T.I. & Iino, M. (1987). An action spectrum in the blue and ultraviolet for phototropism in alfalfa. *Photochem. Photobiol 46*, 127-136.

Baum, G., Long, J.C. Jenkins, G.I. & Trewavas, A.J. (1999) Stimulation of the blue light phototropic receptor NPH1 causes a transient increase in cytosolic Ca2+. *Proc. Nat. Acad. Sci. USA 96*, 13554-13559.

Bernier, G. (1988). The control of flower evocation and morphogenesis, *Annu. Rev. Plant Physiol. Plant Mol. Biol. 39*, 175-219.

Bieza K. & Lois, R. (2001). An *Arabidopsis* mutant tolerant to lethal ultraviolet-B levels shows contsitutively elevated accumulation of flavonoids and other phenolics. *Plant Physiol. 126*, 1105-1115.

Briggs, W.R. & Huala, E (1999). Blue-light photoreceptors in higher plants. *Annu. Rev. Cell Develop. Biol. 15*, 33-62.

Bünning, E. (1964). *The physiological clock.* Springer Verlag, Berlin.

Butler, W.L., Hendricks, S.B. & Siegelman, H.W. (1964). Action spectra of phytochrome *in vitro. Photochem. Photobiol. 3*, 521-528.

Carr-Smith, H.D., Thomas, B. & Johnson, C.B. (1989). An action spectrum for the effect of continuous light on flowering in wheat. *Planta 179*, 428-432.

Casal, J.J. & Sanchez, R.A. (1998). Phytochromes and seed germination. *Seed Sci. Res. 8*, 317-329.

Casal, J.J. (2000). Phytochromes, cryptochromes, phototropin: photoreceptor interactions in plants. *Photochem. Photobiol. 71*, 1-11.

Cashmore, A.R., Jarillo, J.A., Wu, Y.J. & Liu, D.M. (1999). Cryptochromes: Blue light receptors for plants and animals. *Science 284*, 760-765.

Chamovitz, D.A. & Deng, X.W. (1997) The COP9 complex - a link between photomorphogenesis and general developmental regulation. *Plant Cell Environ. 20*, 734-739, 1997 Jun.

Childs, K.L., Miller, F.R., Cordonnier-Pratt, M.M., Pratt, L.H., Morgan, P.W. & Mullet, J.E. (1997). The sorghum photoperiod sensitivity gene, *Ma₃*, encodes a phytochrome B. *Plant Physiol. 113*, 611-619.

Christie, J.M., Salomon, M., Nozue, K., Wada, M. & Briggs, W.R. (1999). LOV (light, oxygen, or voltage) domains of the blue-light photoreceptor phototropin (nph1): Binding sites for the chromophore flavin mononucleotide *Proc. Nat. Acad. Sci. USA 96*, 8779-8783.

Christie, J.M. & Briggs, W.R. (2001). Blue light sensing in higher plants. *J. Biol. Chem.* 276, 11457-11460.

Clack, T., Mathews, S. & Sharrock, R.A. (1994). The phytochrome apoprotein family in *Arabidopsis* is encoded by five genes - the sequences and expression of *PHYD* and *PHYE*. *Plant Mol. Biol.* 25, 413-427.

Clough, R.C. & Vierstra, R.D. (1997). Phytochrome degradation. *Plant Cell Environ. 20*, 713-721.

Colon-Carmona, A., Chen, D.L., Yeh, K.C. Abel S. (2000). Aux/IAA proteins are phosphorylated by phytochrome in vitro. *Plant Physiol. 124*, 1728-1738.

Corbesier, L., Gadisseur, I., Silvestre, G., Jacqmard, A. & Bernier, G. (1996). Design in *Arabidopsis thaliana* of a synchronous system of floral induction by one long day. *Plant Journal. 9*, 947-952.

Coulter M.W. & Hamner K.C. (1964). Photoperiodic flowering response of Biloxi soybean in 72-hour cycles. *Plant Physiol. 39*, 846-856.

Deitzer, G.F., Hayes, R. & Jabben, M. (1982). Phase shift in the circadian rhythm of floral promotion by far-red light in *Hordeum vulgare* L. *Plant Physiol. 69*, 597-601.

Devlin, P.F., Patel, S.R. & Whitelam, G.C. (1998). Phytochrome E influences internode elongation and flowering time in *Arabidopsis. Plant Cell 10*, 1479-1488.

Devlin, P.F., Robson, P.R.H., Patel, S.R., Goosey, L., Sharrock, R.A. & Whitelam, G.C. (1999). Phytochrome D acts in the shade-avoidance syndrome in *Arabidopsis* by controlling elongation growth and flowering time. *Plant Physiol.* 119, 909-915.

Devlin, P.F. & Kay, S.A. (2000). Cryptochromes are required for phytochrome signaling to the circadian clock but not for rhythmicity. *Plant Cell 12*, 2499-2509.

Dieterle, M., Zhou, Y.C., Schäfer, E., Funk, M. & Kretsch, T. (2001). EID1, an F-box protein involved in phytochrome A-specific light signaling. *Genes Develop.* 15, 939-944.

Fairchild, C.D., Schumaker, M.A. & Quail, P.H. (2000). HFR1 encodes an atypical bHLH protein that acts in phytochrome A signal transduction. *Genes Develop.* 14, 2377-2391.

Fankhauser, C., Yeh, K.C., Lagarias, J.C., Zhang, H., Elich, T.D. & Chory, J. (1999). PKS1, a substrate phosphorylated by phytochrome that modulates light signaling in *Arabidopsis. Science 284.* 1539-1541.

Furuya, M. (1989). Molecular properties and biogenesis of phytochrome I and II. *Adv. Biophys. 25*, 133-167.

Garner, W.W. & Allard, A.H. (1920). Effect of the relative length of day and night and other factors of the environment on growth and reproduction in plants. *J. Agric. Res. 18*, 553-606.

Goosey, L., Palecanda, L. & Sharrock, R.A. (1997). Differential patterns of expression of the arabidopsis *PHYB, PHYD*, and *PHYE* phytochrome genes. *Plant Physiol. 115*, 959-969.

Green, R.M. & Tobin, E.M. (1999). Loss of the circadian clock-associated protein I in *Arabidopsis* results in altered clock-regulated gene expression. *Proc. Nat. Acad. Sci. USA 96*, 4176-4179.

Guo, H.W., Mockler, T., Duong, H. & Lin, C.T. (2001). SUB1, an *Arabidopsis* Ca^{2+}-binding protein involved in cryptochrome and phytochrome coaction. *Science*. 291, 487-490.

Guo, H.W., Yang, W.Y., Mockler, T.C. & Lin, C.T. (1998). Regulation of flowering time by *Arabidopsis* photoreceptors. *Science 279*, 1360-1363.

Harper, R.M., Stowe-Evans, E.L., Luesse, D.R., Muto, H., Tatematsu, K., Watahiki, M.K., Yamamoto, K. & Liscum, E. (2000). The *NPH4* locus encodes the auxin response factor ARF7, a conditional regulator of differential growth in aerial *Arabidopsis* tissue. *Plant Cell 12*, 757-770.

Hartmann, K.M. (1967). Ein Wirkungspectrum der Photomorphogenese unter Hochenergiebe-dingungen und seine Interpretation auf der Basis der Phytochroms (Hypokotylwachstumshemmung bei *Lactuca sativa* L.) *Z. Naturforsch. 22b*, 1172-1175.

Hauser, B.A., Cordonnier-Pratt, M.M., Daniel-Vedele, F. & Pratt, L.H. (1995). The phytochrome gene family in tomato includes a novel subfamily. *Plant Mol. Biol. 29*, 1143-1155..

Hauser, B.A., Pratt, L.H. & Cordonnier-Pratt, M.M. (1997). Absolute quantification of five phytochrome transcripts in seedlings and mature plants of tomato (*Solanum lycopersicum* L.). *Planta 201*, 379-387.

Hauser, B.A., Cordonnier-Pratt, M.M. & Pratt, L.H. (1998). Temporal and photoregulated expression of five tomato phytochrome genes. *Plant J. 14*, 431-439.

Hershey, H.P., Colbert, J.T., Lissemore, J.L., Barker, R.F. & Quail, P.H. (1984). Molecular cloning of cDNA for Avena phytochrome. *Proc. Natl. Acad. Sci. U.S.A. 81*, 2332-2336.

Hicks, K.A., Millar, A.J., Carre, I.A., Somers, D.E., Straume, M., Meeks-Wagner, D.R. & Kay, S.A. (1996). Conditional circadian dysfunction of the *Arabidopsis early-flowering 3* mutant. *Science 274*, 790-792.

Hisada, A., Hanzawa, H., Weller, J.L., Nagatani, A., Reid, J.B. & Furuya, M. (2000). Light-induced nuclear translocation of endogenous pea phytochrome A visualized by immunocytochemical procedures. *Plant Cell 12*, 1063-1078.

Hoecker, U., Tepperman, J.M. & Quail, P.H. (1999). SPA1, a WD-repeat protein specific to phytochrome A signal transduction. *Science 284*, 496-499.

Hsieh, H.L., Okamoto, H., Wang, M.L., Ang, L.H., Matsui, M., Goodman, H. & Deng, X.W. (2000). *FIN219*, an auxin-regulated gene, defines a link between phytochrome A and the downstream regulator COP1 in light control of *Arabidopsis* development. *Genes Develop. 14*, 1958-1970.

Huala, E., Oeller, P.W., Liscum, E., Han, I.S., Larsen, E. & Briggs, W.R. (1997). *Arabidopsis* NPH1 - a protein kinase with a putative redox-sensing domain. *Science 278*, 2120-2123.

Hudson, M., Ringli, C., Boylan, M.T. & Quail, P.H. (1999). The *FAR1* locus encodes a novel nuclear protein specific to phytochrome A signaling. *Genes Develop.* 13, 2017-2027.

Hudson, M.E. (2000). The genetics of phytochrome signalling in *Arabidopsis. Sem. Cell Develop.Biol.* 11, 475-483.

Jackson, S.D., Heyer, A., Dietze, J. & Prat S. (1996). Phytochrome B mediates the photoperiodic control of tuber formation in potato. *Plant J.* 9, 159-166.

Jenkins, G.I., Long, J.C., Wade, H.K., Shenton, M.R. & Bibikova, T.N. (2001). UV and blue light signalling: pathways regulating chalcone synthase gene expression in *Arabidopsis. New Phytol. 151*, 121-131.

Johnson, E., Bradley, M., Harberd, N.P. & Whitelam, G.C. (1994). Photoresponses of light-grown phyA mutants of *Arabidopsis* - phytochrome A is required for the perception of daylength extensions. *Plant Physiol. 105*, 141-149.

Kagawa, T., Sakai, T., Suetsugu, N., Oikawa, K., Ishiguro, S., Kato, T., Tabata, S., Okada, K. & Wada, M. (2001). *Arabidopsis* NPL1: A phototropin homolog controlling the chloroplast high-light avoidance response. *Science*. 291, 2138-2141.

Kanegae, H., Tahir, M., Savazzini, F., Yamamoto, K., Yano, M., Sasaki, T., Kanegae, T., Wada, M. & Takano, M. Rice *NPH1* homologues, *OsNPN1a* and *OsNPN1b*, are differently photoregulated. *Plant Cell Physiol.* 41, 415-423.

Kehoe, D.M. & Grossman, A.R. (1996). Similarity of a chromatic adaptation sensor to phytochrome and ethylene receptors. *Science 273*, 1409-1412.

Kerckhoffs, L.H.J., Schreuder, M.E.L., van Tuinen, A., Koornneef, M. & Kendrick, R.E. (1997). Phytochrome control of anthocyanin biosynthesis in tomato seedlings - analysis using photomorphogenic mutants. *Photochem. Photobiol.* 65, 374-381..

Kim, B.C., Tennessen, D.J. & Last, R.L. (1998). Uv-B-induced photomorphogenesis in *Arabidopsis thaliana. Plant J.* 15, 667-674.

Kinoshita, T., Doi, M., Suetsugu, N., Kagawa, T., Wada, M. & Shimazaki, K. (2001). phot1 and phot2 mediate blue light regulation of stomatal opening. *Nature. 414*, 656-660.

Kircher, S., Kozma-Bognar, L., Kim, L., Adam, E., Harter, K., Schafer, E. & Nagy, F. (1999). Light quality-dependent nuclear import of the plant photoreceptors phytochrome A and B. *Plant Cell 11*, 1445-1456.

Kleiner, O., Kircher, S., Harter, K., Batschauer, A. (1999). Nuclear localization of the Arabidopsis blue light receptor cryptochrome 2. *Plant J. 19*, 289-296.

Kohchi, T., Mukougawa, K., Frankenberg, N., Masuda, M., Yokota, A. & Lagarias, J.C. (2001). The *Arabidopsis HY2* gene encodes phytochromobilin synthase, a ferredoxin-dependent biliverdin reductase. *Plant Cell 13*, 425-436.

Koornneef, M., Rolff, E. & Spruitt, C.J.P. (1980). Genetic control of light-inhibited hypocotyl elongation in *Arabidopsis thaliana* L. Heynh. *Z. Pflanzenphysiol. 100*, 147-160.

Kuno, N. & Furuya, M. (2000). Phytochrome regulation of nuclear gene expression in plants. *Sem. Cell Develop. Biol. 11*, 485-493.

Lin, C., Robertson, D.E., Ahmad, M., Raibekas, A.A., Jorns, M.S., Dutton, P.L. & Cashmore, A.R. (1995). Association of flavin adenine dinucleotide with the Arabidopsis blue light receptor cry1. *Science 269*, 968-970.

Lin, C., Yang, H.Y., Guo, H.W., Mockler, T., Chen, J. & Cashmore, A.R. (1998). Enhancement of blue-light sensitivity of *Arabidopsis* seedlings by a blue light receptor cryptochrome 2. *Proc. Nat. Acad. Sci. USA 95*, 2686-2690.

Liscum, E. & Briggs, W.R. (1995). Mutations in the *nph1* locus of *Arabidopsis* disrupt the perception of phototropic stimuli. *Plant Cell 7*, 473-485.

Liscum, E. & Briggs, W.R. (1996). Mutations of *Arabidopsis* in potential transduction and response components of the phototropic signaling pathway. *Plant Physiol. 112*, 291-296.

Liscum, E. Stowe-Evans, E.L. (2000). Phototropism: A "simple" physiological response modulated by multiple interacting photosensory-response pathways. *Photochem. Photobiol. 72*, 273-282.

Malhotra, K., Kim, S.T., Batschauer, A., Dawut. L. & Sancar, A. (1995). Putative blue-light photoreceptors from *Arabidopsis thaliana* and *Sinapis alba* with a high degree of sequence homology to DNA photolyase contain the two photolyase cofactors but lack DNA repair activity. *Biochem. 34*, 6892-6899.

Martinez-Garcia, J.F., Huq, E. & Quail, P.H. (2000). Direct targeting of light signals to a promoter element-bound transcription factor. *Science 288*, 859-863.

Mathews, S. & Sharrock, R.A. (1997). Phytochrome gene diversity. *Plant Cell Environ.* 20, 666-671.

Mazzella, M.A., Magliano, T.M.A. & Casal, J.J. (1997). Dual effect of phytochrome A on hypocotyl growth under continuous red light. *Plant Cell Environ. 20*, 261-267.

McCormac, A.C., Whitelam, G.C., Boylan, M.T., Quail, P.H. & Smith, H. (1992). Contrasting responses of etiolated and light-adapted seedlings to red: far-red ratio: a comparison of wild type, mutant and transgenic plants has revealed differential functions of members of the phytochrome family. *J. Plant Physiol. 140*, 707-714.

McWatters, H.G., Bastow, R.M., Hall, A. & Millar, A.J. (2000). The ELF3 zeitnehmer regulates light signalling to the circadian clock. *Nature 408*, 716-720.

Mockler, T.C., Guo, H. Yang, H., Duong, H. & Lin, C. (1999). Antagonistic actions of *Arabidopsis* cryptochromes and phytochrome B in the regulation of floral induction. *Development 126*, 2073-2082.

Motchoulski, A. & Liscum, E. (1999). *Arabidopsis* NPH3: a NPH1 photoreceptor-interacting protein essential for phototropism. *Science 286*, 961-964.

Mullineaux, P. & Karpinski, S. (2002). Signal transduction in response to excess light: getting out of the chloroplast. *Curr. Opin. Plant Biol. 5*, 43-48.

Muramoto, T., Kohchi, T., Yokota, A., Hwang, I.H. & Goodman, H.M. (1999). The *Arabidopsis* photomorphogenic mutant *hy1* is deficient in phytochrome chromophore biosynthesis as a result of a mutation in a plastid heme oxygenase. *Plant Cell 11*, 335-347.

Mustilli, A.C. & Bowler, C. (1997). Tuning in to the signals controlling photoregulated gene expression in plants. *EMBO J. 16*, 5801-5806.

Nelson, D.C., Lasswell, J., Rogg, L.E., Cohen, M.A. & Bartel, B. (2000). *FKF1*, a clock-controlled gene that regulates the transition to flowering in *Arabidopsis*. *Cell 101*, 331-340.

Osterlund, M.T. & Deng, X.W. (1998). Multiple photoreceptors mediate the light-induced reduction of GUS-COP1 from *Arabidopsis* hypocotyl nuclei. *Plant J. 16*, 201-208.

Osterlund, M.T., Hardtke, C.S., Wei, N. & Deng, X.W. (2000). Targeted destabilization of HY5 during light-regulated development of Arabidopsis. *Nature 405*, 462-466.

Park, D.H., Somers, D.E., Kim, Y.S., Choy, Y.H., Lim, H.K., Soh, M.S., Kim, H.J., Kay, S.A. & Nam, H.G. (1999). Control of circadian rhythms and photoperiodic flowering by the Arabidopsis *GIGANTEA* gene. *Science 285*, 1579-1582.

Parker, M.W., Hendricks, S.B., Borthwick, H.A. & Went, F.W. (1949). Spectral sensitivities for stem and leaf growth of etiolated pea seedlings and their similarity to action spectra for photoperiodism. *Am. J. Bot. 36*, 194-204.

Parks, B.M, & Quail, P.H. (1991). Phytochrome-deficient *hy1* and *hy2* long hypocotyl mutants of Arabidopsis are defective in phytochrome chromophore biosynthesis. *Plant Cell 3*, 1177-1186.

Parks, B.M., Quail, P.H. & Hangarter, R.P. (1996). Phytochrome A regulates red-light induction of phototropic enhancement in Arabidopsis. *Plant Physiol. 110*, 155-162.

Parks, B.M., Cho, M.H. & Spalding, E.P. (1998). Two genetically separable phases of growth inhibition induced by blue light in Arabidopsis seedlings. *Plant Physiol. 118*, 609-615.

Poppe, C., Sweere, U., Drumm-Herrel, H. & Schäfer E. (1998). The blue light receptor cryptochrome 1 can act independently of phytochrome A and B in *Arabidopsis thaliana. Plant J. 16*, 465-471.

Quail, P.H. (1994). Phytochrome genes and their expression. In *Photomorphogenesis in plants*, 2nd ed. (RE Kendrick, GHM Kronenberg, eds), pp 71-104. Kluwer Academic Publishers, Dordrecht.

Quail, P.H. (1997). An emerging molecular map of the phytochromes. *Plant Cell Environ. 20*, 657-665.

Quail, P.H. (2000). Phytochrome-interacting factors. *Sem. Cell Develop. Biol. 11*, 457-466.

Reed, J.W., Nagatani, A., Elich, T.D., Fagan, M. & Chory, J. (1994). Phytochrome A and phytochrome B have overlapping but distinct functions in Arabidopsis development. *Plant Physiol. 104*, 1139-1149.

Robson, P.R.H., McCormac, A.C., Irvine, A.S. & Smith H. (1996). Genetic engineering of harvest index in tobacco through overexpression of a phytochrome gene. *Nature Biotechnol.* 14,995-998.

Ryan, K.G., Swinny, E.E., Winefield, C. & Markham, K.R. (2001). Flavonoids and UV photoprotection in Arabidopsis mutants. *Z. Naturforsch. 56*, 745-754.

Sage, L.C. (1992). *Pigment of the imagination: a history of phytochrome research*. Academic Press, New York.

Sakai, T., Wada, T., Ishiguro, S. & Okada, K. (2000). RPT2: A signal transducer of the phototropic response in Arabidopsis. *Plant Cell 12*, 225-236.

Sakai, T., Kagawa, T., Kasahara, M., Swartz, T.E., Christie, J.M., Briggs, W.R., Wada, M. & Okada, K. (2001). Arabidopsis nph1 and npl1: Blue light receptors that mediate both phototropism and chloroplast relocation *Proc. Nat. Acad. Sci. USA 98*, 6969-6974.

Samach, A. & Coupland, G. (2000). Time measurement and the control of flowering in plants. *Bioessays. 22*, 38-47.

Samach, A., Onouchi, H., Gold, S.E., Ditta, G.S., Schwarz-Sommer, Z., Yanofsky, M.F. & Coupland, G. (2000). Distinct roles of CONSTANS target genes in reproductive development of Arabidopsis. *Science 288*, 1613-1616.

Schaffer, R., Ramsay, N., Samach, A., Corden, S., Putterill, J., Carre, I.A. & Coupland, G. (1998). The *late elongated hypocotyl* mutation of Arabidopsis disrupts circadian rhythms and the photoperiodic control of flowering. *Cell 93*, 1219-1229.

Schneider-Poetsch, H.A.W., Kolukisaoglu, Ü., Clapham, D.H., Hughes, J. & Lamparter, T. (1998). Non-angiosperm phytochromes and the evolution of vascular plants. *Physiol. Plant. 102*, 612-622.

Schultz, T.F., Kiyosue, T., Yanovsky, M., Wada, M. & Kay, S.A. (2001). A role for LKP2 in the circadian clock of Arabidopsis. *Plant Cell 13*, 2659-2670.

Schwechheimer, C., Serino, G., Callis, J., Crosby, W.L., Lyapina, S., Deshaies, R.J., Gray, W.M., Estelle, M. & Deng, X.W. (2001). Interactions of the COP9 signalosome with the E3 ubiquitin ligase SCF/TIR1 in mediating auxin response. *Science 292*, 1379-1382.

Shinomura, T., Nagatani, A., Hanzawa, H., Kubota, M., Watanabe, M. & Furuya, M. (1996). Action spectra for phytochrome A- and B-specific photoinduction of seed germination in *Arabidopsis thaliana. Proc. Nat. Acad. Sci. USA 93*, 8129-8133.

Shinomura, T., Uchida, K., Furuya, M. (2000). Elementary processes of photoperception by phytochrome A for high-irradiance response of hypocotyl elongation in Arabidopsis. *Plant Physiol. 122*, 147-156.

Shirley, B.W. (1996). Flavonoid biosynthesis - new functions for an old pathway. *Trends Plant Sci. 1*, 377-382.

Short, T.W. & Briggs, W.R. (1994). The transduction of blue light signals in higher plants. *Annu. Rev. Plant Physiol. Plant Mol. Biol. 45*, 143-171.

Smith H (1995) Physiological and ecological function within the phytochrome family. *Annu. Rev. Plant Physiol. Plant Mol. Biol. 46*, 289-315.

Smith, H. & Whitelam, G.C. (1997). The shade avoidance syndrome - multiple responses mediated by multiple phytochromes. *Plant Cell Environ. 20*, 840-844.

Smith, H. (2000). Phytochromes and light signal perception by plants - an emerging synthesis. *Nature 407*, 585-591.

Somers, D.E., Devlin, P.F. & Kay, S.A. (1998a). Phytochromes and cryptochromes in the entrainment of the Arabidopsis circadian clock. *Science 282*,1488-1490.

Somers, D.E., Webb, A.A.R., Pearson, M. & Kay, S.A. (1998b). The short-period mutant, *toc1-1*, alters circadian clock regulation of multiple outputs throughout development in *Arabidopsis thaliana. Development 125*, 485-494.

Somers, D.E., Schultz, TF., Milnamow, M. & Kay, S.A. (2000) *ZEITLUPE* encodes a novel clock-associated PAS protein from Arabidopsis. *Cell 101*, 319-329.

Strayer, C., Oyama, T., Schultz, T.F., Raman, R., Somers, D.E., Mas, P., Panda, S., Kreps, J.A. & Kay, S.A. (2000). Cloning of the Arabidopsis clock gene *TOC1*, an autoregulatory response regulator homolog. *Science 289*, 768-771.

Suarez-Lopez, P., Wheatley, K., Robson, F., Onouchi, H., Valverde, F. & Coupland, G. (2001). CONSTANS mediates between the circadian clock and the control of flowering in Arabidopsis. *Nature 410*, 1116-1120.

Takano, M., Kanegae, H., Shinomura, T., Miyao, A., Hirochika, H. & Furuya M. (2001). Isolation and characterization of rice phytochrome A mutants. *Plant Cell 13*, 521-534.

Terashima, I. & Hikosaka, K. (1995). Comparative ecophysiology of leaf and canopy photosynthesis. *Plant Cell Environ. 18*, 1111-1128.

Terry, M.J., Wahleithner, J.A. & Lagarias, J.C. (1993). Biosynthesis of the plant photoreceptor phytochrome. *Arch. Biochem. Biophys. 306*, 1-15.

Thomas, B. & Vince-Prue, D. (1987). *Photoperiodism in Plants* (2nd edn.), Academic Press, London.

Torii, K.U. & Deng, X.W. (1997). The role of COP1 in light control of Arabidopsis development. *Plant Cell Environ. 20*, 734-739.

Toth, R., Kevei, E., Hall, A., Millar A.J., Nagy, F. & Kozma-Bognar, L. (2001). Circadian cloc-regulated expression of phytochrome and cryptochrome genes in Arabidopsis. *Plant Physiol. 127*, 1607-1616.

Vierstra, R.D. & Davis, S.J. (2000). Bacteriophytochromes: new tools for understanding phytochrome signal transduction. *Sem. Cell Develop. Biol. 11*, 511-521.

Vogelmann, T.C., Nishio, J.N. & Smith, W.K. (1996). Leaves and light capture – light propagation and gradients of carbon fixation within leaves. *Trends Plant Sci. 1*, 65-70.

Wang, H.Y., Ma, L.G., Li, J.M., Zhao, H.Y. & Deng, X.W. (2001). Direct interaction of Arabidopsis cryptochromes with COP1 in light control development. *Science 294*, 154-158.

Wang, Z.Y. & Tobin, E.M. (1998). Constitutive expression of the *CIRCADIAN CLOCK-ASSOCIATED 1* (CCA1) gene disrupts circadian rhythms and suppresses its own expression. *Cell 93*, 1207-1217.

Weller, J.L., Murfet, I.C. & Reid, J.B. (1997a). Pea mutants with reduced sensitivity to far-red light define an important role for phytochrome a in day-length detection. *Plant Physiol. 114*, 1225-1236.

Weller, J.L., Reid, J.B., Taylor, S.A. & Murfet, I.C. (1997b). The genetic control of flowering in pea. *Trends Plant Sci. 2*, 412-418.

Weller, J.L., Terry, M.J., Reid, J.B. & Kendrick, R.E. (1997c). The phytochrome-deficient *pcd2* mutant of pea is unable to convert biliverdin IXα to 3(Z)-phytochromobilin. *Plant J. 11*, 1177-1186.

Weller, J.L., Beauchamp, N., Kerckhoffs, L.H.J., Platten, J.D. & Reid, J.B. (2001a). Interaction of phytochromes A and B in the control of de-etiolation and flowering in pea. *Plant J. 26*, 283-294.

Weller, J.L., Perrotta, G., Schreuder, M.E.L., van Tuinen, A., Koornneef, M., Giuliano, G. & Kendrick, R.E. (2001b). Genetic dissection of blue-light sensing in tomato using mutants deficient in cryptochrome 1 and phytochromes A, B1 and B2. *Plant J. 25*, 427-440. Went, F.W. (1941). Effects of light on stem and leaf growth. *Am. J. Bot. 28*, 83-95.

Withrow, R.B., Klein, W.H. & Elstad, V. (1957). Action spectra of photomorphogenic induction and its inactivation. *Plant Physiol. 32*, 453-462.

Yang, H.Q., Wu, Y.J., Tang, R.H., Liu, D.M., Liu, Y. & Cashmore, A.R. (2000). The C-termini of *Arabidopsis* cryptochromes mediate a constitutive light response. *Cell. 103*, 815-827.

Yano, M., Kojima, S., Takahashi, Y., Lin, H.X. & Sasaki, T. (2001). Genetic control of flowering time in rice, a short-day plant. *Plant Physiol. 127*, 1425-1429.

Yanovsky, M.J., Casal, J.J. & Whitelam, G.C. (1995). Phytochrome A, phytochrome B and HY4 are involved in hypocotyl growth responses to natural radiation in Arabidopsis - weak de-etiolation of the *phyA* mutant under dense canopies. *Plant Cell Environ. 18*, 788-794.

Yanovsky, M.J., Mazzella, M.A. & Casal, J.J. (2000a). A quadruple photoreceptor mutant still keeps track of time. *Curr. Biol. 10*, 1013-1015.

Yanovsky, M.J., Izaguirre, M., Wagmaister, J.A., Gatz, C., Jackson, S.D., Thomas, B. & Casal, J.J. (2000b). Phytochrome A resets the circadian clock and delays tuber formation under long days in potato. *Plant J. 23*, 223-232.

Yeh, K.C. & Lagarias, J.C. (1998). Eukaryotic phytochromes - light-regulated serine/threonine protein kinases with histidine kinase ancestry. *Proc. Nat. Acad. Sci. U.S.A. 95*, 13976-13981.

Zhu, Y.X., Tepperman, J.M., Fairchild, C.D. & Quail, P.H. (2000). Phytochrome B binds with greater apparent affinity than phytochrome A to the basic helix-loop-helix factor PIF3 in a reaction requiring the PAS domain of PIF3. *Proc. Nat. Acad. Sci. U.S.A. 97*, 13419-13424.

ANDERS JOHNSSON AND WOLFGANG ENGELMANN

15. THE BIOLOGICAL CLOCK AND ITS RESETTING BY LIGHT

1. OVERVIEW

This chapter presents *the role light plays in synchronizing biological clocks with the 24 hour cycles of the earth*. We will first characterize the different clocks used by organisms. The functions and properties of these clocks are mentioned. In the following section we explain how light synchronizes circadian (that is close to 24 hour) clocks as well as other effects of light on circadian rhythms. Models and mechanisms of circadian clocks are discussed to understand the effects of light. The main part of the chapter presents several examples of organisms and their circadian systems. Man and mice are chosen as representatives of mammals. Furthermore we discuss the insect *Drosophila*, some plants including *Arabidopsis*, the unicellular dinoflagellate *Gonyaulax*, the ascomycetal fungus *Neurospora* and a cyanobacterium. In selecting these examples we want to show the general occurrence of circadian rhythms in almost all organisms and the similarities and differences in the effects of light and the mechanisms of the circadian clocks used by them. It will also become obvious that there are many questions open as is typical for a rapidly progressing research area.

General literature on the subject can be found in Roenneberg and Foster (1997), Dunlap (1999), Lakin-Thomas (2000), King and Takahashi (2000). More detailed reviews are mentioned in the different sections.

2. BIOLOGICAL CLOCKS

The daily revolutions of the earth around its axis and its annual course around the sun are responsible for day and night and for the seasons with their fluctuations in daylength and temperature. Most organisms have adapted to these diurnal and annual cycles. The strategies and mechanisms used are partly quite delicate and complicated. Direct effects of light on organisms such as in photosynthesis, phototaxis, plant morphogenesis, vision, DNA inactivation and others are reported in other chapters of this part B of the book. It is obvious that photosynthesis in green plants occurs during the daytime and understandable that other processes which are inhibited by light or by the products of photosynthesis such as O_2 are restricted to the night. N_2 fixation in *cyanobacteria* is an example.

It came as a surprise, that these and many other events are, however, *in addition controlled by internal (or 'endogenous') clocks*. Thus, photosynthesis fluctuates not only during the daily LD cycle (light-dark change; LD 12:12 means 12 hours of light

335

L.O. Björn (ed.), Photobiology, 335–387.
© 2002 *Kluwer Academic Publishers. Printed in the Netherlands.*

followed by 12 hours of darkness), but also when the plants are kept under LL (continuous light) and constant temperature (Hennessey and Field (1991). The period length (called period for short in the following) of this rhythmic event is typically not exactly 24 hours, but close to it and therefore called *'circadian'* (*circa* Latin for about, *dies* Latin for day). If in the absence of LD- and temperature cycles other 24 hour time cues (also called Zeitgeber, German for time giver) would control the rhythm, it should show an exact 24 hour rhythm. This is not the case, demonstrating the endogenous nature of a controlling clock that is sensitive to light signals.

2.1. Spectrum of Rhythms

Endogenous rhythms of organisms are not only tuned to the daily cycle of 24 hours. The range of rhythms found in organisms covers *ultradian* (with periods of several hours to very short ones), *circadian* and *annual* (with periods of about a year) rhythms. Other rhythms such as tidal, fourteen day and monthly ones are due to influences of the moon on the earth, mainly on the water movements of the oceans. Organisms at the coasts and in the sea have often adapted to them. There are also other long period rhythms.

Annual rhythms are subject to the daylength changes during the year and are the content of a special chapter (see Chapter 14 on photoperiodism). The following discussion of the 'biological clock' is restricted to circadian rhythms with periods around 24 hours. Even they are often not just one clock type, but a 'circadian system' consisting of two or more clocks with different properties which are or are not coupled mutually.

2.2. Function of Clocks

The term 'clock' usually implies a time measuring device or function. For instance, the daylength (or night length) can be determined by an organism. Since daylength is a function of the time of the year (long days in summer, short days in winter), it can be used to time certain events such as tuber formation or flowering of a plant or breeding of birds and mammals during the most appropriate season (see chapter 16).

However, a clock can also be used to set a certain temporal order. For instance, the circadian control of our sleep-wake cycle ensures that we rise in the morning and fall asleep in the evening around a certain preferred time. Food intake and digestion are likewise controlled by this clock and gated to certain times of the day (Forsgren 1935). *Because of its endogenous nature the circadian clock will time these events properly also under constant conditions.*

Furthermore, circadian clocks can serve as alarm clocks. They tell the organism certain times of the day which are important. For instance, the time sense of insects such as bees allows them to visit the flowers of a plant at the time it offers nectar and/or pollen. Or from the standpoint of the plant, attracting certain insects is more efficient if timed to the active period of them. If flowers open at night pollination by specialized butterflies or bats is facilitated. Evolution has worked on the plant and the pollinator to bring about this delicate interplay controlled by circadian clocks.

Alarm clocks might also exist in man. Some humans are able to wake up at a certain time of the night without external help by relying on a 'head clock'. Although not tested yet, it is likely that this alarm device uses the circadian clock.

Circadian clocks can furthermore be used for orientation purposes. If the direction of the sun is used for orientation by insects, birds, fishes and other animals, the changing position of the sun during the day has to be taken into account. Therefore sun compass orientation needs an internal time reference. The circadian clock is used as such.

2.3. Adaptive Significance and Evolutionary Aspects of Circadian Clocks

The different functions of circadian clocks just mentioned are surely not the only reasons why they evolved. Winfree (1986) and others have discussed, that early in evolution circadian clocks might have served to protect organisms from adverse effects of light. *Circadian timing and light reception might have co-evolved* and even preceded the evolution of specialized photoreceptors and eyes (see discussion by Herzog and Block (1999). Homologies between pacemaking molecules and ancient photopigments from fungi to mammals suggest an evolutionary link between modern clock proteins and ancient light sensing proteins (Crosthwait *et al.* 1997). However, this is difficult to prove. It would be interesting to know whether primitive eyes (for instance eye spots) contain circadian pacemaker cells. Among vertebrates, retinal clocks seem to be quite ancient (lamprey, Menaker *et al.* 1997).

The adaptive significance of possessing a circadian clock was demonstrated in *cyanobacteria* by using mutants with different periods in competition with each other and with the wild strain. They were exposed to LD cycles deviating from the normal 24 hour rhythm. Short period mutants crowded out the wild type and the long period mutants in short LD cycles (of 22 hours), long period mutants crowded out the others in long LD cycles (of 26 hours). An arrhythmic mutant was out-competed by the wild strain under all LD cycles (Johnson *et al.* 1998, Ouyang *et al.* 1998).

The vertebrates show a wide evolutionary variety in their circadian system. They possess a 'circadian axis' (retinas, pineal, suprachiasmatic nucleus) with circadian oscillators. In mammals the pineal as part of this axis does not contain a circadian oscillator. Mammals lack also extraretinal and extrapineal circadian photoreceptors in contrast to other vertebrates. Menaker *et al.* (1997) discuss a 'nocturnal bottleneck' that could have led to the evolution of mammals and their exceptional circadian system.

2.4. Properties and Formal Structure of the Circadian System

Besides being found in almost all living beings, from prokaryots to higher organisms, circadian clocks possess a number of formal properties as follows:

• They have a period of roughly 24 hours (about 18 and 28 hours in extremes) under constant conditions.

• They are entrainable by time cues (mainly light and temperature changes) to 24 hours.
• Their period is only slightly changed by temperature (if constant).
• They function at the cellular level and are heritable.
• They are of advantage to the clock bearer.

If, for instance, the plant *Kalanchoe blossfeldiana* is kept under constant weak green light conditions, the period of the opening and closing of the four petals of the flowers amounts to 22 hours at a temperature of 22°C. If exposed to an LD 12:12, the flowers open during the light period and close during the dark period. The period of the cycle is now exactly 24 hours. Under constant conditions the 'free run' is 21.9 hours at 15°C, 22.3 hours at 20°C, and 21.3 hours at 25°C (Oltmanns 1960). The differences in period are quite small compared to the influences temperature normally has on chemical and biochemical reactions.

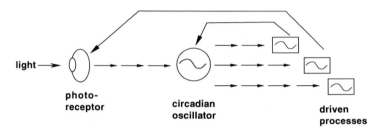

light →

photo-
receptor

circadian
oscillator

driven
processes

Figure 1. Schematic presentation of the circadian system consisting of a photoreceptor unit which receives environmental light, a transduction pathway to the circadian oscillator, output pathways to the driven processes. Some of the clock controlled processes (e.g. locomotor activity in an animal) might feed back to the clock. Other driven processes might feed back to the photoreceptor unit thus modulating its sensitivity in a circadian fashion.

Mutants of *Drosophila* and other organisms are known which differ in clock properties. For instance, the per^8 locomotor activity has a period of 19.5 hours compared to 24.4 for the wild type, and the per^L period amounts to 28.6 hours. Other mutants are arrhythmic (per^0). Any useful model of the circadian system has to take these properties into account and has to offer mechanisms which lead to the long circadian period of about 24 hours, to the low temperature dependence of period, to ways of synchronizing the rhythms to the 24 hour time cues. One of the main aims are to unravel the genetic foundation of the mechanisms underlying these circadian oscillations. The circadian system is often depicted in a way shown in Figure 1. Light excites a photoreceptor. The photochemical changes lead to a signal chain which feeds into a circadian oscillator ('clockwork'). This oscillator in turn controls the output signals via a sequence of reactions. The clock can also affect the photoreceptor system and the oscillator via feedback from an output (Figure 1). In this way they can adapt better to the prevailing light conditions and can protect the receptors from being damaged by strong light. For instance, in numerous plants leaves move in a circadian pattern up and down which continuously changes the amount of light perceived by the photoreceptors in the leaves. Or, to give an example for animals, the retinal cells in the eyes are protected efficiently during the daytime (and also during the subjective daytime under constant conditions) by a circadian clock which controls the movement of shielding pigments (Fleissner and Fleissner (2002a). Renewal of photoreceptor units is also often under circadian control (see disk shedding, subsection 7.3). In models of circadian rhythms *the concepts of feedback and of time delay* are frequently used (see Figure 1 and section 4).

3. SYNCHRONIZATION OF CLOCKS BY LIGHT AND OTHER TIME CUES

A biological clock has to be synchronized to the environmental cycle, which is 24 hours long in the case of circadian clocks. This is necessary, since circadian clocks typically deviate in their period from 24 hours. Time cues of the environment are effective for this synchronization. *The LD cycle is the most frequently used time cue*, but temperature rises or temperature drops can also be Zeitgeber. In the circadian control of conidiation of *Neurospora crassa* cycles of moderate temperature changes are even more effective in synchronizing as compared to LD cycles (Liu *et al.* 1998). In animals social cues and other signals can synchronize.

The dominant role of light in synchronizing might be due to the high reliability of this Zeitgeber, whereas temperature changes during day and night are less reliable. However, the begin of the light period and correspondingly of the dark period does not occur at the same time of the day during the course of the year. During the summer the light period is longer than during the winter. This is quite obvious at higher latitudes. This fact has to be taken into account by the organisms if light is to be used for synchronization.

3.1. Photoreceptors of Circadian Clocks and Effective Light

Photoreceptors are needed to perceive the light as an input signal for resetting the circadian clocks. Depending on the organism these receptors can be quite diverse. In many unicellulars, such as yeast or most algae such as the dinoflagellate *Gonyaulax,* no special receptor structures have been found. Instead pigment molecules in the cell are changed by light and a transduction chain finally resets the clock. In animals specialized light receptive organs are used such as the vertebrate eyes or the compound eyes in insects. But often extraretinal photoreceptors serve to perceive the synchronizing light either in addition to or instead of the usual eyes. For instance, in birds the pineal organ is light sensitive and synchronizes the circadian rhythm if the eyes are obscured or denervated or removed. In *Drosophila* flies, extraretinal structures in the brain (Buchner Hofbauer eye, dorsal neurosecretory cells) have been shown to serve as additional devices for synchronization.

Photopigments like phytochrome, cryptochrome, opsins and others synchronize circadian rhythms. Properties and functions of these pigments are described under the examples for circadian rhythms.

Depending on the kind of photoreceptor, different wavelengths are more or less effective in resetting the circadian clock. Using varying fluence rates of colored light, action spectra can be obtained (see Chapter 6) which tell us how many photons of the different wavelengths are needed in order to evoke the same effect (see next section). The effect of light depends, however, not only on the *wavelength* and *fluence rate*, but *also on the phase of the circadian clock at which the light was given* (see next subsection).

3.2. Entrainment

Light or other Zeitgeber entrain the circadian clock of organisms, provided they are applied in a 24 hour cycle or in a cycle close to 24 hours. The *organism is synchronized*, and the *Zeitgeber has entrained the rhythm*. Organisms kept under constant conditions without synchronizing time cues of the environment show '*free run*' with a period of their circadian rhythm usually deviating from 24 hours. We should now look more closely at this entraining effect of light.

If an organism such as a *Kalanchoe* plant is kept for some days in an air conditioned chamber with 12:12 h LD and after the last 12 hours of light transferred to constant darkness, the circadian opening and closing of the flowers will continue to run with its characteristic period going through subjective day and night cycles.
A light pulse would shift this rhythm or not, depending on the phase of the clock a which the pulse is applied. If given before the subjective midnight point, the rhythm will be delayed, if given after this point, the rhythm will be advanced. During the subjective day period there is normally a time span where a light pulse is without effect on the rhythm (called 'dead zone'). These reactions are to be expected, because an organism receiving light in its early night phase would 'assume' (used of course in a figurative sense) that it is still in the light phase and consequently delay its circadian rhythm in order to adjust it to the 24 hour pattern. On the other hand, receiving light in its late night phase, the organism would 'assume' it is already in the light phase and therefore adjust its clock by advancing. Light during the normal day phase, however, should not shift the clock.

3.3 Phase Response Curves

These phase shifts can be plotted in respect to magnitude and direction by a *phase response curve*. They are based on experiments with light pulses administered at different phases, and examples are given in Figure 2. The phase response curves of the locomotor activity rhythm of a Syrian hamster and of the eclosion rhythm of *Drosophila* flies (a population rhythm) to light pulses are shown and explained in Figure 3. Photic phase response curves are similar in all mammals, nocturnal as well as diurnal, including man. However, they differ in detail such as the amplitude and duration of the advance and delay portion in different species (Rusak and Zucker 1979). This allows to adjust the phase and period of the circadian clock to the 24 hours day.

It is assumed that synchronization of circadian rhythms to an LD cycle is brought about by light effects similar to the light pulse actions, although under normal conditions light is impinging throughout the day phase. A skeleton period where the onset of light is marked by a short light pulse and the end of the light period by another light pulse 12 hours later has the same synchronizing effect as a complete photoperiod consisting of 12 hours continuous light (Pittendrigh 1965).

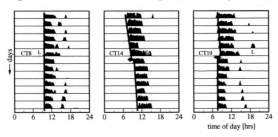

Figure 2. Phase shifting of the locomotor activity rhythm of a Syrian hamster by a brief light exposure at three different phases at CTs 8, 14, 19, (see arrows) (circadian time CT 0 is the time at which in a normal LD 12:12 cycle the light period would have started) is shown in the actograms representing the locomotor activity rhythm of a Syrian hamster. No phase shift is observed at CT8, the subjective day (hamsters are night active). A light pulse at CT 14 delays, a light pulse at CT 19 advances the rhythm. Modified after Rea (1998).

3.4. Range of Entrainment

If LD is applied not in a 24 hour pattern, but for instance in a 20 hour pattern (for instance LD 10:10), most circadian clocks are still entrained. This holds also for driving cycles with periods exceeding 24 hours, such as an LD 13:13 (that is a 26 hour cycle). The *range of entrainment* in which the circadian rhythm of the organism can follow the LD pattern depends on the species and the fluence rate and duration of the light pulse. It usually is between 18 and 30 hours (Enright 1965).

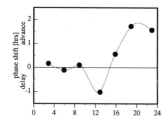 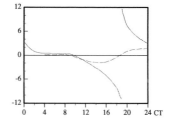

Figure 3. Phase response curves for shifting circadian rhythms with light pulses. Left: The results of phase shifts by light such as shown in Figure 2 lead to a response curve which is of a weak type and typical for Syrian hamster and other rodents. Only small phase shifts are observed even with high fluence rate light pulses. By convention the advanced phase shifts (occurring at late night and early morning) are plotted upward, the delay phase shifts (occurring at early night) downward. Modified after Rea (1998). Right: Phase shifting effect of light pulses in the eclosion rhythm of Drosophila flies (a population rhythm). Cultures of Drosophila larvae were reared under LD 12:12 cycles and after pupation transferred into weak red light (safe light). Portions of the pupae were treated at different times with a short blue light pulse (0.23 mW/cm_). The resulting delay respectively advance phase shifts of the eclosion rhythm are depicted as a function of the phase at which the light pulse was administered. The dashed small amplitude curve is a weak phase response curve with phase shifts up to 2 to 3 hours only and obtained with short light pulses, the solid large amplitude curve a strong phase response curve with phase shifts up to 12 hours and obtained with longer light pulses Note different scales of the vertical axes.

3.5. Arrhythmicity

Circadian rhythms might damp out under certain environmental conditions such as LL and/or DD (continuous darkness) or at too high or too low temperatures (for instance in the rhythm of the crassulacean acid metabolism of *Kalanchoe*). There is apparently a permissive range allowing circadian rhythms to occur.

However, arrhythmicity can also be induced by a special light pulse which is applied at a certain phase. The rhythm disappears after this pulse and the circadian system stays in arrhythmicity until a perturbation (for instance a second light pulse) of sufficient strength starts the oscillation anew. In *Drosophila* eclosion rhythm, *Kalanchoe* petal movement and other cases the phase at which arrhythmicity can be induced is the subjective midnight point, and the strength of the pulse has to be such that it is just between evoking a strong or a weak phase response curve. According to a model of Leloup and Goldbeter (2001) the phases at which arrhythmicity can be induced are, however, more extended.

The underlying mechanism will be treated in the next section under a formal viewpoint.

4. MODELS

4.1. Usefulness and Kind of Models

Models lie in the heart of natural science and their usefulness should not be underestimated. They force us to systematize our knowledge and to quantify our predictions. They could give very useful hints as to the interplay between light and circadian systems. The literature on light effects on circadian rhythms provides many examples of the usefulness of model concepts and predictions. One example is the discovery and the experiments around the concept of light induced arrhythmicity of circadian systems, mentioned above.

Other examples which illustrate the usefulness of models are found in applied fields. For man, the modern jet flights imply rapid changes of time zones and accordingly changed light inputs to the circadian systems. Models are published which facilitate resynchronization to new light conditions. They take the changing environmental light signals into account as well as the circadian system (see subsection 6.3). These models can also be used, at least in principle, in difficulties of the human circadian system to adapt to shift work conditions.

Published models of circadian systems are of different kinds. Some are purely mathematical ones, describing the variables in, usually, differential equations; other models are often presented as a block diagram and use concepts from control theory. Often numerical methods are used to simulate the circadian behavior in the biological system. A third form of models describes reactions in words and figures without deriving or attempting quantitative relations (see the selected examples in section 7 and the following). Ultimately the models should give precise qualitative and quantitative descriptions and predictions at the molecular, the cellular and the organismal level.

The detailed way in which light affects the individual circadian systems is important for a model: The photoreceptors and receptor molecules have to be identified, the details of the signal transduction pathways have to be known, and the way in which the (transformed) light signal enters the circadian clockwork has to be determined. This shows that modelling requires specific knowledge for each circadian system under study.

The circadian system in a cell can be modeled as a single unit model. In a multicellular organism, the concept of a *multi-oscillator model* might be necessary to consider. Furthermore, a particular oscillating system should be modeled by the fewest possible variables that are relevant for the study in question.

4.2. Feedback and Time Delay

In models for circadian rhythms the concepts of *feedback* and of *time delay* are frequently used. Control theory tells that negative feedback in a system does not always lead to oscillations. However, if the signal in the loop is delayed in a suitable way, it may be fed back to reinforce an already existing signal and induce oscillations. In several models of a circadian system, negative feedback with suitable delays are nowadays introduced. In order to increase the tendency to oscillate, the feedback system must also possess sufficient amplification. A simple description can be given.

The rather large time delays which exist in circadian clocks could be due to transcription, translation, transport, and production of clock related components. Specific points will be discussed later in the different examples for circadian rhythms.

4.3 Entrance of Light Signals into the Clockwork

Light signals from the environment are perceived in a receptor organ or in a receptor system, see Figure 1. After the light quanta are absorbed, the excitation energy gives rise to a signal along a transduction chain into the clockwork. Models of this pathway for the light signals will differ widely between organisms.

Feedback models of a circadian oscillator are often used. In such a model the signal $c(t)$ in the loop is delayed in a suitable way before feeding back to reinforce an already existing signal and induce oscillations. In order to increase the tendency to oscillate, the feedback system must also possess sufficient amplification (a non-linear function). If we assume that a substance is produced at time t according to the concentration of the same substance $c(t)$ at a certain earlier time $c(t-t_0)$, we have a simple feedback system with delay. The situation can be expressed, with K as a positive constant, as

production of substance at time $t = -K*($conc. of substance at time $(t-t_0))$

Here 7 is a positive constant and the negative sign indicates that production is decreased if the concentration was high at the earlier time, while it is increased if concentration was low at the earlier time (inhibition occurs if concentration was high, activation occurs if concentration was low at some time units earlier). Transformed into mathematical terms the expression shows sustained oscillations if the delay t_0 and the feed back signal are large enough, that is if K is large enough.

Furthermore, the period of the oscillations will be about 4 times the delay time introduced. Circadian oscillations thus need a delay of about 6 hours in the example in order to end up with a 24 hour period.

Feedback models based on the concepts discussed (but using nonlinearities that are always present in biological systems and needed to limit the maximal concentrations that can be reached) have been used to simulate features of circadian rhythms (*Kalanchoe* petal rhythm, Johnsson and Karlsson 1972), activity rhythm in the New Zealand Weta, (Lewis 1999), and a molecular model by Lema *et al.* (2000). Light could enter the feedback loop at different locations. However, in the case of the *Kalanchoe* petal rhythm, it turned out that in order to simulate experimental results light pulses have to enter the feedback loop at a certain point (Engelmann *et al.* 1978). This might, however, be different in other biological systems.

As mentioned already, the photoreception can be clock controlled by feedback links that change the properties of receptor systems. This makes the modeling and analysis more difficult. In addition, light adaptation and other changes of sensitivity to light might increase the level of complexity in modeling the light induced effects on the circadian clock.

The experimental investigations of the light perception and signal transduction must probably reach a new stage in order to allow modelling the first steps of the light reactions of the circadian system.

4.4. Clock Outputs

Reaction sequences downstream the clock are also important to model. The period of the circadian system will be reflected in the reactions driven by the clock. Amplitude and phase of the driven reactions might change, but the period will be that of the clock. Thus, the final reaction that is observable - a hand of the clock - has the same period as the clock. This is stressed since environmental light signals might affect the clock controlled 'downstream' reactions directly, thereby changing for instance their amplitude. Such changes should not be mistakenly ascribed to light effects on the circadian system itself.

4.5 Some mathematical Properties of circadian Models

General features of circadian systems which must be handled by models are:

• *Phase shifts and phase response curves:* Any model of an oscillating, circadian system should react to external signals - for instance light signals. One often assumes that light pulses given to the organism also enter the circadian system as pulses, affecting a model variable or a model parameter. Detailed modeling of the light signal pathway into the clock is certainly needed.

The light pulse introduced into the model will, in general, cause a phase shift of the rhythm (see subsection 10). Phase delays as well as phase advances will arise in the models.

Small light induced signals to the clockwork will give rise to smaller phase shifts of the circadian oscillator, while stronger signals might cause phase changes of up to 12 hours.

Phase response curves which present the phase shifts in a systematic way have been modeled for different organisms.

• *Entrainment:* Repetitive light pulses given with a period T will, due to successive phase shifts, entrain the circadian rhythm ('entrainment', 'synchonization', 'phase locking'). The external light cycle will function as a synchronizer. This general property of circadian systems will also be simulated by models of circadian systems. The magnitude of phase shifts, the range of entrainment and the amplitude effects on the circadian oscillations will vary with the parameters and the details of the models. Successful descriptions of phase response curves (both of weak and strong type), entrainment regions, etc have been given in several types of models. Such descriptions are, therefore, successful at the system level or the cell level. Refined models are of course required to describe the processes and the dynamics at the molecular level.

• *Stopping the clock by light pulses:* In many models phase shifts and amplitude changes brought about by light pulses are concomitant features. One might ask if an external light pulse could cause a total amplitude reduction of the circadian system, i.e. lead to a 'stop' of the oscillations. Such a feature of the circadian system was found in some circadian systems. Experiments on the eclosion rhythm of *Drosophila* (Winfree 1970), the petal movement rhythm of *Kalanchoe* (Engelmann *et al.* 1978), activity rhythm of the *Culex* mosquito (Peterson 1981a, Peterson 1981b) revealed that the circadian system could be sent into a non-oscillating state by light pulses.

In these experiments, the phase at which the light pulses had to be administered to the organisms was fairly restricted. However, several combinations of irradiance and pulse durations were still effective in *Drosophila* (Chandrashekaran and Engelmann 1973), indicating that a range of parameter choices could be used to reach this special state. This extended range of phases at which pulses could stop a circadian rhythm was theoretically penetrated in a detailed model (Leloup and Goldbeter 2001) of the *Drosophila* rhythm.

A mathematically and biologically interesting question arises: will a circadian system start oscillating spontaneously again after having been sent into the 'non oscillatory state'? Will it then go back to the old oscillation pattern? These problems have attracted much interest and focus on the mathematical structure of the circadian systems (on so called singularities, limit cycles and attractor cycles).

Descriptive models for some circadian systems will be discussed later.

4.6. Single versus Multi-Oscillator Models

Several important features of circadian systems are being modeled on the assumption that one single oscillator is controlling the clock. A one-oscillator model does not preclude the presence of many cellular oscillators - it only assumes that they are so strongly coupled to each other that they (mostly) behave as one single unit (a 'lumped' model).

However, in multi-oscillator models the circadian system has new features that can not be explained under the assumption that the system consists of one single oscillator.

The circadian system of man is an example which is often modeled by two interacting oscillators. One of the oscillators is then assumed to have its strongest influence on (among other rhythms) the *activity rhythm*, the other one on (among others) the *temperature rhythm*. Usually the two oscillators are coupled and oscillate in phase, but the dual nature of the system can show up in, for example, experiments in isolation where the rhythms might display different periods (Wever 1979).

Many circadian systems should be modeled as multi-oscillatory systems. Modeling often starts with a simple one-oscillator assumption, an approach that eventually turns out to be too simple. In the case of *Drosophila,* several oscillators are nowadays implicated in more detailed modeling (Helfrich-Förster and Diez-Noguera (1993), see section 31).

5. MECHANISMS

5.1. Current Concepts and Caveats

In order to understand the exact way of synchronization of the circadian clocks by light and other time cues, the mechanisms of circadian oscillators be known, as well as the photoreceptors and pigments involved in the entrainment. The clock mechanisms are currently under intensive studies. Some of the results will be presented briefly in the following section. The prevailing opinion is, that *feedback loops between clock gene products acting on the promoters of their genes are at the heart of these clocks* (Dunlap 1999, Dunlap 2000, Iwasaki and Dunlap 2000, Sehgal *et al.* 1999).

However, the picture is probably more complicated, and cautions have been raised. For instance, these feedback loops might be also elements which are located before the clock mechanism proper. By virtue of special interactions they are able to affect the proper clock mechanism in such a way that decisive properties, for instance period and entrainment, are influenced (Merrow *et al.* 1999).

There are other cases reported which make it difficult to accept the presently favored concept of a circadian clock mechanism as a general one. For instance, in the large unicellular alga *Acetabularia* it was shown that cutting off the rhizoid containing the nucleus does not prevent the circadian rhythm of oxygen production (Karakashian and Schweiger 1976). As another example, dry seeds of bean plants in which nucleic acid metabolism and synthesis is absent were claimed to still show circadian rhythms in (the extremely low) respiration rate (Bryant 1972). Finally, it was reported, that in human erythrocytes different enzymes such as glucose-6-phosphate dehydrogenase, acid phosphatase and acetyl cholinesterase fluctuate in a circadian way. Mature erythrocytes do not possess nuclei nor nucleic acids, lack most other organelles usually found in cells such as mitochondria, and are therefore not able to perform protein synthesis and respiration. They are tuned to their main task: to bind oxygen to the hemoglobin molecule and deliver it all over the body (Ashkenazi *et al.* 1975). What is common to the two last mentioned systems is the complete lack of nucleic acid metabolism. This is an important issue, since several of the recently proposed models of circadian systems use feedback systems in transcriptional and translational events.

It might therefore be wise to keep an open eye on alternative mechanisms underlying the circadian oscillators. Of course, there is no guarantee that all circadian clocks use the same mechanism, although their properties are often quite similar.

Proteins could, for instance be involved in timing mechanisms. An interesting case was reported by Kai *et al.* (1995). The development of the embryos in the eggs of female silk moths (*Bombyx mori*) is arrested under long days. This *'diapause'* (developmental interruption in a certain stage of insects) is broken by exposure to at least two weeks of low temperature (5°C). The duration of the chilling period is measured by an esterase A4 complexing with another enzyme, PIN. After 14 days the esterase A4 dissociates from PIN, the conformation of the esterase A4 changes and it becomes suddenly active (Kai *et al.* 1999). This enzyme is thus a kind of molecular timer.

5.2. Selected Examples for Light Resetting of Circadian Clocks

Formal models do help in situations where physiological, genetic and molecular studies of circadian systems have not been very extensive. They allow predictions which can be tested experimentally. However, the goal of research into the mechanisms underlying circadian rhythms is a full understanding of the physiological and biochemical mechanisms and their molecular basis and the coding by genes. It is therefore important to study certain organisms and their rhythms in detail using modern approaches. This has fortunately been done successfully in the last decades and we will try to review some of the exciting results by focusing on the light entrainment in these cases.

In discussing the effect of light on circadian rhythms, we have to limit ourselves to a few examples. We have selected several organisms ranging from man to cyanobacteria which have been shown to be of advantage in studying the circadian system and its entrainment mechanisms:

- among rodents the locomotor activity rhythm of hamsters, mice and rats
- among insects the eclosion rhythm and the locomotor activity rhythm of *Drosophila*,
- in plants the circadian expression of several genes in *Arabidopsis*,
- among fungi the conidiation rhythm of *Neurospora crassa*
- among unicellulars the bioluminescence rhythmof the dinoflagellate *Gonyaulax polyedra*
- and in *Synechococcus* and *Synechocystis*, both cyanobacteria, the circadian control of expression of several genes.

For lack of space we could not cover the interesting studies on marine snails such as *Aplysia* and *Bulla*, where the electric output of pacemaker cells in the retina show circadian rhythms. We refer to a review article (Herzog and Block 1999).

In some of these examples we will briefly explain the circadian rhythms and underlying molecular mechanisms and locations of the clock mechanisms. We then present the location and kind of photoreceptors, retinal, extraretinal, and unstructured, and their resetting mechanisms of the clocks. It will become apparent, that the circadian system is quite complicated and often – not only in higher organisms – consists of multiple receptors for resetting light, and of multiple oscillators.

6. LIGHT AND CIRCADIAN CLOCKS IN MAN

In man the circadian system governs not only the sleep-wake cycle, the body temperature, alertness and efficiency, but also many other physiological and metabolic events such as REM sleep, hormonal secretion, enzymatic activities in organs such as the liver or in the red blood cells, and so on. There is a long list of circadian clock driven events (Minors and Waterhouse 1981).

The sleep/wake cycle, the body temperature and the urine production and composition can be monitored easily and have therefore been used as hands of the circadian clock system quite intensively. Often they have also been the subject of modeling the circadian system (Wever (1979), Moore-Ede et al. (1982), Forger et al. 1999). Melatonin concentration in the blood is a particularly useful hand of the circadian clock system because it is rather sturdy and not much disturbed by activities (as is the case with body temperature). However, light has an immediate suppressing effect on melatonin concentration.

The circadian system of man shows up clearly if humans live under isolation from external time cues. In a cave or in a bunker type facility a person who has no information of the outside time will sleep and wake according to his internal circadian clock system. This in most cases runs more slowly than the 24 hour day, as can be shown by continuously measuring the body temperature, the time of going to bed and rising up, and the activities of arm movements and moving around. The average 'free run' period of many such studies is about 25 hours (Wever 1979).

This free run rhythm can be synchronized to the 24 hour day by knowing the time of day, by external time cues such as light, temperature, noise or social contact and so on. Light as a time cue of the circadian system was, in an earlier stage of these studies, thought to be much less effective than had been known from other organisms (Wever 1979). However, it later turned out that the illuminance used in trying to synchronize the subjects in bunker experiments was not high enough (it was below 2500 lux). If stronger light was used, it could easily synchronize the subjects to the 24 hour day and also to other LD cycles if inside the range of entrainment (Czeisler 1995, Wever 1989, Klein et al. 1991). Newer results show that light pulses of only 1 hour duration and 500 lux shift the phase of the circadian melatonin rhythm (Laakso et al. 1993).

6.1. Light Synchronizes the Circadian System

In man, as in other mammals, the eyes seem to be the only photoreceptors which are able to synchronize or phase shift the circadian rhythms. Claims that extraretinal photoreception can phase shift the circadian rhythm of body temperature and melatonin concentration by illuminating the backside of the knees (Campbell and

Murphy 1998) have been questioned in later experiments (recent discussion by Eastman *et al.* 2000). In more than 50% of blind people free run of their circadian rhythms is observed (Miles *et al.* 1977). In the rest either the blindness affects only normal vision of images, but not the capability of cells in the eye that have recently been found to be involved in synchronizing the clock, or other time cues are used for synchronization. Occasionally even in people with intact vision free run is observed, although they live in a normal environmental situation. It is not known why light (and other time cues) are ineffective in these people (Giedke *et al.* 1983).

Single light pulses are able to phase shift circadian rhythms in humans according to a phase response curve (Minors *et al.* 1991). A light pulse 1 to 3 hours before the minimum of the body temperature rhythm delays, a light pulse 1 to 4 hours after it advances the rhythms. The minimum of the body temperature rhythm is in young subjects found 1 to 2 hours before the usual wake-up time. A light pulse during the subjective day does not phase shift the rhythm (see also Figure 3). Besides phase the amplitude of the rhythm is an important parameter. A first light pulse has been claimed to reduce the amplitude of the rhythm, which would render the system more sensitive toward the phase shifting effect of a second light pulse (Czeisler *et al.* 1989). Problems with obtaining phase response curves (see subsection 3.3) in humans such as background light and masking effects of light and more details are discussed by Dijk *et al.* (1995).

6.2. Photoreceptors in the Eye for Circadian Vision, Pathways to the Circadian Centers, SCN

As in other mammals, the hypothalamus at the neuroendocrine crossroad of the brain plays an important role in integrating neurotransmitters and hormones with respect to circadian rhythms. It harbors at the ventral part of the third ventricle above the optic chiasm the paired *suprachiasmatic nucleus* or SCN. They are the centers of circadian pacemakers (Klein *et al.* 1991). Their structure and function will be discussed in the next section.

The eyes are the only photoreceptors in mammals which are used to entrain the circadian system. Which photoreceptors in the eye are responsible for circadian vision is debated (see next section). Retinal rods and cones are not required as shown with mice without these receptors (Freedman *et al.* 1999). It was claimed, that in mammals with their vitamin A-based photopigment opsin for vision, a vitamin B_2-based pigment *cryptochrome* is used for entrainment of the circadian clock (Miyamoto and Sancar 1998). However, this has been debated by Lucas and Foster (1999) and they propose, that circadian vision is also opsin based as is normal vision: The action spectrum does not speak for cryptochrome as the photoreceptor for light entraining the circadian clock, but rather for opsin:retinaldehyde. Cryptochrome might be an additional photoreceptor for shorter wavelengths, but it is more likely that cone photoreceptors mediate these responses (Lucas and Foster 1999).

The phototransduction for circadian vision can be mediated by cone pigments, as shown by exposing human subjects to red light below the rod-, but above the cone sensitivity (Zeitzer *et al.* 1997). The trichromatic visual system is not necessary for circadian vision, as studies showed with color-vision deficient subjects (Ruberg *et al.* 1996). In rodents, 450 to 500 nm wavelengths affect the circadian system maximally, which could involve rods or blue sensitive cones. Newer results speak in favor of specialized cones or a completely novel photoreceptive system (discussed in Rea 1998).

Several genes which encode previously unknown opsin-like proteins were recently cloned in vertebrates. Two of them, VA opsin (Soni *et al.* 1998) and melanopsin (Provencio *et al.* 2000), are expressed in elements of the inner nuclear layer of the mammalian retina and not in the opsin-containing cells of the outer retina that is responsible for normal vision. The inner nuclear layer might be responsible for the regulation of circadian rhythms and for immediate melatonin suppression. But it is not involved in image formation. A special long wavelength sensitive cone opsin found in the atrophied subcutaneous eyes is used by the blind mole rat for entrainment of its circadian clock (David-Gray *et al.* 1998). So far, results suggest that both classical and novel photopigments are involved in synchronizing the circadian system of mammals.

How light information reaches the SCN is presented in the section on mice.

6.3. Significance of Light in Shift Work and Jetlag

The circadian system determines sleep propensity, timing of sleep, sleep structure, and consolidates sleep and wakefulness. Sleep homeostasis interacts with it (Circadian priciple: The longer we are awake, the shorter we sleep. Homeostatic regulation of sleep: The longer we are awake, the deeper our sleep). The effects of light on sleep have been reviewed by Dijk *et al.* (1995).

The circadian rhythm of modern man is often delayed with respect to the natural LD cycle. He uses electric light and can therefore stay up during the winter time much longer than natural day light would otherwise permit (Cardinali 1998). This independence or even insulation from the natural light easily leads to permanent sleep deprivation. In addition, modern society expects full range services throughout the 24 hours. Traffic, economy, health service and security have to rely on shift work or night work by a considerable part of the workers (about 20% in the industrialized nations).

However, shift work clashes with our circadian clock. As a consequence, potential health- and safety problems arise. Many accidents are due to ill-adapted circadian clocks, and sleep disturbances arise from it (review by Eastman *et al.* 1995).

The synchronizing effect of light on the circadian system of man is one of the problems of shift work. For instance, the high fluence rate outdoor light in the morning after a night shift prevents the phase shift of the circadian system of the night worker needed for optimal adjustment of his clock (Czeisler and Dijk 1995). Wearing dark goggles is advisable in this case (Eastman *et al.* 1994). On the other hand, light can be used also for adjusting the clock to the shift work schedule, if properly applied. Models are used successfully for constructing LD cycles which phase shift the rhythm in such a way that they align better with shift work and day sleep schedules (Jewett *et al.* 1999), Martin and Eastman 1998). More empirical data from shift work effects on the circadian rhythms are, however, needed for detailed simulations of this kind (Akerstedt 1998). Other counteractions consist of light exposures at certain times of the circadian cycle (Eastman *et al.* 1995) and of using chronobiotics such as melatonin (Redfern *et al.* 1994). In using combinations of light and melatonin, it should be taken into account that the phase shifting effect of light pulses and melatonin pulses are 180° out of phase (Boulos *et al.* 1995, Redfern *et al.* 1994).

Jet lag is another problem where the circadian clock is suddenly exposed to a new temporal environment and needs some time for adjustment (overviews by Boulos *et al.* 1995, Redfern *et al.* 1994, practical considerations Deacon and Arendt 1996, Samel *et al.* 1995). For a critical review of a potential beneficial effect of light on jet lag syndromes see Samel and Wegmann (1997). The human phase response curve to light pulses tells us that we should after arrival avoid morning light after having flown westward, and expose our self to outdoor light in the evening. Faster adjustment after eastward flights demand exposure to light in the morning and avoiding evening light.

6.4. Seasonal affective Disorders and Endogenous Depressions

In endogenous depression, abnormalities in the circadian system have been observed (Halaris 1987). Furthermore, treatments affecting the circadian rhythms and the sensitivity of the retina to light have a therapeutic effect (Terman and Terman 1999). Depressed patients might suffer under an anomalous phase of the rhythm of light sensitivity due to some defect in the retina (Steiner *et al.* 1987).

A special type of depression is the seasonal affective disorder (SAD). It occurs during the winter as a response to the seasonal shortening of the light period (Graw *et al.* 1999) in adults as well as in children (Swedo *et al.* 1998) in the northern and southern hemisphere (Teng *et al.* 1995). It has been discussed, whether it might reflect some kind of photoperiodic reaction in human (Cardinali 1998).

Besides other abnormalities (increased appetite and carbohydrate craving (Arbisi *et al.* (1996), hypersomnia, an abnormal structure of non-REM sleep, increased need for REM sleep (Palchikov *et al.* 1997), see however Brunner *et al.* (1996) and Wirz-Justice *et al.* (1996), the circadian rhythms show abnormal amplitudes and phases (Bunney and Bunney 2000, Koorengevel *et al.* 2000) and the phase response curve to light is increased (Thompson *et al.* 1997). SAD patients are more sensitive to variations in the length of the natural day (Guillemette *et al.* 1998). They are supersensitive to light during the winter (Terman and Terman 1999), perhaps due to a phase delay of the circadian rhythm (Nathan *et al.* 1999). Furthermore, the serotonergic system (serotonin is an excitatory neurotransmitter) seems to be involved (Partonen 1998).

A light therapy (Lam *et al.* 1997, Partonen and Lonnqvist 1996), outdoor light (Wirz-Justice *et al.* 1996) and administration of serotonin uptake inhibitors (Neumeister *et al.* 1997) or using both methods are effective (Thorell *et al.* 1999), but not in severely ill patients (Schwartz *et al.* 1996). It does not seem to matter whether the light is applied in the morning, at noon or in the evening (Lewy *et al.* 1998, Meesters *et al.* 1995, Thalen *et al.* 1995); see however Leibenluft *et al.* (1996). But the treatment should exceed one week. Preferably it should be applied for three weeks (Eastman *et al.* 1998). Dawn simulation improves the efficiency of the light treatment (Meesters 1998).

However, there are many questions unanswered, before the light/SAD/clock relations are settled (see Levitt *et al.* 1996, Lee *et al.* 1997, Meesters *et al.* 1999). Several types of SAD are known, some of which react poorly to light treatment (Terman *et al.* (1996)). For special literature on SAD see Neumeister *et al.* (1998), two articles in Touitou (1998), and for the practitioners Lam and Levitan (2000), Lam *et al.* (1997), Rosenthal and Oren (1995), and Dalgleish *et al.* (1996).

7. MOUSE-CLOCKS AND LIGHT

Among vertebrates mammals are the best studied class in respect to the circadian timing system. For experimental reasons rodents have been favored (easy rearing, small size, short generation time, simple recording of locomotion) and among these mice and rats are preferred because many mutants are available and genetic and modern molecular biological methods are applilhcle.

Locomotor activity is most frequently measured as a hand of the circadian clock by offering running wheels to the animals. They will run in the inside of the wheel and each turn activates a magnetic switch. The number of contacts per hour is used to construct *actograms* which allow easy determination of the period and phase shifting of the rhythm induced by different treatments (example of actograms in Figure 2). Body temperature can also be monitored with special implanted sensors.

The photic entrainment of these rhythms is reviewed by Rea (1998) and a subject of later subsections. We should first have a look at the pacemaker centers driving these rhythms.

7.1. Circadian Centers

As in other vertebrates a center of circadian control lies in the paired SCN of the anterior part of the hypothalamus. As a main oscillator they control a large number of physiological, endocrine and behavioral rhythms. Among them are locomotor activity, sleep-wake-cycle, thermoregulation, torpor, hibernation, functions of the circulatory system and many endocrine events. The synthesis and secretion of melatonin is also controlled by the SCN. If these nuclei are destroyed, the circadian control of all these events disappears.

The SCN is not just a tissue, which transfers informations of the LD-cycle from the eye to an oscillator. In this case its destruction would have prevented *synchronization* of the various circadian rhythms, but the *rhythms* would not have disappeared. Free run should have occurred instead. However, the animals became arrhythmic. There are further strong indications for the SCN playing a master oscillator role, as reviewed in Silver and Moore (1998) and Klein *et al.* (1991).

In the SCN of all mammals about eight thousand to ten thousand neurons are found. They form two parts which are characterized by the nature of the neurotransmitters of their neurons and the innervation (reviewed in Moore 1997, Esseveldt *et al.* 2000). Retinal information is transmitted through the retinohypothalamic tract (RHT, Figure 4) and additionally, via the intergeniculate leaflet (IGL), through the geniculohypothalamic tract (GHT) to the ventrolateral part (*shell*) containing neurons, the firing of which is under circadian control. The rhythms are light dependent. The dorsomedial part (*core*) receives inputs from non-visual sources and the neurons show rhythms which are light-independent (references given in Ibata *et al.* 1999). The neuronal firing rate shows a circadian rhythm. High firing rate during the subjective day seems to correlate with peptides and the neurotransmitter gamma-aminobutyric acid (GABA) and uses standard synaptic interaction. However, rhythmic information might also be conveyed by a diffusible substance.

The electrophysiological activity of horizontally cut slices of hamster SCN reveal two distinct oscillating components (Jagota *et al.* 2000). They might reflect the activity of *morning* and *evening oscillators* which have been inferred already earlier from behavioral studies (Pittendrigh and Daan 1976, Illnerova and Vanecek 1982). Photoperiodic reactions are supposed to use such evening and morning oscillators, and a long and a short photoperiod do indeed affect the morning and evening peaks of the electrical recordings differently (Jagota *et al.* 2000).

Individual SCN neurons, which were dissociated from each other, were monitored with multi-micro-electrode plates for the electrical activity for extended time spans. Cultures contained cells with different phases and periods, although functional synapses were present (Welsh *et al.* 1995). It is unknown which of the cells are able to oscillate, how the phase shifts are realized and whether the phase response curve to light is brought about by single cells or by a network of cells. Intrinsic pacemaker activity might be restricted to the core. The afferents of the core would then be entrainment pathways, whereas the shell receives inputs from other parts of the brain including light information from the retina (see Figure 4). Efferents of the SCN project to other parts of the hypothalamus and to the pineal gland (Ikonomov *et al.* 1998). The output pathways and target organs of the SCN are not well known except for the projection to the pineal organ. This output modulates melatonin synthesis in a circadian fashion. Melatonin feeds back to the SCN.

After the destruction of the SCN other rhythms are still maintained, such as the anticipatory food uptake behavior (Stephan *et al.* 1979, Mistlberger *et al.* 1996, Marchant and Mistlberger 1997). They must therefore be controlled by another pacemaker center. It is firmly established that the retina of the eye contains circadian pacemaker (see later). So far it is poorly studied how the oscillator in the SCN controls locomotor activity and other events in a circadian fashion.

7.2. Synchronization and Photoreceptors

The synchronization of the circadian clock of mammals is mainly due to the LD-cycle of the environment. Light is perceived via the eyes. Rods and cones are involved, but also an additional component, as discussed before. Blinded rodents can not be synchronized by an LD-cycle but show instead free run (Meijer *et al.* 1996).

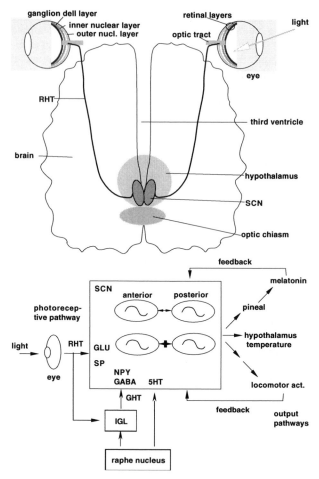

Figure 4: Top: Circadian center of mammals with its photic inputs, circadian outputs and feedbacks. In the eye the three retinal layers are shown schematically with the ganglion cell layer, the inner nuclear layer and the outer nuclear layer containing cell bodies of rods and cones. Temporal information of the environmental light is transferred from the retina to the paired SCN at the base of the third ventricle above the optic chiasm via the retinohypothalamic tract (RHT). Bottom: Details of the mammalian circadian circuitry: A part of the RHT projects directly to the suprachiasmatic nucleus (SCN) using glutamate (GLU) and substance P (SP) as neurotransmitters, another part to the intergeniculate leaflet (IGL) and further to the SCN via the genicohypothalamic tract (GHT) using neuropeptide Y (NPY) and GABA as neurotransmitters. The SCN consists of about 10 000 pacemaker cells (a few shown symbolically, as ~) arranged in a core and a shell (not illustrated). The core mediates the circadian rhythm to other parts of the brain influencing, for instance, the temperature control in the hypothalamus and locomotor activity, but also the pineal and its melatonin secretion. Melatonin feeds back to the SCN, as does locomotor activity. The shell of the SCN receives inputs from the RHT and GHT and from the raphe nucleus with serotonin (5HT) as a neurotransmitter. The anterior and posterior part of the SCN seem to house pacemaker cells responsible for the occurrance of a morning and an evening peak, respectively, in the locomotor activity rhythm. After Moore (1997) and Wollnik, unpublished.

Light information reaches the SCN via the RHT from retinal ganglion cells and via the IGL and the GHT. The neurotransmitters used are probably excitatory amino acids such as glutamate.

The retinal ganglion cells of the RHT are found within a large sampling area across the entire retina (Moore *et al.* 1995). They seem to be specialized to detect average illuminance of the day and integrate it over long intervals. UV-sensitive cones (or their remnants) are also involved. Mutants with degenerated retinas (rd, rds) are available. The circadian rhythm of these mutants is, however, still phase shifted by light. The animals are less sensitive to light as compared to the original strain CBA/N. Either only a few cones without outer segments for light perception are responsible for the phase shifting effect, or there are cells in the retina which are not known yet (Provencio and Foster 1995, Argamaso *et al.* 1995, Lucas and Foster 1999).

The RHT is necessary and sufficient for entraining the circadian clock in the SCN. If the IGLs are destroyed, entrainment of the circadian system by light is not prevented, but modulated (see Morin (1994) for review). The two other nerve bundles to the brain, the optic nerve and the accessory optic system, do not influence synchronization of the circadian system.

RHT neurotransmission is reviewed in Rea (1998). For photic regulation of the circadian pacemakers in the SCN glutamate, serotonin (Bradbury & Dement 1997) and the neuropeptide substance P seem to be involved. Serotonin is an important regulator of circadian phase, probably by inhibiting the release of glutamate, and by modulating GABA-mediated chloride channel activity.

Two distinct kinds of molecules seem to synchronize the circadian rhythm in the SCN after light exposure of the retina (reviewed by Rea 1998): immediate early genes such as transcription regulator proteins and furthermore nitrogen monoxide (NO). Both seem to play a role in the transfer of light-induced signals in the SCN. NO production is required for light induced phase shifts of behavioral rhythms under circadian control. The exact role of these photic entrainment pathways and their connections are not well understood. The advances and delays of the rhythm might be due to different pathways.

Some of the neuropeptides in the SCN serving as transmitters or modulators under LD and DD conditions (vasopressin, somatostatin) show marked circadian rhythms. They are restricted to the 'core′ of the SCN which mediates the rhythm to other parts of the brain. Other neuropeptides show rhythms only under LD, but not under constant conditions. They are found in the ventrolateral part ('shell′) of the SCN and convey information on environmental light conditions to the pacemakers (Inouye 1993).

7.3. The Eye Clock

Besides the SCN mammals possess another circadian pacemaker: It is located in the retina of the eye (Tosini and Menaker 1996). The rhythmicity is apparently generated by photoreceptor cells (Cahill and Hasegawa 1997). The pacemaker cells are responsible for a circadian output of melatonin from the eye. This seems to change the sensitivity of the retina to light. The genetic basis of this oscillator is the same as that of the SCN: Period mutants show altered periods also in the output of

356

the eye clock. The effects of this clock are probably restricted to the eye. These pacemaker cells are responsible for the rhythmic adaptation of phototransduction and for the recycling of biochemical components in the retina. It demonstrates the tied connection between pacemakers and photoreceptors which are known also from invertebrates and *Neurospora* (see subsection 8.4 and 10.2, respectively). Other rhythmic events in the eye such as visual resolution (Tassi *et al.* 2000), ERG, intraocular pressure (Nickla *et al.* 1998), choroid thickening and eye growth might be or are also under circadian control. The shedding of outer segment tips and the phagocytosis of the shed membranes of rods and cones are controlled by the circadian clock (Young 1976, Grace *et al.* 1999). This *internal* renewal is important for the functional integrity of rods and cones, which can not be replaced by new cells. Shedding of rod segments occurs in the morning, shedding of cones in the evening. In this way it does not interfere too much with the times the rod or cones are mainly used.

7.4. Pineal Organ and Melatonin

Reproduction of different rodents is controlled by day length (Baker and Ranson 1932). The pineal and its hormone melatonin transmit the photoperiodic signals of the environment to the neuroendocrine axis (review: Steinlechner 1992). These signals can be either stimulating or inhibiting (Hoffmann 1981). Melatonin passes the placenta and is in this way able to convey circadian and seasonal informations to the fetus. Melatonin does not only affect the gonads and other physiological events, but also the SCN. It inhibits the neuronal activity and shifts the phase of the circadian rhythm. If properly timed and illuminated with a certain amount of light the melatonin rhythm can disappear by sending the oscillator into a singular state (Shanahan *et al.* 1997). It is not known, whether the human reproductive system responds to the melatonin message. The duration of melatonin synthesis, which depends on night length, might help man to adjust his circadian rhythm to the seasonally varying night lengths. It plays a role in perceiving and processing morning light (Wehr 1997).

In contrast to the pineal of other vertebrates that of mammals does not react to light and the pinealocytes do not posses circadian clocks. Instead, the light is perceived by the retina and the signals are transmitted via the retinohypothalamic tract to the SCN. From the SCN neuronal signals are sent via sympathetic nerves passing the superior cervical ganglion to the pineal. The pineal organ is in this way synchronized to the environmental LD cycles. It synthesizes and secretes melatonin transmitting it to the brain and the body. The melatonin mRNA in the pineal is rhythmically transcribed and therefore the melatonin synthesis occurs in a circadian way (Foulkes *et al.* 2000).

Melatonin also provides a neuroendocrine feedback mechanism by affecting the SCN and modulating the activity and other processes. The nature of this feedback is unknown. Although pinealectomy in rats does not influence the circadian rhythm in the LD-cycle and in continuous darkness, the running wheel activity is heavily disturbed in LL. Either the feedback from the pineal to the SCN influences the light sensitivity of the SCN or the circadian output of the SCN (Cassone 1993).

Light affects melatonin synthesis not only via its resetting effect on the circadian system. It also has an immediate effect. Illumination of the nasal part of the retina is the most effective in reducing melatonin synthesis during the night (Visser *et al.* 1999). Under optimal conditions light of 200 lux and single light pulses are already sufficient.

Melatonin is produced not only in the pineal gland, but also in the Harderian gland, the retina, and the gastrointestinal tract. Besides its photoperiodic effect and its phase shifting of the circadian system (by changing the functional state of the SCN (Illnerova and Sumova 1997), it modulates sleep, reproductive behavior (review Luboshitzky and Lavie 1999), mood and immune responses. It has been used to cure sleep disturbances and insomnia (for instance in elderly people), depression, jet lag and shift work related sleep cycle disorders. Skin protection against UV relies on the antioxidant properties of melatonin (free radical scavenger).

With administered melatonin pulses (orally, injected), the circadian rhythms can be phase advanced or delayed, depending on the phase of application. The phase response curve to melatonin pulses is similar to that of light pulses, but displaced by 180°. Therefore melatonin can be used in a similar way as light pulses – if properly phased to shift the circadian system. Circadian phase disorders can be treated in this way (Lewy and Sack 1997).

Melatonin receptors (G-protein-coupled) with high affinity have been found in the eyes, kidneys, in the gastrointestinal tract, blood vessels and in the brain. Melatonin effects are furthermore modulated by adenylat cyclase, guanyl cyclase, phospholipase C and potassium channels (Kokkola and Laitinen 1998).

8. *DROSOPHILA'S* CIRCADIAN CLOCKS AND LIGHT

Drosophila has been used extensively as an experimental animal for studying circadian rhythms, especially *eclosion* of the flies out of the puparium and *locomotor activity* of the adults. It is also one of the best studied objects for the effects of light on the circadian clock. *Drosophila* is well known genetically and amenable to methods of molecular biology. Quite a number of clock mutants are available ('per'). In addition many mutants affecting photoreception are known. General reviews on the circadian rhythms of *Drosophila* are those by Hall (1998), Young (1998), special reviews on the organization of the circadian system by Helfrich-Förster *et al.* (1998), and on the effect of light on the circadian rhythm by Helfrich-Förster & Engelmann (2002) and Foster and Helfrich-Förster (2001). A mini-review (Reppert and Weaver 2000) compares the circadian system of *Drosophila* and mice.

8.1. Circadian Eclosion and Locomotor Activity

After several larval stages *Drosophila* forms a puparium in which pupation and metamorphosis into the adult stage takes place. Eclosion from the puparium occurs under the daily LD cycles during the early morning hours. The eclosion rhythm can be observed in a population of flies only, since it happens only once during the development of a fly. If a culture of *Drosophila* flies is transferred into constant conditions of darkness, eclosion is still rhythmic. This shows that eclosion is not just the response to the onset of light, but under control of a circadian clock. The eclosion rhythm can be entrained by an LD cycle and shifted by a single light pulse. Therefore light receptors must exist which transfer the light signal to the oscillator which controls eclosion.

An action spectrum for phase shifting the eclosion rhythm with a single light pulse shows a broad maximum in the blue (457 nm) and further maxima at 375, 435 and 473 nm. Wavelengths beyond 540 nm are ineffective. Only one photopigment is involved in the advance and delay effects.

The photoreceptor pigment is very likely a flavoprotein and not a carotenoid derivative (rhodopsin): Whereas the visual sensitivity of the compound eyes of flies reared on a carotenoid free diet had decreased by three orders of magnitude, the photosensitivity of the circadian eclosion rhythm was not affected. Furthermore, the eclosion rhythm of mutants lacking compound eyes was still synchronized by light. The compound eyes in the metamorphosed fly in the pupae are thus not needed to phase shift and entrain the eclosion rhythm. The responsible photoreceptor(s) must be located somewhere else.

The larval eyes are involved in entraining the eclosion rhythm. Functional larval eyes are needed to entrain the molecular rhythms in the pacemaker cells in mutants which lack extraretinal photoreception.

The circadian clock controls also the locomotor activity of *Drosophila*. In an LD cycle the flies are only active during the light period. In constant darkness the rhythm free-runs with a period close to but usually not exactly 24 hours. Temperature compensation keeps period (almost) constant at different environmental temperatures. In continuous light the period is influenced by the fluence rate. At higher light intensities the flies become arrhythmic.

8.2. Multioscillatory System of Drosophila and the Underlying Mechanisms

Drosophila flies, like other insects, possess a *multioscillatory system* to control different events in a circadian way. Circadian oscillators seem to be widespread throughout the different tissues and cells: Using a construct in which the luciferase gene *lux* is fused to per, which is under circadian control, the luminescence of the whole fly, of parts of the fly, and of cultured tissue could be monitored. Under LD-cycles it shows diurnal, and under continuous darkness circadian fluctuations. The cultures could be synchronized to a new LD-cycle. This shows that the circadian system in *Drosophila* is cell-autonomous and photoreceptive. Light might serve as the master coordinator of these cellular clocks.

However, behavior such as eclosion of the flies out of the puparium and locomotor activity is driven by circadian centers in the brain. Where they are localized, how they are coupled, connected with inputs and outputs and organized is being intensively studied (Helfrich-Förster *et al.* 1998 and Figure 5).

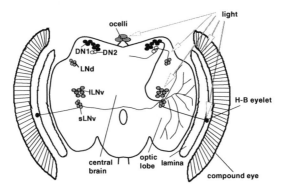

Figure 5: Neuroanatomy of the circadian pacemakers of locomotor activity and eclosion in Drosophila: *Looking toward the front of the brain from the neck, rhythm relevant neurons and photoreceptors are shown. Light for synchronization is received by several photoreceptors, the compound eye, the Hofbauer-Buchner eye (H.B.eye), ocelli, and by ventrolateral neurons (LNv). The LN are deep brain cells and consist of two groups: The smaller LNv (black) and the larger LNv (circles). The smaller ones are responsible for the circadian control of adult behavior: They project to dorsolateral neurons (LNd) and to dorsal neurons (DN), which are other clock-gene expressing cells. They furthermore project to the contralateral side across the midline of the brain in the posterior optical tract. And finally they project with highly ramified neurites to the optic lobes (see right side). The smaller LNv secrete the pigment dispersing factor PDF into the hemolymph. Secretion is under circadian control. After Helfrich-Förster et al. (2001).*

The circadian oscillators which control activity and eclosion consist of a molecular feedback loop. It is generated by interactions of several 'clock genes', namely period (per), timeless (tim), Clock (Clk), cycle (cyc) and doubletime (dbt), and their products PER, TIM, CLK, CYC, and DBT. The eclosion and locomotor activity rhythms are affected by mutations in these genes: The flies become arrhythmic or change period of the circadian rhythm. A present model is shown in Fig. 6. For reviews see Hall (1998), Young (1998), Dunlap (1999), Edery (1999), and Reppert andWeaver (2000).

More components and clock factors are probably involved. For instance, the cAMP response element binding protein (CREB) participates in the feedback loop. It promotes oscillations of PER and TIM. *vrille* is a further candidate of a clock gene (Blau and Young 1999). It codes for a transcription factor, but its precise role in the clockwork is not clarified yet.

8.3. Photoreceptors for the Synchronization of the Locomotion Clock

The locomotor activity of mutants which lack compound eyes can still be entrained by LD cycles. But compound eyes are also involved in circadian photoreception: Eyeless flies (*sine oculis*) are less sensitive to light by 2 orders of magnitude and show a more narrow action spectrum as compared to wildtype flies. Red light has no effect on the circadian eclosion of wild types and the circadian

360

activity of blind mutants. But in the wild type the activity rhythm is entrained by rhodopsin as the photopigment. A spectral response curve for phase shifts of the activity rhythm (maximum at 500 nm with some sensitivity to red light) resembles more the action spectrum for entrainment of activity than the phase shifting of eclosion. The action spectra for eclosion and adult activity differ mainly in the red part of the spectrum, where eclosion is insensitive. They are probably due to the different compositions of circadian photoreceptors in larvae and adults.

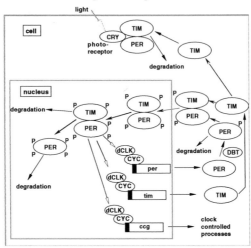

Figure 6: Model of the circadian Drosophila *clock which is based on a feedback loop involving transcriptional and translational events in the pacemaker cells. The feedback loop has the following structure: CLK and CYC are basic helix-loop helix-PAS transcription factors. They form heterodimers and bind to an element in the promoters of the per and tim genes and other clock controlled genes (ccg) in the nucleus activating their transcription. per and tim mRNA levels increase until they reach high levels early in the evening. The proteins PER and TIM reach maximal levels with a lag in late evening. This delay is due to post-transcriptional regulation of per mRNA and of PER. DBT in the cytoplasm phosphorylates PER and destabilizes it in this way. But PER becomes stabilized via dimerization with TIM, while its synthesis is rising. The PER:TIM heterodimers move into the nucleus and repress their own transcription by direct interaction of PER and TIM with CLK-CYC transcriptional activators. The phorphorylated TIM:PER is now converted to TIM and PER monomers which are degraded. This stops transcriptional repression at the end of the circadian cycle. Due to the lag between mRNAs and proteins, this negative feedback results in a stable cycling in per and tim mRNA and protein levels. Entrainment of the clockwork by light functions by affecting the blue sensitive cryptochrome (CRY). It stimulates the CRY:TIM interaction, which triggers TIM degradation. TIM is phosphorylated, ubiquinated and degraded in proteosomes. Other photoreceptors (rhodopsin, see text) and their interaction with the molecular gears of the clock are not shown, since they are so far unknown. After Edery (1999) and Dunlap (1999).*

The rhythm of activity in adult flies can be entrained by several photoreceptors (see Figure 5, Helfrich-Förster *et al.* 2001): The compound eyes, extraretinal photoreceptors located underneath the compound eye (Hofbauer-Buchner eyelets, Hofbauer and Buchner 1989) and chryptochrome in the pacemaker neurons (LNv) which control eclosion and activity (see Figure 5 and reviews by Hall 1998 and Helfrich-Förster *et al.* 1998). The entrainment is completely abolished in glass/cry *b* double mutants. They lack cryptochrome and all external and internal eye structures.

8.4. Synchronization and Entrainment of the molecular Feed-Back Oscillator

If the feedback loop described before is indeed the basis of the circadian rhythm, it must explain the following points:
1. attenuation of the rhythm by continuous light,
2. phase shifting by light pulses,
3. entrainment by LD cycles.

According to the model (Figure 6), all these effects are achieved through light dependent degradation of TIM. Continuous light keeps the TIM level permanently low. Because cytoplasmic PER degrades if not protected by TIM, its level is low, too while per and tim mRNA remain at a median level. Therefore under LL the clock genes and proteins do not oscillate. This explains point 1.

Delaying and advancing phase shifts of the rhythm by light pulses are brought about in the following way: Light is perceived by the circadian photoreceptor cryptochrome. Photochemical changes in the chromophore allow CRY to interact with TIM in the cytoplasm and the nucleus. As a result TIM is degraded and the PER/TIM heterodimeric complex is disrupted before it enters the nucleus. If a light pulse hits the phases of the rising TIM concentration, TIM is reduced and builds up again after the end of the pulse. The following peaks in TIM concentration are thus delayed. If the light pulse hits at peak TIM concentrations or afterward, the degradation of TIM is enhanced and the subsequent buildup is earlier. Thus the following peaks in TIM are advanced. This explains point 2.

The entrainment by LD cycles is the result of advancing and delaying phase shifts. They will keep the circadian oscillation in a certain phase relationship to the LD cycle. This explains point 3.

Thus, the model can explain the three light effects mentioned. But is there evidence? Yes, there is: TIM degradation induced by light pulses can be measured in the LNs (see Figure 5). It correlates well with the amount of phase shifts of the activity rhythm elicited by light pulses. Furthermore, the spectral response curves for TIM degradation and for phase shifts of the activity rhythm display a maximum between 400 and 450 nm. This shows that both events are causally related. TIM degrades also in the absence of PER (that is: no functional clock). Finally, TIM degradation does not depend on functional compound eyes. This implies, that the clock can be reset by an extraretinal pathway. All these results indicate that TIM degradation is crucial in circadian light perception. TIM is not light-sensitive by itself. The light signal must be transduced on its way to TIM. The blue light absorbing photopigment *Drosophila*-cryptochrome (DCRY) seems to be responsible. Cryptochromes are flavoproteins (review: Cashmore *et al.* 1999) and the absorption spectrum corresponds to the action spectra of the light effects on the *Drosophila* rhythm (Selby and Sancar 1999).

DCry gene transcription is clock-controlled, but the DCRY protein level is controlled by light and independent of the clock molecules (Egan *et al.* 1999, Ishikawa *et al.* 1999). The DCry gene dose correlates with the phase shifting effect of light in the activity rhythm: a low dose phase shifts slightly, over-expression of DCry shifts strongly. At higher light intensities the system is saturated, because DCRY exceeds a certain level. The over-expression is more pronounced in the phase delaying part of the cycle because the DCRY level is rather low and can be increased considerably by over-expression. At the phase advancing part of the cycle, DCRY is already close to its maximal level and almost saturated. Therefore none or only small behavioral responses are found. In the delay zone the phase shifts with low light fluence rate pulses (high fluence rate pulses saturate!) should therefore be more pronounced than in the advance zone, due to the different levels of available DCRY. This was found experimentally in behavioral studies in the eclosion rhythm but interpreted wrongly (Chandrashekaran and Engelmann 1973).

DCRY is thus an important player in circadian photoreception. TIM is probably the direct target of DCRY. Light changes the conformation of DCRY which allows it to enter the nucleus and to interact directly with the TIM/PER complex. The complex is now not able anymore to participate in the negative feed back loop (Ceriani et al. 1999). Degradation of TIM is thus a consequence of DCRY blocking PER/TIM and not the first step in phototransduction.

A mutation in the DCry gene was isolated (cryb) which affects binding of one of the two cofactors needed for cryptochrome functioning. Consequently the clock of the mutant does not react to light pulses anymore, and TIM and PER levels are not entrainable by LD-cycles. They do, however, oscillate in a quite normal way under temperature cycles (Stanewsky et al. 1998).

DCry seems to be expressed and effective in the LNv (Egan et al. (1999) and Figure 5). The small LNv are already present in the first larval instar and the level of PER and TIM cycles. This cycling is still entrainable in cryb mutants as well as in glass and norpA mutants that impair the function of the larval photoreceptors (Emery et al. 2000). The eclosion rhythm of all these mutants is entrainable (Helfrich-Förster, unpublished results). The larval small LNv are apparently entrained by cryptochrome *and* the larval eyes. They do not entrain when both pathways are impaired.

8.5. Why are multiple Photoreceptors used?

Drosophila uses several photoreceptors for entraining its circadian system. Larvae and adults utilize the blue-light photopigment cryptochrome which probably affects the pacemaker neurons in the brain directly. For entrainment compound eyes are used in addition. Furthermore, extraretinal eyes adjust the activity of the adults to the environmental light cycles. The roles of the different photoreceptors in affecting the circadian system of *Drosophila* are not yet completely clarified.

Why do *Drosophila* and other organisms use multiple photoreceptors for setting their circadian clocks? There are several reasons (Roenneberg and Foster 1997):
- The signal-to noise ratio is reduced if several inputs are used.
- During twilight at dusk and dawn not only the intensity of light changes, but also its spectral composition. Different qualities of the environmental light can be used by a set of different photoreceptors (see Roenneberg and Foster 1997).
- Natural LD cycles do not simply consist of light steps. Instead, light is increasing and decreasing slowly during the twilight of the day. If organisms use certain light intensities during twilight as the begin respectively end of the day, the daylength can be measured accurately and reliably, and independently of daily weather conditions.

• Entraining by dawn and dusk is more effective than lights-on /off programs in all animals tested so far including man (Fleissner and Fleissner 2002b).

9. LIGHT EFFECTS ON CIRCADIAN CLOCKS IN PLANTS: *ARABIDOPSIS*

Arabidopsis thaliana plants are well suited for studying circadian rhythms and their genetic and molecular background (McClung 1993, Millar 1999). In the 'Drosophila of the botanists' quite a number of circadian rhythms have been described, among them petal movements, circadian inhibition of elongation of the hypocotyl and the flower stalk (Jouve *et al.* 1998), leaf movement (Dowson-Day and Millar 1999), stomatal closure and opening (Webb 1998), coordination of metabolism (Kreps and Kay 1997) such as photosynthesis, activity of enzymes (Jones *et al.* 1998), expression of a wide variety of genes (overview by McClung *et al.* 2000), post-transcriptional rhythms, hormone production and responsiveness, photoperiodic control (see articles by Thomas, Lumsden, Jackson and Thomas, Coupland, Carre in Lumsden and Millar 1998). Furthermore, many mutants are known, among them several which affect the clock work, clock inputs and clock outputs. Besides mutants in which the function of the photoreceptors is affected, others are known, in which the *transfer* of the light induced signals is changed. Such mutants are known both for phytochromes and for the blue light receptors.

Arabidopsis has been used successfully for studying the molecular basis of the circadian clock(s) in this plant. For continuous recording it was of much advantage to use a construct of firefly luciferase gene with a promoter of the lhc2 gene which is under control of the circadian clock (Millar *et al.* 1995a). The method allows to monitor circadian rhythms in whole plants, but also in different tissues of the plant. It also makes screening of mutations in the clock easy by looking for aberrant temporal patterns of luciferase expression.

9.1. Zeitgeber

Light is, as in most other organisms, the most important time cue for synchronizing the circadian rhythms of *Arabidopsis*. As mentioned before, synchronization by light under natural conditions has to take into account, that the daylength changes systematically during the course of a year: Under short photoperiods, the pattern of entrainment ensures that at dawn the plant is highly responsive to light. Under long days the light response at dawn is reduced (Millar and Kay 1996). In addition to resetting the phase, light modulates also the period of the *Arabidopsis* clock. In continuous light the period of the luciferase rhythm is 24.5 hours, but in continuous darkness it is 30 to 35 hours (Millar *et al.* 1995a). Non-photic time cues are also used by the plant. Temperature cycles entrain the clock. Imbibition of the seeds sets a circadian clock which is insensitive to light during the first 60 hours. From the 36th hour onward light initiates a second rhythm which runs independently of the imbibition rhythm (that is, the output, namely LHC2 and CAT2, shows the two rhythms superimposed). Light after the 60th hour synchronizes the two rhythms (Kolar *et al.* 1998, Zhong *et al.* 1998).

9.2. Photoreceptors

In plants multiple photoreceptors measure the quality and quantity of light in the environment. Phytochromes, blue and UV-A light photoreceptors (cryptochromes) and a UV-B photoreceptor are used (Millar *et al.* 1995a, Whitelam and Devlin 1998, Batschauer 1998). This complex system of different photoreceptors with partially antagonistic functions and overlapping action spectra detects different wavelengths over a wide spectral range.

Whereas photoreceptors and the light regulated responses including gene expression have recently been intensively studied, the *signal transduction components* are much less known. A large number of signaling components must exist, which are affected by external and internal factors. Both genetical and biochemical approaches are used to clarify these transduction pathways and modes.

Many processes in plants are *directly* affected by light. But light affects also processes which are *under circadian control* by entraining these rhythms (see Figure 7). The immediate and circadian responses are genetically separable (Anderson *et al.* 1997). Phytochromes, cryptochromes (Briggs and Huala 1999) and a UV-B photoreceptor are used by the plants to entrain circadian rhythms. Phytochrome A is used for low fluence rate red light and low fluence rate blue light to control the circadian clock, phytochrome B is a high fluence rate red light photoreceptor controlling the circadian clock, cryptochrome 1 serves at low fluence rate of blue light for the circadian clock control and at high fluence rate it influences the period of the clock. A mutant lacking functional phytochromes and cryptochromes is still entrainable by visible light (Yanovsky *et al.* 2000). This indicates that another photoreceptor besides the phytochromes and cryptochromes is present in *Arabidopsis* which synchronizes the circadian clock to the LD cycle.

Mutations in the perception of light can be detected by screening for altered periods of the circadian clock (since period depends on the fluence rate). hy1 is one of these mutations (see Figure 7). det1 exhibits a rather short period (18 hours). It apparently mimics the effect of very high irradiance. cop1 works in the same direction, although to a lesser extent (Millar and Kay 1997).

Mutants are known which affect the transduction pathway between photoreceptors and the clock. The elf3 mutation causes arrhythmicity in lhc expression in LL and abolishes the leaf movement rhythm and the photoperiodic sensitivity for flower induction (early flowering) (Hicks *et al.* 1996). Since the rhythms are still observable under LD-cycles and since they occur also under continuous darkness, ELF3 is not a part of the circadian oscillator. Instead, it seems to be a component between the photoreceptor and the clock. ZEITLUPE (ZTL) is another example. It plays a primary role in the photo-control of circadian period (Somers *et al.* 2000).

Newly discovered mutants of circadian rhythmicity suggest a close link between photoreception, the circadian oscillator and transcriptional regulation. Light-regulated elements are rather closely interwoven with promoter sequences of clock controlled genes (Millar 1998). A few examples:

• The promoter of Lhcb (light-harvesting chlorophyll-binding) genes contains light-regulatory elements directly implicated in circadian responses and in phytochrome and blue light activity. Blue light responses utilize regions of the

promoter which are independent of the regions modulating the phytochrome and circadian responses (Folta and Kaufman 1999).
- The observed rapid damping of the rhythm of CAT3 mRNA in extended dark conditions requires phytochrome A and cryptochrome 1 but not phytochrome B. If light perception is disrupted by mutations, strong circadian cycling for many cycles in extended darkness is observed (Zhong *et al.* 1997).
- The gene which encodes Rubisco activase (RCA) is organ-specific, light-responsive and regulated by the circadian clock. It is expressed throughout the green plant but not in petals and roots (Liu *et al.* 1996).

9.3. Clock Mechanism

To identify clock genes (i.e. genes which are directly involved in the construction and functioning of the clockwork) several characteristics are used: mutations in these genes shorten or lengthen the period of the rhythm; null mutations lead to arrhythmicity; gene expression (mRNA production) is rhythmic; keeping the activity at high or low levels stops the clock from functioning; induced changes in component activities should shift or reset the phase of the rhythm (Aronson *et al.* 1994) A toc1 mutant (timing of lhc expression) specifically shortens the period of several 'hands´ (LHC, stomatal movement, circadian rhythmof photoperiodic control of flowering) to 21 hours. It has been cloned (Strayer *et al.* 2000). Temperature entrainment is normal, the period changes with light fluence rate as in the wild type, except that it is generally 2-3 hours shorter. All this speaks in favor of TOC being an essential element of the circadian clock and not an element in the input signal pathway (Somers *et al.* 1998). Other mutants lengthen the period. This points, together with other indications given by Aronson *et al.* (1994), toward a central role of these genes in the clock mechanism.

As in other circadian systems, the *Arabidopsis* clock consists of auto-regulatory feedback circuits which involve transcription and translation of clock genes. Figure 7 shows a current model of the *Arabidopsis* clock (Staiger and Heintzen 1999). Details and the inputs and outputs to overt rhythms are shown in the Figure. This regulated network integrates photoreceptor signals of the environmental light conditions to the clock, thus setting its phase. It further shows the output of the clock information to affect gene expressions. In addition, light affects gene expressions also directly and not only via the clock (immediate light effects). There is furthermore a developmental program which contributes information to the system.

366

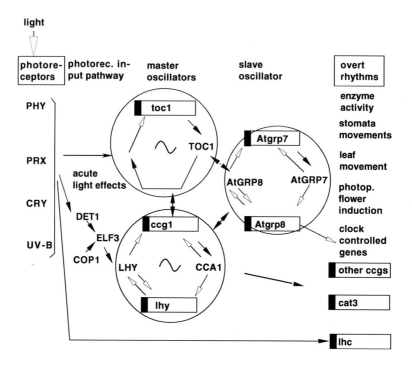

light

| photore-ceptors | photorec. in-put pathway | master oscillators | slave oscillator | overt rhythms |

Figure 7: Model of the circadian system of Arabidopsis *at the molecular level. It consists of two feedback loops as master oscillators (~), one involving TOC1 (timing of cab expression), the other LHY (late elongated hypocotyl) and CCA1 (circadian clock associated). Both loops influence several physiological ('overt rhythms') and molecular processes, among them a slave oscillator (a subordinated feedback loop) which consists of AtGRP7 and AtGRP8 (*Arabidopsis thaliana *glycine-rich protein). There is furthermore an outer feedback loop (not shown) which is required to maintain amplitude and period of the oscillation. Light is perceived by several photopigments, including phytochromes (PHY), cryptochromes (CRY), a UV-B sensitive pigment and an unknown pigment (PRX: the circadian rhythm of a mutant without PHY and CRY can still be synchronized Millar et al. 1995). The signals are transferred from the photoreceptors via input pathways (the gene products DET1, COP1 andELF3 are involved) to the master oscillators and also as immediate light effects to promoters of clock controlled genes such as lhc. Modified after* Staiger and Heintzen (1999) *and* Park et al. (1999).

It is debated, whether a single clock underlies circadian rhythms. The toc mutation changes period and affects several hands to the same extent (Somers *et al.* 1998). But in the wild type leaf movement occurs with a 25 hour period, whereas the lhc rhythm has only a 24.7 hour rhythm. The toc mutant shortens the period of the lhc rhythm to 20.7 hours, and the rhythm of the leaf movement to 23.3 hours. The simplest explanation is, that one clock drives the diverse rhythms, but the different cellular environments modify the speed of the oscillator. For instance, if coupling between oscillators is affected, oscillations can occur with different periods.

Alternatively more than one clock might control different activities in the *Arabidopsis* plant or even in a single cell (as was shown in *Gonyaulax*, see Roenneberg and Hastings (1988). In *Phaseolus* the rhythm of leaf movements differs from the one driving stomatal opening and photosynthetic activity (Hennessey and Field 1992). Using a luciferase reporter gene, LD cycles differing in phase were applied to different restricted tissue areas. The phases were maintained after the entraining treatments ended. This shows that the circadian oscillators in intact plants are autonomous without localized pacemakers and not affected by the entrainment of the rest of the plant. Phase-resetting signals are thus also autonomous (Thain *et al.* 2000).

The chloroplasts may harbor a separate clock. The Ca^{2+} level in the chloroplasts do not oscillate in constant light, in contrast to the calcium levels in the cytosol. Apparently the levels are regulated independently in the chloroplasts and the cytosol. The circadian clock in the chloroplasts stops oscillating under continuous light (Johnson *et al.* 1995). Transcription in the chloroplasts can be regulated simultaneously by light acting on topoisomerases. They affect super-coiling in the chloroplasts and in this way form a general way of regulating transcription of chloroplast genes (Salvador *et al.* 1998).

9.4. Clock controlled Genes

The output of the circadian system is visible in quite a number of cellular activities. Transcription is in many cases controlled by the circadian clock (Piechulla 1999). Many of these genes are related to photosynthesis. They are expressed early in the subjective day, such as the lhc gene (Millar and Kay 1996). cat2 is also expressed in the morning (CAT2 is involved in the glyoxysomal and photo-respiratory degradation of H_2O_2), but cat3 in the late afternoon (CAT3 is probably scavenging H_2O_2 at night, Zhong and McClung 1996). cat 1 does not cycle (McClung 1997). This shows that even in a gene family the timing of expression may differ. Other gene expressions occur at the same circadian time such as chloroplastic (Rubisco small subunit and Rubisco activase), peroxisomal (catalase) and mitochondrial components of the photorespiratory pathway (McClung *et al.* 2000). The isolation and study of these clock controlled genes will provide new tools for understanding the circadian system (Kreps *et al.* 2000).

Circadian variations in cytosolic Ca^{2+} levels were found by using the Ca sensitive luminescent aequorin protein (Johnson *et al.* 1995). Finally, post-transcriptional regulation by the clock is found in nitrate reductase (Pilgrim *et al.* 1993).

9.5. Conclusion

In plants the molecular basis of circadian oscillations is not yet well understood. The number of clock composing gene products and clock controlled genes is increasing constantly. It is to be expected, that a model of the interwoven feedback loops composing the clock work, the outputs to the driven processes and clock controlled genes and the inputs from the time cues will emerge.

There are multiple photoreceptors used by the plants to entrain the circadian clock. They are apparently quite closely linked to the clock mechanism. Whereas in animals a hierarchy of clock units with pacemaker centers is the rule, in plants the different tissues and organs are equipped with cellular circadian clocks which are autonomously running and synchronized by the light/dark environment. Plant hormones might serve as coordinators of these local clocks.

10. FUNGAL CLOCKS AND LIGHT RESETTING: *NEUROSPORA*

The ascomycete *Neurospora crassa* is originally a tropical fungus, but nowadays found all over the world. A circadian rhythm of macroconidia formation can easily be measured by using the bands which are formed while the mycelium grows over the agar surface in 'running tubes'. The period and phase shifts can be determined simply by using a ruler and time markings at the growth front. *Neurospora* is coenocytic, that is, many nuclei share a common cytoplasm. But it is compartmentalized.

The circadian rhythm is not only manifested in conidiation. Biochemical rhythms, metabolites, enzyme activity, heat shock and other proteins, mRNA´s of certain genes ('*clock controlled genes*') are also under circadian control. As a rule, these genes are not only clock controlled, but respond also directly to light (Nakashima and Onai 1996). For instance, expression of the clock-controlled gene ccg-2 (=eas) is circadian but regulated also directly by light in wild-type strains. In white collar mutants (wc-1 and wc-2) which are 'blind' to blue light, the ccg-2 mRNA does not cycle. However, ccg-2 mRNA is induced by light in a mutant without functional clock (frq-9). The circadian clock is thus not needed for the direct light induction of ccg-2 (Arpaia *et al.* 1993).

Different approaches have been used to unravel the circadian clock that underlies overt rhythms such as conidiation. In the pharmacological approach inhibitors are used that are known to interfere with certain parts of the metabolism. It is then checked whether they affect also the circadian clock by changing its speed and amplitude or by stopping the clock. Pulses of those substances might phase shift the rhythm if interfering with the clock, and this may depend on the phase of application. In the genetic approach biochemical mutants with known effects in metabolism or sensitive toward certain drugs are used or clock mutants which have altered clock properties.

A number of reviews on circadian rhythms in *Neurospora* are available (e.g. Bell-Pedersen 2000, Bell-Pedersen *et al.* 1996b, Brody 1994, Lakin-Thomas *et al.* 1990) and on light effects (Linden *et al.* 1997, Linden *et al.* 1999, Macino *et al.* 1998).

10.1. Effects of External Cues on the Circadian Clock

Usually light is the strongest time cue for entraining circadian rhythms. However. *Neurospora* is an exception: moderate temperature changes can dominate LD-cycles in phase shifting the rhythm (1-2°C are already sufficient; Liu *et al.* 1998). The amount of FRQ (see later) depends on the phase of the oscillator and on the environmental temperature. It is higher at higher temperatures (3 times higher at 28°C as compared to 21°C). Changing temperature corresponds to shifts in clock time, because the amount of FRQ changes immediately. In contrast to light, which triggers a cell response outside the clock, temperature resetting seems to be the result of changes brought about directly within the clock.

Continuous light suppresses the circadian modulation of conidiation. Instead, conidia are formed all the time. In continuous darkness or safelight such as red, conidiation occurs in a circadian pattern. A single pulse of light phase shifts the conidiation rhythm either by advancing or by delaying it. A light pulse applied at late subjective day and early subjective night delays the rhythm, a light pulse at late subjective night and early subjective morning advances the rhythm (Figure 8). It will be shown later how these effects of light, the rhythm annihilating one of continuous light and the phase shifting one of pulses, can be understood in terms of molecular events.

10.2. Photoreceptors and Transduction of Light Signal to Clock

In contrast to plants, *Neurospora* senses light only in the blue range. It is blind to light beyond 520 nm. The light that sets the clock and has immediate effects is perceived by flavin-containing substances (a flavin/b-type cytochrome complex?) and not by carotenoids (Russo 1986), but also not by nitrate reductase, a flavin containing substance (Ninnemann 1984). It is not yet known, whether one or several flavin-type photoreceptors are involved. An action spectrum of light phase shifting the rhythm shows maximal effects at 465 nm (Dharmananda 1980). Blue light induces the circadian rhythm. The double mutant white color wc-1 and wc-2 is blind (Russo 1988). The role of WC is discussed below.

The transduction of this blue light signal to the clock involves protein synthesis (protein synthesis inhibitors interfere with phase shifting by light pulses and phase shift the rhythm in a phase-dependent way if administered as a pulse) and mRNA expression of specific genes (see next subsection). Ca^{2+} is also involved, but probably not cAMP, and definitely not the IP3 cycle. The blue light seems to activate WC-1 rapidly by phosphorylating it via a protein kinase C, a light specific, positively acting element. WC-1 is thus the substrate for protein kinase C (Macino *et al.* 1998). WC-1 and WC-2 interact with each other via a dimerization domain 'PAS' which recognizes binding sites in light regulated promoters. A white color complex WCC is formed. WCC has a dual role: It activates the expression of the frq gene in darkness and it transfers light signals to light responsive and clock-controlled genes (arrows to frq, wc-1 and ccg in Figure 8).

It has been discussed whether cryptochrome is involved in the light responses of the circadian clock of *Neurospora*. CRY1 is present in *Neurospora*, but does not function as a photoreceptor in carotenogenesis and in light regulated transcription.

Although the photoreceptor for synchronizing the circadian rhythm seems to be close to the clock mechanism, it can be separated from the clock: light input pathways can be disrupted (mutation) without preventing the circadian clock from running (see next subsection).

10.3. Molecular Mechanism of the circadian Oscillator

The formation of aerial hyphae and asexual macroconidia is under circadian control. Genetic (Feldman 1982, Dunlap 1996) and molecular biological studies (Dunlap 1999, Lakin-Thomas 1998) show this to be a complicated interplay of a circadian clock, direct control by light, metabolic and developmental controls.

The following picture of the molecular mechanism of the circadian clock is presently favored: The product FRQ of the frq gene is an essential part of the circadian clock mechanism. The mRNA and the FRQ proteins of the frq gene are parts of a feedback system where FRQ regulates its own expression via the white color complex WCC (Lee *et al.* 2000). FRQ would thus be a state variable in the circadian system (Aronson *et al.* 1994). Temperature compensation was supposed to be the result of expressing two different kinds of FRQ in different amounts at higher and lower environmental temperatures (Liu *et al.* 1997). Furthermore, the effect of light as the most important time cue was modeled. For more details see Figure 8.

In synchronizing the rhythm, light affects this circadian system as a function of the phase at which it is seen. frq is rapidly induced by light pulses via WCC activation. Speed and magnitude of this induction suggests that this is the first clock-specific event in resetting the clock by light. If the synthesis of FRQ is blocked, light does not reset the clockwork anymore (Crosthwaite *et al.* 1995).

The phase shifting effects of light pulses occur, according to this model, in the following way: Light, received by the photoreceptor, quickly induces frq expression. if given during late night or early morning (the circadian activation would take several hours). The amounts of frq mRNA and FRQ increase earlier as without light pulse and advance the rhythm. Light in the late day and early evening again immediately increases the expression of frq and FRQ synthesis. This postpones the FRQ turnover and the rhythm is delayed. There is a close correlation between the dose response of frq mRNA and phase shifting in response to the fluence rate.

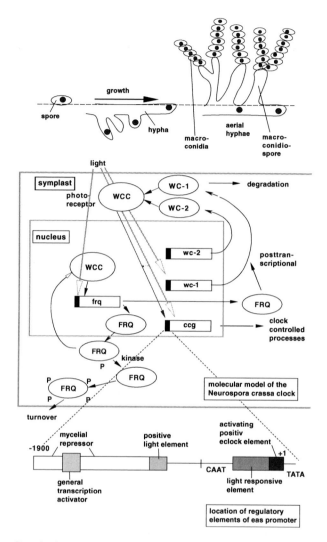

Figure 8: Top: Growth of Neurospora crassa *hypha on agar medium. In the late (real or subjective) night aerial hyphae are formed, at the tips of which macroconidiophores are formed. They produce asexual macroconidia. Center: A recent model of the molecular feedback loop responsible for the circadian oscillation and the effects of light on the different gears of the clock. The product FRQ of the frq gene is an essential part of the circadian clock mechanism. The mRNA and the FRQ proteins of the frq gene are parts of a feedback system where FRQ regulates its own expression via blocking (arrow with open head) the WCC heterodimeric complex. FRQ, on the other hand, activates the synthesis of the WC-1 protein. WC-1 and WC-2 form WCC. FRQ promoted synthesis of WC-1 is balanced by degradation of WC-1. FRQ is exported into the cytoplasm, where it is phosphorylated by a kinase and degraded (turnover). WCC has a dual role: It activates the expression of the frq gene in darkness and it transfers light signals to light responsive and clock-controlled genes (arrows to frq. wc-1 and ccg). Not shown is the temperature compensation mechanism. After Edery (1999). Bottom: The promoter of a clock controoed genes, eas, with a mycelial repressor, a general transcription activator, a positive light element, a light responsive element and an activating positive element (after Bell-Pedersen et al. 1996a).*

The wc-1 and wc-2 genes have both been cloned. wc-mutants have a low frq expression in darkness and show no circadian rhythm. Temperature is also unable to induce rhythmicity. This shows, that WC-1 and WC-2 are components and/or clock associated factors. They encode transcription factors with PAS domains putatively associated with protein-protein interactions. PAS domains are found in many regulatory proteins with functions in signal transduction and reception of different stimuli (light, chemical compounds, oxygen). These domains may serve to interact between receptors and signal transduction components. They are found also in phytochrome and in clock proteins (PER).This might indicate that clock proteins have evolved from proteins with photo-responses (Crosthwait et al. 1997). There is furthermore a sequence similarity between WC and PER which apparently extents also to the mouse clock (Antoch et al. 1997, King et al. 1997).

There are still many questions left regarding FRQ, its activity and stability, the regulators of WC and their targets. It is interesting that not all clock outputs are required for or associated with the daily cycling of development. The ccg-7 for instance codes glyceraldehyde-3-phosphate of the glycolytic pathway which is found in all organisms. It is thus a core enzyme of the metabolism. Other clock controlled genes are involved in development, stress responses, intermediary metabolism (Dunlap et al. 1998).

This Neurospora model might have to be modified in view of more recent findings where FRQ, WC-1 and WC-2 deficient mutants can be restored to exhibit robust rhythmicity under lipid-deficient conditions. Instead of being elements of the core oscillator, frq, wc-1 and wc-2 gene products may be elements of an input unit to the core oscillator. This input unit transduces the light signal into the core oscillator and sustains its oscillation. It can be bypassed under altered lipid metabolism (Lakin-Thomas and Brody 2000, Merrow et al. 1999, Lakin-Thomas 2000). There seems to be a FRQ-less oscillator coupled to the feedback loop shown in the Figure (Loros and Feldman 1986, Iwasaki and Dunlap 2000).

11. CLOCKS IN THE UNICELLULAR DINOFLAGELLATE GONYAULAX

Circadian rhythms are also found in unicellular organisms such as algae. Euglena, Chlamydomonas and the giant cell alga Acetabularia could serve as examples (Roenneberg and Mittag 1996). The synchronizing and phase shifting effect of light has been studied particularly in Chlamydomonas (Kondo et al. 1991).

The marine dinoflagellate Gonyaulax has been used extensively to study the molecular mechanisms of the circadian rhythms and the processes controlled by it (Roenneberg 1996), but also with respect to light affecting the circadian rhythms (review Roenneberg and Foster 1997). This alga shows a circadian rhythm of bioluminescence and of aggregation of the cells. The bioluminescence consists of two phenomena: a series of flashes caused by a mechanical or a chemical disturbance and a much weaker glow which is observable in an undisturbed culture. The bioluminescence of the flash rhythm peaks during the middle of the dark period and lasts a few hours only each day. The glow rhythm peaks toward the end of the dark period.

The bioluminescence rhythm and the aggregation rhythm are, as is the rule for circadian rhythms, relatively independent of the environmental temperature. At higher temperatures the period is slightly longer. The clock is in this way buffered against changes in the environmental temperature. This temperature compensation has been explained by two chemical reactions with similar temperature-dependencies, in which one reaction product inhibits the other reaction (Hastings and Sweeney 1957).

As in most other organisms light is the strongest time cue for synchronizing the *Gonyaulax* rhythms, but nutrients also act as Zeitgeber and interact with light. Light pulses administered at different phases of the cycle shift the rhythm.

Under conditions of constant light the bioluminescence rhythm continues in a circadian fashion. The period depends on the illuminance and amounts to 24.4 hours at 1200 lux, to 22.8 hours at 3800 lux with damping of the rhythm. Beyond 10000 lux the bioluminescence rhythm disappears. In continuous darkness the period is 23.0 to 24.4 hours and the rhythm damps out. Period of the bioluminescence rhythm depends furthermore on the light quality: Under continuous red light it is longer than 24 hours and will increase further in stronger light. Under continuous blue light it is shorter and will further shorten under higher fluence rates. Furthermore, in red light the flash is stronger than the glow and both kinds of luminescence are stronger as compared to white light (further information and references in Roenneberg and Foster 1997).

Light pulses given during weak continuous light shift the rhythm. The phase response curve is asymmetric with small delays and stronger advances. For day-active organisms this makes sense and allows for a better adaptation to the varying light periods in the course of the year. The action spectrum shows maxima in the blue and red spectral regions: Light around 475 and around 650 nm are the most effective wavelength components (Hastings and Sweeney 1960). This could indicate chlorophyll as the responsible photoreceptor. However, this was experimentally excluded for the bioluminescence rhythm. Phytochrome also does not participate in synchronization. In contrast to the visible light, UV-light shifts the rhythm of the bioluminescence to earlier times only (advances the rhythm). No delay shifts are induced by UV.

Two photoreceptors influence the circadian system of *Gonyaulax*. One of them responds mainly to blue light and is highly sensitive, inducing strong phase responses and eliciting advance phase shifts during the subjective night (Morse *et al.* 1994). This receptor is furthermore under circadian control: It is activated after CT 15. The other photoreceptor is sensitive to blue *and* red light, delays the rhythm during subjective day (before CT 15) and does not lead to strong phase shifts even at higher fluence rates. The dependency of period on wavelength and fluence rate and the phase responses to light of different wavelengths explain the asymmetry of the phase response curve (Roenneberg and Hastings 1988).

Gonyaulax furthermore exhibits a swimming activity rhythm which is synchronized by LD cycles and is circadian under constant conditions. The rhythm can be followed by measuring the aggregation of the cells. It is under the control of a circadian clock, the *A oscillator*, which is separate from the one controlling bioluminescence, the *B oscillator*. This can be shown under certain experimental conditions where the two rhythms dissociate from each other by exhibiting different periods. The phase shifts brought about by light pulses differ also for these two rhythms: The B oscillator is mainly blue light sensitive, whereas the A oscillator is sensitive to both blue and red light.

The phase shifts by light pulses can be influenced by different substances. Creatine shortens the period in blue and white light, but not in red. Phase advances by blue are apparently affected (Roenneberg and Taylor 1994). Under creatine the asymmetrical phase response curve to light is converted to the more conventional symmetric type 0 phase response (Roenneberg and Taylor 1994). Allopurinol (which inhibits xanthine oxidase, a flavoprotein) inhibits blue light phase shifting after CT 15. Inhibition of protein synthesis and clock resetting are also correlated (Olesiak *et al.* 1987).

Each of the two oscillators control classes of proteins, the synthesis of which are differently timed (Roenneberg and Mittag 1996). The kinase inhibitor DMAP (inhibiting protein phosphorylation) inhibits phase shifting of light pulses at all phases (Comolli *et al.* 1994). Creatine and allopurinol seem to act close to the photoreceptor, whereas DMAP acts closer to the clock. Drugs inhibiting mitochondria and/or electron transport partially inhibit the phase shift by light.

12. CLOCKS AND LIGHT IN CYANOBACTERIA

In cyanobacteria the expression of many genes is under circadian control. To facilitate the monitoring of these rhythms, a bacterial luciferase luxAB gene set was used as a reporter in *Synechococcus* and *Synechocystis* allowing continuous video camera recording of the amount of emitted light from many clones on a medium in Petri dishes (Kondo *et al.* 1993). The lux gene reporter construct was originally fused to the psbA1 gene (encoding D1, a major component of the photosystem II reaction center) and afterward to a large number of promoters. This method offers an easy to read 'hand' of the underlying clock. It turned out that almost every gene and thus the entire metabolism is under circadian control (Liu *et al.* 1995). More than 100 mutants have been selected with altered properties of the clock and they were used to shed light on the possible functioning of the circadian system (Kondo *et al.* 1994).

All rhythmic strains were entrained by dark pulses given during the continuous light of the recording conditions. Their period was not affected by different temperatures. The results are reviewed in several articles (Golden *et al.* 1997, Johnson *et al.* 1996, Kondo and Ishiura 2000).

12.1. Molecular Clock Model

The circadian system of cyanobacteria consists of a negative feedback loop where the products of a gene cluster of three open reading frames kaiA, kaiB and kaiC influence the transcription of their genes (Ishiura *et al.* 1998). A negative feedback control of kaiC expression by KaiC generates a basic oscillation, and KaiA sustains as a positive factor the oscillation by enhancing kaiBC expression (kaiB and kaiC are under control of the same promoter, see Figure 9). The Kai proteins physically interact to form a complex which is critical for the generation of the circadian rhythm (Iwasaki *et al.* 1999). There is no rhythm if one or several of the kai genes are deleted or inactivated. Re-introduction restores rhythmicity. This shows, that all three genes are important for the generation of the circadian rhythm. Furthermore, they represent the key gears of the clock, since clock mutants are complemented by introducing the kai gene cluster DNA (Ishiura *et al.* 1998). Finally, the kai genes are especially dedicated to generate the rhythm: Mutants show no other changes besides altered clock properties. The KaiC binds ATP and GTP and this plays an important role in the rhythm generation (Nishiwaki *et al.* 2000). Details of the clock mechanism are not yet fully understood. Some clock-output factors control the kaiA promoter. In addition they control genes ('clock controlled genes') which in turn lead to circadian expression of their products (Ishiura *et al.* 1998). For more details see Figure 9 and Kondo and Ishiura (2000).

The period is mainly determined by the KaiC, since period mutations consist of single amino acid substitutions in the KaiC protein. The periods of the mutants range from 14 hours to 60 hours. The three Kai proteins show various hetero- and homotypic associations in vitro. A long period mutation kaiA1 enhances these interactions. The biochemical function of the Kai proteins is unknown. KaiC contains two ATP / GTP binding domains which plays an important role in the rhythm generation (Nishiwaki *et al.* 2000). The histidine kinase SasA interacts with KaiC. Its disruption lowered kaiBC expression, reduces amplitude of the rhythm and shortens the period. Continuous over-expression eliminates the rhythm, while limited over-expression shifts the phase. SasA is necessary for a robust circadian rhythm (Iwasaki *et al.* 2000).

12.2. Temporal Orchestration of Gene Expression

The extensive circadian control of gene expression in *Synechococcus* consists of at least two groups with respect to phase: As with PsbA1, about 80 % of the assayed promoters are active during the day with a maximum near the end of day. In the smaller group activity is maximal at dawn and minimal at dusk (see lower right in Figure 9). These genes may encode for instance oxygen sensitive enzymes and they perform best at night, where photosynthesis is absent.

Some genes involved in photosynthesis in *Synechococcus* and *Synechocystis,* such as the psbAI gene, which drives luxAB in the reporter strain, are regulated by light (Golden *et al.* 1995, Kumar *et al.* 1999). Other studies indicate that nitrogenase activity during the dark cycle results from de novo synthesis of the enzyme. During the light phase it is quickly degraded, perhaps by a protease. Transcription of the rbcL gene, which encodes the catalytic subunit of the key enzyme of CO_2 fixation (a light-dependent process), is enhanced in the light (Chow and Tabita 1994).

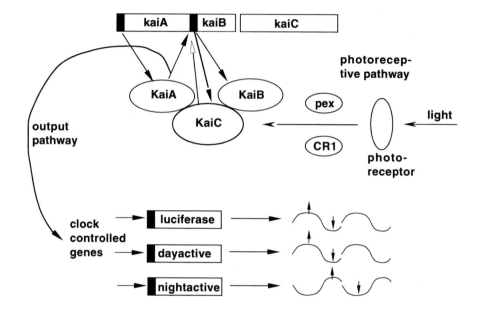

Figure 9: *Model of the circadian clock of cyanobacteria: The circadian system of cyanobacteria consists of a feedback loop where the products of three kai-genes kaiA, kaiB and kaiC influence the transcription of their genes. kaiB and kaiC are controlled by the same promoter. The different kai products interact with each other in the following way: The Kai proteins physically interact to form a complex KaiA KaiC KaiB. A negative feedback control of the kaiC expression by KaiC generates a basic oscillation, and KaiA generates or sustains the oscillation by enhancing kaiB and kaiC expression. Since the Kai proteins are without DNA binding motifs, transcription factors must be involved in the feedback loop. They are so far unknown. Some clock-output factors control the kaiA promoter (not shown). In addition they control clock controlled genes which in turn lead to circadian expression of their products. The phase of the overt rhythms belongs either to the day active type or to the night active type (bottom right, arrow up maximum, arrow down minimum of oscillation). The output and phase relationship of the luciferase reporter are also shown. Light is perceived by a photoreceptor (right) and a signal transduced to the oscillator via the kai-complex. Factors such as CR1 and PEX are probably involved in the transduction of the light signal. After Ishiura et al. (1998).*

12.3. Zeitgeber and Photoreceptors

A circadian rhythm is induced in *Synechococcus* by lowering the fluence rate of continuous light (Chen *et al.* 1993).

The circadian clock of *Synechocystis* is synchronized by LD-cycles and light pulses (Aoki *et al.* 1997). The rhythm continues if the cultures are transferred to LL or DD conditions. The period differs: It amounts to 25 hours if the culture is kept under DD, and to 22.6 hours under LL conditions (Chen *et al.* 1991, Kondo *et al.* 1993). Three hour light pulses administered during the DD free run leads to advancing or delaying phase shifts of the rhythm depending on the phase of the application. The circadian rhythm is thus, as expected, light responsive. In *Synechococcus* it was shown that temperature in addition to light entrains the circadian clock, but light has priority over temperature (Lin *et al.* 1999). The clock is temperature compensated (Aoki *et al.* 1995).

Synechocystis, with its rhythm expressed in DD and with its genome entirely sequenced, will help to unravel the photo-transduction pathway leading to phase shifts and synchronization of the circadian rhythm. Several candidates for photoreceptors are already known from the sequence of the genome, including a phytochrome, two-component systems and adenylate cyclases, among others. The bacteriophytochrome CikA seems also to provide light input to the oscillator (Schmitz *et al.* 2000). Furthermore, a bacterial cryptochrome has been found (Hitomi *et al.* 2000). Disrupting or over-expressing these genes will allow to determine whether some of these candidates are involved in the transduction of the light signal to the circadian clock. An action spectrum of phase shifting light has been determined (Inouye *et al.* 1998).

ACKNOWLEDGEMENTS AND REMARK

We acknowledge help of Franziska Wollnik, Charlotte Förster and Till Roenneberg for providing literature. For lack of space the references had to be restricted. This is especially the case for the *Drosophila* section. However, two papers on this topic can be referred to (Helfrich-Förster & Engelmann 2002 and Foster and Helfrich-Förster 2001). If more detailed references are needed, contact the second author at engelmann@uni-tuebingen.de.

REFERENCES

Åkerstedt, T. (1998). Is there an optimal sleep-wake pattern in shift work?. *Scandinavian J. Work, Environment & Health,* 24 *(Suppl. 3),* 18–27.

Anderson, S., Somers, D., Millar, A., Hanson, K., Chory, J. & Kay, S. (1997). Attenuation of phytochrome A and B signaling pathways by the *Arabidopsis* circadian clock. *Plant Cell, 9,* 1727–1743.

Antoch, M., Song, E., Hang, A., Vitaterna, M., Zhao, Y., Wilsbacher, L., Sangoram, A., King, D., Pinto, L. & Takahashi, J. (1997). Functional identification of the mouse circadian clock gene by transgenic BAC rescue. *Cell, 89,* 655–667.

Aoki, S., Kondo, T. & Ishiura, M. (1995). Circadian expression of the dnaK gene in the cyanobacterium *Synechocystis* sp. strain PCC 6803. *J. Bacteriology, 177,* 5606–5611.

Aoki, S., Kondo, T., Wada, H. & Ishiura, M. (1997). Circadian rhythm of the cyanobacterium *Synechocystis sp.* strain PCC 6803 in the dark. *J. Bacteriology, 179,* 5751–5755.

Arbisi, P., Levine, A., Nerenberg, J. & Wolf, J. (1996). Seasonal alteration in taste detection and recognition threshold in seasonal affective disorder: The proximate source of carbohydrate craving. *Psychiatry Res., 59*, 171–182.

Argamaso, S., Froehlich, A., McCall, M., Nevo, E., Provencio, I. & Foster, R. (1995). Photopigments and circadian systems of vertebrates. *Biophys. Chem., 56*, 3–11.

Aronson, B., Johnson, K., Loros, J. & Dunlap, J. (1994). Negative feedback defining a circadian clock: autoregulation of the clock gene frequency. *Science, 263*, 1570–1572.

Arpaia, G., Loros, J., Dunlap, J., Morelli, G. & Macino, G. (1993). The interplay of light and the circadian clock: Independent dual regulation of clock-controlled gene ccg-2(eas). *Plant Physiology, 102*, 1299–1305.

Ashkenazi, I., Hartman, H., Strulovitz, B. & Dar, O. (1975). Activity rhythms of enzymes in human red blood cell suspension. *J. interdiscipl. Cycle Res., 6*, 291–301.

Baker, J. & Ranson, R. (1932). Factors affecting the breeding of the field mouse (*Microtus agrestis*). I. Light. *Proc.Roy.Soc., Series B, 110*, 113–332.

Batschauer, A. (1998). Photoreceptors of higher plants. *Planta, 206*, 479–492.

Bell-Pedersen, D. (2000). Understanding circadian rhythmicity in *Neurospora crassa*: from behavior to genes and back again. *Fungal Genetics & Biology, 29*, 1–18.

Bell-Pedersen, D., Dunlap, J. & Loros, J. (1996a). Distinct cis-acting elements mediate clock, light, and developmental regulation of the *Neurospora crassa* eas (ccg-2) gene. *Molec. & Cellular Biol., 16*, 513–521.

Bell-Pedersen, D., Garceau, N. & Loros, J. (1996b). Circadian rhythms in fungi. *J. Genet., 75*, 387–401.

Blau, J. & Young, M. (1999). Cycling vrille expression is required for a functional *Drosophila* clock. *Cell, 99*, 661–671.

Boulos, Z., Campbell, S., Lewy, A., Terman, M., Dijk, D. & Eastman, C. (1995). Light treatment for sleep disorders: Consensus report. VII. Jet lag. *J. Biological Rhythms 10*, 167–176.

Bradbury, M.J. & Dement, WC, E. (1997). Serotonin-containing fibers in the suprachiasmatic hypothalamus attenuate light-induced phase delays in mice. *Brain Research, 768*, 125–134.

Briggs, W. & Huala, E. (1999). Blue-light photoreceptors in higher plants. *Annual Review of Cell & Developmental Biology, 15*, 33–62.

Brody, S. (1994). Circadian rhythms in microorganisms. *Res. Microbiol., 145*, 499–501.

Brunner, D., Kraucht, K., Dijk, D., Leonhardt, G., Haug, H. & Wirz-Justice, A. (1996). Sleep electroencephalogram in seasonal affective disorder and in control women: Effects of midday light treatment and sleep deprivation. *Biological Psychiatry, 40*, 485–496.

Bryant, T. (1972). Gas exchange in dry seeds: Circadian rhythmicity in the absence of DNA replication, transcription, and translation. *Science, 178*, 634 –636.

Bunney, W. & Bunney, B. (2000). Molecular clock genes in man and lower animals: Possible implications for circadian abnormalities in depression. *Neuropsychopharmacology, 22*, 335–345.

Cahill, G. & Hasegawa, M. (1997). Circadian oscillators in vertebrate retinal photoreceptor cells. *Biological Signals, 6*, 191–200.

Campbell, S. & Murphy, P. (1998). Extraocular circadian phototransduction in humans. *Science, 279*, 396–399.

Cardinali, D. (1998). The human body circadian: How the biological clock influences sleep and emotion. *Ciencia e Cultura (Sao Paulo), 50*, 172–177.

Cashmore, A., Jarillo, J., Wu, Y. & Liu, D. (1999). Cryptochromes: Blue light receptors for plants and animals. *Science, 284*, 760–765.

Cassone, V. (1993). Melatonin in vertebrate circadian rhythm. *Chronobiology International, 15*, 457–473.

Ceriani, M., Darlington, T., Staknis, D., Mas, P., Petti, A., Weitz, C. & Kay, S. (1999). Light-dependent sequestration of TIMELESS by CRYPTOCHROME. *Science, 285*, 553–568.

Chandrashekaran, M. & Engelmann, W. (1973). Early and late subjective night phase of the *Drosophila* rhythm require different energies of blue light for phase shifting. *Zeitschr. Naturforschung 28c*, 750–753.

Chen, T., Chen, T., Hung, L. & Huang, T. (1991). Circadian rhythm in amino acid uptake by *Synechococcus* RF-1. *Plant Physiology, 97*, 55–59.

Chen, T., Pen, S. & Huang, T. (1993). Induction of nitrogen-fixing circadian rhythm *Synechococcus* RF-1 by light signals. *Plant Science, 92*, 179–182.

Chow, T. & Tabita, F. (1994). Reciprocal light-dark transcriptional control of nif and rbc expression and light-dependent posttranslational control of nitrogenase activity in *Synechococcus* sp. strain RF-1. *J. Bacteriology, 176*, 6281–6285.

Comolli, J., Taylor, W. & Hastings, J. (1994). An inhibitor of phosphorylation stops the circadian oscillator and blocks light-induced phase shifting in *Gonyaulax polyedra*. *J. Biological Rhythms, 9,* 13–26.

Crosthwait, S., Dunlap, J. & Loros, J. (1997). *Neurospora* wc-1 and wc-2: Transciption, photoresponses, and the origin of the circadian rhythmicity. *Science, 276,* 763–769.

Crosthwaite, S., Loros, J. & Dunlap, J. (1995). Light-induced resetting of a circadian clock is mediated by a rapid increase in frequency transcript. *Cell, 81,* 1003–1012.

Czeisler, C. (1995). The effect of light on the human circadian pacemaker. *In* Chadwick, D. & Ackrill, K. (eds), *Circadian clocks and their adjustment,* pp. 254–302. Chichester: John Wiley.

Czeisler, C. & Dijk, D. (1995). Use of bright light to treat maladaptation to night shift work and circadian rhythm sleep disorders. *J. Sleep Research, 4 (Suppl. 2),* 70–73.

Czeisler, C., Kronauer, R., Allan, J., Duffy, J., Jewett, M., Brown, E. & Ronda, J. (1989). Bright light induction of strong (type 0) resetting of the human circadian pacemaker. *Science, 244,* 1328–1333.

Dalgleish, T., Rosen, K. & Marks, M. (1996). Rhythm and blues: The theory and treatment of seasonal affective disorder. *British J. Clinical Psychology, 35,* 163–182.

David-Gray, Z., Janssen, J., DeGrip,W., Nevo, E. and Foster, R. (1998). Light detection in a 'blind' mammal. *Nature Neuroscience, 1,* 655–656.

Deacon, S. & Arendt, J. (1996). Adapting to phase shifts. I. An experimental model for jet lag and shift work. *Physiology & Behavior, 59,* 665–673.

Dharmananda, S. (1980). *Studies on the circadian clock of Neurospora crassa: Light induced phase shifting.* PhD thesis. University of California, Santa Cruz.

Dijk, D., Boulos, Z., Eastman, C., Lewy, A., Campbell, S. & Terman, M. (1995). Light treatment for sleep disorders: Consensus report. II. Basic properties of circadian physiology and sleep regulation. *J. Biological Rhythms. 10,* 113–125.

Dowson-Day, M. & Millar, A. (1999). Circadian dysfunction causes aberrant hypocotyl elongation patterns in *Arabidopsis. Plant J. 17,* 63–71.

Dunlap, J. (1996). Genetic and molecular analysis of circadian rhythms. *Annu. Rev. Genetics., 30,* 579–601.

Dunlap, J. (1999). Molecular bases for circadian clocks. *Cell, 96,* 271–290.

Dunlap, J. (2000). An end in the beginning. *Science, 280,* 1548–1550.

Dunlap, J., Loros, J., Crosthwaite, S., Liu, Y., Garceau, N., Bell-Pedersen, D., Shinohara, M., Luo, C., Collett, M., Cole, A. & Heintzen, C. (1998). The circadian regulatory system in *Neurospora. In* Chaddick, M., Baumberg, S. D., Hodgson, D. & Phillips-Jones, M. (eds), *Microbial Responses to Light and Time,* pp. 279–295. Cambridge: University Press.

Eastman, C., Boulos, Z., Terman, M., Campbell, S., Dijk, D. & Lewy, A. (1995). Light treatment for sleep disorders: Consensus report. VI. Shift work. *J. Biological Rhythms, 10,* 157–164.

Eastman, C., Martin, S. & Hebert, M. (2000). Failure of extraocular light to facilitate circadian rhythm reentrainment. *Chronobiol. Intern., 17,* 807–826.

Eastman, C., Stewart, K., Mahoney, M., Liu, L. & Fogg, L. (1994). Dark goggles and bright light improve circadian rhythm adaptation to night-shift work. *Sleep, 17,* 535–543.

Eastman, C., Young, M., Fogg, L., Liu, L. & Meaden, P. (1998). Bright light treatment of winter depression: A placebo-controlled trial. *Archives General Psychiatry, 55,* 883–889.

Edery, I. 1999. Role of postranciptional regulation in circadian clocks: Lessons from *Drosophila. Chronobiol. Internat. , 16,* 377–414.

Egan, E., Franklin, T., Hilderbrand-Chae, M., McNeil, G., Roberts, M., Schroeder, A., Zhang, X. & Jackson, F. (1999). An extraretinally expressed insect cryptochrome with similarity to the blue light photoreceptors of mammals and plants. *J. Neurosci., 19,* 3665–3673.

Emery, P., Stanewsky, R., Hall, J. & Rosbash, M. (2000). A unique circadian-rhythm photoreceptor.. *Nature, 404,* 456–457.

Engelmann,W., Johnsson, A., Kobler, H. & Schimmel, M. (1978). Attenuation of the petal movement rhythm of *Kalanchoe* with light pulses. *Physiol. Plant., 43,* 68–76.

Enright, J. (1965). Synchronization and ranges of entrainment. *In* Aschoff, J. (ed.), *Circadian clocks. Proceedings of the Feldafing summer school, 7-18 September 1964,* pp. 112–124. Amsterdam: North-Holland Publishing Co.

Esseveldt, L., Lehman, M. & Boer, G. (2000). The suprachiasmatic nucleus and the circadian time-keeping system revisited. *Brain Research Reviews, 33,* 34–77.

Feldman, J. (1982). Genetic approaches to circadian clocks. *Annu. Rev. Plant Physiol., 33,* 583–608.

Fleissner, G. & Fleissner, G. (2002a). Retinal circadian rhythms. *In* Kumar, V. (ed.), *Biological rhythms,* pp. 71-82. New Delhi: Narosa Publ. House.

Fleissner, G. & Fleissner, G. (2002b). Perception of natural Zeitgeber signals. *In* Kumar, V. (ed.), *Biological rhythms,* pp. 83-93. New Delhi: Narosa Publ. House.

Folta, K. and Kaufman, L. (1999). Regions of the pea Lhcb1*4 promoter necessary for blue-light regulation in transgenic *Arabidopsis. Plant Physiology, 120,* 747–755.

Forger, D., Jewett, M. & Kronauer, R. (1999). A simpler model of the human circadian pacemaker. *J. Biological Rhythms, 14,* 532–537.

Forsgren, E. (1935). *Über die Rhythmik der Leberfunktion, des Stoffwechsels und des Schlafes.* Göteborg: Gumperts Bokhandel

Foster, R. & Helfrich-Förster, C. (2001). Photoreceptors for circadian clocks in mice and fruit flies. *Phil. Trans. Roy. Soc. Sci. London. B, 356,* 1-11..

Foulkes, N., Cermakian, N., Whitmore, D. & Sassone-Corsi, P. (2000). Rhythmic transcription: The molecular basis of oscillatory melatonin synthesis. *Novartis Foundation Symposium, 227,* 5–14.

Freedman, M., Lucas, R., Soni, B., von Schantz, M., Munoz, M., David-Gray, Z. & Foster, R. (1999). Regulation of mammalian circadian behavior by non-rod, non-cone, ocular photoreceptors. *Science, 284,* 502–504.

Giedke, H., Engelmann,W. & Reinhard, P. (1983). Free running circadian rest-activity cycle in normal environment. A case study. *Sleep Research, 12,* 365.

Golden, S., Ishiura, M., Johnson, C. H. & Kondo, T. (1997). Cyanobacterial circadian rhythms. *Annu. Rev. Plant Physiol. Plant Mol. Biol., 48,* 327–354.

Golden, S., Tsinoremas, N., Liu, Y., Kutsuna, S., Lebedeva, N., Andersson, C., Aoki, S., Johnson, C., Ishiura, M. & Kondo, T. (1995). The quest for the cyanobacterial circadian clock. *Plant Physiology, 108 (2 Suppl.),* 15.

Grace, M., Chiba, A. and Menaker, M. (1999). Circadian control of photoreceptor outer segment membrane turnover in mice genetically incapable of melatonin synthesis. *Visual Neuroscience, 16,* 909–918.

Graw, P., Recker, S., Sand, L., Krauchi, K. & Wirz-Justice, A. (1999). Winter and summer outdoor light exposure in women with and without seasonal affective disorder., *J. Affective Disorders 56,* 163–169.

Guillemette, J., Hebert, M., Paquet, J. & Dumont, M. (1998). Natural bright light exposure in the summer and winter in subjects with and without complaints of seasonal mood variations. *Biological Psychiatry, 44,* 622–628.

Halaris, A. (1987). *Chronobiology and psychiatric disorders.* New York, Amsterdam, London: Elsevier.

Hall, J. (1998). Molecular neurogenetics of biological rhythms. *J. Neurogenetics, 12,* 115–181.

Hastings, J. & Sweeney, B. (1957). On the mechanism of temperature independence in a biological clock. *Proc. Natl. Acad. Sci. USA, 43,* 804–811.

Hastings, J. & Sweeney, B. (1960). The action spectrum for shifting the phase of the rhythm of luminescence in *Gonyaulax polyedra. J. Gen. Physiol., 43,* 697–706.

Helfrich-Förster, C. & Diez-Noguera, A. (1993). Use of a multioscillatory system to simulate experimental results obtained for the period-mutants of *Drosophila melanogaster. J. Interdiscipl. CycleRes., 24,* 225–231.

Helfrich-Förster, C. & Engelmann,W. (2002). Photoreceptors for the circadian clock of the fruitfly. *in* Kumar, V. (ed.), *Biological Rhythms, pp. 94-106.* New Delhi: Narosa Publ. House. p. in press.

Helfrich-Förster, C., Stengl, M. & Homberg, U. (1998). Organization of the circadian system in insects. *Chronobiol. Internat., 15,* 567–594.

Helfrich-Förster, C., Winter, C., Hofbauer, A., Hall, J. and Stanewsky, R. (2001). The circadian clock of fruit flies is blind after elimination of all known photoreceptors. *Neuron, 30,* 249–261.

Hennessey, T. & Field, C. (1991). Circadian rhythms in photosynthesis. *Plant Physiol., 96,* 831–836.

Hennessey, T. & Field, C. (1992). Evidence for multiple oscillators in bean plants. *J. Biological Rhythms, 7,* 105–113.

Herzog, E. & Block, G. (1999). Keeping an eye on retinal clocks. *Chronobiology International 16,* 229–247.

Hicks, K., Millar, A., Carre, I., Somers, D., Straume, M., Kay, S. & Meeks-Wagner,

D. (1996). Conditional circadian dysfunction of the Arabidopsis early flowering 3 mutant. *Science, 274,* 790–792.

Hitomi, K., Okamoto, K., Daiyasu, H., Miyashita, H., Iwai, S., Toh, H., Ishiura, M. & Todo, T. (2000). Bacterial cryptochrome and photolyase: Characterization of two photolyase-like genes of *Synechocystis sp.* PCC6803. *Nucleic Acids Research, 28,* 2353–2362.

Hofbauer, A. and Buchner, E. (1989). Does *Drosophila* have seven eyes? *Naturwiss., 76*, 335–336.

Hoffmann, K. (1981). The role of the pineal gland in the photoperiodic control of seasonal cycles in hamsters. *In* Follett, B. & Follett, D. (eds), *Biological clocks in seasonal reproductive cycles,* pp. 237–250. Bristol: Wright.

Ibata, Y., Okamura, H., Tanaka, M., Tamada, Y., Hayashi, S., Iijima, N., Matsuda, T., Munekawa, K., Takamatsu, T., Hisa, Y., Shigeyoshi, Y. & Amaya, F. (1999). Functional morphology of the suprachiasmatic nucleus. *Frontiers in Neuroendocrinology, 20*, 241–268.

Ikonomov, O., Stoynev, A. & Shisheva, A. (1998). Integrative coordination of circadian mammalian diversity: neuronal networks and peripheral clocks. *Progress in Neurobiology, 54*, 87–97.

Illnerova, H. and Sumova, A. (1997). Photic entrainment of the mammalian rhythm in melatonin production. *J. Biological Rhythms, 12*, 547–555.

Illnerova, H. and Vanecek, J. (1982). Two-oscillator structure of the pacemaker controlling the circadian rhythm of N-acetyltransferase in the rat pineal gland. *J. Comp. Physiol., A145*, 539–548.

Inouye, C. (1993). Circadian rhythms in peptides and their precursor messenger RNAs in the suprachiasmatic nucleus. *In* Nakagawa, H., Oomura, Y. & Nagai, K. (eds). *Internat. Symp. Osaka: New functional aspects of the suprachiasmatic nucleus of the hypothalamus,* pp. 219–233. London: John Libbey & Co.

Inouye, C., Okamoto, K., Ishiura, M. and Kondo, T. (1998). The action spectrum of phase shift by light signal in the circadian rhythm in cyanobacterium. *Plant Cell Physiol., 39 (Suppl.),* S82.

Ishikawa, T., Matsumoto, A., Kato, T., Togashi, S., Ryo, H., Ikenaga, M., Todo, T., Ueda, R. & Tanimura, T. (1999). DCRY is a *Drosophila* photoreceptor protein implicated in light entrainment of circadian rhythm. *Genes to Cells, 4*, 57–65.

Ishiura, M., Kutsuna, S., Aoki, S., Iwasaki, H., Andersson, C., Tanabe, A., Golden, S., Johnson, C. & Kondo, T. (1998). Expression of a gene cluster kaiABC as a circadian feedback process in cyanobacteria. *Science, 281*, 1519–1523.

Iwasaki, H. & Dunlap, J. (2000). Microbial circadian oscillatory systems in *Neurospora* and *Synechococcus*: models for cellular clocks. *Curr. Opinion Microbiology, 3*, 189–196.

Iwasaki, H., Taniguchi, Y., M., I. & Kondo, T. (1999). Physical interactions among circadian clock proteins KaiA, KaiB and KaiC in cyanobacteria. *EMBO J., 18*, 1137–1145.

Iwasaki, H., Williams, S., Kitayama, Y., Ishiura, M., Golden, S. & Kondo, T. (2000). A kaiC-interacting sensory histidine kinase, SasA, necessary to sustain robust circadian oscillation in cyanobacteria. *Cell , 101*, 223–233.

Jagota, A., de la Iglesia, H. & Schwartz, W. (2000). Morning and evening circadian oscillations in the suprachiasmatic nucleus in vitro.. *Nature Neuroscience, 3*, 372–376.

Jewett, M., Kronauer, R. & Megan, E. (1999). Interactive mathematical models of subjective alertness and cognitive throughput in humans.. *J. Biological Rhythms, 14*, 588–597.

Johnson, C., Golden, S. & Kondo, T. (1996). Circadian clocks in prokaryotes. *Molecular Microbiology, 21*, 5–11.

Johnson, C., Golden, S. & Kondo, T. (1998). Adaptive significance of circadian programs in cyanobacteria. *Trends Microbiology, 6*, 407–410.

Johnson, C., Knight, M., Kondo, T., Masson, P., Sedbrook, J., Haley, A. & Trewavas, A. (1995). Circadian oscillations of cytosolic and chloroplastic free calcium in plants.. *Science, 269*, 1863–1865.

Johnsson, A. & Karlsson, H. (1972). A feedback model for biological rhythms. I. Mathematical description and basic properties of the model. *J. Theoretical Biology, 36*, 153–174.

Jones, T., Tucker, D. & Ort, D. (1998). Chilling delays circadian pattern of sucrose phosphat synthase and nitrate reductase activity in tomato. *Plant Physiology, 118*, 149–158.

Jouve, L., Greppin, H. & Degli Agosti, R. (1998). *Arabidopsis thaliana* floral stem elongation: Evidence for an endogenous circadian rhythm. *Plant Physiol. Biochem. (Paris), 36*, 469–472.

Kai, H., Arai, T. & Yasuda, F. (1999). Accomplishment of time-interval activation of esterase A4 by simple removal of pin fraction. *Chronobiology International, 16*, 51–58.

Kai, H., Kotani, Y., Miao, Y. & Azuma, M. (1995). Time Interval Measuring Enzyme for Resumption of Embryonic Development in the Silkworm, *Bombyx mori*. *J. Insect Physiol., 41*, 905–910.

Karakashian, M. & Schweiger, H. (1976). Circadian properties of the rhythmic system in individual nucleated and enucleated cells of *Acetabularia mediterranea*. *Exper. Cell Res., 97*, 366–377.

King, D. & Takahashi, J. (2000). Molecular genetics of circadian rhythms in mammals. *Annu. Rev. Neuroscience, 23*, 713–42.

King, D., Zhase, Y., Sangoram, A., Wilsbacher, L., Tanaka, M., Antoch, M., Stewes, T., Vitaterna, M., Kornhauser, J., Lowry, P., Turek, F. & Takahashi, J. (1997). Positional cloning of the mouse circadian clock gene. *Cell, 89*, 641–652.

Klein, D., Moore, R. & Reppert, S. (1991). *Suprachiasmatic nucleus: The mind's clock*. New York & Oxford: Oxford University Press.

Kokkola, T. & Laitinen, J. (1998). Melatonin receptor genes. *Ann. Medicine., 30*, 88-94.

Kolar, C., Fejes, E., Adam, E., Schaefer, E., Kay, S. & Nagy, F. (1998). Transcription of *Arabidopsis* and wheat Cab genes in single tobacco transgenic seedlings exhibits independent rhythms in a developmentally regulated fashion. *Plant J., 13*, 563–569.

Kondo, T. & Ishiura, M. (2000). The circadian clock of cyanobacteria. *BioEssays, 22*, 10–15.

Kondo, T., Johnson, C. & Hastings, J. (1991). Action spectrum for resetting the circadian phototaxis rhythm in the CW15 strain I: Cells in darkness. *Plant Physiology, 95*, 197–205.

Kondo, T., Strayer, C., Kulkarni, R., Taylor, W., Ishiura, M., Golden, S. & Johnson, C. (1993). Circadian rhythms in prokaryotes: Luciferase as a reporter of circadian gene expression in cyanobacteria. *Proc. Natl Acad. Sci. USA, 90*, 5672–5676.

Kondo, T., Tsinoremas, N., Golden, S., Johnson, C., Kutsuna, S. & Ishiura, M. (1994). Circadian clock mutants of cyanobacteria. *Science, 266*, 1233–1236.

Koorengevel, K., Beersma, D., Gordijn, M., den Boer, J. & van den Hoofdakker, R. (2000). Body temperature and mood variations during forced desynchronization in winter depression: A preliminary report. *Biological Psychiatry, 47*, 355–358.

Kreps, J. & Kay, S. (1997). Coordination of plant metabolism and development by the circadian clock. *Plant Cell, 9*, 1235–1244.

Kreps, J., Muramatsu, T., Furuya, M. & Kay, S. (2000). Fluorescent differential display identifies circadian clock-regulated genes in *Arabidopsis thaliana*. *J. Biological Rhythms, 121*, 208–217.

Kumar, A., Munehiko, A. & Mukato, S. (1999). Light-dependent and rhythmic psbA transcripts in homologous/heterologous cyanobacterial cells. *Biochem. Biophys. Res. Comm., 255*, 47–53.

Laakso, M., Hätönen, T., Stenberg, D., Alila, A. & Smith, S. (1993). The human circadian response to light -strong and weak resetting. *J. Biological Rhythms, 8*, 351–360.

Lakin-Thomas, P. (1998). Choline depletion, frq mutations, and temperature compensation of the circadian rhythm in *Neurospora crassa*. *J, Biological Rhythms, 13*, 268–277.

Lakin-Thomas, P. (2000). Circadian rhythms: new functions for old clock genes. *Trends Genetics, 16*, 135–142.

Lakin-Thomas, P. & Brody, S. (2000). Circadian rhythms in *Neurospora crassa*: lipid deficiencies restore robust rhythmicity to null frequency and white-collar mutants. *Proc. Natl Acad. Sci. USA, 97*, 256–261.

Lakin-Thomas, P., Cote, G. & Brody, S. (1990). Circadian rhythms in *Neurospora crassa*: biochemistry and genetics. *Crit. Revs Microbiology, 17*, 365–416.

Lam, R. & Levitan, R. (2000). Pathophysiology of seasonal affective disorder: a review.. *J. Psychiatry Neurosci., 25*, 469–480.

Lam, R., Terman, M. and Wirz-Justice, A. (1997). Light therapy for depressive disorders: Indications and efficacy. *In* Rush, A. (ed.), *Mood disorders: Systematic medication management. Modern Problems of Pharmacopsychiatry*, pp. 215–234. Basel & London: Karger.

Lee, K., Loros, J. & Dunlap, J. (2000). Interconnected feedback loops in the *Neurospora* circadian system. *Science, 289*, 107-110.

Lee, T., Chan, C., Paterson, J., Janzen, H. & Blashko, C. (1997). Spectral properties of phototherapy for seasonal affective disorder: A meta-analysis. *Acta Psychiatrica Scandinavica, 96*, 117–121.

Leibenluft, E., Turner, E., Feldman-Naim, S., Schwartz, P., Wehr, T. & Rosenthal, N. (1996). Light therapy in patients with rapid cycling bipolar disorder: Preliminary results. *Psychopharmacology Bull., 31*, 705–710.

Leloup, J. & Goldbeter, A. (2001). A molecular explanation for the long-term suppression of circadian rhythms by a single light pulse.. *Amer. J. Physiol. (Regulatory, Integrative and Comparative Physiology)*.

Lema, M., Golombek, D. & Echave, J. (2000). Delay model of the circadian pacemaker. *J. theor. Biol., 204*, 565 – 573.

Levitt, A., Wesson, V., Joffe, R., Maunder, R. & King, E. (1996). A controlled comparison of light box and head-mounted units in the treatment of seasonal depression. *J. Clin. Psychiatry, 57*, 105–110.

Lewis, R. (1999). Control system models for the circadian clock of the New Zealand Weta, *Hemideina thoracia* (Orthoptera: Stenopelmatidae). *J. Biological Rhythms, 14*, 480 – 485.

Lewy, A. & Sack, R. (1997). Exogenous melatonin's phase-shifting effects on the endogenous melatonin profile in sighted humans: A brief review and critique of the literature. *J. Biological Rhythms, 12*, 588–594.

Lewy, A., Bauer, V., Cutler, N., Sack, R., Ahmed, S., Thomas, K., Blood, M. and Latham-Jackson, J. (1998). Morning vs evening light treatment of patients with winter depression. *Arch. General Psychiatry, 55*, 890–896.

Lin, R., Chou, H. and Huang, T. (1999). Priority of light/dark entrainment over temperature in setting the circadian rhythms of the prokaryote *Synechococcus* RF-1. *Planta, 209*, 202–206.

Linden, H., Ballario, P. & Macino, G. (1997). Review: Blue light regulation in *Neurospora crassa*. *Fungal Genetics Biology, 22*, 141–150.

Linden, H., Ballario, P., Arpaia, G. & Macino, G. (1999). Seeing the light: News in *Neurospora* blue light signal transduction. *Adv. in Genetics* **41**, 35–54.

Liu, Y., Garceau, N., Loros, J. & Dunlap, J. (1997). Thermally regulated translational control of FRQ mediates aspects of temperature responses in the *Neurospora* circadian clock. *Cell* **89**, 477–486.

Liu, Y., Golden, S., Kondo, T., Ishiura, M. & Johnson, C. (1995). Bacterial luciferase as a reporter of circadian gene expression in cyanobacteria. *J. Bacteriology, 177*, 2080–2086.

Liu, Y., Merrow, M., Loros, J. & Dunlap, J. (1998). How temperature changes reset a circadian oscillator. *Science, 281*, 825–829.

Liu, Z., Taub, C. & McClung, C. (1996). Identification of an *Arabidopsis thaliana* ribulose-1,5-bisphosphate carboxylase/oxygenase activase (RCA) minimal promoter regulated by light and the circadian clock. *Plant Physiology, 112*, 43–51.

Loros, J. & Feldman, J. (1986). Loss of temperature compensation of circadian period length in the frq-9 mutant of *Neurospora crassa*. *J. Biological Rhythms, 1*, 187–198.

Luboshitzky, R. & Lavie, P. (1999). Melatonin and sex hormone interrelationships - A review. *J. Pediatric Endocrinology Metabolism, 12*, 355–362.

Lucas, R. & Foster, R. (1999). Photoentrainment in mammals: a role for cryptochrome?. *J. Biological Rhythms, 14*, 4–10.

Lumsden, P. & Millar, A. (1998). *Biological rhythms and photoperiodism in plants*. Environmental Plant Biology. Oxford, Washington, D.C.: Bios Scientific Publishers.

Macino, G., Arpaia, G., Linden, H. & Ballario, P. (1998). Responses to blue light in *Neurospora crassa*. *In* Chaddick, M., Baumberg, S., Hodgson, D. & Phillips- Jones, M. (eds), *Microbial Responses to Light and Time*, pp. 213–224. Cambridge: Cambridge University Press.

Marchant, E. & Mistlberger, R. (1997). Anticipation and entrainment to feeding time in intact and SCN-ablated C57BL/6j mice. *Brain Res., 765*, 273–282.

Martin, S. and Eastman, C. (1998). Medium-intensity light produces circadian rhythm adaptation to simulated night-shift work. *Sleep, 21*, 154–165.

McClung, C. (1993). The higher plant *Arabidopsis thaliana* as a model system for the molecular analysis of circadian rhythms. *In* Young, M. (ed.), *Molecular genetics of biological rhythms*. Vol. 4 of *Cellular Clocks Series* (Edmunds, L. N., ed.), pp. 1–35. New York, Basel, Hong Kong: Marcel Dekker.

McClung, C. (1997). Regulation of catalases in *Arabidopsis*. *Free Radical Biology & Medicine, 23*, 489–496.

McClung, C., Hsu, M., Painter, J., Gagne, J., Karlsberg, S. and Salome, P. (2000). Integrated temporal regulation of the photorespiratory pathway. Circadian regulation of two *Arabidopsis* genes encoding serine hydroxymethyltransferase. *Plant Physiology, 123*, 381–391.

Meesters, Y. (1998). Case study: Dawn simulation as maintenance treatment in a nine-year-old patient with seasonal affective disorder. *J. American Acad. Child & Adolescent Psychiatry, 37*, 986–988.

Meesters, Y., Beersma, D., Bouhuys, A. and van den Hoofdakker, R. (1999). Prophylactic treatment of seasonal affective disorder (SAD) by using light visors: Bright white or infrared light?. *Biological Psychiatry, 46*, 239–246.

Meesters, Y., Jansen, J., Beersma, D., Bouhuys, A. & van den Hoofdakker, R. (1995). Light therapy for seasonal affective disorder: The effects of timing. *British J. Psychiatry, 166*, 607–612.

Meijer, J., Watanabe, K. & Detari, L. (1996). Light entrainment of the mammalian biological clock. *Progr. Brain Research, 111*, 175–190.

Menaker, M., Moreira, L. & Tosini, G. (1997). Evolution of circadian organization in vertebrates. *Brazilian J. Medical Biological Res. 30*, 305–313.

Merrow, M., Brunner, M. & Roenneberg, T. (1999). Assignment of circadian function for the *Neurospora* clock gene frequency. *Nature, 399*, 584–586.

Miles, L., Raynal, D. & Wilson, M. (1977). Blind man living in normal society has circadian rhythm of 24.9 hours. *Science, 198*, 421–423.

Millar, A. (1998). Molecular intrigue between phototransduction and the circadian clock. *Ann. Botany, 81*, 581–587.

Millar, A. (1999). Biological clocks in *Arabidopsis thaliana. New Phytologist, 141*, 175–197.

Millar, A. and Kay, S. (1996). Integration of circadian and phototransduction pathways in the network controlling CAB gene transcription in Arabidopsis. *Proc. Natl Acad. Sci. USA, 93*, 15491–15496.

Millar, A. and Kay, S. (1997). The genetics of phototransduction and circadian rhythms in *Arabidopsis. Bioessays, 19*, 209–214.

Millar, A., Carre, I., Strayer, C., Chua, N. & A., K. (1995a). Circadian Clock Mutants in *Arabidopsis* identified by luciferase imaging. *Science, 267*, 1161–1163.

Millar, A., Straume, M., Chory, J., Chua, N. & Kay, S. (1995b). The Regulation of Circadian Period by Phototransduction Pathways in *Arabidopsis. Science, 267*, 1163–1166.

Minors, D. & Waterhouse, J. (1981). *Circadian rhythms and the human.* Bristol, London, Boston: Wright.

Minors, D., Waterhouse, J. & Wirz-Justice, A. (1991). A human phase response curve to light. *Neuroscience Lett., 133*, 36–40.

Mistlberger, R., de Groot, J. & Marchant, E. (1996). Discrimination of circadian phase in intact and suprachiasmatic nuclei ablated rats. *Brain Res., 96*, 12–18.

Miyamoto, Y. & Sancar, A. (1998). Vitamin B_2-based blue-light photoreceptors in the retinohypothalamic tract as the photoactive pigments for setting the circadian clock in mammals. *Proc. Natl Acad. Sciences USA, 95*, 6097–6102.

Moore-Ede, M., Sulzman, F. & Fuller, C. (1982). *The clocks that time us.* Cambridge, Mass. & London: Harvard Univ. Press.

Moore, R. (1997). Chemical neuroanatomy of the mammalian circadian system. *In* P. Redfern, P. & Lemmer, B. (eds), *Physiology and Pharmacology of biological rhythms*, chapter 4, pp. 79–93. Berlin, Heidelberg, New York: Springer.

Moore, R., Speh, J. & Card, J. (1995). The rhd originates from a distinct subset of retinal ganglion cells. *J. Comp. Neurol., 352*, 351–366.

Morin, L. (1994). The circadian visual system. *Brain Research Rev., 67*, 102–127.

Morse, D., Hastings, J. & Roenneberg, T. (1994). Different phase responses of two circadian oscillators in *Gonyaulax. J. Biological Rhythms, 9*, 263–274.

Nakashima, H. & Onai, K. (1996). The circadian conidiation rhythm in *Neurospora crassa. Semin. Cell Developm. Biol. 7*, 765–774.

Nathan, P., Burrows, G. and Norman, T. (1999). Melatonin sensitivity to dim white light in affective disorders. *Neuropsychopharmacology, 21*, 408–413.

Neumeister, A., Praschak-Rieder, N., Hesselmann, B., Rao, M., Glueck, J. & Kasper, S. (1997). Effects of tryptophan depletion on drug-free patients with seasonal affective disorder during a stable response to bright light therapy. *Arch. Gen. Psychiatry, 54*, 133–138.

Neumeister, A., Turner, E., Matthews, J., Postolache, T., Barnett, R., Rauh, M., Vetticad, R., Kasper, S. & Rosenthal, N. (1998). Effects of tryptophan depletion vs catecholamine depletion in patients with seasonal affective disorder in remission with light therapy. *Arch. Gen. Psychiatry, 55*, 524–530.

Nickla, D., Wildsoet, C. & Wallman, J. (1998). The circadian rhythm in intraocular pressure and its relation to diurnal ocular growth changes in chicks. *Exp. Eye Res., 66*, 183–193.

Ninnemann, H. (1984). The nitrate reductase system. *in* H. Senger (ed.), *Blue light effects in biological systems*, pp. 95–109. Berlin: Springer.

Nishiwaki, T., Iwasaki, H., Ishiura, M. & Kondo, T. (2000). Nucleotide binding and autophosphorylation of the clock protein KaiC as a circadian timing process of cyanobacteria. *Proc, Natl Acad. Sci. USA, 97*, 495–499.

Olesiak, W., Ungar, A., Johnson, C. & Hastings, J. (1987). Are protein synthesis inhibition and phase shifting of the circadian clock in *Gonyaulax* correlated?. *J. Biological Rhythms.*

Oltmanns, O. (1960). Über den Einfluss der Temperatur auf die endogene Tagesrhythmik und die Blühinduktion bei der Kurztagpflanze *Kalanchoe blossfeldiana. Planta, 54*, 233–264.

Ouyang, Y., Andersson, C., Kondo, T., Golden, S. & Johnson, C. (1998). Resonating circadian clocks enhance fitness in cyanobacteria. *Proc. Natl Acad. Sci. USA, 95*, 8660–8664.

Palchikov, V., Zolotarev, D., Danielenko, K. & Putilov, A. (1997). Effects of the seasons and of bright light administered at different times of day on sleep EEG and mood in patients with seasonal affective disorder. *Biological Rhythm Res., 28*, 166–184.

Park, D., Somers, D., Kim, Y., Choy, Y., Lim, H., Soh, M., Kim, H., Kay, S. & Nam, H. (1999). Control of circadian rhythms and photoperiodic flowering by the *Arabidopsis* GIGANTEA gene. *Science, 285*, 1579–1582.

Partonen, T. (1998). Short note: Extrapineal melatonin and exogenous serotonin in seasonal affective disorder. *Medical Hypotheses, 51*, 441–442.

Partonen, T. & Lonnqvist, J. (1996). Prevention of winter seasonal affective disorder by bright-light treatment.. *Psychological Medicine, 26*, 1075–80.

Peterson, E. (1981a). Dynamic response of a circadian pacemaker. I. Recovery from extended light exposure. *Biol. Cybern., 40*, 171–179.

Peterson, E. (1981b). Dynamic response of a circadian pacemaker. II. Recovery from light pulse perturbations. *Biol. Cybern., 40*, 181–194.

Piechulla, B. (1999). Circadian expression of the light-harvesting complex protein genes in plants. *Chronobiol. Internat., 6*, 115–128.

Pilgrim, M., Caspar, T., Quail, P. & McClung, C. (1993). Circadian and light-regulated expression of nitrate reductase in *Arabidopsis. Plant Molecular Biology 23*, 349–364.

Pittendrigh, C. (1965). On the mechanism of the entrainment of a circadian rhythm by light cycles. *In* Aschoff, J. (ed.), pp. 277–297. *Circadian clocks.* Amsterdam: North-Holland Publishing Co.

Pittendrigh, C. & Daan, S. (1976). A functional analysis of circadian pacemakers in nocturnal rodents. *J. Comparative Physiology, A106*, 333–355.

Provencio, I. & Foster, R. (1995). Circadian rhythms in mice can be regulated by photoreceptors with cone-like characteristics. *Brain Res., 694*, 183–190.

Provencio, I., Rodriguez, I., Jiang, G., Hayes,W., Moreira, E. & Rollag, M. (2000). A novel human opsin in the inner retina. *J. Neuroscience, 20*, 600–605.

Rea, M. (1998). Photic entrainment of circadian rhythms in rodents. *Chronobiol. Internat., 15*, 395–423.

Redfern, P., Minors, D. & Waterhouse, J. 1994. Circadian rhythms, jet lag, and chronobiotics: An overview. *Chronobiol. Internat., 11*, 253–265.

Reppert, S. & Weaver, D. (2000). Comparing clockworks: Mouse versus fly. *J. Biological Rhythms, 15*, 357–364.

Roenneberg, T. (1996). The complex circadian system of *Gonyaulax polyedra. Plant Physiology, 96*, 733–737.

Roenneberg, T. & Foster, R. (1997). Twilight times: light and the circadian system. *Photochem. Photobiol., 66*, 549–61.

Roenneberg, T. & Hastings, J. (1988). Two photoreceptors influence the circadian clock of a unicellular alga. *Naturwiss., 75*, 206–207.

Roenneberg, T. and Mittag, M. (1996). The circadian program of algae. *Seminar Cell. Developm. Biology, 7*, 753–763.

Roenneberg, T. & Taylor, W. (1994). Light induced phase responses in *Gonyaulax* are drastically altered by creatine. *J. Biological Rhythms 9*, 1–12.

Rosenthal, N. & Oren, D. (1995). Light therapy. *In* Gabbard. G. (ed.), *Treatments of psychiatric disorders.* 2nd ed. Vol. 1 and 2 of *Proc. Internat. Congr. Chronobiol. Paris 7-11 September 1997*, pp. 1263–1273. Washington, D.C.: Am. Psych. Press.

Ruberg, F., Skene, D., Hanifin, J., Rollag, M., English, J., Arendt, J. & Brainard, G. (1996). Melatonin regulation in humans with color vision deficiencies. *J. Clin. Endocrin. Metabol., 81*, 2980–2985.

Rusak, B. and Zucker, I. (1979). Neural regulation of circadian rhythms. *Physiol. Rev., 59*, 449–526.

Russo, V. (1986). Are carotenoids the blue light photoreceptor in the photoinduction of protoperithecia in *Neurospora crassa?. Planta, 168*, 56–60.

Russo, V. 1988. Blue light induces circadian rhythms in the bd mutant of *Neurospora*: double mutants bd,wc-1 and bd,wc-2 are blind. *J. Photochem. Photobiol. B 2*, 59–65.

Salvador, M., Klein, U. & Bogorad, L. (1998). Endogenous fluctuations of DNA topology in the chloroplast of *Chlamydomonas reinhardtii. Molecul. Cellular Biol., 18*, 7235–7242.

Samel, A. & Wegmann, H. (1997). Bright light: A countermeasure for jet lag? *Chronobiol. Internat., 14*, 173–183.

Samel, A., Wegmann, H. & Vejvoda, M. (1995). Jet lag and sleepiness in aircrew. *J. Sleep Res., 4 (Suppl. 2)*, 30–36.

Schmitz, O., Katayama, M.,Williams, S., Kondo, T. & Golden, S. (2000). CikA, a bacteriophytochrome that resets the cyanobacterial circadian clock. *Science, 289*, 765–768.

Schwartz, P., Brown, C., Wehr, T. & Rosenthal, N. (1996). Winter seasonal affective disorder: A follow-up study of the first 59 patients of the National Institute of Mental Health Seasonal Studies Program. *American J. Psychiatry, 153*, 1028–1036.

Sehgal, A., Ousley, A., Yang, Z., Chen, Y. & Schotland, P. (1999). What makes the circadian clock tick: genes that keep time? *Recent Progr. Hormone Res., 54*, 61–84.

Selby, C.P. & Sancar, A. (1999). A third member of the photolyase/blue-light photoreceptor family in *Drosophila*: A putative circdain photoreceptor. *Photochem. Photobiol.. 69*, 105–107.

Shanahan, T., Zeitzer, J. and Czeisler, C. (1997). Resetting the melatonin rhythm with light in humans. *J. Biological Rhythms, 12*, 556–567.

Silver, R. & Moore, R. (1998). Special issue on suprachiasmatic nucleus. *Chronobiol. Internat., 15*, VII–X and 395 ff.

Somers, D., Schultz, T., Milnamow, M. & Kay, S. (2000). ZEITLUPE encodes a novel clock-associated PAS protein from *Arabidopsis*. *Cell, 101*, 319–329.

Somers, D., Webb, A., Pearson, M. & Kay, S. (1998). The short-period mutant, toc1-1, alters circadian clock regulation of multiple outputs throughout development in *Arabidopsis thaliana*. *Development 125*, 485–494.

Soni, B., Philp, A., Knox, B. & Foster, R. (1998). Novel retinal photoreceptors. *Nature, 394*, 27–28.

Staiger, D. & Heintzen, C. (1999). The circadian system of *Arabidopsis thaliana*: Forward and reverse genetic approaches. *Chronobiology Internat., 16*, 1–16.

Stanewsky, R., Kaneko, M., Emery, P., Beretta, B., Wager-Smith, K., Kay, S., Rosbash, M. & Hall, J. (1998). The cryb mutation identifies cryptochrome as a circadian photoreceptor in *Drosophila*. *Cell, 95*, 681–692.

Steiner, M.,Werstiuk, E. & Seggie, J. (1987). Dysregulation of neuroendocrine crossroads: Depression, circadian rhythms and the retina – a hypothesis. *Progr. Neuro-Psychopharmacol. Biological Psychiatry, 11*, 267–278.

Steinlechner, S. (1992). Melatonin: an endocrine signal for the night lenght. *Verh. Dtsch. Zool. Ges., 85*, 217–229.

Stephan, F., Swann, J. & Sisk, C. (1979). Entrainment of circadian rhythms by feeding schedules in rats with suprachiasmatic lesions. *Behav. Neural Biol., 25*, 545–554.

Strayer, C., Schultz, T., Raman, R., Somers, D., Mas, P., Panda, S., Kreps, J. & Kay, S. (2000). Cloning of the *Arabidopsis* clock gene TOC1, an autoregulatory response regulator homolog. *Science, 289*, 768–771.

Swedo, S., Lowe, C. & Rosenthal, N. (1998). Case series: Pediatric seasonal affective disorder. A follow-up report. *J. American Academy of Child & Adolescent Psychiatry, 37*, 218–220.

Tassi, P., Pellerin, N., Moessinger, M., Hoeft, A. & Muzet, A. (2000). Visual resolution in humans fluctuates over the 24h period. *Chronobiol. Internat., 17*, 187–195.

Teng, C., Akerman, D., Cordas, T., Kasper, S. & Vieira, A. (1995). Seasonal affective disorder in a tropical country: A case report. *Psychiatry Res., 56*, 11–15.

Terman, J. & Terman, M. (1999). Photopic and scotopic light detection in patients with seasonal affective disorder and control subjects. *Biological Psychiatry, 46*, 1642–1648.

Terman, M., Amira, L., Terman, J. & Ross, D. (1996). Predictors of response and nonresponse to light treatment for winter depression. *Am. J. Psychiatry, 153*, 1423–1429.

Thain, S., Hall, A. & Millar, A. (2000). Functional independence of circadian clocks that regulate plant gene expression. *Curr. Biology, 10*, 951–956.

Thalen, B., Kjellman, B., Morkrid, L., Wibom, R. & Wetterberg, L. (1995). Light treatment in seasonal and nonseasonal depression. *Acta Psychiatrica Scandinavica, 91*, 352–360.

Thompson, C., Childs, P., Martin, N., Rodin, I. and Smythe, P. 1997. Effects of morning phototherapy on circadian markers in seasonal affective disorder. *British J. Psychiatry, 170*, 431–435.

Thorell, L., Kjellman, B., Arned, M., Lindwall-Sundel, K., Walinder, J. & Wetterberg, L. (1999). Light treatment of seasonal affective disorder in combination with citalopram or placebo with 1 year follow-up. *Internat. Clin. Psychopharmacol., 14 (Suppl.2)*, S7–S11.

Tosini, G. & Menaker, M. (1996). Circadian rhythms in cultured mammalian retina., *Science 272*, 419–421.

Touitou, Y. (1998). *Biological clocks: Mechanisms and applications*. Proc. Internat. Congr. Chronobiology Paris 7 - 11 September 1997. Amstedam: Elsevier.

Visser, E., Beersma, D. & Daan, S. (1999). Melatonin suppression by light in humans is maximal when the nasal part of the retina is illuminated. *J. Biological Rhythms, 14*, 116–121.

Webb, A. (1998). Stomatal rhythms. *In* Lumsden, P. & Millar, A. (eds), *Biological rhythms and photoperiodism in plants*, pp. 69–79. Bios Scientific Publishers.

Wehr, T. (1997). Melatonin and seasonal rhythms. *J. Biological Rhythms, 12*, 518–527.

Welsh, D., Logothetis, D., Meister, M. & Reppert, S. (1995). Individual neurons dissociated from rat suprachiasmatic nucleus express independently phased circadian firing rhythms. *Neuron, 14,* 697–706.

Wever, R. (1979). *The circadian system of man*. New York: Springer.

Wever, R. (1989). Light effects on human circadian rhythms: A review of recent Andechs experiments. *J. Biological Rhythms, 4*, 161–186.

Whitelam, G. & Devlin, P. (1998). Light signalling in *Arabidopsis*. *Plant Physiol. Biochem. 36*, 125–133.

Winfree, A. (1970). Integrated view of resetting a circadian clock. *J. Theor. Biol., 28*, 327–374.

Winfree, A. (1986). *The timing of biological clocks*. New York: Scientific American Books, Inc.

Wirz-Justice, A., Graw, P., Krauchi, K., Sarrafzadeh, A., English, J., Arendt, J. & Sand, L. (1996). 'Natural' light treatment of seasonal affective disorder. *J. Affective Disorders, 37,* 109–120.

Yanovsky, M., Mazzella, M. & Casal, J. (2000). A quadruple photoreceptor mutant still keeps track of time. *Curr. Biology, 10*, 1013–1015.

Young, M. (1998). The molecular control of circadian behavioral rhythms and their entrainment in *Drosophila*. *Annu. Rev. Biochem. 67*, 135–152.

Young, R. (1976). Visual cells and the concept of renewal. *Investigative Ophthalmol., 15*, 700–725.

Zeitzer, J., Kronauer, R. & Czeisler, C. (1997). Photopic transduction implicated in human circadian entrainment. *Neuroscience Letters, 232*, 35–138.

Zhong, H. and McClung, C. (1996). The circadian clock gates expression of two *Arabidopsis* catalase genes to distinct and opposite circadian phases. *Molecul. Gen. Genetics, 251*, 196–203.

Zhong, H., Painter, J., Salome, P., Straume, M. & McClung, C. (1998). Imbibition, but not release from stratification, sets the circadian clock in *Arabidopsis* seedlings. *Plant Cell, 10*, 2005–2017.

Zhong, H., Resnick, A., Straume, M. & McClung, C. (1997). Effects of synergistic signaling by phytochrome A and cryptochrome1 on circadian clock-regulated catalase expression. *Plant Cell 9*, 947–955.

LARS OLOF BJÖRN

16. BIOLUMINESCENCE

1. INTRODUCTION AND DEFINITIONS

Apart from the fluorescence and phosphorescence introduced in Chapter 1, there are three kinds of light emission that may take place from a living organism:

(1) Photosynthetic *delayed light emission*, also called *delayed fluorescence* or *afterglow*. This is weak red light emitted by all green plants and algae. The intensity is so low, and the light of such long wavelength that we cannot see it, but it is easily measured. It is due to reversion of the first steps of photosynthesis.

(2) *Ultraweak light emission* takes place from all organisms. It is due to various processes, mostly (but not always) involving molecular oxygen. It is regarded as a by-effect of metabolic activity, and has no biological function in itself. It is even weaker than the previous kind of light emission, and although it is often of shorter wavelength it cannot be seen. Rather sophisticated equipment is needed for its measurement. It can be exploited for studying what is going on in cells in a non-invasive and non-destructive way.

(3) *Bioluminescence*. This is the best-known of the biological luminescence phenomena, mostly because it can be observed using one's eyes only. We shall devote most of this chapter to bioluminescence. Although photosynthetic delayed light emission and ultraweak light emission in our terminology is not bioluminescence, sections at the end of the chapter will deal with these phenomena.

In addition to the more recent literature cited in this chapter, we would like to mention the book by Harvey (1952) as an excellent summary of the older bioluminescence literature. Extensive reviews of marine bioluminescence are provided by Tett & Kelly (1973) and Herring (1982).

2. EVOLUTION AND OCCURENCE AMONG ORGANISMS

Although most species are non-bioluminescent, most phyla have bioluminescent representatives. Among the exceptions are true plants and higher vertebrates (i.e., amphibians, reptiles, birds, and mammals).

Comparison of the biochemical systems involved clearly shows that bioluminescence has evolved multiple times. About 30 independently evolved systems are still extant (Hastings 1983, Wilson & Hastings 1998). Still, most bioluminescence systems share some common features, as we shall see in the section on biochemistry.

Thus bioluminescence occurs among bacteria, fungi, dinoflagellates, protozoa, sponges, cnidaria (coelenterates), ctenophores (comb jellies), molluscs, annelids,

389

L.O. Björn (ed.), Photobiology, 389–410.

crustaceans, insects, bryozoa, echinoderms, and fish. The majority of bioluminescent species live in the sea, although there are also many terrestrial bioluminescent insects, especially beetles. It has been estimated that 60-80% of the fish species in the deep sea are bioluminescent. Table 1 (adapted from Campbell 1988) shows the distribution in more detail.

Table 1. Systematic distribution of bioluminescence

Phylum	Approximative number of genera with bioluminescence
Bacteria	5
Pyrrophyta	11
Protozoa	9
Porifera	1
Cnidaria	66
Ctenophora	15
Rhyncocoela	1
Nematoda	1
Mollusca	74
Annelida	40
Arthropoda	207
Bryozoa	1
Echinodermata	47
Chordata	208

Since bioluminescent microorganisms exist, one must be careful not to confuse microbial luminescence with luminescence of the host. Many fish and mollusc species which have been regarded as bioluminescent organisms have been shown to glow by the light of symbiotic bacteria. There are, however, also cases of true fish and mollusc bioluminescence.

It is probable that bioluminescence first appeared during the "cambrian explosion" when the evolution of eyes had made it meaningful. Molecular oxygen is required for all known bioluminescence mechanisms, but the required oxygen partial pressure (or equivalent chemical activity) is much lower than that of the contemporary atmosphere.

3. BIOLOGICAL ROLES: WHAT IS BIOLUMINESCENCE GOOD FOR?

In some cases the advantage to the organisms of bioluminescence is quite clear, in other cases quite obscure. An old hypothesis for explaining bioluminescence in cases where no other explanation could be found is that it takes care of big energy quanta which could act destructively and converts them to harmless photons (McElroy & Seliger 1962, Seliger & McElroy 1965). The energy in a bioluminescence photon is an order of magnitude greater than the energy bound in a high energy phosphate bond in, e.g., ATP. The evolution bioluminescence could have been triggered by the appearence of free oxygen, causing formation of dangerous peroxides. As we shall see, peroxides play a role in most bioluminescent systems. It has also been pointed out (Seliger & McElroy 1965) that bioluminescent reactions, although depending on the presence of oxygen, require only very low partial pressures, corresponding to conditions in the distant past when organisms would first have had to adapt to this dangerous triplet molecule. Barros & Bechara (2000) further discuss the protective effect of luciferase with special regard to a beetle larva.

In those cases when we can clearly see a present-day biological role for bioluminescence, we can divide the advantages gained into three main categories: (1) reproduction, (2) protection (defence, camouflage or aposematic signalling), and (3) food acquisition. We shall give examples of each of them.

3.1 Reproduction

The best known examples of bioluminescence having a role in the propagation of the species is found among the beetles of the family Lampyridae (true fireflies and glowworms), although bioluminescence occurs also in several other beetle families. In glow worms only the female glows brightly (with a steady light), and by this attracts the male, and the same is the case in some firefly species. In other fireflies a sophisticated "light conversation" between males and females has evolved, with a different "language" for each species. Males send out an "interrogation" flash, and females respond. Species specificities are obtained both by the time course of the flash on a half-second time scale, and on the time delay between "interrogation" and "answer". Some details of this with references are given by Seliger & McElroy (1965).

The competitive value of various flashing abilities has been investigated by Branham & Greenfield (1996) and Vencl & Carlson (1998). For one species Branham & Greenfield (1996) found that the flash rate, rather than the flash length or flash intensity determined the female's preference.

An intriguing phenomenon is the synchronous flashing of the males of some fireflies. These males often collect in a tree, and the whole tree flashes "in step" (Buck & Buck 1976, Buck 1988). Species differences, various ideas about how this synchronised flashing comes about as well as how it aids the reproduction of the species are discussed with many references by Buck (1988). For newer investigations on the phenomenon we refer to Moiseff & Copeland (2000).

There are also a multitude of deep sea animals which use bioluminescence for finding a mate in the dark abyss, and here new discoveries are certainly going to be made for some time to come. Among the more interesting cases is the dragonfish, which uses light of a wavelength so long (maximum at 702 nm) that it cannot be perceived by other organisms.

There is no solid information on the role of bioluminescence in fungi. Bioluminescence has been reported in about 40 species of fungi, of which nearly two thirds belong to the genus *Mycena*. Other genera with luminescent members are *Panellus*, *Armilariella*, *Lampteromyces*, *Pleurotus*, *Omphalia* and *Omphalotus*. *Panellus stypticus* is a brightly luminescent fungus common in North America, which has served as material for several investigations. It has been speculated that luminescent fungi attract insects which aid in dispersal of spores (O'Kane et al. 1990a, Bermudes et al. 1992, and many others). However, in many fungi only the mycelium, and not the fruiting bodies, luminesce.

3.2. *Protection*

One of the best-known examples of bioluminescence, described already by Aristotle, is the "fire of the sea" caused by dinoflagellates such as *Noctiluca* and *Gonyaulax*. Its survival value remained obscure for a long time, but it has now been shown that it protects from grazing by copepods (Esaias & Curl 1972, Buskey & Swift 1983).

In different animal groups there are examples of how bioluminescence can protect by diverting the attacker's attention away from the prospective prey. Some squids, when attacked, give off a luminescent secretion which confuses the attacker, and luminescent secretion from a shrimp (Inouye et al 2000) may serve a similar purpose. A kind of marine annelid called a scale worm is covered on the ventral side by scales which first emit flashes when the animal is attacked and then are shed, still glowing.

By aposematic colouration we mean easily recognised bright colour patterns like the black-yellow banding of wasps, spots on ladybirds, and stripes on coral snakes, which warn a predator of nasty consequences of an attack (and frequently are mimicked by species which do not have any other protection). It was shown by Underwood et al. (1997) that bioluminescence of firefly larvae serve a similar purpose.

Somewhat surprisingly, bioluminescence can also be used for camouflage, in two different ways. Fish can be either luminescent by themselves, or can harbour luminescent bacteria. Some fish of both categories use bioluminescence for counterillumination and "disruptive illumination" (McFall-Ngai & Morin 1991), to avoid perception of their shape and size. Most fish are lighter on the ventral than on the dorsal side, and this can be regarded as minimising their visibility: from above they look dark like the background depth, and from below they look bright like the sky above. This cannot give complete protection; they still look rather dark against a bright sky. But some fish, by bioluminescence, do match both the intensity and the angular distribution of the downwelling surface light (Denton et al. 1972, Figure xx). Also the spectral match is very good (Denton et al. 1984), and the intensity is regulated according to ambient light (McFall & Morin (1991).

3.3. *Food acquisition*

The deep sea angler fish *Linophryne arborifera* uses a bait with luminescent bacteria. Female fireflies of the genus *Photuris* reply to the "interrogating flashes" from males of other fireflies, lure them to approach, and eat them (Lloyd 1984a). Most sly and cunning of them all are the females of *Photuris versicolor*, who know firefly languages sufficiently well to be able to prey on 11 different species (Lloyd 1984b). In other cases luminescent fireflies rely for their catch on the more unspecific attraction of insects to light.

In addition to beetles, among bioluminescent insects are fungus gnats, members of the order Diptera. About a dozen of more than 3000 species in the family have larvae which use bioluminescence in different ways to catch prey. Particularly famous are the larvae in the Te Ana-au caves on the South Island of New Zealand.

The gnat larvae sit on the roof and deploy luminescent and sticky "fish lines", which attract other insects which are caught and devoured.

In many cases, marine animals are aided in their vision by their own luminescence, which functions mainly in the service of food acquisition. Several fish species, like *Aristomonias scintillans* in the deep sea, and *Photoblepharon palpebratus* and *Anomalops katoptron* in shallow waters, have luminescent organs in proximity to their eyes. The dragonfish, which are so remarkable that we shall return to them later, also belong to those for which vision is aided by own bioluminescence, as are the shrimplike euphausiids.

3.4. *Role of bioluminescence in bacteria*

As for the role of bioluminescence in bacteria, there is more speculation than solid fact. Those bacteria which live in symbiosis with animals certainly have the benefits of shelter, food and a future life that the animals offer. An interesting observation that has recently been made is that bacteria luminesce only when they are crowded.

Luminous bacteria occur in the genera *Vibrio, Photobacterium, Lucibacterium, Alteromonas* and *Xenorhabdus*. Most investigations deal with either the first or second of these.

4. MECHANISMS OF LIGHT PRODUCTION

As was mentioned before, bioluminescence has evolved several times, ant it is not surprising therefore that there are many different mechanisms. The different mechanisms do, however, have that in common that they all require oxygen at some stage. Many of them also involve a peroxide, either a hydroperoxide or a cyclic peroxide. They involve catalysis by an enzyme called luciferase, but luciferases from different organisms are different (the outcome of comparisons of luciferase sequences is the main reason for the statement that bioluminescence in extant taxa probably has evolved about 30 times). Luciferase action on a relatively low molecular organic compound called luciferin results in an excited state of a pigment, which either emits light directly, or transfers excitation energy to another emitter.

Bacterial luminescence is based on peroxidation of flavin mononucleotide and oxidation of a long-chain aldehyde to carboxylic acid, so flavin mononucleotide can be said to be the luciferin in this case. The reaction scheme is shown in Figure 1. The flavin molecule is here drawn isolated, but is, in fact, bound to the luciferase. The emission may be either the blue-green emission from FMN, or from an accessory chromoprotein to which the excitation is transferred. There are several such proteins known for different bacteria (Eckstein et al. 1990, Lee et al. 1991), but the mechanisms for light emission are not completely understood. Some bacteria can emit radiation with different spectra.

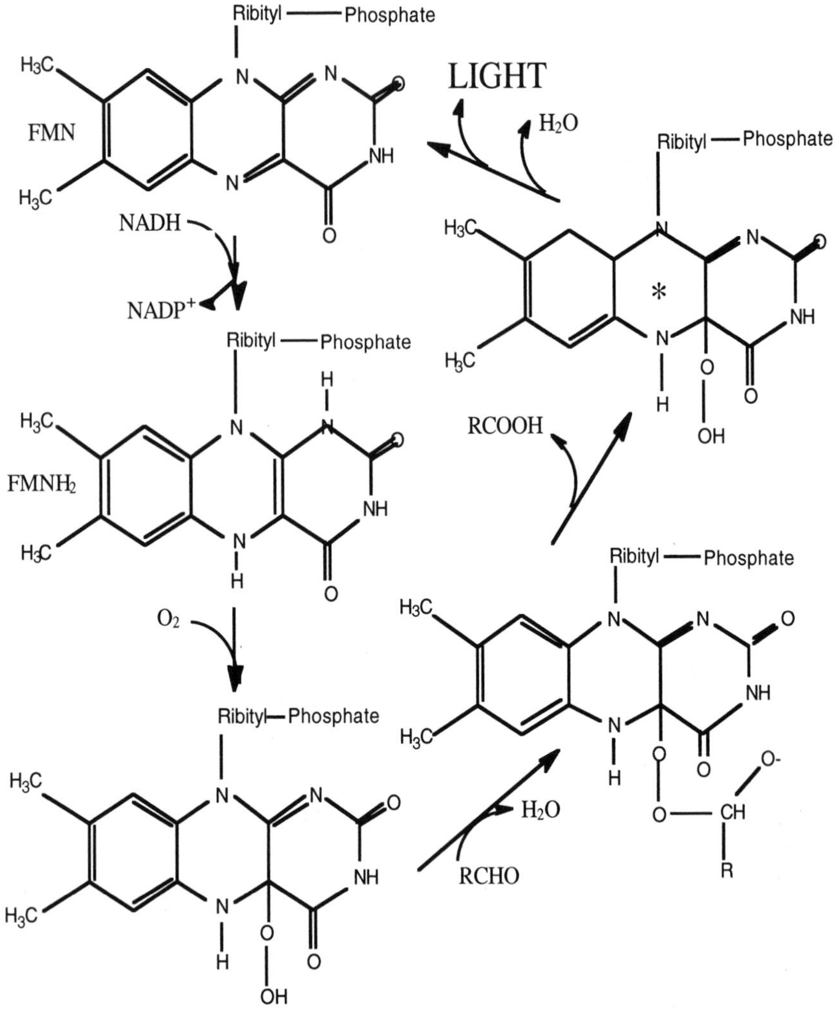

Figure 1. Reactions involved in bacterial bioluminescence.

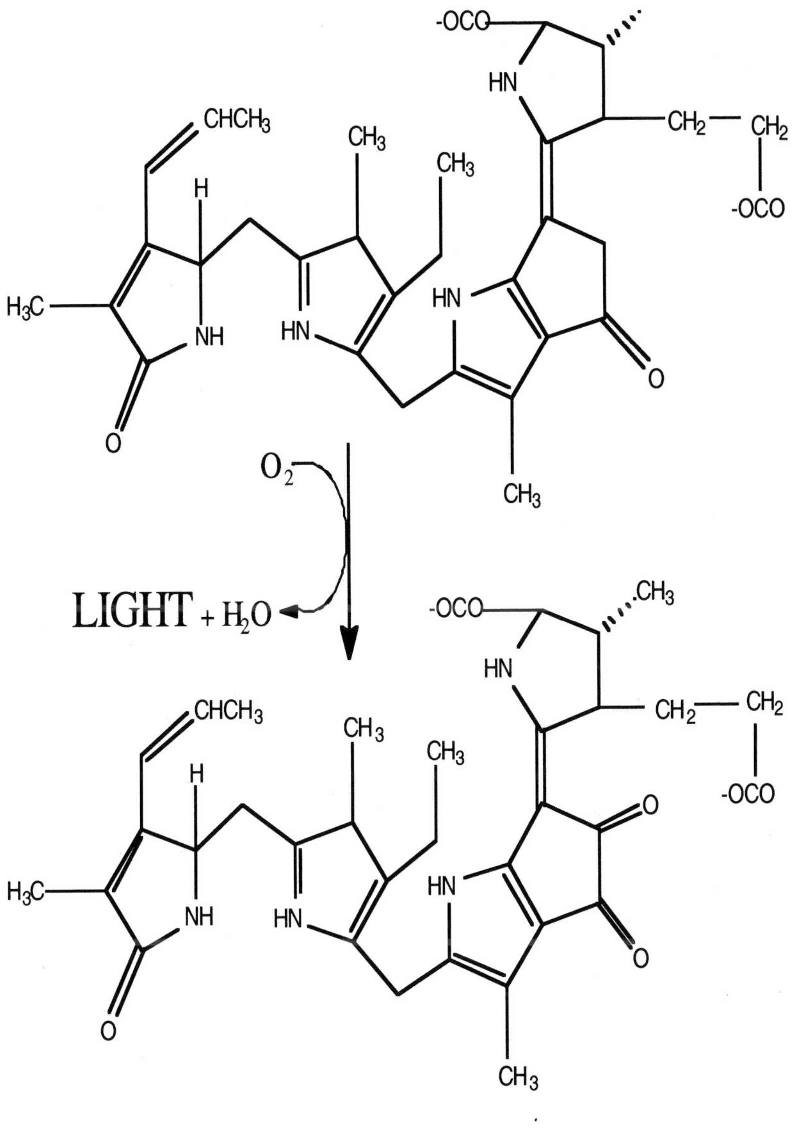

Figure 2. Bioluminescence reaction in the dinoflagellate Gonyaulax.

Fungal bioluminescence is relatively little explored, and in most investigations spectral analysis of the emitted light suffers from deficient methods and equipment. The most reliable spectrum so far was published by O'Kane et al. (1990b). They found an emission maximum (on a photons per wavelength interval basis) of about 525 nm. Shimomura (1989, 1992) favours the the view that the emission is caused by a reaction between hydrogen peroxide, a low molecular amine, and panal (sesquiterpene aldehyde). However, as O'Kane et al. (1990b) point out, such a conclusion cannot be based on spectral data, since chemiluminescence of the latter system in vitro has emission maxima ranging from 485 to 570 nm depending on conditions. Several other emitters have been proposed in fungal bioluminescence; one of them is riboflavin (Isobe et al 1987).

Dinoflagellates have different luciferins depending on species. In the most studied organism, *Gonyaulax polyedra*, it is a tetrapyrrol-like substance with an extra ring (Nakamura et al. 1989), clearly derived from chlorophyll (Fig. 2), but in another species, *Pyrocystis lunula*, it is quite different (Nakamura et al. 1989). The bioluminescence of *Gonyaulax* differs from that of most bioluminescent organisms in that no peroxide seems to be involved. Another remarkable thing is that the spectrum of luminescence agrees with the fluorescence spectrum of unreacted luciferin (Hastings 1978), while the postulated emitter is non-fluorescent. Therefore the mechanism sketched in Fig. 2 must be regarded as tentative.

For light emission by the firefly luciferin/luciferase reaction (Fig. 3) prior adenylation of the luciferin by reaction with ATP is required. Both luciferin and luciferase are located in the peroxisomes in one part of the cell, while another part of the cell is full of ATP-generating mitochondria.

Although different beetles can produce light with different colours from green to red, they all seem to possess the same kind of luciferin. The differences in wavelength distribution are probably due to differences in luciferase. Possibly there are two different molecular species involved as emitters, as sketched in Fig. 3.

As for colour of emitted light, the most remarkable animal is the larvae in the beetle family Phengodidae (*Phrixothrix vivianii* and *Euryopa* species), so-called railroad worms (Viviani & Bechara 1997, Viviani & Ohmiya 2000). They carry both red (on the head) and yellow-green (on the sides) lanterns on the same individual. Colour pictures of these magnificent animals are available at several Internet sites, e.g., http://lifesci.uscb.edu/-biolum/forumvviviani2.html.

The mechanism of bioluminescence in dipterans differs from that of beetles, but is until now little explored.

In addition to *Vargula* (see below), among crustaceans the euphausiids are worth special mention because of their strong light and sophisticated lantern optics. They are shrimp-like animals, but distinct from true shrimps and not members or the group Decapoda. Like the dinoflagellate *Gonyaulax* they have a tetrapyrrole chlorophyll derivative as a light emitting chromophor, but the macrocycle of the chlorophyll molecule is split open at another site. Their lanterns are equipped both with a reflective backing and a lens system to direct the light. At least those lanterns which are located on the eye stalks just above the eyes certainly serve as an aid to vision.

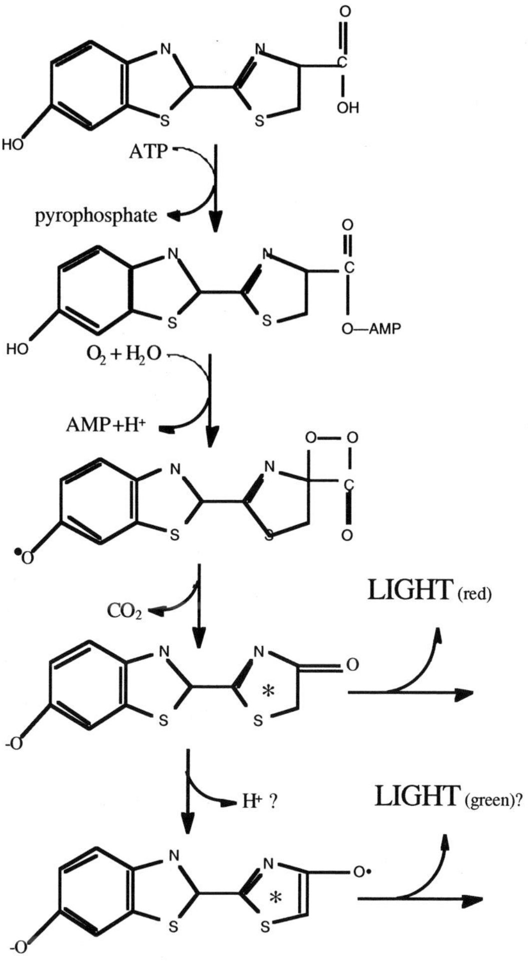

Figure 3. Reactions leading to luminescence in fireflies and other beetles. The luciferin and its derivatives remain bound to luciferase throughout the reaction sequence. The asterisk indicates excited state. The diagram is essentially according to Wilson & Hastings (1998).

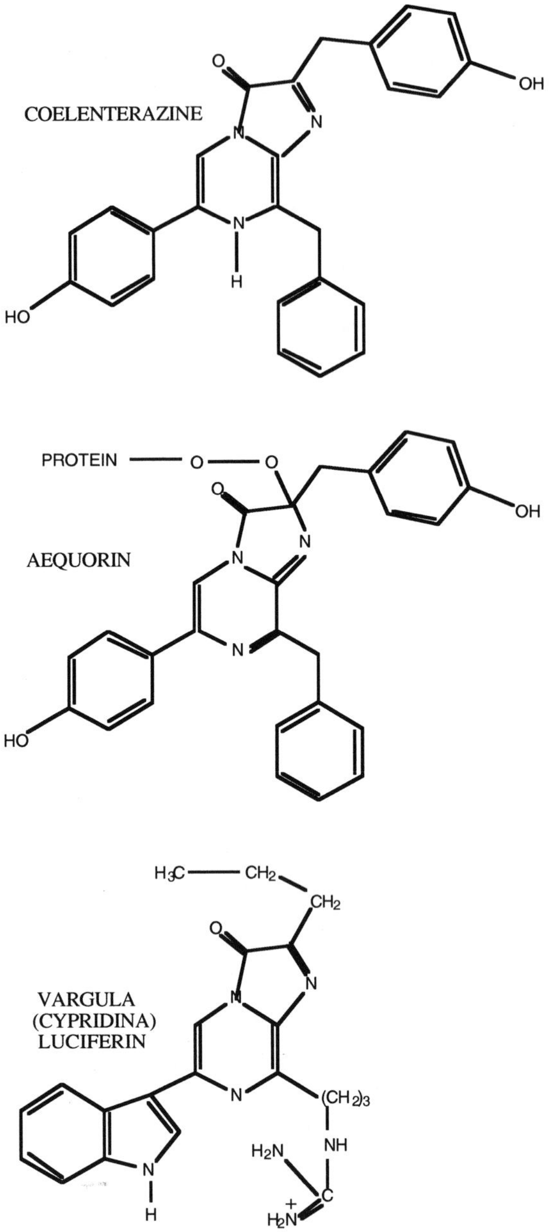
COELENTERAZINE
O
N
N
N
H
HO
OH
PROTEIN
O
O
O
N
AEQUORIN
N
N
HO
OH
H_3C
CH_2
CH_2
O
N
VARGULA
(CYPRIDINA)
LUCIFERIN
N
N
$(CH_2)_3$
NH
H_2N
N
H
C
+
H_2N

Fig. 4. Coelenterazine and related chromophores.

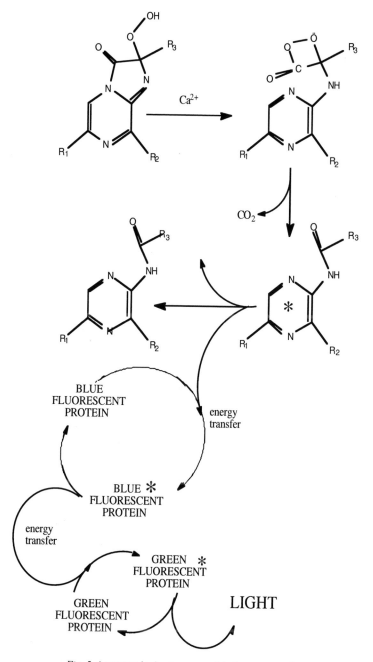

Fig. 5. Aequorea *bioluminescence.* * indicates excited state.

The structure of the luciferin typical of the coelenterates (coelenterazine) has great similarities to the prostethic group of a light-generating chromoprotein, aequorin, of the jellyfish *Aequorea*. The luciferin of the squid *Watasenia*, is coelenterazine with the hydroxy groups replaced by sulfate groups. Also the luciferin of the crustacean (ostracod) *Vargula* (formerly known as *Cypridina*) is structurally related to these chromophores (Fig. 4).

The luciferins of coelenterates and of *Vargula* function together with luciferases and also require oxygen for the light emitting reactions to take place. Earlier it was thought that the bioluminescence of *Aequorea* was something in principle different, since the chromoprotein aequorin extracted from it would glow *in vitro* when calcium ion was added, without the need for molecular oxygen. However, it is now realised that aequorin is an enzyme-substrate (luciferase-luciferin) complex which requires oxygen for formation, and is stable in the absence of calcium ions.

Just as some bacteria, also some animals have accessory light-emitting chromoproteins, proteins different from the luciferin-luciferase complexes. This is the case with *Aequorea*, and also the sea pansy (Pennatulacea, relatives of corals) *Renilla*. Excitation energy is transferred to the covalently bound chromophores of these accessory proteins from the excited reaction product of the luciferin by the Foerster mechanism.

5. DRAGONFISHES: LONG-WAVE BIOLUMINESCENCE AND LONG-WAVE VISION

Even the dark deep-sea harbours creatures of great interest to photobiologists. Dragonfishes are worth our attention for two reasons: Their bioluminescence and their vision. Both operate at the long-wavelength limit of our own perceptual abilities, at wavelengths of around 700 nm.

Three genera, *Aristomias* , *Malacosteus*. and *Pachystomias* emit both blue light from organs behind the eyes and far-red light from organs below the eyes (Widder et al. 1984, Douglas et al. 2000). It is possible that the basic luminescence system is the same in both blue-emitting and far-red-emitting organs, and that the latter are equipped with secondary emitters to which excitation energy is transferred from the luciferin. If this is the case, probably also an intermediary pigment is required to span the spectrum from the 479 nm (*A. scintillans*) or 469 nm (*M. niger*) primary emitter to the final 703 nm (*A. scintillans*) or 660 nm (*M. niger*) emitter, to make possible sufficiently large overlap integrals for Foerster energy transfer to take place. *M. niger* has a short wavelength cut-off filter in front of the light emitter to shift the emission maximum from 660 nm to 702 nm (Widder et al. 1984, Denton et al. 1985).

What could the pressure be to drive evolution of emission maxima in both these fishes to such long wavelength, using different methods? This for a long time was an enigma. Water absorbs such radiation rather efficiently, so its range cannot be very large. And, above all, what could its use be? It was hard to believe that any visual pigment could exist to permit the fish to see such long-wave radiation.

But just in the difficulty in constructing a retinal-based visual pigment with such long-wavelength absorption lies the explanation for the advantage. The dragonfish can use its far-red torch without being observed by other animals. But still - to have any use for the light, it must be able itself to perceive it. And how it manages to do that is the most remarkable fact about this fish.

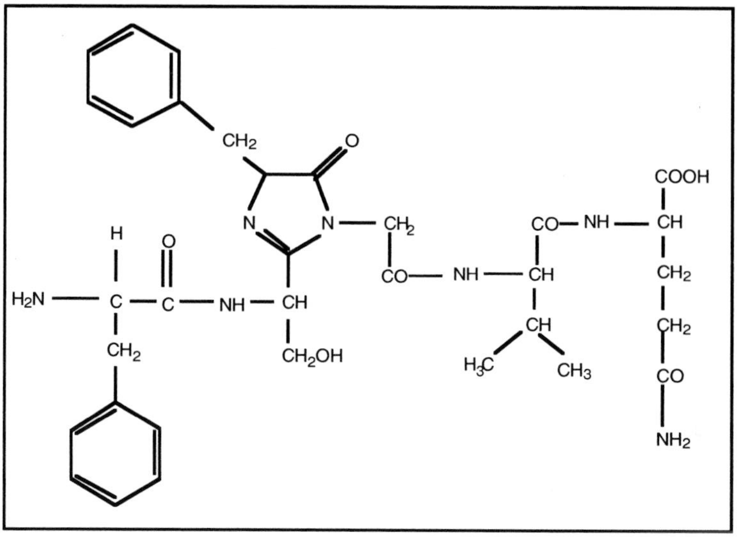

Fig. 6. Structure of the chromophore of the green fluorescent protein from Cody et al. (1993). A slightly different structure has been published by Shimomura (1979). The fluorescence spectrum has a narrow band peaking at 509 nm.

The genera *Aristomias* and *Pachystomias* use a "conventional" method. It has managed to tune a rhodopsin by "protein engineering" to get an absorption band peaking at 588-595 nm (Douglas et al. 1998, 2000). This is quite a feat, considering that essentially the same chromophore, in a different protein environment, is used by other animals for UV receptors with sensitivity peaks around 360 nm. The tail of the 588 nm pigment extends above 700 nm, and would give some sensitivity overlapping the bioluminescence spectrum. However, in addition to rhodopsins (retinal based visual pigments), fishes (and also dragonfishes) usually contain porphyropsins (3,4-dehydroretinal based pigments) with the same opsin-type protein moiety. Although Douglas et al. (2000) were unable to find the porphyropsin analogue of the 588 nm rhodopsin, they speculate that *A. tittmanni* is equipped with it. 'They calculate the absorption peak to be at 669 nm, and show that it would match the bioluminescence almost perfectly.

But *Malacosteus* has not succeeded with this "protein engineering". It has only two "conventional" visual pigments with peaks at 520 and 540 nm. But it has another trick up its sleeve.

Not only is it difficult to construct a retinal-based pigment with an absorption maximum above 650 nm, but there are, in fact, very few types of organic substances which absorb at such long wavelengths. One well-known type wide-spread in the biosphere is chlorophyll. Bacterial variants of this may have absorption peaks even above 1 micrometre.

And this type of pigment is just what *M. niger* uses for its vision. Chlorophyll from photosynthetic bacteria at the base of the food chain has been converted to a mixture of pheophorbides and deposited in the outer segments of the photoreceptor cells in close contact with the 520 and 540 nm pigments. It seems that either in some way energy can be transferred from pheophorbide to these pigments (Douglas et al. 1998 speculate that this could take place via the triplet state of the rhodopsin or porphyropsin), or that pheophorbide can act in place of ordinary visual pigments.

6. CONTROL OF BIOLUMINESCENCE

Even in bacteria bioluminescence is regulated; bacteria do not glow unless the cell density is high, as it is for instance in the organs where some squids and fishes harbour luminescent bacteria.

A common feature for many bioluminescent systems is that the light intensity decreases when ambient light increases. This indicates both that the light has some function in itself (and not primarily serves to divert excess energy) and that bioluminescence is energetically and metabolically demanding and that an organism cannot afford to waste the resources. In one case an action spectrum has been constructed for the inhibition of luminescence (Fig. 6), but the chromophore corresponding to this spectrum is unknown, as is the mechanism of inhibition.

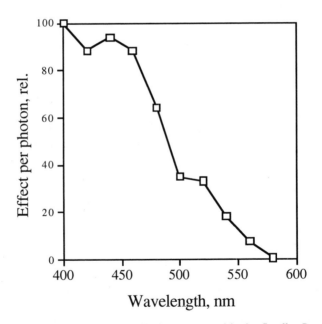

Figure 7. Action spectrum for the inhibition of bioluminescence of the dinoflagellate Protoperidinium depressum. *Redrawn from Li et al. (1996).*

There are many interesting findings regarding regulation of bioluminescence. These range from the rapid flashing of fireflies and dinoflagellates and the circadian rhythms to long-term effects of environment and nutrient status, and from the organismal to the subcellular levels. We must refrain from descriptions of most of this here, and the reader is referred to the treatises by Campbell (1988), Ulitzur & Dunlap (1995), Hosseini & Nealson (1995), and Wilson & Hastings (1998). A few words will be devoted to exciting recent findings about one of the best investigated cases, firefly bioluminescence.

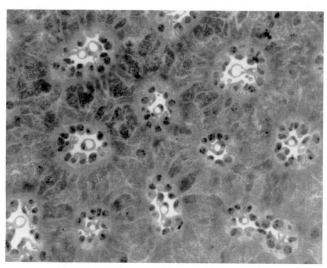

Figure 8. A section through the light emitting organ of the firefly Photuris sp. *showing tracheae surrounded by cylindrical tracheal end organs and photocytes. The photocytes stretch rosette-fashon from one cylinder to the next. Cf. the following figures. From Ghiradella (1998).*

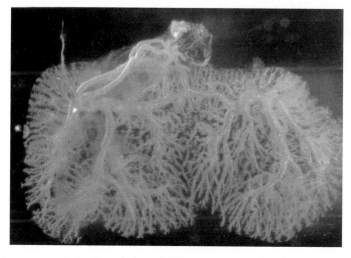

Figure 9. Preparation of isolated tracheae from a light-emitting organ fo a firefly, showing the repeated ramification. From Ghiradella (1998).

Most fireflies flash with precise timing in a species-specific way, and rapidly respond to flashes from other fireflies. Obviously they must have very tight control of their so-called photocytes, the light-emitting cells. How this control is possible has been an enigma, since in some case they have no direct nerve connections. As we have seen, the light emitting process requires oxygen, and the photocytes are located in close proximity to the the profusely branched tracheae (Figs 8-10). It has turned out (Trimmer et al. 2001) that the signalling takes place along the same path as the supply of oxygen, using the gaseous hormone nitrogen monoxide (NO). This hormone, coined "molecule of the year" by Science in 199?, is important also in human physiology, but in man and most other animals it is transported in the dissolved state. The fireflies take advantage of the fact that it is a small molecule diffusing very rapidly in the gaseous phase. The nitrogen monoxide is produced in the tracheal end organs (Fig. 10) in which the nerve projections end, and is transported to the photocytes along the tracheoles (branches of the tracheae).

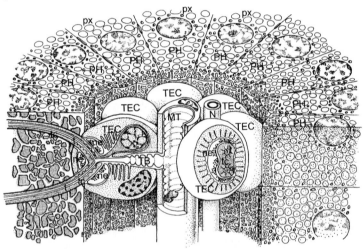

Figure 10. Diagrammatic view of part of a firefly lantern. At centre the main trachea (parallel to a nerve) connected to a cluster of tracheal end organs. At left a longitudinally sectioned end organ with its tracheolar branches and the webbed intercellular border of an associated photocyte. To the right a transversely sectioned end organ. In the upper part of the picture a half-circle of cross sections of photocytes. From Ghiradella (1998).

7. HUMAN EXPLOITATION OF BIOLUMINESCENCE

Warfare is perhaps not an application one would guess at for bioluminescence. Japanes soldiers are said to have used dried *Vargula* for reading maps during World War II. When wetted these dried animals emitted sufficient light for the purpose.

The cloned gene of the fluorescent accessory protein of *Aequorea* ("green fluorescent protein") has found wide application in molecular biology as a reporter gene. It has even been improved by genetic engineering (Blinks 1989, Cubitt et al. 1995a, Heim et al. 1995, Heath 2000, Deo & Daunert 2001).

The "arrested" luciferin-luciferase complex of *Aequorea* can be used for extremely sensitive assays of calcium ion. Using this protein it is even possible to map the concentration of calcium ions inside cells. The protein can be injected into the cells under study by different methods. If the aqueorin gene is cloned into the organism under study, it is deposited in the cytoplasm of that organism. The cytoplasm has a low calcium ion concentration most of the time, but one step in many signal transduction chains consists in a sudden elevation of the concentration, and this can be studied by this method (Knight et al. 1991, 1993, Cubitt et al. 1995b, Wood et al. 2000, 2001).

Also various other luciferase genes, especially the firefly luciferase gene, are also used for the study of gene regulation. The luciferase gene is fused to the regulator gene under study. A disadvantage is that the organism has to be killed and treated in such a way that ATP and luciferin can be added. This can be circumvented by using instead bacterial luciferase, which can be activated by addition of the vapour of an aldehyde.

The classic use of bioluminescence, however, is the use of a luciferin-luciferase mixture from fireflies. The sample to be analysed is mixed with a luciferin-luciferase mixture or crude firefly lantern extract. Since the luciferase preparations usually also contain an enzyme capable of converting two molecules of ADP to one molecule of AMP and one of ATP (a slower reaction than the luciferase reaction), AMP in the sample will also produce light, but with much slower kinetics. Therefore the intensity of the initial flash upon mixing the sample with the reagents is taken as a measure of ATP.

Thje firefly assay for ATP has found very wide use, and kits for assay are commercially available. This application forms a natural bridge to the next section.

8. PHOTOSYNTHETIC AFTERGLOW

In the 1950's many people were involved in the discovery and study of photosynthetic phosphorylation, and the present author had the privilege of working as an assistant to D.I. Arnon during this exciting time. Already early in the decade Strehler and Arnold (1951) attempted to demonstrate the formation of ATP in a preparation of isolated chlorplasts by mixing the chloroplast suspension with firefly extract. After a period of illumination to let the chloroplasts synthesise ATP, the experimenters shut off all external light and tried to measure bioluminescence from the mixture. They were pleased to find a clear light signal, which rapidly decayed with time (as they expected, because the tiny amount of ATP would soon be consumed by the luciferase reaction).

Then they did a control experiment without firefly extract, and to their surprise found that the light was as strong as before. This was the discovery of a phenomenon to which several names have been given: Afterglow, delayed light emission, delayed fluorescence. Briefly, it is due to the reversion of early steps in photosynthesis: Light energy recently converted to chemical, electrical and proton gradient energy is reconverted to light. Chlorophyll serves as the emitter.

Figure 11. Part of a bean leaf imaged in different ways by its own emitted light: Left fluorescence light, centre fast luminescence (about 0.25 s after the cessation of light incident on the leaf), right slow luminescence (integrated between 30 and 60 s after the cessation of incident light). The leaf veins had been injected with DCMU, a substance that interrupts electron transfer between the photosystems, before the picture was taken. This makes photosystem 2 dissipate energy faster, so fluorescence and fast luminescence become stronger, but slow luminescence weaker than from unpoisoned cells further away from the veins. From Björn & Forsberg (1979).

The delayed light emission is due almost exclusively to emission by photosystem 2, and consists of several kinetic components. The most rapidly decaying component is due to return of electrons from pheophytin and the quinone called Q_A to the chlorophyll ion in the reaction centre, resulting in excited chlorophyll. The slowest component involves also photosystem 1, and is best excited using long wavelength light preferentially absorbed by this system. It is also dependent on molecular oxygen (Björn 1971a). Energy for this emission is stored as proton gradient and as ATP, and is inhibited not only by electron transport blockers as DCMU, but also by uncouplers. Even before the acceptance of the chemiosmotic theory of Mitchell, study of this long-lived component demonstrated that the membrane system in a chloroplast stores energy as a unit (Björn 1971b). Several reviews have been written on delayed light emission; one of the best and most comprehensive is still that by Lavorel (1975). Apart from its use in studies of the mechanism of photosynthesis, delayed light emission has been used for study of plant damage by disease, frost etc. A special technique for this is "phytoluminography", whereby pictures of plants are produced using only the delayed light (Sundbom & Björn 1977, Björn & Forsberg 1979, see Figs 11).

9. ULTRAWEAK LIGHT EMISSION

So far we have treated special organisms which emit bioluminescence, and photosynthetic organisms which produce delayed light emission due to reversal of photosynthesis. But all other organisms and all cells with active metabolism which have been studied emit very weak light. This is called ultraweak light emission. The ultraweak light emission from green leaves can be studied when the delayed light emission has decayed for about 4 hours (at room temperature).

Ultraweak light emission probably has several components with different causes. The main emission from green leaves is of a wavelength exceeding 600 nm, and probably is emitted from chlorophyll. In most other cases the emission is of shorter wavelength. It is believed that a main component stems from peroxidation of unsaturated membrane lipids. However, most cells contain also other components which can emit light. For instance, when hydrogen peroxide is decomposed by catalase in the presence of various organic compounds (the plant hormone auxin is one example, pyrogallol another), light is emitted. A number of other causes of ultraweak light emission are listed by Campbell (1988).

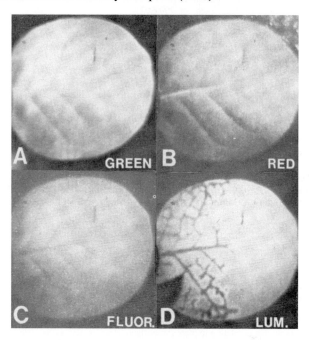

Figure 12. A leaf on a tobacco plant on which another leaf was inoculated with tobacco mosaic virus 6 days before the picture was taken. The virus has spread through the veins to the depicted leaf. Pictures were taken using reflected green (A) and red (B) light, red fluorescence evoked by blue light (C) and light emitted by the leaf after cessation of external illumination (D). Only in the last case does the infection become visible at this early stage. From Björn & Forsberg (1979).

If plants are irradiated with ultraviolet-B or ultraviolet-C radiation, the intensity of ultraweak light emission increases after a long lag period (about 2 days). The reason is probably that the ultraviolet radiation initiates a chain reaction leading to peroxidation of membrane lipids. An action spectrum for this ultraviolet effect has been determined (Cen & Björn 1994).

Various medical and other applications of measurements of ultraweak light emission are described by Campbell (1988) and Jezowska-Trzebiatowska et al. (1990).

REFERENCES

Barros, M.P. & Bechara, E.J.H. (2000). Luciferase and urate may act as antioxidant defeneses in larval Pyrearinus termitilluminans (Elateridae: Coleoptera) during natural development and upon 20-hydroxyecdysone treatment. *Photochem. Photobiol., 71,* 648-654.

Bermudes, D., Petersen, R.H. & Nealson, K.H. (1992). Low-level bioluminescence detected in *Mycena haematopus* basidiocarps. *Mycologia, 84,* 799-802.

Björn, L.O. (1971). Far-red induced, long-lived afterglow from photosynthetic cells. Size of afterglow unit and paths of energy accumulation and dissipation. *Photochem. Photobiol., 13,* 5-20.

Björn, L.O. & Forsberg, A.S. (1979). Imaging by delayed light emission (phytoluminography) as a method for detecting damage to the photosynthetic system. *Physiol. Plant., 47,* 215-222.

Blinks, J.R. (1989). Use of calcium-regulated photoproteins as intracellular Ca^{2+} indicators. *Meth. Enzymol., 172,* 164-203.

Bowmaker, J.K., Dartnall, H.J.A. & Herring, P.J. (1988). *J. Comp. Physiol., A163,* 685-698.

Branham, M.A. & Greenfield, M.D. (1996). Flashing males win mate success. *Nature, 381,* 745-746.

Buck, J. & Buck, E. (1976). Synchronous fireflies. *Sci. Am., 234,* 74-85.

Buck, J. (1988). Synchronous rhythmic flashing of fireflies. II. *Quart. Rev. Biol., 63,* 265-289.

Buskey, E.J. & Swift, E. (1983). Behavioural responses of the coastal copepod Acartia hudsonica to simulated dinoflagellate bioluminescence. *J. Exp. Biol. Ecol., 72,* 43-58.

Campbell, A.K. (1988). *Chemiluminescence: Principles and applications in biology and medicine,* pp. 608. Chichester: Ellis Horwood. ISBN 3-527-26342-X

Cen, Y.-P. & Björn, L.O. (1994). Action spectra for enhancement of ultraweak luminescence by ultraviolet radiation (270-340 nm) in leaves of *Brassica napus*. J. Photochem. Photobiol. B: Biol. 22:125-129.

Cody, C.W., Prasher, D.C.. Westler, W.M., Prendergast, F.G. & Ward, W.W. (1993). Chemical structure of the hexapeptide chromophore of the *Aequorea* gfreen-fluorescent protein. *Biochemistry, 32,* 1212-1218.

Cubitt, A.B., Heim, R., Adams, S.R., Boyd, A.E., Gross, L.A. & Tsien, R.Y. (1995a). Understanding, improving and using green fluorescent proteins. *Trends Biochem. Sci., 20,* 448-455.

Cubitt, A.B., Firtel, R.A., Fischer, G., Jaffe, L.F. & Miller, A.L. (1995b). Patterns of free calcium in multicellular stages of *Dictyostelium* expressing jellyfish apoaqueorin. *Development, 121,* 2291-2301.

Denton, E.J., Gilpin-Brown, J.B. & Wright, P.G. (1972). The angular distribution of the light produced by some mesopelagic fish in relation to their camouflage. *Proc. R. Soc. Lond., B182,* 145-158.

Denton, E.J., F.R.S., Herring, P.J., Widder, E.A., Latz, M.F. & Case, J.F. (1985). The roles of filters in the photophores of oceanic animals and their relation to vision in the oceanic environment. *Proc. R. Soc. Lond., 225,* 63-97.

Deo, S.K. & Daunert, S. (2001). Luminescent proteins from *Aequerea victoria*: applications in drug discovery and in hight throughput analysis. *Fresenius J. Anal. Chem., 369,* 258-266.

Douglas, R.H., Partridge, J.C., Dulai, K.S., Hunt, D.M., Mullineaux, C.W Tauber, P.H. & Hynninen, P.H. (1998). Dragon fish see using chlorophyll. *Nature, 393,* 425.

Douglas, R.H., Partridge, J.C., Dulai, K.S., Hunt, D.M., Mullineaux, C.W & Hynninen, P.H. (1999). Enhanced retinal longwave sensitivity using a chlorophyll-derived photosensitiser in *Malacosteus niger,* a deep-sea dragon fish with far red bioluminescence. *Vision Res., 39,* 2817-2832.

Douglas, R.H., Mullineaux, C.W. & Partridge, J.C. 2000. Long-wave sensitivity in deep-sea stomiid dragonfish with far-red bioluminescence: evidence for a dieatry origin of the chlorophyll-derived retinal photosensitizer of *Malacosteus niger. Phil. Trans. R. Soc. Lond., B 355,* 1269-1272.

Eckstein, J., Cho, K.W., Colepicolo, P., Ghisla, S., Hastings, J.W. & Wilson, T. 1990. A time-dependent bacterial luminescence emission spectrum in an in vitro singel turnover system: energy transfer alone cannot account for the yellow emission of *Vibrio fischeri* Y-1. *Proc. Natl. Acad. Sci. USA, 87,* 1466-1470.

Esaias, W.E. & Curl, H.C. (1972). Effect of dinoflagellate bioluminescence on copepod ingestion rates. *Limnol. Oceanogr., 17,* 901-906.

Esaias, W.E., Curl, H.C., Jr & Seliger, H.H. (1973). Action spectrum for a low intensity rapid photoinhibition of mechanically stimulable bioluminescence in the marine dinoflagellates *Gonyaulax catenella, Gonyaulax acatenella* and *Gonyaulax tamarensis. J. Cell Physiol,. 82,* 363-372.

Ghiradella, H. (1998). The anatomy of light production: The fine structure of the firefly lantern. *In* Harrison, F.W. & Locke, M. (eds) *Microscopic anatomy of invertebrates*, vol. 5 *Insecta*, pp. 363-381.New York: Wiley-Liss. ISBN 0-471-56118-5.

Harvey, E.N. (1952). *Bioluminescence*. New York: Academic Press.

Hastings, J.W. (1978). Bacterial and dinoflagellate luminescent systems. *In* Herring, P.J., (ed.) *Bioluminescence in Action*, pp. 129-170. London: Academic Press.

Hastings, J.W. (1983). Biological diversity, chemical mechanisms, and the evolutionary origins of bioluminescent systems. *J. Mol. Evol., 19*, 309-321.

Hastings, J.W. (1996). Chemistries and colors of bioluminescent reactions: a review. *Gene, 173*, 5-11.

Hastings, J.W. & Tu, D. (eds) (1995). Symposium-in-print: Molecular mechanisms in bioluminescence. *Photochem. Photobiol., 62*, 597-673.

Heath, M.C. (2000). Advances in imaging the cell biology of plant-microbe interactions. Annu. Rev. Phytopath. 443-459.

Heim, R., Cubitt, A.B. & Tsien, R.Y. (1995). Improved green fluorescence. *Nature, 373*, 663-664.

Herring, P.J. (1982). Aspects of the bioluminescence of fishes. *Oceanogr. Mar. Biol. Ann. Rev., 20*, 415-470.

Hosseini, P. & Nealson, K.H. (1995). Symbiotic luminous soil bacteria: Unusual regulations for an unusual niche. *Photochem. Photobiol., 62*, 633-640.

Inouye, S., Watanabe, K, Nakamura, H. & Shimomura, O. (2000). Secretional luciferase of the luminour shrimp *Opiophorus gracilirostris*: cDNA cloning of a novel imidazopyrazinone luciferase. *FEBS letters, 481*, 19-25.

Isobe, M., Uyakul, D. & Goto. T. (1987). *Lampteromyces* bioluminescence -1. Identification of riboflavin as the light emitter in the mushroom *L. japonicus*. *J. Biolum. Chemilum., 1*, 181-188.

Knight, M.R., Campbell, A.K., Smith, S.M. & Trewavas, A.J. (1991). Transgenic plant aequorin reports the effects of touch and cold-schock and elicitors on cytoplasmic calcium. *Nature, 352*, 524-526.

Knight, M.R., Read, N.D., Campbell, A.K. & Trewavas, A.J. (1995). Imaging dynamics in living plants using semisynthetic recombinandt aequorins. J. *Cell Biol., 121*, 83-90.

Lavorel, J. (1975). Luminescence. *In* Govindjee (ed.) Bionergetics of photosynthesis, pp. 223-317. New York: Academic Press, pp. xii+698. ISBN 0-12-294350-3.

Lee, J., Matheson, I.B.C., Müller, F., O'Cane, D.J., Vervoort, J. & Visser, A.J.W.G. (1991). The mechanism of bacterial bioluminescence.*In* Muller, F. (ed.) *Chemistry and Biochemistry of Flavins and Flavoenzymes*, Vol. 2, FL, pp. 109-151. Orlando: CRC Press.

Li, Y., Swift, E. & Buskey, E.J. (1996). Photoinhibition of mechanically stimulable bioluminescence in the heterotrophic dinoflagellate *Protoperidinium depressum* (Pyrrophyta). *J. Phycol., 32*, 974-982.

Lloyd, J.E. (1980). Male *Photuris* mimic sexual signals of their females' prey. *Science, 210*, 669-671.

Lloyd, J.E. (1984a). On deception, a way of all flesh, and firefly signaling and systematics. *Oxford Surveys in Evolutionary Biology.* (Oxford University Press), 1, 49-84.

Lloyd, J.E. (1984b). Evolution of a firefly flash code. *Florida entomologist, 67*, 368-376.

Lloyd, J.E. & Wing, S.R. (1993). Nocturnal aerial predation of fireflies by light-seeking firefles. *Science, 222*, 634-635.

McFall-Ngai, M. & Morin, J.G. (1991). Camouflage by disruptive illumination in leiognathids, a family of shallow-water, bioluminescent fishes. *J. Exp.Biol., 158*, 119-137.

McElroy, W.D. & Seliger, H.H. (1962). Origin and evolution of bioluminescence. *In* Kasha, M. & Pullman, B. (eds) *Horizons in Biochemistry*, pp. 91-101. New York: Academic Press.

Moiseff, A. & Copeland, J. (2000). A new type of synchronied flashing in a North American firefly. *J. Insect Behavior, 13*, 597-612.

Nakamura, H., Kishi, Y., Shimomura, O., Morse, D. & Hastings, J.W. (1989). Structure of dinoflagellate luciferin and its enzymatic and non-enzymatic air-oidation products. *J. A,. Chem. Soc., 111*, 7607-7611.

O'Kane, D.J., Lingle, W.L., Porter, D. & Wambler, J.E. (1990a). Localization of bioluminescent tissues during basidiocarp development in *Panellus stypticus*. *Mycologia, 82*, 595-606

O'Kane, D.J., Lingle, W.L., Porter, D. & Wambler, J.E. (1990b). Spectral analysis of bioluminescence of Panellus stypticus. *Mycologia, 82*, 607-616.

Partridge, J.C. & Douglas, R.H. (1995). Far-red sensitivity of dragon fish. *Nature, 375*, 21-22

Shimomura, O. (1979). Structure of the chromophore of *Aequorea* green fluorescent protein. *FEBS Lett., 1054*, 220-222.

Shimomura, O. (1980). Chlorophyll-derived bile pigment in bioluminescent euphausiids. *FEBS Lett., 116*, 203-206.

410

Shimomura, O. (1989). Chemiluminescence of panal (a sesquiterpene) isolated from the luminous fungus *Panellus stipticus*. *Photochem. Photobiol., 49,* 355-360.

Shimomura, O. (1992). The role of superoxide dismutase in regulating the light emission of luminescent fungi. *J. Exp. Bot., 43(256),* 1519-1525.

Seliger, H.H. & McElroy, W.D. (1965). *Light: Physical and biological action,* pp. 417. New York: Academic Press.

Sundbom, E. & Björn, L.O. (1977). Phytoluminography: Imaging plants by delayed light emission. *Physiol. Plant., 40,* 39-41.

Tett, P.B. & Kelly, M.G. (1973). Marine bioluminescence. *Oceanogr. Mar. Ann. Rev., 11,* 89-173.

Trimmer, B.*A., Aprille, J.R., Dudzinski, D.M., Lagace, C.J., Lewis, S.M., Michel. T., Qazi, S. & Zayas, R.M. (2001). Nitric oxide and the control of firefly flashing. Science, 292, 2486-2488.

Ulitzur, S. & Dunlap, P.V. (1995). Regulatory circuitry controlling luminescence autoinduction in *Vibrio fischeri. Photochem. Photobiol., 62,* 625-632.

Vencl, F.V. & Carlson, A.D. (1998). Proximate mechanisms of sexual selection in the firefly *Photinus pyralis* (Coleoptera: Lampyridae). *J. Insect Behavior, 11,* 191-207.

Viviani, V.R. & Bechara, E,J.H. (1997). Bioluminescence and biological aspects of Brazilian railroad worms (Coleoptera: Phengodidae). *Ann. Entomol. Soc. Am., 90,* 389-398.

Viviani, V.R. & Ohmiya, Y. (2000). Bioluminescence and color determinants of *Phrixothrix* railroad worm luciferases: Chimeric luciferases, site-directed mutagenesis of Arg 215 and guanidine effect. *Photochem. Photobiol., 72,* 267-271.

Vencl, F.V., Blasko, B.J. & Carlson, A.D. (1994). Flash behavior of female *Photuris versicolor* fireflies (Coleoptera: Lampyridae) in simulated courtship and preadatory dialogs. *J. Insect Behavior, 7,* 843-858.

Widder, E.A., Latz, M.I., Herring, P.J. & Case, J.F. (1984). Far red bioluminescence from two deep-sea fishes. *Science , 225,* 512-513.

Wilson, T. & Hastings, J.W. (1998). Bioluminescence. *Annu. Rev. Cell Dev. Biol., 14,* 197-230.

Wood, N.T., Allan, A.C., Haley, A., Viri-Moussaid, M. & Trewavas, A.J. (2000). The characteristics of differential calcium signalling in tobacco guard cells. *Plant J., 24,* 335-344.

Wood, N.T. Haley, A., Viri-Moussaid, M., Johnson, C.H., van der Luit, A.H. & Trewavas, A.J. (2001). The calcium rhythms of different cell types oscillate with different circadian phases. *Plant Physiol., 125,* 787-796.

LARS OLOF BJÖRN

17. HINTS FOR TEACHING EXPERIMENTS AND DEMONSTRATIONS

1. INTRODUCTION

The following is not aimed to represent a complete set of practicals for a photobiology course. I shall try to do just what the title says, give some hints. As a general reference book for photobiological teaching experiments Valenzeno et al. (1991) merits special mention. Those planning a general photobiology course should consult that book also in addition to the present one. One disadvantage with several of the experiments in the book by Valenzeno et al. is that they need advanced equipment which may not be available anywhere, and I have tried to concentrate below on less demanding experiments.

Experiments and demonstrations must, of course, always be adapted to the audience. In the following you will find some which, with suitable adaptations, can be used from elementary school to graduate student levels. From some of them you may even learn something yourself.

I am grateful to former students, Drs Björn Sigfridsson, Gunvor Björn and Susanne Widell for testing some of the descriptions below.

Reference

Valenzeno, D.P., Pottier, R.H., Mathis, P. & Douglas, R.H. (eds) (1991). *Photobiological techniques*, pp. xiii+381. New York: Plenum Publishing Corp. ISBN 0-306-43778-3.

2. A GOOD START

I shall start with my own pet demonstration which I have used as an introduction to many courses, and which makes students start thinking, and can be a starting point for discussions on the interaction of light with matter, vision, photosynthesis and phytochrome. It requires a lecture hall which can be efficiently darkened, an overhead projector, a beaker and an acetone extract of plant leaves. You should also make a cover of cardboard or masonite for the projector, so the light can emerge only from a hole slightly smaller than the bottom of the beaker.

Before the demonstration you should prepare the leaf extract. It is essential that it is very concentrated. The easiest way, if you have the time available, is to put leaves into a bottle together with the acetone and let it stand in the cold and dark. A faster way is to grind the leaves with some sand and acetone in a mortar and filter the

L.O. Björn (ed.), Photobiology, 411–439.

slurry. You will require about one litre of extract for a two litre beaker. Make sure that the extract is clear.

Ask your audience what colour plant leaves have, and whether they know what makes them have that colour. They are likely to answer that plant leaves are green and in most audiences at least somebody is likely to answer that it is chlorophyll that makes them green. Explain that in your bottle (which should preferably be brown or opaque not to show the colour of the extract) you have an extract with the pigments of plant leaves, and that you are now going to demonstrate the colour.

Switch on the overhead projector, put the cardboard mask on and place the beaker so that it covers the hole in the mask. Turn off the room light completely. Pour a small amount of your extract into the beaker. As expected a green colour will be projected onto the screen.

Pour more and more of the extract into the beaker and let the (usually very surprised) audience see how the colour on the screen goes through a dirty brown to a clear red. Then point to the beaker to make them see the brilliant red chlorophyll fluorescence and ask them again what the colour of chlorophyll is, and to think about what colour really is: a property of an object, or a sensation in our brains.

A few explanations:

(1) The red colour to be seen on the screen is something quite different from the red light radiating in all directions from the solution. The light on the screen, which appears red to us is really far-red, i.e. light of very long wavelength which chlorophyll cannot absorb. This is the kind of light dominating at the bottom of a dense forest or inside a wheat canopy. It is the kind of light driving phytochrome (a light-sensitive pigment in plants, see Chapter 14) from the Pfr form to the Pr form.

(2) The red light radiating from the solution is energy which has been absorbed by, and excited chlorophyll molecules, and been reradiated as fluorescence when the chlorophyll molecules reverted to the ground state. Chlorophyll in solution fluoresces much more intensely than chlorophyll in the plant, because the plant uses more of the absorbed energy for photosynthesis.

(3) For us the colour on the screen and the light radiated by the solution look the same although the wavelength composition is different, because we do not have any light sensitive pigment in our eyes absorbing at longer wavelength than the one which has an absorption maximum at 552 or 557 nm (depending on the person, see Chapter 7).

You can do a similar demonstration, except that there will be no fluorescence, using coloured Perspex or Plexiglass instead of the chlorophyll solution. You start with one layer of green acrylate, showing a green colour on the screen, and add additional layers of green acrylate until only far-red light goes through. The version with chlorophyll solution is no doubt best, but if you are travelling around or have to arrange a demonstration in a hurry, it may be good to have the coloured plastic version available. You can also use blue acrylate; in fact any colour will end up far-red if you add a sufficient number of acrylate.

3. WHAT IS COLOUR? BENHAM'S DISK

Now that we have touched upon colour and colour perception it is appropriate to continue with a simple demonstration which has impressed me very much: The Benham disk or Benham's disk or Benham top. This was originally a toy, not even evented by Benham, but described by him in a paper in Nature (Benham 18...). Since then it many learned papers have been written about it. A review with 67 references was published by von Campenhausen & Schramme (1995), and publications continue on the subject until the present day, e.g. Le Rohellec & Vienot (2001). There is no real consensus about how it works, but in any case it affords a clear demonstration that colour and spectral composition of light are not the same. It is a disk with a black-and white pattern. Surprisingly, you can change the apparent colour in diffferent sectors of it just by changing the direction of rotation.

A template for the disk is shown in Fig. 1. Copy it and glue the copy to cardboard, cut it out and attach it to an electric or hand-driven drill so you can rotate it around its centre. An electric drill with variable speed and possibility to reverse the direction is ideal. The disk in Fig. 1 is an example; other version can be found on the Internet, but one half of the disk should always be black. In my experience, the most vivid colours are obtained if the black sectors in the other half are narrow, like the one shown. It can be instructive to make a disk with stripes of various widths. It is very important that you can adjust the speed. If the rotation is too slow, or too fast, you will see no colour.

One of the first explanations for appearance of the colour phenomena was that receptors in the eye for different spectra regions have different time constants. This explanation, however, is no longer regarded to be correct. The contrast between different parts of the image projected on the retina is regarded as an essential ingredient.

Literature

Le Rohellec, J. & Vienot, F. (2001). Interaction of luminance and spectral adaptation upon Benham subjective colours. *Color Res. Appl.* S174-S179. Supplement.

von Campenhausen, C. & Schramme, J. (1995). 100 years of Benham top in color science. Perception, 24, 695-717.

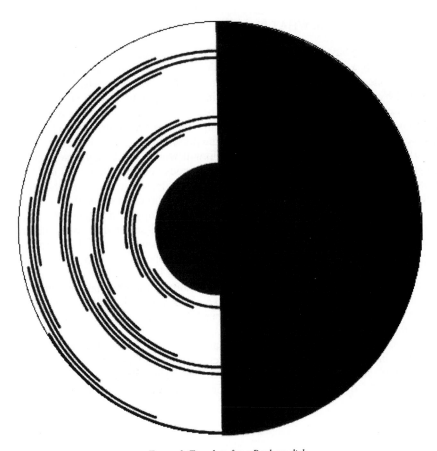

Figure 1. Template for a Benham disk.

4. THE WAVE-NATURE OF LIGHT

The laser-pointers now available on the marked make demonstrations of diffraction and interference much easier than they used to be. Pointers with laser diodes can be obtained very cheaply now. I have bought one for an equivalent of U.S. $ 2 in China, while the price in Sweden is about three times that. Even laser pointers of the same kind emit light with slightly different wavelengths, but with the same indistinguishibly red colour. If you have a couple of such pointers with different spectral tuning you can therefore easily arrange one more demonstration of the fact that wavelength and colour is not the same thing. It is possible to find some coloured plastic, for instance blue plexiglass, which transmits light from one pointer but not from the other, although both give light of the same red colour.

More importantly, you can easily demonstrate both single slit diffraction and Young's famous double-slit experiment with using laser-pointers. You may be able to cut out sufficiently thin slits in a (not too thick) cardboard. Alternatively you can use black adhesive tape that you stick either on a glass slide or over a hole in a piece of cardboard. Better precision can be obtained by drawing one black line (or for the double slit experiment two parallel black lines with a small spacing in between) on white paper, photograph the paper with ordinary negative black-and-white film, and use the negative. A number of other items for optical experiments can be purchased from Edmund Scientific Company, Mass.

If you want to demonstrate how a grating works and have not got one from an instrument, you can use a recordable compact disk (CD). The curvature of the grooves will not matter with the small spot from a laser-pointer If you have sufficient money for teaching your course, you may like to buy also the (much more expensive) laser pointers with green light and UV-A radiation now available, to show how the behaviour of the beam depends on wavelength.

5. SINGLET OXYGEN

Singlet oxygen for demonstration purposes can be easily generated by mixing hydrogen peroxide (30%) with solid sodium hypochlorite. If you do this in a darkroom you will see a rapidly fading red glow. This is the so-called dimol emission of singlet oxygen, which has half the wavelength of emission from isolated singlet oxygen molecules. Dimol emission takes place when the concentration of singlet oxygen is sufficiently high, and the probability is high that two singlet oxygen molecules collide and the excitation energy of both are added to form one photon.

Concentrated hydrogen peroxide is extremely caustic, and mixing it with sodium hypochlorite will cause heating and gas production, so you have to take great care to avoid splashing, and especially to protect your eyes. Try it out in a lighted room before you do it in the dark. One good way is to use a Pasteur pipette with a rubber baloon for the hydrogen peroxide, fill it in the light and put it in a test-tube that you hold in one hand together with a test-tube with the hypochlorite. To see the dimol emission well, you should first dark-adapt your eyes for five minutes.

6. ULTRAVIOLET RADIATION DAMAGE AND ITS PHOTOREACTIVATION

This exercise can be carried out as a demonstration, but is recommended as a practical for students. In its simplest form it is a direct repeat of the classical first demonstration of photoreactivation by Hausser & Oehmcke (1933) and is evaluated visually. With relatively cheap equipment, also useful for other experiments, it can be evaluated quantitatively.

You will need a low pressure (germicidal) mercury lamp, a strong lamp, preferably medium pressure mercury with fluorescent coating (or other lamp with sufficient emission in the blue and UV-A) for photoreactivating light, a big jar or (preferably) small plastic aquarium, and a few unripe, green, bananas. You should also have goggles for protecting your eyes from the UV-C radiation from the mercury lamp.

416

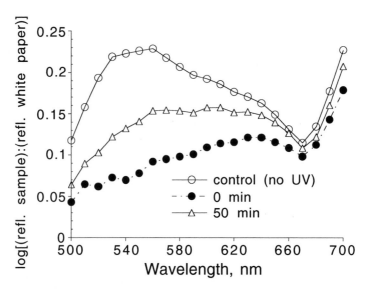

Figure 2. Reflectance spectra of bananas measured with Colortron. The top curve is an average spectrum fo rfive bananas which were not exposed to ultraviolet radiation, the two other for bananas exposed to ultraviolet-C radiation for two minutes. The bananas for the middle curve, in addition, received 50 minutes of white light immediately after UV-C irradiation. In this case spectra were measured two days after irradiations, when the bananas were still green. If the UV-C irradiation is decreased, more complete approach to the control curve can be achieved by white light irradiation.

Mount the germicidal lamp so you can irradiate the bananas at a distance of about 20 to 30 cm (not critical, but suitable exposure times vary with distance). Mount the photoreactivation lamp with the aquarium below it, and some space below the aquarium. The bananas can be irradiated for photoreactivation in the space below the aquarium. The purpose of the aquarium (which should be filled with clear water) is to remove infrared radiation and avoid heating of the bananas by the strong lamp. If you find it more convenient it is also possible to mount the bananas in the water in the aquarium, but there should be at least 10 cm of water above them.

If green bananas are exposed to the bactericidal radiation (mainly 253.7 nm) radiation only (about 1-2 minutes, but depending on the lamp, the distance and the bananas), and then left to ripen during a couple of days at room temperature, they will turn brown instead of yellow. If they are exposed to photoreactivating light (15-60 minutes required depending on lamp etc.) immediately after the ultraviolet radiation, the browning is prevented, and they will turn yellow during ripening. You can use, e.g., aluminium foil to shade different parts of the banana during exposures, and so expose different parts for different times and get many combinations of damaging and reactivating exposures on the sama banana.

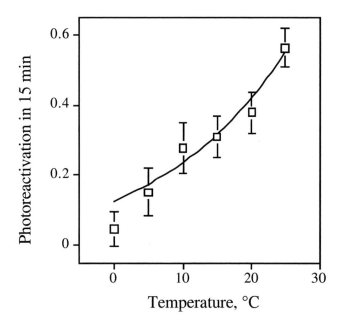

Figure 3. A study of the temperature dependence of photoreactivation of the UV-C-induced darkening of bananas. The experiment was carried out as in Fig. 2, except that the bananas were kept in water of different temperature during white light irradiation. The UV-C irrqadiation was for 1 min, and the reactivation time was only 15 min. The reflectance at 570 nm was monitored, and photoreactivation expressed as fractional approach to the control curve. The curve in the diagram here is an Arrhenius plot fitted to the data with the zero degree data omitted.

A simple spectrophotometer for reflectance measurements, such as Colortron (see section 13 below), can be used to evaluate "colour changes" (or more correctly: reflectance changes) quantitatively. You can also adapt a spectrophotometer intended for transmission measurements by attaching a Y-shaped light conductor to it. One arm of the conductor picks up light from the monochromator, the other arm delivers it to the photomultiplier, and the base of the Y is placed at a distance of about 5 mm from the banana. Light conductors of this kind can be obtained, e.g., from Oriel. It is best to plot the logarithm of the reflectance of the sample relative to a white standard.

Another educational photoreactivation experiment is described by Delpech (2001).

Literature

Delpech, R. (2001). Using *Vibrio natriegens* for studying bacterial population growth, artificial selection, and the effects of UV radiation and photo-reactivation. *J. Biol. Educ., 35*, 93-97.
Hausser, K.E. & v. Ohmcke, H.V. Lichtbräunung an Fruchtschalen. *Strahlentherapie, 48*, 223-229.

7. ULTRAVIOLET DAMAGE TO MICROORGANISMS

Any actively swimming microorganism can be used to repeat in a simplified form Hertel's (1905) classical experiment, using UV-C radiation (254 nm) from a germicidal lamp to stop the movement. Note that a visible effect takes time to develop, so it is in many cases sufficient to irradiate for a minute, and then wait. Take care to protect your eyes! We have used *Euglena gracilis* for the purpose. In addition to stopping the movement, the radiation causes this organisms to change shape, and with sufficient radiation the colour will also eventually change from green to yellow. This is a very simple experiment to carry around if you are a "travelling teacher" as I have sometimes been. See also Delpech (2001) listed in the previous section, and either Valenzeno et al. (1991) or Smith (1977)..

Literature

Valenzeno, D.P., Pottier, R.H., Mathis, P. & Douglas, R.H. (eds) (1991). *Photobiological techniques*, pp. xiii+381. New York: Plenum Publishing Corp. ISBN 0-306-43778-3.
Hertel, E. (1905). Ueber physiologische Wirkung von Strahlen verschiedener Wellenlänge. - *Zschr. Allgem. Physiologie, 5,* 95-122.
Smith, K.C. (1977). *The Science of photobiology* (first edition). New York: Plenum.

8. COMPLEMENTARY CHROMATIC ADAPTATION OF CYANOBACTERIA

Some cyanobacteria form no phycoerythrin when grown under red light, and then appear blue-green in colour from chlorophyll, phycocyanin and allophycocyanin. When grown under green light they change to almost black, because of the presence of phycoerythrin.

The most difficult part of this is to obtain a suitable culture of cyanobacteria. We recommend either *Tolypothrix tenuis* or *Fremyella diplosiphon*. The American Type Culture Collection (http://www.atcc.org) has *Tolypthrix tenuis* Kutzing 20335 and many other strains, but at high cost. From Carolina Science and Math (http://www,carolina.com/#) you can buy a set of six cyanobacterial cultures including a *Tolypothrix* sp. (not tested) for US$ 29.70 (early 2002). Sammlung Algenkulturen at the University of Göttingen sells *Tolypothrix tenuis* (accession number 94.79) and *Fremyella diplosiphon* (accession number 1429-1b) to non-commercial customers for 10.50 Euro per culture (early 2002). the most complete collection of chromatically adapting cyanobacteria is available at http://www.pasteur.fr/recherche/banques/PCC/help/htm (the Institut Pasteur in Paris). The following culture medium is suitable for *Tolypothrix tenuis* (concentrations in g l^{-1}): KNO_3 (3.0), $MgSO_4 \cdot 7H_2O$ (0.5), Na_2HPO_4 (0.2, or $Na_2HPO_4 \cdot 12H_2O$ 0.5), $CaCl_2$ (0.02, or $CaCl_2 \cdot 2H_2O$ 0.027), $FeSO_4 \cdot 7H_2O$ (0.02), and 5 ml per litre of the following trace element solution: Recipies for other suitable media are listed on the Internet at the above culture collection sites.

Light to adapt to is generated by green and red fluorescent lamps (e.g., Philips TLD 17 and TLD 15, respectively). The red lamps can be wrapped in red (or yellow) cellulose acetate foil to filter off small amounts of blue light given off (particularly at the ends), the green lamps with green (or yellow) foil. If coloured fluorescent lamps are hard to obtain, also white ones can be used if several layers of appropriate coloured cellulose acetate are wrapped around them. The cultures grow more rapidly if they are bubbled with 5% carbon dioxide in air, but ordinary air will also do. With added carbon dioxide one week should be allowed for sufficient growth, without added carbon dioxide two weeks.

For evaluation of the result you may be content just looking at the cyanobacteria with the naked eye and in the microscope. You can also do in vivo absorption spectra using a "Shibata plate", i.e. a scattering plate, just behind the cuvettes in the spectrophotometer. For quantitative estimation of the phycobiliprotein pigments proceed as follows:

Separate the organisms from the culture medium by low-speed centrifugation and resuspend them in a small amount of 0.01 M phosphate buffer (pH 7.0) and sonicate them. Avoid overheating by precooling the suspension, keeping it in ice during treatment, and sonicating in several short pulses. If you have a French press or equivalent you can disintegrate the cells more completely by using it. Only grinding in a mortar does not give a very good result. Centrifuge the slurry at 10,000xg for 5 min and then the supernatant for 1 h at 80,000xg to obtain a clear phycobiliprotein extract.

Measure the absorbance of the extract (if necessary after dilution) at 565, 620, and 650 nm. The specific absorption coefficients of the phycobiliproteins, in $g^{-1} l^{-1} cm^{-1}$, are as follows:

	phycoerythrin	phycocyanin	allophycocyanin
565 nm	12.6	3.12	1.81
620 nm	0.22	6.77	4.39
650 nm	0.15	1.75	6.54

The absorbance you measure at each wavelength will be the sum of the products of concentration (c) and absorption coefficient (ε) for each of the three pigments, times the pathlength (l), i.e. $A_\lambda = (c_E*\varepsilon_E+c_C*\varepsilon_C+c_A*\varepsilon_A)*l$, where subscripts E, C, and A stand for phycoerythrin, phycocyanin, and allophycocyanin, respectively. You will thus have an equation system with three equations and three unknowns, which can be solved for the pigment concentrations.

Variations: Instead of continuous light treatments you can try inductive pulses according to descriptions in the literature. The coloured polypeptides can be separated into very beautiful bands by isoelectric focussing.

Literature

Björn, G.S. & Björn, L.O. (1976). Photochromic pigments from blue-green algae: Phycochromes a, b, and c. *Physiol. Plant., 36,* 297-304.

Diakoff, S. & Scheibe, J. (1973). Action spectra for chromatic adaptation in Tolypothrix tenuis. *Plant Physiol., 51,* 382-385.

Fujita, Y. & Hattori, A. (1960a). Formation of phycoerythrin in preilluminated cells of *Tolypothrix tenuis* with special reference to nitrogen metabolism. *Plant Cell Physiol., 1,* 281-292.

Fujita, Y. & Hattori, A. (1960b). Effect of chromatic lights on phycobilin formation in a blue-green alga, *Tolypothrix tenuis. Plant Cell Physiol., 1,* 293-220.

Scheibe, J. (1972). Photoreversible pigment in a blue-green alga. *Science, 176,* 1037-1039.

9. PHOTOMORPHOGENESIS IN PLANTS AND RELATED TOPICS

For those readers who know German I recommend the book by Schopfer (1970), in which a number of class experiments on plant photomorphogenesis are described in detail. I do myself have three favourite experiments, of which the first two are in principle similar to experiments in Schopfer's book. Many experiments, such as basic observations on phototropism, solar tracking leaves, phototaxis etc. are so simple that they do not need to be described here. For those who feel they need some guidance reference made to Valenzeno et al. (1991) for phototropism and phototaxis, and to Vogelmann & Björn (1993) for sun-tracking.

9.1. Photomorphogenesis of bean plants

Prepare sources for red and far-red light as follows. For red light, wrap a red fluorescent lamp in red cellulose acetate film to remove traces of blue light. If you cannot obtain a red fluorescent lamp, use a white one with red acetate film. For far-red light, use an incandescent lamp (25-40 W) behind one layer (3 mm) of red and one layer (3 mm) of blue plexiglass (many other coloured plastic filters will also work). You can use a standard photographic dark-room lamp in which you change the filter for the far-red irradition, or you can build your own lamp- and filter-holder; just make sure that only far-red light escapes to your experimental dark-room or cabinet. Do not leave the incandescent lamp on when not in use, since gives off heat which may damage the filter.

Soak red beans (*Phaseolus vulgaris*) overnight in water, peel of the seed-coats and sow in (at least) five pots with soil (or sow seeds directly without soaking and peeling. although the result will be more variable then). Place one pot in a greenhouse or other suitable place where the plants can continue development under normal light conditions. Put the other pots in a completely dark place, where you can inspect them using dim green light from a green fluorescent lamp (prefereably wrapped in green or yellow cellulose acetate film to remove traces of blue light) and water when necessary (not too much, which may result in mould development). Do not unnecessarily expose the plants even to the green safelight. It is not ideal to use an incandescent lamp with a green filter as safelight, since it will emit a lot of far-red light, to which the plants are sensitive, even if you do not see it yourself. When the seedlings have reached a few cm in height, expose one pot to red, one pot to far-red,, one pot to red followed by far-red light, and keep one pot as dark control. Each exposure should be for ten minutes. Let all bean plants continue their development in darkness, except that you repeat the light treatments on each of the two following days. After the three days of light treatment, let the plants develop for two more days, after which they can be inspected in full roomlight.

Measure internode lengths and plumular angle, and compare pigmentation visually.

As a variant, you can do a similar experiment with peas.

9.2. Regulation of seed germination by phytochrome

The red and far-red light sources for this experiment can be the same as in the bean plant experiment. As plant material most educators have used lettuce seeds of a variety called Grand Rapids, available from many sources (see the Web). However, even different lots of seeds obtained from a particular source do not react uniformly, depending on temperature prehistory and other factors. If you find that seed germinate independently of light treatment, try one of the following pretreatments:

(1) Hydrate the seeds in the dark, expose them for 10 min to far-red light, and dry them again for class use (recommendation by John Hoddinott, University of Alberta).

(2) Try doing the experiment at a rather high temperature (25-28°C) rather than normal room temperature (recommendation by Brad Goodner, University of Richmond).

When you have got a good lot, keep it in a closed container in your freezer for future use, preferably in small portions, so you do not have to refreeze.

The above advice was found on an Internet chat site at http://www.clemson.edu/biolab/phyt.html. One participant (Jon Monroe), who points out that Grand Rapids is no longer as good for the experiment as it used to be, because suppliers treat the seeds to ensure complete germination, seeds from the Harris Seed Company (http://commercial.harrisseeds.com/), while Ross Koning at Eastern Connecticut State University recommends "Salad Bowl" lettuce from Agway (http://www.seedway.com/catalog). Dan Tennessen at Cornell University suggests trying seeds of *Poa pratensis* (bluegrass) and *Lepidium virginianum* (peppergrass).

For the experiment, cover the bottoms of 12 small petri dishes with a double layer of filter paper, cut to size so it is flat on the bottom. Pour distilled water into the dishes and pour it off again, so the paper is moist but no longer dripping. Place 50 seeds in rows in each dish. Cover the dishes with aluminium foil. After letting the seeds imbibe for one to three hours in the dark, irradiate with your red (R) and far-red (FR) as follows; each time for five minutes, or leave as dark control. Do duplicates of each treatment.

Irradiation sequences: R, FR, R+FR, R+FR+R, R+FR+R+FR.

Leave in complete darkness (wrapped in aluminium foil) for two days, then evaluate the result.

As an alternative to seeds, fern spores can be used for a red/far-red germination experiment. The procedure for preparing and sowing spores described in the next experiment can be used for this,

9.3. Effects of blue and red light on development of fern prothallia.

Collection of fern spores: When fern spores are about to be released (late August in northern Europe), collect spore-bearing leaves and put them on a white paper, sporangia side down. We have used *Dryopteris filix-mas*, but some other species should also work. After a couple of days most spores have fallen out on the paper and form a beautiful imprint of the leaf. By holding the paper at a slight angle and gently tapping it, it is easy to separate spores from empty sporangia and other debris. This is easily done manually, but if you like gadgets you can use the handle of an electric toothbrush to vibrate the paper.

Nutrient medium (according to Etzold 1965, amounts in g per liter): NH_4NO_3 (0.2), K_2HPO_4 (0.1), $MgSO_4\cdot 7H_2O$ (0.1), $CaCl_2\cdot 2H_2O$ (0.1); add a couple of drops of a solution of $FeCl_3$ (1% w/w) per liter. Adjust pH to 6.0-6.3 with HCl. Add 5 g of agar per liter and heat to dissolve the agar. Transfer the medium to erlenmayer flasks, plug with cotton and autoclave at for 10 min at 1 atmosphere overpressure. Let cool to about 40°C and pour the medium into sterile 9 cm plastic petri dishes. The dishes can be stored in the dark at 5°C until use.

Sterilization and sowing of spores: Pour a few (about twenty) drops of dilute (5% w/v) sodium hypochlorite solution (or corresponding concentration of commercial bleach) into a small glass cup. Put a small amount of spores (as much as the outermost mm of the tip of a knife or spatula can hold) on the solution and stir with a glass-rod until all spores are in the solution. After three to four minutes you can start to sow your petri plates with the suspension using a sterile inoculation loop. Put the loop under the spores and lift, so you get a drop with spores in the loop. Streak out the drop on a plate, and repeat with more plates (you need at least four plates). The spores can remain in the hypochlorite solution for some time without being damaged. After transfer to the agar the hypochlorite will be sufficiently diluted and need not be removed.

Irradiation and development: Put two plates under blue light, two under red light (best to use coloured fluorescent lamps, but white fluorescent lamps with coloured cellulose acetate film will probably also work well). After about six days the spores should have germinated. Study the development during one week. When you see that the protallia have started to grow, you can transfer one plate from red light to blue light, and one from blue light to red light. Under red light the prothallia will grow as filaments, under blue light as plates.

If you run a dark control you will find that the spores do not germinate. You can do red/far-red reversion experiments as on the seeds in the previous experiment.

Literature

Björn, L.O. & Virgin, H.I. (1958). The influence of red light on the growth of pea seedlings. An attempt to localize the perception. *Physiol. Plantarum 11,* 363-373.

Cone, J.E. & Kendrick, R.E. (1986). Photocontrol of seed germination. *In* Kendrick, R.E. & Kronenberg, G.H.M. (eds) *Photomorphogenesis in plants,* pp. 443-463. Dordrecht: Martinus Nijhoff/Junk Publishers.

Etzold, H. (1965). Der Polarotropismus und Phototropismus der Chloronemen von *Dryopteris filix-mas* (L.) Schott. *Planta, 64,* 254-280.

Frankland, B. & Taylorson, R.B. (1983). Light control of seed germination. *In* Shropshire, Jr, W. & Mohr, H. (eds), *Encycl. Plant Physiol., New Series, 16A,* pp. 428-456. Berlin: Springer.

Haupt, W. & Björn, L.O. (1987). No action dichroism for light-controlled fern-spore germination. *J. Plant Physiol., 129,* 119-128.

Schopfer, P. (1970). *Experimente zur Pflanzenphysiologie,* pp. 418. Freiburg: Rombach Verlag.

Valenzeno, D.P., Pottier, R.H., Mathis, P. & Douglas, R.H. (eds) (1991). *Photobiological techniques,* pp. xiii+381. New York: Plenum Publishing Corp. ISBN 0-306-43778-3.

Vogelmann, T.C. & Björn, L.O. (1983). Response to directional light by leaves of a sun-tracking lupine (*Lupinus succulentus*). *Physiol. Plant., 59,* 533-538.

Withrow, R.B., Klein, W.H. & Elstad, V.B. (1957). Action spectra of photomorphogenetic induction and its inactivation. *Plant Physiol., 32,* 453-462.

10. SPECTROPHOTOMETRIC STUDIES OF PHYTOCHROME *IN VIVO*

Photobiology practicals in our own photobiology courses have included also purification of phytochrome and in vitro experiments with the purified phytochrome, but this may be to laborious to set up in a laboratory not doing research on phytochrome. The following is a simple experiment to carry trough, but requires a good dual wavelength spectrophotometer such as Aminco DW-2. It may be possible to adapt the experiment for an ordinary double-beam single wavelength computerized spectrophotometer of good quality, but I have not tried this.

The aim of the experiment is to demonstrate spectrophotometrically the *in vivo* transformations between the red-absorbing (P_r) and far-red absorbing (P_{fr}) forms of phytochrome: $P_r ==> P_{fr}$ (under red light) and $P_{fr} ==> P_r$ (under far-red light).

Plant material: Oats or wheat is sown in moist Vermiculite (mica) in plastic or metal trays (a total of 0.25 m^2 required). The trays are incubated in complete darkness at 20-25°C for about four days. The when the seedlings have emerged, and preferably before the leaves have broken through the coleoptiles, the shoots are harvested. The harvesting and the following manipulations should be carried out using dim green working light. About 50 plants are required for an experiment.

Sample preparation: The coleoptiles are cut open so the leaves in them can be removed. This procedure is rather time-consuming and not absolutely necessary if you just want to demonstrate the presence of phytochrome. For quantitative experiments it is of advantage to get rid of the leaves, which are rich in protochlorophyllide. This is also photoconvertible, which obscures the phytochrome signal.

The coleoptiles are collected in a petridish with water, so they do not dry out. When you have enough coleoptiles, order them into a bundle which you cut across with a razor or scalpel in the middle to get two bundles with a sharp delimitation. They are joined with the cut ends in the same direction and pressed into a water-filled spectrophotometer cuvette in such a way that you avoid air bubbles between the coleoptiles. There is no need to crush an expensive cuvette trying to squeeze in the coleoptiles; you may as well use a cheap plastic one. We have also with good result used small round glass test-tubes which fit the cuvette holder of the spectrophotometer.

Measurement: Seitch the spectrophotometer to dual-wavelength mode, set the monochromators to measure the absorbance difference between 660 and 730 nm (or, preferably, between 660 and 800 nm in one series of measurement and between 730 and 800 nm in another series of experiments). If you have not removed the leaves it is best to focus on measuring the P_{fr} changes only by recording the absorbance difference between 730 and 800 nm. Using the Aminco DW-2 and DW-2a spectrophotometers we have set full scale to 0.01 absorbance units and the damping to medium.

After taking a reading, irradiate the sample with strong red light for one minute and take a new reading. Then do the same with far-red light and take a new reading. If this works you can continue the experiments in a variety of ways, studying reaction kinetics and action spectra, for instance. We have also let students take coleoptiles outside to different environments, above and below tree canopies etc., and then bring the coleoptiles back to the laboratory (on ice) for evaluation of the phytochrome state.

The phytochrome transforming irradiations can either be carried out with the sample cuvette still in the spectrophotometer, using a xenon lamp-monochromator combination, or after removal from the spectrophotometer. In the former case the photomultiplier should be protected with a piece of sheet metal during irradiations, and if the spectrophotometer does not switch off the photomultiplier voltage automatically this should be done manually. In the latter case, we have used light sources similar to those used for photomorphogenesis and germination experiments. In this case it is important to handle the cuvette carefully so the coleoptiles or any air bubble present do not change position, not to touch the places where the measuring beam is to enter and exit, and to put the cuvette back with the original orientation.

As an alternative to coleoptiles, the epicotyls (internodes between cotyledons and first two leaves) of dark-grown pea plants serve very well. This circumvents the time-concuming trouble of removing leaves.

11. PHOTOCONVERSION OF PROTOCHLOROPHYLLIDE

Background: The synthesis of chlorophyll *a* proceeds via 5-aminolevulinic acid (ALA), protochlorophyllide and chlorophyllide *a*. When angiosperms are grown in darkness, the synthesis is arrested at the protochlorophyllide stage. However, there is no major buildup of protochlorophyllide, because the synthesis of ALA is exposed to feedback inhibition by protochlorophyllide, and is resumed only when protochlorophyllide is used up. This is the reason that plants do not become visibly green in darkness, even if protochlorophyllide in itself has a green colour. When the plant is illuminated, protochlorophyllide is transformed to chlorophyllide by photoreduction, and it is the protochlorophyllide itself, bound to NADPH:protochlorophyllide oxidoreductase (POR, EC 1.6.99.1). The reductant is NADPH, which in darkness is bound to the enzyme together with protochlorophyllide.

The transformation of protochlorophyllide to chlorophyllide a can be monitored spectrophotometrically directly in living leaves, since the weak light beam in the spectrophotometer does not noticeably affect the process.A single flash from an electronic photoflash, on the other hand, transforms a large part of the protochlorophyllide present.

Spectrophotometer: You will need a good split-beam spectrophotometer (or a single beam spectrophotometer with computer that can store a reference signal). For this you should make a special leaf holder. If you use bean leaves (see below) the leaf holder should consist of two glass plates and some kind of clamp holding the plates together, so you can position the leaves between the plates, covering the measuring beam. The details depend on the spectrophotometer. One of the glass plates should be scattering (ground surface, or milky glass), or you must add a special "Shibata plate" to scatter light after passage of the sample and reference.

If you use grass leaves, you can use the standard cuvette holder (except that you need to add a "Shibata plate"). Instead of cuvette you use a square rod of Plexiglas or Perspex (1 cm x 1 cm x 4 cm) in which you have drilled nine holes (1.5 mm diameter) along the 4 cm direction. Five of the holes are in a row, with 1 mm in between. The remaining four holes are in another row, such that the holes are in between the holes in the first row. Fig. 4 gives an idea of what the grass leaf holder looks like from above. You may need to adjust the dimension to the kind of grass you use; maize needs larger holes.

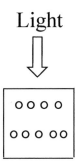

Figure 4. Sketch of holder for grass leaves in spectrophotometer.

Experimental procedure: You can use either bean or wheat plants (or some other grass). All steps in which you handle your plants or leaves after sowing should be carried out in dim green working light. Your eyes, fortunately, are most sensitive to this light, while the process you want to study has a sensitivity minimum in the green spectral region.

If you use bean plants, soak the beans in dilute hydrogen peroxide (initial concentration 1-2%) overnight and then remove the seed coats. The hydrogen peroxide serves both to kill some mold spores, and to facilitate peeling by producing gas under the seed coats. Sow the beans directly in Vermiculite or sand that has been heated to kill microorganisms, or pregerminate on filter paper and then sow under green light. Let the plants grow (about 5-7 days) until the first true leaves (between the cotyledons) have become a little longer than the cotyledons. If you wait until the plants have become much older, only part of the protochlorophyll(ide) will be immediately phototransformable.

If you use grass leaves, after sowing a few tens of caryopses allow the plants to grow in darkness until you (by inspection under green light) see that the leaves have just emerged from the coleoptiles. They will still remain tightly rolled together unless you have used too much of your inspection light (unrolling is a phytochrome-controlled reaction). Collect nine leaves, cut out 5 cm of the middle portion and put them in the holes in your leaf holder.

Irrespective of what kind of leaf you use, it is important that the leaves completely cover the whole measuring beam in the spectrophotometer.

Record a spectrum for the unirradiated leaves from 500 to 720 nm. Do not worry about absolute absorbance; it is the shape of the spectrum that is of interest. To balance the split-beam spectrophotometer with the leaves on the sample side, you may need to put something "neutral" in the reference beam. One or a few pieces of filter paper serve well.

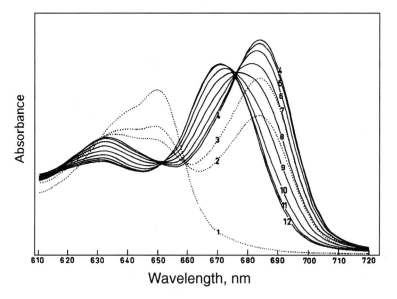

Figure 5. Absorption spectra for dark-grown bean leaves, before illumination (curve 1), after 1, 2 and 10 exposures to a photographic flash (curves 2- 4), and after incubation for various times in darkness after the light exposures (curves 5-12). The first curve represents mainly the absorption of the protochlorophyllide-NADPH-enzyme complex, which is changed to a chlorophyllide a-NADP+-enzyme complex by illumination. During the following dark incubation the chlorophyllide is released from the enzyme, attaches to another protein, and is esterified to chlorophyll a. More sophisticated experiments reveal several spectral shifts.

When you have recorded the "dark" spectrum, irradiate in a suitable way, for instance with a photoflash as I did for Fig. 2 above. Record a new spectrum. If you want to produce a good-looking series of spectra as in Fig. 5, it is important that you work fast, since the "Shibata shift" from the initial chlorophyllide a-enzyme complex starts at once (rate depending on age of leaves and other factors).

This experiment can be modified in a number of ways. The protochlorophyllide-enzyme complex ("protochlorophyllide holochrome") can be extracted and the conversion studied in solution, and fluorescence spectroscopy can be used in place of absorption spectroscopy. See, for instance, Björn (1969) in the literature list.

Literature

Björn, L.O. (1969). Action spectra for transformation and fluorescence of protochlorophyll holochrome from bean leaves. *Physiol. Plant., 22*, 1-17.

12. SEPARATION OF CHLOROPLAST PIGMENTS

There are many ways of carrying out this experiment, but I have my definite favourite. It is to use a cylindrical chromatography paper, as described below.

Pigment extraction: Choose a plant species which does not have extremely acid cell sap, which will result in conversion of chlorophyll to pheophytin. We have used, among others, spinach, bean, and stinging nettle. Extract a leaf using a small amount of 100% acetone. We have usually done this in a mortar with sand, and then one has to filter or centrifuge to get a clear extract. I have read that other people with good result have just left the leaf with acetone in a bottle in the dark for a day or two, so if you do not want to do the whole experiment in one day this saves some work and dishwashing.

The acetone extract may be applied as it is to the chromatogram, but then you cannot put on very much of it without getting into trouble. It is better to mix it with a little (one tenth the volume of the extrac or so) light petroleum (or hexane) and add water (a about the same volume as the original extract) to separate into two layers. Discard the lower (aqueous) layer (the most convenient way is to suck up the upper layer using a Pasteur pipette and transfer the liquid to a clean test tube), add water again, shake and let separate. If you have difficulty getting a clean phase boundary, add some sodium chloride. Pipette the organic, dark green phase to a clean test-tube.

Chromatography: Get a glass jar with tightly fitting lid (it is very important that it fits tighly). You may find one in your home that has contained mayonnaise or some other food product. A suitable size is 150 mm high and 70 mm wide. Cut out a rectangular paper of a size that can produce a cylinder that fits inside the jar without touching the walls or the lid, but do not fold it to a cylinder yet. Avoid touching the paper with your fingers.

Pour into the jar a mixture of 90% (v(v) petroleum ether and 10% (v/v) acetone. You should have a layer about 10 mm deep at the bottom of the jar. Put on the lid.

Use a Pasteur pipette to apply your pigment extract along a line parallel to and 15 to 20 mm from a side of the paper which will form the base of your cylinder. The paper edge should rest on a test tube or glass rod, or stick out over the edge of the lab bench, so it is free in the air (or put the paper in a book, with the edge sticking out). With some practice you will be able to produce a nice green line. You should let the tip of the pipette move quickly across the paper to avoid big blobs. Try not to scratch the paper with the tip of the pipette. If the extract is in light petroleum or hexane it will dry quickly, and you can in a short time repeat the application until you have quite a lot of pigment on the line. But do not apply so much that you clog the pores in the paper.

Now shape the paper into a cylinder with the green line at the bottom. Staple together along the edges perpendicular to the bottom edge in such a way that the edges do not touch (if they do, the liquid will rise in the paper in an irregular way). Open the jar briefly and put down your paper cylinder, green line down. Put on the lid immediately, to avoid liquid evaporating from the paper. Do not move the jar until you finish the experiment. Wait and enjoy the result!

This chromatography is mainly a liquid/liquid distribution of substances according to lipophily/hydrophily between the moving hydrocarbon and stationary water molecules hydrogen bonded to the OH groups in the cellulose.

Within a few minutes you can see the pigments separating into several bands: A yellow band almost at the rising liquid front are the carotenoids. They do not stick to the water adsorbed on the paper because they contain only carbon and hydrogen. Next comes another yellow band, the oxygen-containing xanthophylls. Lówer down follow chlorophylls *a* and *b*, in that order because chlorophyll *a* has a lipophilic methyl group where chlorophyll *b* has an aldehyde group.

Spectrophotometry: When the liquid front has almost reached the top of the paper after half an hour or so, take it out and let it dry. This experiment is easily done as a demonstration during a lecture, and in that case you can conclude it by cutting strips of the paper and sending them around in the audience for inspection. If it is done as a student experiment, it can be concluded by determining the absorption spectra of the various pigments. If you have got a hand-held reflection spectrophotometer such as the Colortron (see next section) it this can be done within a couple of minutes directly on the paper. Otherwise the chromatogram can be eluated with acetone. Cut out the strips with the various pigments. If you dip such an end into acetone, the acetone will rise and carry all of the pigment with the front, so you can easily concentrate the pigment and dissolve it in a small amount of acetone. You will easily get a concentration high enough and sufficient volume for a standard spectrophotometer cuvette from a single chromatogram.

13. LIGHT ACCLIMATION OF LEAVES. THE XANTHOPHYLL CYCLE

We have used a simple hand-held reflectance spectrophotometer (Colortron, and a later model, Colortron II) for a number of biological experiments. It has already been mentioned above in connection photoreactivation (section 6) and spectrophotometry of paper chromatograms (section 12). These excellent and relatively cheap (about US$ 1000) instruments (originally marketed by Light Source, Inc.) are currently available from but X-rite (http://store.xrite.com). Other companies sell other instruments of a similar type that should also work.

Such instruments can certainly also be used to evaluate complementary chromatic adaptation (section 8), changes in skin colour, changes in leaf colour during senescence or nutrient deficiency, and a large number of other experiments. A very simple but interesting experiment, relating to the light acclimation of leaves and the xanthophyll cycle, can also be carried out.

Introduction to the xanthophyll cycle

Since the xanthophyll cycle is not covered elswhere in the book, a more thoroough introduction is given here for readers who are not familiar with it.

Daylight is highly variable, and the plant must be able to adjust to strong as well as weak light. When the rate of light absorption by the photosynthetic system exceeds the rate with which carbon dioxide can be assimilated, or other assimilatory reactions can be carried out, there is risk for damage to the plant. One thing that may happen when carbon dioxide cannot be reduced by the electrons transported through the electron transport chain is that *electrons* may end up on molecular oxygen and reduce it to superoxide anion. The plant has superoxide dismutase to take care of this. However, the electron transport chain itself may be overloaded, so that the excitations in the pigment system cannot be used up for electron transport at all. What may then happen is that the *energy* is transferred to oxygen molecules, resulting in a form of excited oxygen called singlet oxygen (actually there is more than one kind of singlet oxygen). Singlet oxygen is very reactive and may cause damage. The plant needs a way of disposing of excitation energy, but this should operate only when there is an excess. In weak light the plant needs all the energy it can collect.

One way for the plant of regulating the dissipation of excess energy absorbed by the photosynthetic system is by adjusting the amount of violaxanthin and zeaxanthin through the reactions of the xanthophyll cycle. In this cycle violaxanthin is converted to zeaxanthin via the intermediate antheraxanthin, a reaction (so-called deepoxidation) catalyzed by the enzyme violaxanthin-deepoxidase. The reverse reaction (epoxidation of zeaxanthin to violaxanthin) can also take place (Fig. 6). Deepoxidation takes place in strong light, epoxidation in weak light or darkness.

Both zeaxanthin and violaxanthin have long conjugated double bond systems. The important difference between them is that the zeaxanthin molecule has a longer conjugated system (11 double bonds vs 9 for violaxanthin). This causes the lowest excited state in the zeaxanthin molecule to be at a lower level than the lowest excited state of violaxanthin. It so "happens" that the lowest energy levels of the chlorophyll a molecule are just between that of zeaxanthin and that of violaxanthin.

I have here to introduce a little complication. If you do not understand this paragraph, do not worry, since it is not important for an understanding of the plant's physiology. Xanthophylls like zeaxanthin and violaxanthin are yellow because they have energy levels low enough for blue light photons to be absorbed and excite them to an excited state. These energy levels cause the bands in the blue part of the absorption spectrum. That the xanthophylls possess these energy levels is because they have long conjugated bond systems. However, these are *not* the energy levels we are talking about in explaining the xanthophyll cycle. There are even lower energy levels, corresponding to red light. They do not, however, show up in the absorption spectra, because the transition to this state from the ground state is "forbidden". Like most of quantum mechanics, we shall, as biologists specializing in another branch of science, just have to accept this.

Let's now return to the main line of thought. Since the lowest excited state of violaxanthin is slightly higher than the lowest excited state of chlorophyll a, energy from an excited violaxanthin molecule may be transferred to a chlorophyll a molecule, provided it is close enough. In other words, violaxanthin can act as an antenna pigment.

Zeaxanthin, on the other hand, cannot act as an antenna pigment for chlorophyll a, since its lowest energy level is lower than the lowest energy level of chlorophyll *a*. On the contrary, it can accept energy from an excited chlorophyll a molecule, provided it is close enough.

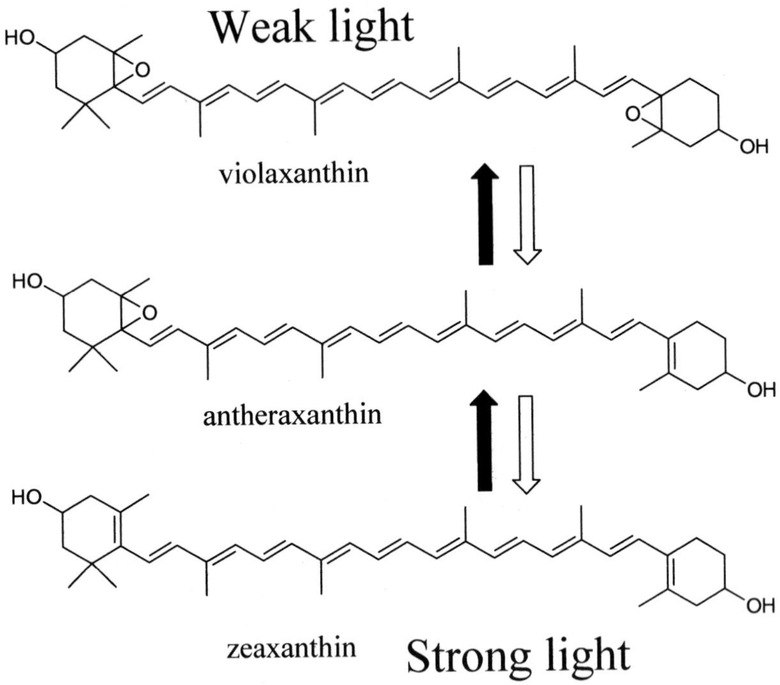

Figure 6. The xanthophyll cycle in higher plants (also called the violaxanthin cycle). In some algae an analogous cycle with other carotenoids takes place.

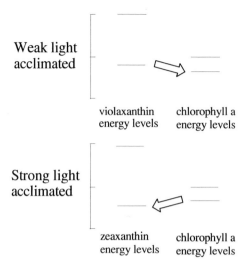

Weak light acclimated

violaxanthin   chlorophyll a
energy levels   energy levels

Strong light acclimated

zeaxanthin   chlorophyll a
energy levels   energy levels

Figure 7. Energy levels in violaxanthin, chlorophyll a and zeaxanthin. For violaxanthin the lowest excited state is higher, for zeaxanthin lower than the lowest (singlet) excited state in chlorophyll. The arrows show direction of energy transfer. Simplified after Frank et al. (1994)

We are more accustomed to describing the photosynthetic pigments in terms of the positions of their absorption maxima than in terms of energy levels. This is OK, as long as we remember that energy is inversely proportional to wavelength. The reaction centre pigment in photosystem I is referred to as P700, because its (long wavelength) absorption peak is at 700 nm, and that of photosystem II is called P680 for the corresponding reason. These wavelengths correspond to energies of 1.77 and 1.82 eV, respectively. The antenna pigments have peaks at shorter wavelength, corresponding to higher energy. This is the reason that energy can flow from antenna pigments to reaction centres, while some of the energy is degraded to heat, giving an overall positive entropy change.

The lowest energy level (1.89 eV) of violaxanthin corresponds to 655 nm, allowing this pigment to donate energy to any pigment with an absorption peak at longer wavelength. This includes all forms of protein-bound chlorophyll a. The lowest energy level (1.76 eV) of zeaxanthin corresponds to 704 nm. It allows it to accept energy from any antenna chlorophyll a.¨

For the intermediate xanthophyll, antheraxanthin with 10 conjugated double bonds, the lowest energy level corresponds to 680 nm. It therefore can accept energy from photosystem II antenna pigments on the same terms as the reaction centre, but donate energy to photosystem I.

The xanthophyll cycle can thus function as a safety valve for the photosynthetic system when it is overloaded with energy. The valve opens when zeaxanthin is formed from violaxanthin. The energy received by zeaxanthin is degraded to small quanta (heat), which is less dangerous than the large energy quanta that can break bonds and cause chemical reactions.

This is just half the truth, though. There are other energy sinks than zeaxanthin in the photosynthetic system. Violaxanthin has a third important function in addition to being an antenna pigment and being a source of zeaxanthin. It is able to inactivate the other (not yet characterized) energy sinks. The xanthophyll cycle thus regulates the electron pumping by the photosystems in a threefold way.

Violaxanthin deepoxidase has a pH optimum of about 5, and it is likely that deepoxidation is activated by the proton uptake into the thylakoids that occurs in light, and the "trigger point" is the pH that is just a little bit lower than that required for ATP synthesis.

The acidification of the thylakoids and the conversion of violaxanthin to zeaxanthin that take place when the plant is subjected to excess light cause secondary changes in the antenna pigments. The structure is changed, which can be seen as increased light scattering power. This is a contributing factor to the signals that we shall monitor in our experiment.

The xanthophyll cycle pigments are present both in photosystem I and photosystem II antennas (chlorophyll a/b binding proteins), and also dissolved in the thylakoid membrane lipid. In addition, violaxanthin occurs in the envelope membrane. All violaxanthin in the plant is not available to the deepoxidase, and the available fraction seems to increase with increasing reduction of the plastoquinone pool. This may be an important point for us who are interested in effects of UV-B radiation on plants. UV-B inhibits photosystem II more than photosystem I and thus presumably leads to a more oxidized plastoquinone pool, resulting in less substrate for the xanthophyll cycle. UV-B may also inhibit the deepoxidase, and both these changes may lead to a decrease in the ability of the plant to protect the photosystem agains overloading by excess light. One of the experiments we can carry out is to look just at how UV-B affects the deepoxidation.

The pool of xanthophylls that can participate in the xanthophyll cycle also increases when the plant prepares for the winter. There is a big need for safety valve function during the spring, when the plant may be subjected to strong light at the same time as the temperature is so low that carbon dioxide assimilation impaired. The pool is larger in plants adapted or acclimated to strong light than in shade plants.

Finally, a different role has also been proposed for zeaxanthin and the xanthophyll cycle: as a blue light sensing system in stomatal regulation and phototropism.

Experiment

Gamon et al. (1990) described a principle for sensing of the xanthophyll cycle state by monitoring reflectance at 530 nm. The method described here employs the same principle. The advantage with our variant is that it uses equipment that is cheap, battery-operated and portable. Although leaves picked from plants were used for the examples shown here, the method has potential of being non-destructive, as there seems to be nothing preventing measurements to be made on attached leaves.

A leaf is kept in darkness for at least ten minutes. The reflectance spectrum is then recorded. If a Colortron instrument is used, this simply means pressing the instrument, connected to a Macintosh or PC with the appropriate software running, against the leaf. In a few seconds the spectrum from 390 to 700 nm is then automatically stored in the computer, using the two lamps built into the instrument as light sources. We have built a simple leaf holder, so a leaf can be placed in front of the instrument in a repeatable way. After the first spectrum is stored, the leaf is exposed to strong light (PAR) for a 0.5-7 min, and the spectrum again recorded. The spectra can be viewed with the original software, but in this case they are best transferred to another programme (we have used CricketGraph III or KaleidaGraph for the purpose), in which the light-dark difference spectra can be computed and displayed. An alternative is to not follow the directions for calibrating the spectrophotometer using the "100% reflectance standard" supplied, but to use the dark-adapted leaf as reference, against which the illuminated leaf is compared. One can easily investigate the kinetics using a series of irradiations. Best is to plot the log of the ratio of reflectances, to have an analogue to the absorbance used in transmission measurements.

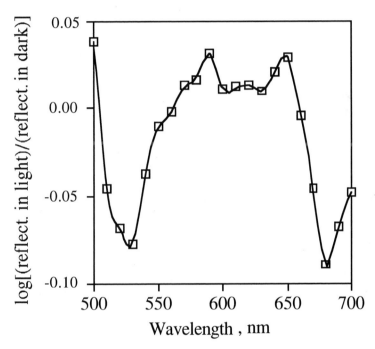

Figure 8. Light/dark difference reflectance spectrum of a Prunus laurocerasus *lea measured with Colortron. The dip at 530 nm signals the xanthophyll changes, the dip at 680 nm the quenching of chlorophyll fluorescence associated with it. CricketGraph was used for plotting the data.*

The result of this experiment is that the reflectance in the green region, especially near 530 nm, is decreased by the radiation. This change is caused by conversion of violaxanthin to zeaxanthin, which causes rearrangements in the thylakoid membranes with cocomitant changes in light scattering. One can also see an apparent reflectance decrease in the red region. This, however, is not a true reflectance decrease, but signals the quenching of fluorescence associated with it.

This measurement is in principle similar to that carried explored by Gamon et al. (1990) as a method for remote sensing of photosynthetic efficiency of plants, and later in more detail by Gamon et al. (1992, 1997), Gamon & Surfus (1999), Penuelas

434

et al. (1995, 1997), Filella et al. (1996), Nichol et al (2000), and Barton & North (2001).

The Colortron has been used in a few other scientific investigations (Andrews & Freeman 1996, Hill 1998, Okubo et al. 1998).

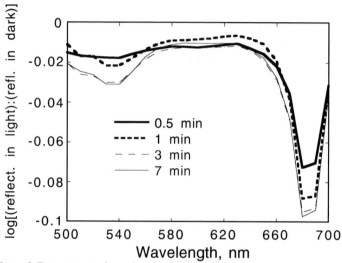

Figure 9. Experiment similar to that in Fig. 8, but with several irradiation times and with leaves of Vaccinium vitis-idaea as experimental material. As can be seen from a comparison of the two figures, the relation between the scattering and the fluorescence change varies among species. It can also be seen here that the change is complete after 3 min in strong light.

Literature

Andrews, J.T. & Freeman, W. (1996). The measurement of sediment color using the Colortron (TM) spectrophotometer. *Arctic Alpine Res. 28*, 524-528.

Barton, C.V.M. & North, P.R.J. (2001). Remote sensing of canopy light use efficiency using the photochemical reflectance index: Model and sensitivity analysis. *Remote Sensing Environ., 78*, 264-273.

Filella, I., Amaro, T., Araus, J.L. & Penuelas, J. (1996). Relationship between photosynthetic radiation-use efficiency of barley canopies and the photochemical reflectance index (PRI). *Physiol. Plant., 96*, 211-216.

Frank, H.A., Cua, A., Chynwat, V., Young, A., Gosztola, D. & Wasielewski, M.R. (1994). Photophysics of the carotenoids associated with the xanthophyll cycle in photosynthesis. *Photosynthesis Res., 41*, 389-395.

Gamon, J.A. & Surfus, J.S. (1999). Assessing leaf pigment content and activity with a reflectometer. *New Phytol., 143*, 105-117.

Gamon, J.A., Peñuelas, J. & Field, C.B. (1992). A narrow-waveband specral index tht tracks diurnal changes in photosynthetic efficiency. *Remote Sensing Envir., 41*, 35-44.

Gamon, J.A.m Serrano, L. & Surfus, J.S. (1997). The photochemical reflectance index: an optical indicator of photosynthetic radiation use efficiency across species, functional types, and nutrient levels. *Oecologia, 12*, 492-501.

Hill, G.E. (1998). An easy, inexpensive means to quantify plumage coloration. *J. Field Ornithol. 69*, 353-363.

Nichol, C.J., Hümmrich, K.F., Black, T.A., Jarvis, P.G., Walthall, C.L., Grace, J. & Hall, F.G. (2000). Remote sensing of photosynthetic light-use efficiency of boreal forest. Agric. Forest Meteorol., 101, 131-142.

Okubo, S.R., Kanawati, A., Richards, M.W. & Childress, S. (1998). Evaluation of visual and instrument shade matching. J. Prosthet. Dent., 80, 642-648.

Peñuelas, J., Filella, I. & Gamon, J.A. (1995). Assessment of photosynthetic radiation-use efficiency with spectral reflectance. *New Phytol., 131*, 291-296.

Peñuelas, J., Llusia, J., Pinol, J. & Finella, I. (1997). Photochemical reflectance index and leaf photosynthetic radiation-use efficiency assessments in Mediterranean trees. *Int. J. Remote Sensing, 18*, 2863-2868.

Srivastava, A. & Zeiger, E. (1995). Guard cell zeaxanthin tracks photosynthetically active radiation and stomatal apertures in *Vicia faba* leaves. *Plant Cell Environm., 18*, 813-817.

14. PHOTOCONVERSION OF RHODOPSIN

Background: Visual pigments in vertebrate eyes are concentrated in cells called rods and cones. The rods are active in "dusk vision", while cones are used for colour vision in stronger light. The light-sensitive pigment in rods is rhodopsin, a protein with 11-cis retinal as a chromophore (Chapters 7 and 8). Absorption of light triggers a multistep isomerisation process, resulting in the separation of protein and chromophore, and to bleaching of the initially red colour. The first part of the rhodopsin conversion, from rhodopsin to metarhodopsin I, is photoreversible.

Chemicals required for the experiment:
Sodium phosphate buffer (0.15 M, pH 6.5)
Sucrose solution (40% w/v dissolved in the phosphate buffer)
Potassium-aluminium sulfate ($KAl(SO_4)_2$ (4% w/v in water)
Hydroxylamine hydrochloride ($NH_2OH \cdot HCl$, 0.05 M in the
        phosphate buffer).
Digitonin (2% w/v dissolved in the phosphate buffer in a heated water bath)

Biological sample: Obtain several (at least eight required) cattle eyes from a slaughterhouse, and leave them in the dark for an hour at room temperature. Dissect out the retinas under red light. The retinas can be frozen and preserved for use later.

Extraction of rhodopsin: The preparation is carried out in a cold-room under red light. Avoid too much of the working light on the retinae. Put eight retinae in a cold mortar and grind very thoroughly for 20 min. Then add five to ten ml of the sucrose solution (about three times the volume of the ground retinae). Mix well and transfer the suspension to a transparent centrifuge tube. Carefully layeer phosphate buffer (about the same volume) on top using, for instance, a syringe with a short tubing. Try to keep a sharp interface between the two liquids. Centrifuge at high speed for 30 min.

The rod outer segments have about the same density as the sucrose solution, will be enriched at the interface between the liquids.

Carefully suck the rod outer segments into a syringe. Transfer them to a new centrifuge tube and fill it to half with buffer. Centrifuge at high speed for ten minutes. Discard the supernatant and repeat the washing procedure. Then add three ml KAl(SO4)$_2$, which will fix the rhodopsin. Let stand for ten minutes. Precipitate the segments again by ten min high-speed centrifugation and wash twice more, but now with water.

Add four ml of the digitonin solution to the rod segments. Stir thorougly, and let stand in the cold overnight. This procedure will solubilize the rhododpsin. Centrifuge just before the experiment, which is carried out on the clear supernatant.

Experiment 1. Low fluence rate treatment: Under red working light, pipet 600 microlitres if rhodopsin solution and 150 microlitres of NH$_2$OH·HCl into a 10 mm spectrophotometer cuvette. Prepare a reference cuvette with 600 microlitres of buffer plus 150 microlitres of NH$_2$OH·HCl. Determine the absorbance difference between the cuvettes throughout the range 500 – 600 nm.

Remove the sample cuvette from the spectrophotometer and put it about half a metre from a small incandescent lamp such as a microscope lamp. Illuminate for 30 s and let stand for two to three minutes. In this time the retinal released from the opsin protein will react with the hydroxylamine, which results in a product with absorption maximum at much shorter wavelength than rhodopsin. Disappearance of intermediates will also be speeded up by the hydroxylamine. Again determine the difference absorption spectrum between sample and reference cuvettes. Repeat the procedure until 80-90% of the rhodopsin has been bleached. Then switch on normal room light and bleach the remaining rhodopsin.

Repeat the irradiation experiment with one quarter of the fluence rate. If you cannot measure the light, and have not calibrated filter, and provided you have a small light source (not a fluorescent tube), and a not-to-white or shiny lab bench, you can approximate this by doubling the distance between lamp and cuvette.

Compute the ratio $(A_i-A_t)/(A_i-A_e)$ as a measure of converted rhodopsin for each time-point A_i = initial absorbance difference, A_e = final absorbance difference, A_t absorbance difference for irradiation time t. Compare the result for the two fluence rates and figure out whether the process follows the Bunsen-Roscoe law ("reciprocity law"). What is the reaction order?

Experiment 2. High fluence rate illumination: Repeat the same, but using a photoflash instead as light source. Does this alter the validity of the reciprocity law, and if so, why?

Literature

Johnson, R.H. & Williams, Th.P. (1970). Action of light upon the visual pigment rhodopsin. *J. Chem. Edu., 47*, 736-739.

15. PHOTOSYNTHESIS OF PREVITAMIN D

An experiment to determine the quantum yield of provitamin D_3 to previtamin D_3 conversion is described by Pottier & Russell (1991). Part of their description concerns the use of the ferrioxalate actinometer, which is already covered elswhere (Chapter 4) in the present book, as is the theoretical introduction. One difficulty in carrying out the experiment as described by Pottier & Russell (1991) is that the pure previtamin D_3 used for calibration in their experiment is not commercially available. The present description is modified with this in mind.

I shall also take the opportunity to show an unusual way of determining the quantum yield, which sometimes has advantages.

For determination of quantum yield of a photochemical reaction, one usually determines the number of molecules converted and the number of photons absorbed, and takes the ratio between the two quantities. Björn (1969), when faced by the problem to determine the quantum yield of inactivation of an enzyme, could not directly measure the amount of light absorbed by the enzyme, because the enzyme solution was dilute, and the amount of light absorbed by it very small. He showed, instead, that the quantum yield equals $^{10}\log(a_o/a_t)/(\varepsilon t\ I)$, where a_o is the concentrations at the start and a_t that after irradiation time t, I the photon fluence rate, and ε the molar absorption coefficient.

In our present case, the photoconversion of provitamin D we have to keep in mind that the primary product, previtamin D, can undergo a number of photochemical reactions which could disturb our measurement. We avoid them in two ways: by limiting the amount of radiation to keep the amount of product (previtamin D) much lower than the amount of substrate (provitamin D), and by choosing a wavelength at which the substrate absorbs much more strongly than the product, i.e. 294 nm. For an ethanol solution containing 5 µg provitamin D per ml (which is a suitable concentration for experimentation) the absorbance at 294 nm in a 1 cm layer is 0.088. The absorbance of a corresponding solution of previtamin D is only 0.034.

We choose the initial provitamin concentration with the following in mind: it should be high enough to allow accurate spectrophotometry, yet low enough to keep the fluence rate within the sample reasonably uniform. If the absorbance in a 1 cm cuvette is 0.088, then the irradiance at the exit side of the cuvette is $10^{-0.088} = 81.7\%$ of that on the illuminated side, i.e. transmission T=0.817 for a thickness x=1 cm. The average irradiance is $\int T^x dx = (T-1)^{10}\log e/^{10}\log T = -0.183/\ln 0.817 = 0.905$. Use of such an average irradiance in the calculations requires that the solution is stirred. For an unstirred solution it is better to mentally divide the solution into, say, ten (or one hundred) layers, calculate for each layer separately how the chemical change (and the change in irradiance) takes place, and finally add up the chemical change for the layers. I leave this as an exercise for the reader.

The molecular weight for provitamin D_3 is 384.65 (for provitamin D_2 it is 400.70; this can be looked up in the Handbook of Chemistry and Physics, where the compounds are called $\Delta^{5,7}$-cholestadien-3β-ol and ergosterol, respectively). Thus the molar absorption coefficient ϵ (for the D_3 form) is $384.65 \cdot 10^6 \cdot 100/5/1000$ m^{-1} M^1 = $7.7 \cdot 10^4$ m^{-1} M^1 (the factor 10^6 is for converting microgram to gram, 100 for converting cm to m, and 1000 for converting millilitre to litre). Thus the quantum yield in our special case is $^{10}\log(a_o/a_t)/(\epsilon t \, I \, 0.905) = {}^{10}\log(a_o/a_t)/(7.7.10^4 t \, I \, 0.905)$ with t and I in SI units.

Literature

Björn, L.O. (1969). Photoinacivation of catalases from mammal liver, plant leaves and bacteria. Comparison of inactivation cross sections and quantum yields at 406 nm. *Photochem. Photobiol., 10*, 125-129.

Pottier, R.H. & Russell, D.A. (1991). Quantum yield of a photochemical reaction. *In* Valenzeno, D.P., Pottier, R.H., Mathis, P. & Douglas, R.H. (eds) (1991). *Photobiological techniques*, pp. 45-52. New York: Plenum Publishing Corp. ISBN 0-306-43778-3.

16. BIOLUMINESCENCE

Fireflies

If you do not have bioluminescent insects where you live, or if the season is not the right one, you can buy dried fireflies from one of several suppliers, such as Sigma Chemical Co. (St. Louis, M.O.) or Worthington Chemical Corp. (Freehold, N.J.). The simplest experiment you can do is to wet the abdomen of such a dead insect with ATP solution (dissolve 10 mg ATP in 10 ml 0.1 M phosphate buffer, pH 7.6, containing 1 mM $MgCl_2$) and watch it glow. If you remove oxygen by flushing with nitrogen the glow disappears. Other experiments can be carried out with extracts of the fireflies containing luciferin and luciferase. For extraction of the fireflies in a mortar with sand use a solution of 0.4 g glycine and 0.1 g ammonium bicarbonate in 100 ml distilled water. You can also buy ready-made, dried firefly extract which only needs reconstitution with water.

If you can get living bioluminescent insects, it would be interesting to try repeating the new observation of the role of nitrogen monoxide (Trimmer et al. 2001), but I have not done this myself.

Bacteria

A culture of luminescent *Photobacterium phosphoreum* can be purchased from a culture collection, such as the American Type Culture Collection (http://www.atcc.org). but the cost is considerable. It is, however, not very difficult to get luminescent bacteria starting to grow on old decaying fish (if you can stand the smell), preferably fish from the sea. See Lee (1977, 1991) and

Literature

Lee, J. (1977). Bioluminescence. *In* Smith, K.C. (ed). The *Science of Photobiology,* Chapter 14, section 14.2. Reprinted in Valenzeno, D.P. et al. (1991). *Photobiological Techniques,* pp. 359-360. New York: Plenum.

Lee, J. (1991). Experiment 33: Bactereial bioluminescence. In Valenzeno, D.P. et al. (1991). *Photobiological Techniques,* pp. 317-320. New York: Plenum.

Trimmer, B.A., Aprille, J.R., Dudzinski, D.M., Lagace, C.J., Lewis, S.M., Michel. T., Qazi, S. & Zayas, R.M. (2001). Nitric oxide and the control of firefly flashing. *Science, 291,* 2486-2488.

SUBJECT AND ORGANISM INDEX